Environmental geotechnics

Robert Sarsby

Thomas Telford

Published by Thomas Telford Publishing, Thomas Telford Ltd, 1 Heron Quay,
London E14 4JD.
URL: http://www.thomastelford.com

Distributors for Thomas Telford books are
USA: ASCE Press, 1801 Alexander Bell Drive, Reston, VA 20191-4400, USA
Japan: Maruzen Co. Ltd, Book Department, 310 Nihonbashi 2-chome, Chuo-ku,
Tokyo 103
Australia: DA Books and Journals, 648 Whitehorse Road, Mitcham 3132,
Victoria

First published 2000

Also available from Thomas Telford Books

Contaminated soil 2000 (2 vols). ConSoil. ISBN: 0 7277 2954 3
Design applications of raft foundations. J. A. Hemsley. ISBN: 0 7277 2765 6
Environmental assessment. R. Singleton, P. Castle & D. Short. ISBN: 0 7277 2612 9
Geoenvironmental engineering: ground contamination. R. N. Yong & H. R.
Thomas (eds). ISBN: 0 7277 2840 7
Green 2: contaminated and derelict land. R. Sarsby (ed.). ISBN: 0 7277 2633 1

A catalogue record for this book is available from the British Library

10035244 75

ISBN: 0 7277 2752 4

This book is published on the understanding that the author is solely responsible
for the statements made and opinions expressed in it and that its publication does
not necessarily imply that such statements and/or opinions are or reflect the views
or opinions of the publishers. While every effort has been made to ensure that the
statements made and the opinions expressed in this publication provide a safe and
accurate guide, no liability or responsibility can be accepted in this respect by the
author or publishers.

Typeset by MHL Typesetting Ltd, Coventry
Printed and bound in Great Britain by MPG Books, Bodmin, Cornwall

Preface

The construction industry has always served to aid society and has contributed significantly to its development and mankind's well-being through, for example, the provision of clean water and sanitation, the building of transport networks and, more recently, the engineering of waste disposal and the rehabilitation of derelict and contaminated land. These latter contributions are prime examples of the inevitable interaction that occurs between construction and the environment. This interaction can be classified as incidental, accidental, opportunistic or deliberate. While much of this interaction has been beneficial, there have of course been instances of detrimental interaction, for a variety of reasons. In recent years growing environmental awareness, within both society and the construction industry, has led to the formal statement and definition of attitudes and approaches to construction and its environmental consequences so as to prevent harm.

The construction industry interacts broadly with the environment in two ways:

- as a service industry to environmentally orientated operations
- as a major consumer of materials (and thereby being partner to the associated environmental consequences).

From the nature of the foregoing activities it is readily apparent that the major construction–environment interface is geotechnical in nature, primarily by virtue of the need to use the ground for the support of works of construction and the bulk use of geotechnical materials in construction. The recognition of environmental geotechnics as an identifiable facet of civil engineering is the *raison d'être* for this text.

This book is directed towards the application of geotechnical principles, processes and techniques in situations where there is a major environmental component, such as engineered waste disposal by landfill or landraise, rehabilitation of derelict and contaminated land and use of waste materials in construction.

The first part of the book contains an outline of those elements of soil mechanics which I feel are required knowledge for persons studying or working within the general area of environmental geotechnics. My intention was not to produce another textbook on soil mechanics, but to incorporate my academic and practical experience in a presentation of those aspects of soil mechanics that are most important for environmental geotechnics. In this respect it is believed that the key to the development of environmental geotechnics as an analytical and design tool, as it continues to grow and become more 'refined', is to understand the basic principles of geotechnics. I regard understanding the concept of effective stress and the ability to select relevant behavioural mechanisms and appropriate representation of the strength behaviour of materials as vital. In this context I hope that the reader will see the connection between geotechnics and the hovercraft, aquaplaning, ice skating, etc.

It is intended that this text is read as a whole book rather than as a step-by-step guide to obtaining solutions to grossly simplified, artificial questions, and hence it contains no worked examples of exam-type questions. To my mind, engineers have to be problem-solvers (not just identifiers of problems), and a prime requirement is for them to have knowledge and judgement to approximate a previously unseen, atypical, seemingly intractable problem into a situation that can be solved with an appropriate degree of accuracy.

In presenting information on the various environmental topics I have exercised my personal judgement, based on 25 years of practical experience in environmental geotechnics, in giving data, information and suggestions for tackling common situations and problems. Others may have different views on data values or analytical methods and they may believe that I have ignored or overlooked important facets or aspects. Hence particular questions that I would like readers to reflect upon when they are dealing with practical situations in environmental geotechnics are:

- How confident are they of the value of relevant parameters?
- What is the anticipated spatial variability of their material (both inanimate and animate)?
- What are the inherent, unconscious simplifications in their approach?
- What are the immediate and long-term consequences of making a mistake?

We have to undertake construction in a way that produces an overall benefit to society and, as we become more aware of ways and means of reducing environmentally negative outcomes, our obligations keep increasing. Environmental protection legislation encourages and supports engineers in reducing environmentally negative effects from construction operations. I fervently believe that engineers must take up this challenge and incorporate environmental considerations as a fundamental criterion in all their work.

Bob Sarsby

Acknowledgements

Many sources of information have been used in the preparation of this book and these sources are acknowledged within the text and the bibliography. In addition I would like to thank the following organisations and persons for granting permission to use specific copyrighted material: A A Balkema (Rotterdam), Ashgate Publishing Ltd (Gower Publishing Ltd), BDA (British Drilling Association), Dr L Beeuwsaert, Prof G E Blight, BSI (British Standards Institution), Dr A K Chakroborty, CIRIA (Construction Industry Research and Information Association, London), CIWEM (Chartered Institute of Water and Environmental Management), Corus Construction, ENPC (Ecole Nationale des Ponts et Chaussees, Paris), Prof D Ellis, Elsevier Science, The Engineering Council, Foundation for Water Research, Dr J Gettinby, HM Stationery Office, ICE (The Institution of Civil Engineers), International Atomic Energy Agency, International Thomson Publishing Services Ltd, John Wiley & Sons Ltd (Chichester and New York), Kluwer Academic Publishers, Prof G S Littlejohn, M Neden, Plenum Publishing Corporation, The Royal Society, Soil Instruments Ltd, Swiss National Co-operative for the Disposal of Radioactive Wastes, Thomas Telford Publishing, W F Thompson, US Bureau of Mines, US Geological Survey, WRc-NSF Ltd, Dr L Wu.

My greatest acknowledgement (and thanks) must go to my wife, Irene, for her patience, encouragement, fortitude and strength in driving me into my study ('the torture chamber') at times when I did not feel like working on this book — kocham cie bardzo.

Notation

This notation only contains those symbols which have more-or-less universal meaning within the construction industry. Other terms which are used within this book are defined as they occur, within the section to which they are specifically applicable.

Symbol	Parameter	Units
A	Skempton's pore pressure parameter (for shearing)	/
B	Skempton's pore pressure parameter (for isotropic compression)	/
$\bar{B}$	Pore pressure parameter for vertical loading	/
c'	Effective cohesion	kN/m^2 or kPa
C_c	Compression index	/
c_r'	Effective residual cohesion	kN/m^2 or kPa
c_w	Wall adhesion	kN/m^2 or kPa
c_u	Undrained cohesion	kN/m^2 or kPa
C_v	Coefficient of consolidation (vertical drainage)	m^2/yr
C_h	Coefficient of consolidation (horizontal drainage)	m^2/yr
C_α	Coefficient of secondary compression	strain/log(time)
CU	Coefficient of uniformity	/
CZ	Coefficient of curvature	/
e	Voids ratio	/
G or G_s	Specific gravity of soil solids	/
i	Hydraulic gradient	/
I_P	Plasticity index	%
k	Coefficient of permeability (or hydraulic conductivity)	m/s
k_h	Coefficient of permeability (horizontal flow)	m/s
k_v	Coefficient of permeability (vertical flow)	m/s
K_A	Coefficient of Active earth pressure	/
K_A'	Effective coefficient of Active earth pressure	/
K_o	Coefficient of earth pressure at-rest	/
K_P	Coefficient of Passive earth pressure	/
K_P'	Effective coefficient of Passive earth pressure	/
L_{eq}	Equivalent notional uniform sound level	dB(A)
L_p	Sound level	dB(A)
L_w	Sound power level	dB(A)
m	Volumetric moisture content	/
m_v	Coefficient of volume compressibility	m^2/MN
n	Porosity	%
r_u	Pore pressure ratio	/
S_r	Degree of saturation	%
T_v	Time factor (vertical drainage)	/

T_R	Time factor (radial drainage)	/
U	Degree of consolidation	%
U_v	Degree of consolidation (vertical drainage)	%
U_R	Degree of consolidation (radial drainage)	%
w	Moisture (or water) content	%
w_L	Liquid limit	%
w_P	Plastic limit	%
w_S	Shrinkage limit	%
α	Angle	Degrees or radians
δ	Angle of wall friction	Degrees
ε	Axial strain	/
ε_v	Volumetric strain	/
ϕ'	Effective friction angle	Degrees
ϕ'_r	Effective residual friction angle	Degrees
ϕ_u	Undrained friction angle	Degrees
γ	Bulk unit weight	kN/m^3
γ'	Buoyant unit weight	kN/m^3
γ_d	Dry unit weight	kN/m^3
ρ	Bulk density	Mg/m^3
ρ'	Buoyant density	Mg/m^3
ρ_d	Dry density	Mg/m^3
σ	Total direct stress	kN/m^2 or kPa
σ'	Effective direct stress	kN/m^2 or kPa
σ_n	Normal stress	kN/m^2 or kPa
σ'_n	Effective normal stress	kN/m^2 or kPa
σ_1	Major principal stress	kN/m^2 or kPa
σ'_1	Effective major principal stress	kN/m^2 or kPa
σ_2	Intermediate principal stress	kN/m^2 or kPa
σ'_2	Effective intermediate principal stress	kN/m^2 or kPa
σ_3	Minor principal stress	kN/m^2 or kPa
σ'_3	Effective minor principal stress	kN/m^2 or kPa
τ	Shear stress	kN/m^2 or kPa

Contents

1   Geotechnics and the environment

1.1. Introduction

The interaction of economic and social development and the natural environment, and the reciprocal impacts between human actions and the biophysical world, are universally recognized. For hundreds of years man's socio-economic development has been marked by his modification of the environment (in its widest sense). The construction industry has served as a major agent of this change through the designing and building of projects to benefit mankind. Civil engineers can be rightly proud of the dams, roads, water and sanitation systems, power plants, structures, etc., that have been built to meet the needs of society and to enhance the quality of life. However, at the same time, the construction industry must be acutely aware that its efforts can be a two-edged sword and that it has a responsibility to ensure that its activities and products are consistent with environmental policies and good environmental practice.

An example of the negative impact of construction is the disastrous effect of the Canadian Pacific Railway on the salmon stock of the Fraser River. Pacific salmon spawn only once, upriver, after feeding and growing in the sea. They invariably migrate back to a single spawning ground in a particular river, stream or lake and rarely wander to another. If a run of such salmon is blocked one year it does not spawn and that year's run is lost forever. If a run is blocked on two successive years, then all the stocks which live on a 2-year life cycle are gone forever. At the height of the migration period up to 1.5 million salmon pass up the Fraser River (Western Canada) in one week.

The Fraser River is about 1400 km long and is one of the world's major rivers, with a discharge range of about 300–15 000 m^3/s. The river has several rapids but the most constraining bottleneck is the Hell's Gate gorge, which is just 32 m wide. At this location the river level can change by as much as 6 m from day to day and by 1 to 2 m from minute to minute due to surges and turbulence. It was here that an environmental disaster was caused by the construction industry as the Canadian Pacific Railway was built along one of the river banks.

Because of space constraints at Hell's Gate gorge, rock was dumped into the river to increase the width of one bank to provide a bed for the railway. The rock was tipped into the river, so that it came to rest at its angle of repose, and rock dumping continued until the required railway bed width had been achieved. Unfortunately, the result of this accumulated rock tipping onto the slopes of the gorge was narrowing of the river and the velocity of flow through Hell's Gate increased. The water velocity increased so much that in 1913 Pacific salmon on their upriver spawning migration could not pass this point. The next year, 1914, a massive rock slide (of tipped stone and river bank) occurred and once again salmon could not pass through Hell's Gate. The next 14 years saw a massive decline in the west coast salmon fishery (Figure 1.1). These environmental impacts were totally unexpected, as controlling potential landslides along salmon rivers and the risk of causing environmental impacts by construction were not perceptions of the civil engineers at that time. The emergency responses were to try to remove rock and transfer salmon by net over the blockages. Eventually a permanent engineering solution (fishways) was invented (Ellis, 1989).

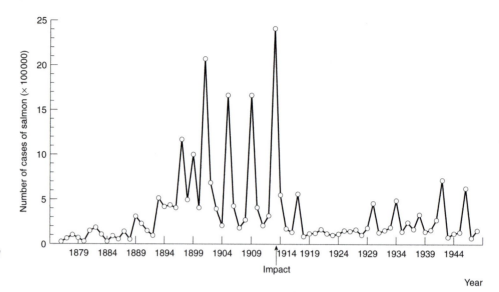

Figure 1.1. The effect of construction at Hell's Gate on salmon catches (from Ellis, 1989)

The Hell's Gate disaster illustrates the environmental risks from construction at major engineering sites. Over a period of 2 years (1913–14) construction-induced rock slides virtually destroyed fisheries with a value of about $2.5 billion (in today's values). That was the risk that was unknowingly taken when the railway was built. It was decades before there was much recovery of the impacted stocks, although the eventual cost of assessing and remedying the impact at the site was relatively trivial (by comparison to the cost to the fishing industry).

Ecosystems are dynamic and are always changing. Natural recovery will eventually occur — with time concrete cracks and crumbles, bacteria in the ground digest pollutants, waste tips are colonized by vegetation, etc. Unfortunately the rate of increase in man's development and population does not now allow the time for natural recovery of damaged environments, which may take decades, centuries or even longer. Hence it is incumbent upon the construction industry to minimize the negative environmental effects of any of its works.

1.2. Construction and the environment

Construction projects range from the very major (e.g. the Channel Tunnel, costing at least £6 billion) to small, local works (e.g. repairs to short stretches of road). Construction (civil engineering and building) is one of the largest industries in the UK employing, in one way or another, about 10% of the working population. Overall, construction covers large areas and employs many workers and invariably it generates a complex array of organizational activity during the various stages of the life of a project. While most building and construction activities will lead to long-term benefits to the community and society as a whole, the location of these developments, and the way they are planned, designed, contracted and operated, can also have wide-ranging, long-term, significant negative impacts on the environment (Table 1.1).

According to the original charter of the Institution of Civil Engineers the profession is 'the art of directing the great resources of power in nature for the use and convenience of man'. Many important public and private works and services, which contribute to the well-being of mankind, are dependent on civil engineers. In recent years it has been stated, and emphasized, that increasingly civil engineers must employ their professional knowledge and judgement to make the best use of scarce resources and to care for the environment.

However, civil engineers have been practising some form of environmental engineering for many years. For example, engineers worked to make major improvements in public health and water supply engineering because of the

Table 1.1. General environmental impacts from construction

Area	Positive achievements	Negative aspects
Water management	Control of floods and storage of water to ensure reliable supply. Energy production, fisheries, navigation, recreation and general development to satisfy current basic social and economic needs	A negative feature of tropical dams and irrigation schemes has been the enhancement of conditions for the establishment of water-related diseases in endemic areas
Sewage management	Systems handle not only human wastes but also a variety of discharges (trade effluents) from commercial and manufacturing activities	Sewage sludge is often heavily contaminated with heavy metals and thus cannot be applied to land and has to be landfilled. Many sludge-holding lagoons exist that have not been engineered for stability or containment
Transport	Improved communications and access to facilities (medicine, work, etc). Improvement of atmosphere and environment by moving traffic away from residential areas	Habitats are affected, both at the construction site and in the region, and tracts of forest, fields, farms, water, etc., may be eliminated. Changed water-flow patterns will affect the sites and rates of soil erosion. Blasting and earth movement can cause rock or soil slides
Mineral extraction	Supports improved standard of living, development of society and its infrastructure, provision of work for people	Impacts from ore extraction, dumping of waste rock and trace metal contaminants. Production of liquid wastes with a high oxygen demand on the ecosystem. The construction and related industries are conspicuous consumers of resources
Waste disposal	Landfills are properly engineered to safely contain waste, thereby preventing uncontrolled egress of gases and contaminated water. Civil Engineers have developed lining systems, gas venting and protection measures	Encouragement of waste disposal philosophy as opposed to recycling, recovery and re-use of waste materials. The construction industry has been slow to develop technology appropriate for long-term waste containment
Energy	In developing alternative energy sources the problem-solving capabilities of the engineer have come to the fore (e.g. wave energy, estuary barriers, wind energy, energy from waste)	Traditional systems of energy protection have given rise to large quantities of dumped by-products and waste materials

appalling unsanitary conditions in London in early Victorian times (Braithwaite, 1995). In fact it could be said that environmental engineering is synonymous with civil engineering because everything constructed has an impact on the environment. This is recognized in the Institution of Civil Engineers' Mission Statement (1995), wherein promotion of the art and science of civil engineering, for the well-being of mankind, is coupled with the need for works of construction to be in sympathy with the environment and to ensure sustainable development. The 1987 Report of the World Commission on Environment and Development (the Brundtland Report — United Nations, 1987) defined sustainable development as 'development which meets the needs of the present generation without compromising the ability of future generations to meet their own needs'. Sustainable development means handing down to future generations not only 'man-made capital' such as roads, schools and historic buildings and 'human capital' such as knowledge and skills, but also 'natural/environmental capital' such as clean air, fresh water, rain forests, the ozone layer and biological diversity (Glasson *et al*, 1994). This is exactly what the civil engineering profession strives to achieve.

In 1993, The Engineering Council issued a 'code of conduct' for engineers with the aim of fostering greater awareness, understanding and effective management of environmental issues. It was intended that the code would encourage engineers to take the lead in proposing and implementing sound engineering solutions to safeguard the future. Furthermore, it would enable them to progress the debate over how to achieve sustainable development. The generic components of an engineer's work were identified in the code and for each component an overall environmental aim, and specific suggestions on how to work towards this aim, was made (Table 1.2).

1.3. Environmental impact from construction

Major projects have a planning and development life cycle with a variety of stages. It is important to recognize such stages because impacts can vary considerably between the phases. The major stages in the life cycle of a project are outlined in Figure 1.2. There may be variations in timing between each stage, and internal variations within each phase, but there is a broadly common sequence of events.

The construction industry has a major impact on the natural and built environments at local, national and global levels. Environmental impact can be caused by the construction industry in several ways:

- that arising from the built environment (e.g. sewage waste).
- pollution caused during the manufacture of materials and products (e.g. dust from quarries, slag from steelworks).
- pollution and hazards from the handling and use of materials on the site itself (e.g. noise, contaminated soil).
- depletion of non-renewable resources (e.g. quarrying of stone for aggregate, use of land for the dumping of waste).
- other construction-related activities (e.g. loss of heritage and amenity areas because of new development rather than rehabilitation).

The design and construction phases involve the specification of materials and the use of plant, processes and techniques. Most phases of the works involve extensive disturbance to the existing environment, whether on greenfield or previously developed sites. Each activity poses a risk of introducing pollutants into the environment, which can affect the workers on site, the neighbourhood or the local ground, water and air quality. Similar impacts can occur during the operational phase of the development — society's use of the resources of land, water and air takes two forms, i.e. as a resource and as a repository or conduit for waste.

The design profession can, and should, have a very significant influence on the environmental impact of the projects they design and develop. Currently there is a wealth of information available for a designer who wishes to pursue a policy of minimizing the environmental impact of a project. Those responsible for the construction phase must also seek to reduce the impact of their own operations on the environment. From the wide range of environmental issues facing the construction industry, a few stand out as very important (CIRIA, 1994b):

- proper functioning of pollution-control measures
- effective conservation of natural resources
- bringing derelict or contaminated land into beneficial use
- selection of materials on the basis of environmental criteria
- energy conservation and reduction of emissions of 'greenhouse gases'
- the move towards a sustainable environment.

Currently all major works of construction must be subjected to an Environmental Impact Assessment (EIA) and the procedure is enshrined within European Commission (EC) Directives (CEC, 1985). The procedure applies to the assessment of the environmental effects of public and private projects where

Table 1.2. Engineers and the environment — a code of conduct (after The Engineering Council, 1993)

Component	Overall aim and specific suggestions
Role	*To work to enhance the quality of the environment* Be aware of the wide variety of uses of natural resources with which you may be concerned — e.g. human, flora and fauna, air, water and land — and the interactions between these. Seek ways to change, improve and integrate designs, methods, processes, operations, raw materials and products to enhance the environment. Use the body of knowledge generally available to the engineering profession at the time to anticipate environmental problems that could arise from your professional activities. Assess projects to ensure that the products and wastes can be re-used, recycled or rendered harmless and the discharges are controlled to minimise environmental impact.
Approach	*To maintain a balanced and comprehensive approach to environmental issues* Be aware of the interaction of your work with that of others involved in the same activity. Recognize that the impact on the environment: • might be so great that a project should be avoided altogether • could be so insignificant that the project could proceed without formal assessment • may lie between the above, and action should be taken to minimise environmental effects as far as reasonably practicable. Be aware of the need to balance reliance on regulations with project-specific environmental reviews. Be mindful of all aspects of the project in an environmental review, including all stages of design, manufacturing, construction, operation, decommissioning and disposal of the product, process or system as well as the energy and materials utilised.
Cost benefit	*To balance economic, environmental and social benefits* Seek to balance costs with the net benefit to the environment and to human society, to achieve the best practical environmental option. Use the best available technology not entailing excessive cost (BATNEEC).
Management	*To encourage management to follow positive environmental policies* Encourage a top-level commitment to an environmental policy, which includes public environmental statements and monitoring systems. Seek personnel policies, which provide for education, training and open communications on environmental issues.
Conduct	*To act in accordance with the code of conduct* Recognize the general duties to avoid creating danger or damage or waste of resources.
Law	*To know about and comply with the law* Accept the duty of care and do whatever is reasonably practicable to respond to environmental issues including, where necessary, going beyond specific standards or codes of practice.
Professional development	*To keep up to date by seeking education and training* Take every opportunity to contribute towards the advancement of knowledge of environmental matters relevant to each engineering discipline. Influence, where possible, the initial education of engineers and technicians to include awareness of their role in protecting and enhancing the environment.
Communication and public awareness	*To encourage understanding of environmental issues* Discuss environmental issues, developing technology and regulatory requirements with others. Bring major or potential environmental damage to the attention of those in authority in a responsible manner. Seek to educate others and encourage public awareness of environmental issues, and to join debate over drafting and implementation of legislation.

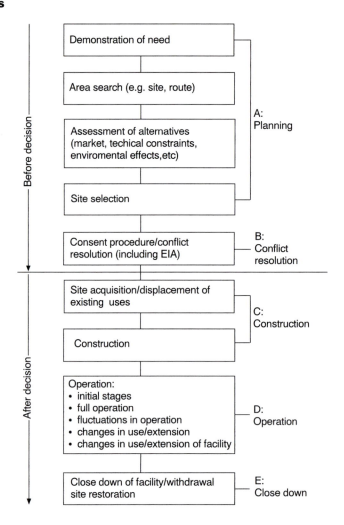

Figure 1.2. The stages in the whole life of a project (from Glasson et al., *1994)*

projects are defined as 'interventions in the natural surroundings and landscape including those involving the extraction of mineral resources'. Environmental Impact Assessments determine the best options for a project, weighing up adverse effects against the economic benefits it will bring. The assessment must consider the direct and indirect effects of the project on human beings, flora and fauna, soil, water, air, climate, the landscape, material assets and cultural heritage. The potential construction impacts are clearly influenced by the overall design of the scheme and the building and engineering methods employed in its construction. The assessment of potential impacts can only be achieved through an understanding of the operations employed and their associated risks. However a project will involve a wide range of different activities, some of which are listed in Box 1.1.

This need to take account of environmental issues from the very earliest stages of a project is placing new demands on all the main parties — the developer, the tenderers and ultimately the successful contractor. There is a growing tendency for both developers and contractors to call on environmental expertise as part of the day-to-day construction management team. To become recognized as being at the forefront of environmental awareness it is necessary to change the perception of civil engineers as primarily engineering problem-solvers. Furthermore, it is necessary for the construction industry to undertake its work in close association with planners, environmental scientists and ecologists. There is no reason why civil engineers should not lead multidisciplinary teams to make major EIAs of projects.

Box 1.1. Construction activities (adapted from CIRIA, 1994b)

Construction activity	Specific components
• Site investigation.	Formation of trial holes, geotechnical sampling, in situ tests, physical tests.
• Demolition and site clearance.	Removal of buildings and structures, stripping of topsoil and vegetation, provision of temporary access.
• Geotechnical activities.	Modification of ground profile and properties of the ground.
• Tunnelling.	Drilling, blasting, removal of spoil.
• Earthworks.	Excavations, placement and compaction of fills, dredging.
• Piling.	Bored, driven, preformed, cast in situ, temporary ground support.
• Superstructure works.	Construction of walls, floors, columns, etc., using steelwork, concrete, brickwork, etc. Installation of cladding and internal services.
• Services.	Drainage systems, water/gas/electricity supplies.
• Roads.	Permanent access roads, hardstanding and car parking.
• Landscaping.	Topsoil placement, planting and seeding.

1.4. Environment–construction interactions

Construction interacts with, and impacts on, the environment in both beneficial and detrimental ways. These interactions can be broadly classified as; deliberate, incidental, opportunist and accidental (Sarsby, 1991).

Deliberate interaction

Deliberate interaction relates to construction works that are specifically designed to prevent pollution or protect/improve the environment. Usually the initiative for this work does not come from the construction industry but it is the duty of the builders to ensure that the finished product functions correctly. With works that are designed to protect the environment (particularly in the long-term) there must be a high degree of confidence in every part of the final product.

Examples of this type of interaction are: engineered landfilling of refuse (see Chapter 12), remediation of contaminated land (see Chapter 13) and nuclear waste disposal (see Chapter 18). For instance, with landfilling of refuse the deliberate interaction is the collection and treatment of the wastes in specific locations to ensure that groundwater is not polluted. At the same time there is the potential for engineering the waste disposal to maximize landfill gas yield to conserve other energy resources and reduce the 'greenhouse effect'. The waste facility must protect the environment both during filling and for a long time afterward. Serious considerations have to be given to the fact that the composition of the waste (whether decomposing or not) may give rise to chemical and biological attack of the components of the system and could lead to eventual failure of the works.

Incidental interaction

Occasionally, construction will give rise to adverse environmental effects that are an inevitable consequence of the works but, prior to construction, there has been a failure to foresee or appreciate the severity of the effects. This is likely to be because the designer has little or no knowledge in relevant specialist areas (e.g. ecology, health and medical matters, demographic aspects, hydrogeology). This type of adverse environmental impact has been designated as incidental interaction.

Examples of this type of interaction include; silting-up of man-made dams, irrigation systems which provide breeding grounds for parasites, and the effect of piling vibrations on poorly-maintained buildings (see Chapter 17). As a specific example consider the spread of Bilharzia in rural Africa. This disease is caused by parasitic worms which use human beings as a primary host and snails as secondary hosts and world-wide it affects some 200 million people. Chronic stages of the disease affect the body's vital organs, leading to liver damage, stone formation in the bladder and bacterial infections of the urinary tract. Irrigation schemes and reservoir construction have greatly increased the potential habitats of the host snails and have also provided ideal play areas (because of the residual/ standing shallow water) for people in hot climates, thereby promoting contact between humans and infected snails. The water-supply systems were developed for the best of reasons (i.e. increased food production for the population), and were engineered for efficient distribution of water using materials and technology appropriate to the area. Unfortunately, the engineers failed to appreciate the severity of a problem that could have been largely avoided by changing or modifying the design of the schemes.

Opportunistic interaction

Opportunistic interaction occurs when a deliberate decision is made to undertake construction in such a way, or using particular materials, as to benefit the environment. With opportunistic interaction the prime motive is usually financial benefit, but the resultant conservation of existing resources and the environmental amelioration due to recycling must not be forgotten.

Prime examples of opportunistic interaction are: the development of 'second-hand' and derelict land (see Chapter 14), the use of waste materials/industrial by-products and low-quality fills as building materials (see Chapter 16), and the controlled production and utilization of gas from landfill sites. A specific example of this type of interaction would be the utilization of Pulverised Fuel Ash (PFA), which is a waste product from coal-fired electricity-generating stations. In the UK approximately 50% of this waste is re-used, most of it by the construction industry. The ash is used as bulk fill and as a replacement for cement in concrete and grouts. It has advantageous engineering properties such as low bulk density, self-hardening and pozzolanic behaviour, and a low sedimentation rate when suspended in water. The re-use of PFA avoids having to use land to store the waste (typically in lagoons) and saves the energy that would otherwise be needed for extraction of conventional aggregates and fill materials. Hence any assessment of the costs of using PFA should include not only the basic cheapness of the ash but also the value of the benefit to the environment.

Accidental interaction

In most construction works minor errors or malfunctions can be tolerated (and these do occur despite extensive supervision and checking by all parties to the works) without detriment to the performance of the finished product. This is because the errors usually compensate one another (i.e. a minor deficiency in one part is balanced by a better-than-designed part elsewhere). However, sometimes all the errors are deficiencies and any small disturbance to the system will induce failure, thereby causing accidental interaction with the environment.

Accidental interaction is usually brought about by: viewing construction works from a narrow perspective, a lack of experience, failure to consider the long-term performance of the works, lack of thoroughness (in both design and construction), lack of technical rigour, and failure to 'design' major works because they are not perceived as engineered construction. The dramatic failure of the Aberfan coal-mining spoil tip (see Chapter 14) falls within the 'accidental interaction' category of environmental damage. However, less high-profile events such as tailings dams collapses (see Chapter 15), the escape of methane gas from refuse disposal sites, and the failure of effluent and sewage containment dykes adjacent to water courses have caused environmental pollution on a much greater scale. Many of the foregoing situations involve, in terms of their dimensions, major works of construction, but most have not been regarded as structures that need to be properly designed, constructed and monitored. As a result, failure of this type of system is a periodic event.

1.5. Environmental geotechnics

Construction of the Pacific Railway in Canada caused a major environmental disaster and almost wiped out the west coast salmon-fishing industry. Seventy years later environmental awareness had increased so much that when the Channel Tunnel was being built invitations to promoters of a Channel Fixed Link required that a full EIA was made (Kershaw & McCulloch, 1993). A check-list was provided of matters to be addressed and some potential issues were highlighted, including tunnel spoil disposal, environmental pollution and the assimilation of the design within its environmental location. The general check-list of environmental issues in the invitation was arranged in 18 groups of topics (Box 1.2).

The EIA was based upon the then draft EC Directive which required the description of the likely effects on the environment 'to cover the direct effects and any indirect, secondary, cumulative, short, medium and long-term, permanent and temporary, positive and negative effects of the project'. In particular, it was stated that the environmental effects of the development during the construction phase were to be considered separately from those during operation. Furthermore, environmental mitigation measures were to be incorporated from the start of the design process. Specialist contributors to the EIA process were identified within the construction companies of the Channel Tunnel Group or from consultants already working on the proposals. At the peak of design and construction activity the in-house environmental team comprised an environmental manager, an environmental health officer and a graduate environmental scientist. Over a dozen specialist consultants were also employed, either to carry out specific tasks or to provide technical advice when necessary (Box 1.3). Remaining topics were addressed by recognized 'external' authorities. The total cost of the initial EIA, including the Environmental Assessors' overview, was around £250 000 (at today's prices).

The recommendations and issues identified by the initial EIA were fully incorporated in the construction contract. The contractor was specifically required to design and construct the works to reduce and, where practical, to avoid all harmful effects on the environment as a consequence of the design, construction, commissioning and maintenance of the works. This was elaborated

Box 1.2. The Channel Tunnel: specialist environmental reports (from Kershaw & McCulloch, 1993)

1. Landform evaluation
2. Groundwater and hydrogeology
3. Soils, land quality and agriculture
4. Terrestrial ecology
5. Coastal hydrography
6. Marine ecology and fisheries
7. Archaeological features
8. Architectural heritage
9. Population, housing and recreation
10. Electricity infrastructure and telecommunications
11. Water and gas infrastructure
12. Energy consumption
13. Design principles and visual impact
14. Transport networks
15. Residues and emissions — sound and vibration
16. Residues and emissions — air
17. Residues and emissions — water
18. Residues and emissions — spoil and waste

Box 1.3. The Channel Tunnel: environmental design input (from Kershaw & McCulloch, 1993)

Acoustics
Numerical prediction of road and rail noise levels, sound levels in buildings, operational areas and off-site, vibration.

Animal control
Site hygiene and preventive measures for rabies control.

Air quality
Computer modelling of vehicle and stationary sources to predict on-site and off-site air quality.
Atmospheric dispersion, assessment from underground sources.

Ecology
Advice on implications of design options on terrestrial, freshwater and marine environments.

Meteorological studies
Site observations and numerical analyses.

Landscape and planting design
Coordination of visual aspects of the works, functional analyses.

Seed bank determination
Planning woodland topsoil relocation.

Vegetation establishment experiments and trials
Consideration of tunnel spoil ameliorants and appropriate native seed mixes.

Beach transport and cliff recession
Historical database assessment and computer modelling.

Sediment transport
Hydrographic surveys and computer modelling.

Tunnel spoil characteristics
Sediment and chemical laboratory analyses and spoil placement/settlement studies.

with particular reference to; the control of noise, dust and other emissions, disposal of liquid and solid waste, protection of water bodies, remedial measures for any uncontrolled emissions, and woodland habitat creation.

Soil is widely used as a construction material in various civil engineering projects, and ultimately all forms of construction are supported by soil and/or rock. Geotechnics is the subdiscipline of civil engineering that relates to the natural materials found close to the surface of the Earth. It includes the application of the principles of soil and rock mechanics to the design of earth structures, retaining systems and foundations. Since much of the environmental content of construction involves the ground and earthworks, a thorough knowledge of geotechnics and its practical applications is essential. The following chapters in this book are intended to provide this knowledge.

2 Environmental basics

2.1. Introduction

Pollution of the environment is due to the release of substances that are capable of causing harm (including offence to any senses) to man or any other living organisms supported by the environment. A narrow view of the components of environmental interaction would focus primarily on the physical environment, for instance all media susceptible to pollution (air, water, soil, flora and fauna, human beings, landscape, urban and rural conservation and the built heritage). However, there are important economic and socio-cultural dimensions to the environment, such as economic structure, labour markets, demography, housing, services (education, health, police, fire, etc.), life-styles and values, and a wider check-list for environmental interaction might be more appropriate (Box 2.1). Furthermore, the environment can also be analysed at various stages and at different scales. Many impacts affect only the 'local' environment (although this may vary according to the stage in the life of a project), but some impacts may have a regional and/or global dimension. The environment also has a time aspect.

The process of assessing the environmental impact of intended construction work is well illustrated by a study undertaken in support of a proposed redevelopment of an old cement works (covering 11 ha) which was surrounded by former quarry areas (Mallet, 1996). The development was adjacent to general industrial, warehousing and leisure-use areas. The old quarry areas (half of which had been landfilled) were not subject to environmental assessment because they were not intended for redevelopment. Nevertheless, the presence of landfill

Box 2.1. Check-list for environmental interaction (from Glasson et al., 1994)

Physical environment
- Air and atmosphere: air quality.
- Water resources and water bodies: water quality and quantity.
- Soil and geology: classification, risks (e.g. erosion, contamination).
- Flora and fauna: birds, mammals, fish, etc.; aquatic and terrestrial vegetation.
- Human beings: physical and mental health.
- Landscape: characteristics and quality of landscape.
- Cultural heritage: conservation areas; built heritage; historic and archaeological sites.
- Climate: temperature, rainfall, wind.
- Energy: light, noise, vibration, etc.

Socio-economic environment
- Economic base — direct: direct employment; labour market characteristics; local/non-local trends.
- Economic base — indirect: non-basic/services employment; labour supply and demand.
- Demography: population structure.
- Housing: supply and demand.
- Local services: supply and demand of services: health, education, police, etc.
- Socio-cultural: life-styles/quality of life; social problems, community stress and conflict.

within the old quarry areas was taken account of by recognizing the potential effects that the redevelopment could have on the landfill areas, and vice versa. The assessment was undertaken in several stages:

- Identification of the primary issues from consideration of the very wide range of environmental aspects which could be investigated (i.e. scoping). The key issues identified were: landscape and visual amenity, ecology, surface water, landfill gas, ground contamination and traffic.
- Provision of information on the existing conditions at the site for each of the preceding key issues (i.e. baselining). The scale of the investigations reflected the potential magnitude of environmental impacts identified during the scoping stage. Thus there was a considerable programme of site investigation concerned with the determination of the ground condition, the presence of leachates and the soil-gas regime.
- Definition of potential environmental effects of the development and identification of the need for, and scope of, any mitigation measures. Potential design solutions were assessed against their impacts to lead to a redevelopment proposal which would be a feasible and economic scheme while minimizing environmental effects.
- Production of a formal statement of the findings of the assessment. In this case the process resulted in the following main mitigation measures: construction practices which ensured that no adverse effects arose from existing land contamination, drainage to control leachate migration and landfill-gas control measures.
- The environmental report was evaluated by the local authority. This led to the commissioning of some further baseline work on ecological matters and discussion and negotiation regarding the proposed mitigation measures.

A major conclusion arising from the case history was that environmental assessment and the preparation of a rigorous environmental statement is a positive aid to the development of brown-field sites. In such situations it is vital to identify mitigation measures that promote redevelopment while minimizing the resultant environmental impact.

2.1.1. Environmental protection

In recent years there has been an immense growth of interest in environmental issues and better management of development in harmony with the environment. Associated with this growth of interest has been the introduction of new legislation, emanating from national and international sources (such as the European Commission), that seeks to influence the relationship between development and the environment. Much of the driving force behind the development of a consistent UK environmental policy and its associated legislation has stemmed from the European Community. Directives are the instrument most used for environmental matters because they allow European Member States the flexibility to encompass new obligations within existing national procedures. Directives are binding only with regard to the results that must be achieved, but Member States are left to choose their own methods. A guiding principle behind the controlling legislation is that the polluter should pay the costs of rectifying the effects of pollution. However, in practice it is often far from easy to actually measure effects and so attribute blame.

Many new construction projects that have significant geotechnical components and potential major environmental interaction will need an environmental assessment (EA). EA is a technique and a process by which information and data about the environmental effects of a project are collected, both by the developer and from other sources. Environmental analysis may indicate ways in which the project can be modified to anticipate possible adverse effects, for example through the identification of a better practicable environmental option or by considering alternative processes.

Since the UK's implementation of CEC Directive 85/337/EEC (the assessment of the effects of certain public and private projects on the environment — CEC, 1985) there has been increasing application of EA to development projects. The effect of the Directive is to require environmental assessment to be carried out, before development consent is granted, for certain types of major project that are judged likely to have significant environmental effects (e.g. new transportation links, open-cast mining, dam erection, sites for disposing of sludge, installations for disposal of mine and quarry wastes, landfill sites). The EC Directive contains a small list (Schedule 1) of major projects for which an EA is required in every case and a much longer list (Schedule 2) of projects which potentially have significant environmental impact. The recommendations of the Directive appertaining to geotechnically-related situations are summarized in Table 2.1. Significantly, redevelopment of previously developed land is unlikely to require EIA unless the proposed use is one of the specific types of development listed in Schedules 1 or 2 or the project is on a very much greater scale than the previous use of the land.

Of the 2300 environmental statements published between 1988 and 1994, 35% were for waste management and road projects, with infrastructure, extraction and energy projects being the next most common categories. Since EA is undertaken when the project proposal is in its infancy, this presents many opportunities to influence the planning and design in order to minimize potentially significant environmental impacts by highlighting them at an early stage. Indeed there is evidence that substantial changes to project design have arisen as a result of EA (CIRIA, 1996).

Control of pollution within the UK is exercised by a number of different agencies and departments of the Government under a wide variety of legislation.

Table 2.1. Assessment of construction-related projects affecting the environment (adapted from CEC Directive 85/337/EEC (CEC, 1985))

Project area	Comments
Disposal of radioactive waste	EIA mandatory
Industrial facilities	EIA mandatory for integrated chemical installations, crude-oil refineries, major thermal power stations
	EIA usually required for other large facilities
General waste disposal	EIA mandatory for facilities dealing with toxic and dangerous wastes
	EIA may be required for facilities dealing with; industrial and domestic waste, residues from sewage treatment and other sludges
Infrastructure	All new motorways and trunk roads which are over 10 km in length require the publication of an environmental statement for the preferred route. EIA may be required for roads, harbours, airfields, canals and dams
Extractive industries	Whether or not EIA is required depends on: sensitivity of the location, size of the site, working methods, proposals for waste disposal, nature/extent of processing, means of transporting minerals from the site, and duration of the proposed workings

EIA, Environmental Impact Assessment.

The Environmental Protection Act brought together many previously separate pieces of legislation and thereby presented a more unified picture of environmental responsibility in a legislative setting. The Act addresses the question of contaminants/pollutants and states the objective of ensuring that, in carrying out a prescribed process, the best available techniques not entailing excessive cost (BATNEEC) will be used to prevent pollution. The Environment Agency of England and Wales was established in 1996 to provide an integrated approach to the protection and management of the environment. The Agency's principal aim is to make a contribution towards achieving sustainable development by protecting and enhancing the environment as a whole. The Agency works at the local level with Local Environment Agency Plans. These are detailed documents which are prepared with local consultation to identify, assess and prioritize the work of the Environment Agency. Local planning authorities are obliged to consult the Agency on a range of planning applications, and this may result in conditions being attached to planning applications or an objection to a proposal being registered.

2.1.2. Aims of assessment

The ecosystem includes: individuals and populations of all plant and animal species; communities of different species; terrestrial and aquatic habitats; places where flora and fauna feed, water, rest or breed, or pathways of travel and migration. Both naturally occurring systems and those 'created' for commercial purposes (such as agriculture, fish farming or woodland) are included. A number of components are vital to the maintenance of ecological systems, these include: soil-water chemistry, the water table, relevant geology, geomorphological processes such as erosion and deposition, air quality and climate, and also management. A development could impact on any, or all, of the preceding components, not only at the site itself but also in the surrounding area.

Some of the potential impacts of a development are obvious to the lay person (e.g. visual intrusion, traffic congestion, loss of agricultural land, noise). Other impacts require the services of specialists to measure and assess them (e.g. employment, effect on local services, ecological effects, natural resource depletion) (Fortlage, 1990). To undertake the work a multidisciplinary team is required that brings together specialist knowledge but comprises a group both able and willing to understand other people's points of view. A typical environmental management team might comprise specialists in the fields of: air pollution, acoustics and vibrations, wastes management, soil and groundwater pollution, ecology, occupational hygiene and public health, land resources planning and energy management. Supporting disciplines would cover other requirements, in particular the need for computer modelling and data processing.

EA is the name given to the whole process of gathering information about a project and analysing the data in order to assess the possible and probable effects on the environment and mankind's health and welfare of undertaking particular activities. The purpose of the exercise is usually to provide decision-makers with information of the likely consequences of their actions so that a decision can be made on whether or not a project should be allowed to go ahead. Many of the impacts of a proposed development may be trivial or of no significance to the decisions that have to be taken. It is the combination of project and location which determines the magnitude and significance of any environmental impact. In practice a decision (regarding a course of action) will depend on only a small subset of issues of overriding importance. Environmental Impact Assessment (EIA) is a systematic process that examines the environmental consequences of development actions in advance. The process is illustrated in Figure 2.1. The emphasis, compared with many other mechanisms for environmental protection, is on prevention of problems. Prediction and evaluation are essential within the

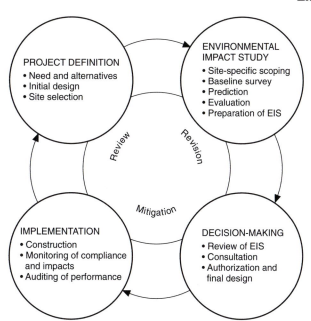

Figure 2.1. The general process of environmental assessment (after Petts & Eduljee, 1994)

assessment process, requiring the developer to appreciate and assess both the negative and positive aspects of a project. Environmental analysis is likely to indicate ways in which a project can be modified to forestall potential adverse effects.

The question of what is a significant effect (i.e. the disturbance or alteration of the existing environment to a measurable degree), is one of the most difficult items to resolve. The evaluation of significance for any specific impact can be based on one or more of the following criteria (Petts & Eduljee, 1994):

- comparison with regulations, standards or other pre-set criteria
- comparison with tolerable risk criteria
- consultation with relevant decision-makers
- consistency with policy or plan objectives
- acceptability to the local community or general public.

The existing physical, social and financial environment must obviously be established before any assessment of future effects can be made. The work can be divided into two categories: fieldwork (which includes surveys, trial holes, photographs and interviews) and recorded data (which include all information obtained from records held by various organizations). Environmental systems are not static, but change over the course of time, even without the influence of man. Some systems are highly dynamic while others only change imperceptibly. To assess the impact of a project it is necessary to establish both the initial state of the site and the changes that would occur naturally over the duration of the project. The environmental impacts of a project are those resultant changes in environmental parameters, in space and time, compared with what would have happened had the project not been undertaken. Figure 2.2 provides a simple illustration of the concept. The time-span for the prediction of the future state of the environment should be comparable with the life of the proposed development, which may necessitate prediction forward for several decades. Spatial coverage may focus on the local situation, but with reference to the wider region and beyond for some environmental elements.

The outcome of the assessment process is the Environmental Statement (ES), which contains all the required information and is submitted with the application for planning permission. The Environmental Statement must be published and the public and interested parties given the opportunity to comment. The authority will

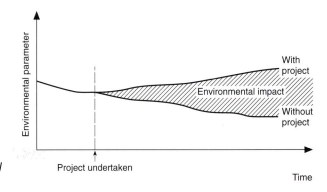

Figure 2.2. The environmental impact concept

take the findings of the assessment and any representations received into account in deciding whether or not to authorize the development.

There are two general approaches to EA, i.e., EIA and Environmental Risk Assessment (ERA) — both of which are concerned with the likely consequences of environmental change. In concept, EIA and ERA have evolved as parallel and sometimes overlapping procedures for rational development of policy-making (Wathern, 1988). Both are intended to provide reasoned predictions of the possible consequences of particular decisions, and thus to permit wiser choices among alternative courses of action. EIAs often exhibit crude and simplistic estimates of the magnitude, likelihood and time distribution of impacts. Prediction is typically limited to judgements that particular consequences are 'likely' or 'unlikely', although exceptions do exist (Wathern, 1988). On the other hand, ERA stresses formal quantification of probability and uncertainty, and typically includes a determination of the types of hazard posed, together with estimates of the probability of their occurrence, the population at risk of exposure and the ensuing adverse consequences. ERAs tend to be numerical appraisals (statistical analyses of likely events) and their purpose is to provide a scientific basis for making public decisions. However, it must be remembered that the results of the investigation are judgements made within constraints of time, money and existing knowledge. ERAs may be based on quite tenuous or debatable assumptions and the predictions ultimately may be unreliable, despite their apparent quantitative rigour.

2.1.3. Environmental impact

There are four main stages in EIA: project definition, environmental impact study, decision-making and implementation and monitoring. The EIA process can be broken down into a number of steps, as outlined in Figure 2.3 and Table 2.2. It should be noted that EIA should be a cyclical activity with feedback and interaction between the various steps.

Scoping is the process of determining which issues are likely to be important for a specific site. The scoping exercise serves to identify those parts of the assessment for which specialist assistance may be required. The types of impact can be divided into three broad categories: physical/chemical, human and nature conservation. For instance, physical effects include ground stability problems and changes in groundwater pattern. Chemical effects are primarily those involving discharges to land, water or air. Disturbances to human beings may arise from noise, odours, traffic congestion, vibrations, etc. Social and cultural impacts can also arise due to effects on features of geological or historical interest. Nature conservation may be affected by pollution, disruption of wildlife conditions and breeding patterns. In addition to the direct loss of vegetation during site preparation and operation, the works may upset the natural balance at a site in various ways, such as by favouring certain plants at the expense of others, by inducing germination of seeds which were previously buried and dormant, or by

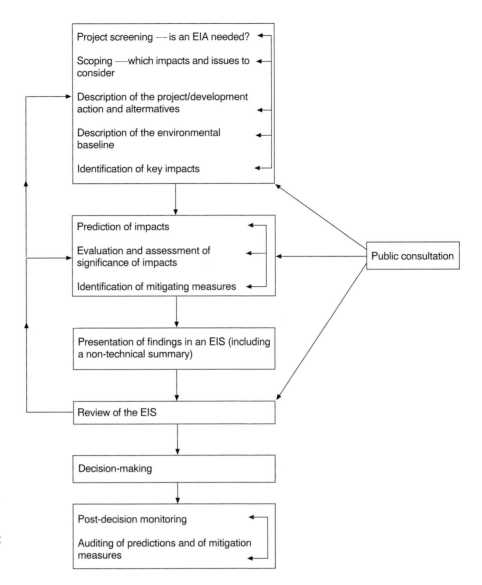

Figure 2.3. The steps in the EIA process (from Glasson et al., 1994). EIS, environmental impact statement

the introduction of 'non-native' flora. Redevelopment of land previously left abandoned and derelict and infilling of land not previously thought suitable for use (such as marshes and boggy areas) can result in the elimination of delicate and specialist ecosystems. The general types of impact arising from some geotechnical-related activities are indicated in Table 2.3, while one particular activity (waste disposal) is covered in detail in Table 2.4.

To assess environmental impacts it is necessary to establish the initial conditions at a site. *Baseline studies* are designed to provide the relevant information on the issues and questions raised during the scoping exercise. It should be remembered that consideration must be given to the construction, operational and post-operational phases of all activities.

2.1.4. Risk assessment

Risk can be defined as the probability of suffering harm or loss under specific circumstances. ERA is a process in which prediction and evaluation are combined to estimate the probability or frequency of environmental harm (risk) for a given hazard (i.e. an event that has the potential to be harmful). 'Harm' includes harm to man, living organisms, offences to any of man's senses, and damage to property. For a risk to exist there must be a hazard, a pathway for transport and a

Component	Purpose
Project screening	To narrow the application of assessment to those projects that may have significant environmental impacts
Scoping	To identify at an early stage, from all of a project's possible impacts and from all the alternatives that could be addressed, those that are the key significant issues
Consideration of alternatives	To ensure that the proponent has considered various feasible approaches (e.g. alternative location/scale/processes) and the 'no action' option
Project description	To clarify the purpose and rationale of the project and to identify its various characteristics (including stages of development, location and processes)
Environmental baseline description	To establishment both the present and future state of the environment (in the absence of the project), taking into account changes resulting from natural events and from other human activities
Key impacts	To bring together the previous steps so that all potentially significant environmental impacts (adverse and beneficial) are identified and taken into account
Impact prediction	To identify the magnitude and other dimensions of any identified change in the environment due to the project
Assessment	To assess the relative importance of the predicted impacts and thus focus on key adverse impacts
Mitigation	The introduction of measures to avoid/reduce/remedy or compensate for any significant adverse impacts
Public consultation	To ensure the quality, comprehensiveness and effectiveness of the EIA, as well as to ensure that the public's views are taken into consideration
Environmental statement	To present the outcomes of the assessment process and the proposed mitigation measures. If done badly, much good work in the EIA may be negated
Review	To provide a systematic appraisal of the quality of the ES, as a contribution to the decision-making process
Decision-making	Consideration by the relevant authority of the ES (including the consultation process), together with other material considerations
Monitoring	Recording the outcomes associated with development impacts, after a decision to proceed
Auditing	To compare actual outcomes with predicted outcomes to assess the quality of predictions and the effectiveness of mitigation. It provides a vital step in the EIA learning process

EIA, Environmental Impact Assessment; ES, Environmental Statement.

	Activity	Potential environmental impacts
Table 2.3. Environmental impacts from geotechnical activities	Raw materials extraction	Noise, dust, loss of visual amenity, alteration of groundwater regime, ground vibrations, effect on habitats of flora and fauna, creation of new habitats, creation of waste tips, employment, traffic
	Piling	Noise, ground vibrations, reclamation and re-use of derelict land, materials for disposal
	Demolition	Noise, ground vibrations, waste materials for disposal, recycling of land
	Earthworks	Noise, alteration of groundwater regime, effect on habitats of flora and fauna, effect on visual amenity, re-use of waste materials
	Contaminated land work	Rehabilitation of polluted zones, use of raw materials for site reinstatement, spreading of contamination, traffic, employment, effect on visual amenity
	Waste disposal	Safe containment of refuse, possible escape of polluted liquid/gas/radiation, restoration of land surface due to infilling, loss of old habitats for flora and fauna, effect on visual amenity, odours, traffic

receptor which can be harmed at the exposure point; if any of these are absent there is no risk. The hazard–pathway–target scenario must also be plausible. Examples of hazards related to environmental geotechnics include the presence of toxic substances in the ground, uncharted mineshafts and abandoned mineral workings, and 'unengineered' tips and lagoons of waste material. Of prime importance in carrying out a risk assessment is the need to identify what or who is at risk. Likely target groups are listed in Table 2.5. In essence, ERA provides a structured approach for ascertaining the nature and extent of the relationship between cause and effect.

ERA can be viewed as consisting of four key stages (Petts & Eduljee, 1994):

1. *Hazard identification*. This is the primary stage of scoping the sources and components of a hazardous event. The stage involves:
   - identifying the hazardous 'substances' present
   - determining their quantity, form and location
   - selecting the indicator 'substances' (i.e. those that could contribute the greatest proportion of the risk).
2. *Hazard analysis*. This is the process of determining:
   - release probabilities, rates and quantities
   - routes or pathways by which released substances could reach targets
   - the fate of substances in the environmental media through which they move and the characteristics of the receptors.

   The objective is to calculate the doses and resultant intakes at the targets. If the frequency and/or doses received by the target are insignificant, no further assessment is necessary.

3. *Dose–response assessment*. The basis of the assessment of harm is to determine the dose–response relationship between a pollutant and a target. Factors include toxicity (the potential of a material to produce injury to biological systems) and hazard (the nature of the adverse effect posed by the toxic material). The general assumption is that there is a relationship

Table 2.4. The impacts of waste disposal

Activity	Examples	Waste treatment	Final disposal
Construction phase			
Site preparation	Land clearance, diversion constructions	C	C
Ground works	Placement of materials, foundations	O	C
Building	Factory premises, accommodation for personnel	C	O
Ancillary works	Temporary access, power supplies, water, etc.	C	C
Raw materials demand	Concrete, soil (effect on source)	O	C
Man-made materials demand	Geomembranes, machinery (effect on manufacturing base)	O	C
Site deliveries	Materials, plant, employees, etc.	C	C
Noise	Increased traffic, construction operations	O	C
Vibrations	Highway traffic, earth-moving, piling	O	O
Dust and mud	Earth-moving, site deliveries	O	C
Gaseous emissions	On-site burning during clearance, exhaust fumes	O	O
Liquid discharges	Site dewatering, spillages	O	O
Solid waste disposal	Excavation spoil, surplus construction materials	O	—
Immigration	Skilled, semi-skilled work force	O	O
Employment	Local trades people, unskilled labour	O	—
Local expenditure	Services, local materials, accommodation	O	O
Accidents/hazards	Traffic, vandalism, trespass	O	C
Operational phase			
Physical changes	Progressive site development	—	C
Earthworks	Stage placement of materials, bund construction	—	C
Raw materials demand	Liner construction (effect on source)	—	O
Traffic	Import of raw materials, import and export of goods/materials	C	C
Site noise	Placing and compacting operations, factory machinery, continuous pumping	C	C
Vibrations	Highway traffic	C	C
Dust and mud	Import and export of materials	O	C
Gaseous emissions	Exhaust fumes, flue gases, landfill gas	C	C
Liquid discharges	Leachate, run-off from stockpiled materials	O	C
Solid waste disposal	Residues from incineration and recycling operations	C	—
Employment	Long-term job creation	C	O
Immigration	Permanent work-force	O	—
Local expenditure	For services	O	O
Accidents/hazards	Traffic, vandalism, trespass	C	C
Site control	Mobility of litter, rodent infestation	O	C
Odours	Waste decomposition, flue gases, waste spreading	O	C
Energy generation/conservation	Landfill gas, incineration heat, waste spreading	C	C
Land reclamation	Controlled landfilling, clearance of industrial sites	—	C
Post-operation			
Cessation of activity	Traffic, site noise, employment	C	C
Physical changes	Demolition, capping, ground movement	C	C
Raw materials demand	Capping materials, topsoil	O	C
Traffic	Import of raw materials, removal of demolition waste	O	O
Noise	Traffic, demolition, cap construction	O	O
Gaseous emissions	Landfill gas, flue dust	O	C
Liquid discharges	Leachate, run-off from site	O	C
Solid waste disposal	Demolition wastes	C	—
Odours	Landfill gas, leachate chimneys	—	C
Energy generation	Landfill gas	—	C
Land reclamation	Landscaping, treatment of contaminated ground	C	C
Accidents/hazards	Leakage from site, access to leachate chimneys	—	O

C, Commonly causes significant impact; O, occasionally causes significant impact

Table 2.5. Potential targets/ receptors of hazards

Target group	Members
People	Adults and children, site workers, permanent residents, visitors, neighbours
Water resources	Surface water, groundwater
Flora and fauna	Sites of special scientific interest, livestock and wild creatures, landscaping
Buildings	Foundations, services, structures, roads, paths

between the concentration of a pollutant and the probability of both an effect occurring and the magnitude of impact. For many pollutants the nature of this relationship remains uncertain; however, there is usually a threshold concentration below which no effect is likely to occur. The three important parameters are:

- the exposure of an individual to the hazard
- the dose received
- the level of toxicity or effect associated with the hazard.

4. *Risk evaluation.* All human activities carry some degree of risk and various measures of human risk are used, including annual risk, lifetime risk, or risk of loss of life-expectancy. Some risks are known with relative certainty because sufficient data have been collated to establish their occurrence; for example there is a 1 in 10 900 risk of death at work for a construction worker and a 1 in 560 000 risk of death through a gas incident (fire, explosion, poisoning, etc.). The final task in ERA is to establish whether the estimated level of risk is tolerable. In the UK this refers to a willingness to live with a risk so as to secure certain benefits and in the confidence that the risk is being properly controlled.

Risk assessment can be used on a site-specific basis to:

- determine whether any further action is required
- establish the level of intervention required to control or reduce risks to an acceptable level
- evaluate different risk control/reduction options
- demonstrate to third parties that a proposed form of action is the best way to proceed.

The risk may be expressed in quantitative terms, with values ranging from nominally 0 (expressing certainty that harm will not occur) to 1 (indicating the certainty that harm will occur). In many cases, however, it may be possible to describe the risk only in qualitative terms (e.g. high, medium, low, negligible, chronic, acute). Several phases of investigation may be required before sufficient data are available to characterize fully the hazards, pathways and targets of concern and to estimate (even qualitatively) the risks involved.

Risk management provides an objective, iterative process for identifying, describing and evaluating risks and deciding the best way of controlling or reducing these risks and implementing strategies to achieve acceptable levels of risk. Risk management equates to risk assessment plus risk reduction, and therefore comprises the components identified previously (hazard identification, analysis, etc.), together with risk control.

The principal advantages of risk management are that it is structured, objective and comprehensive and that it explicitly considers uncertainties with a rational and defensible basis for discussion of a prepared course of action. Risk can also be defined as the product of hazard and vulnerability (the amount of damage that a given hazard can inflict). Thus risk can be managed by either controlling the hazard or reducing vulnerability until a tolerable level of risk is reached.

2.1.5. Life-cycle assessment

The concept of life-cycle assessment (LCA) is to evaluate the environmental effects associated with any given activity from the initial gathering of raw material from the Earth until the point at which all residuals are returned to the Earth (USEPA, 1993b). There are potential adverse environmental effects at all stages of the life cycle of a product, beginning with raw-material acquisition and continuing through materials manufacture and product fabrication. These effects also occur during product distribution, consumption and a variety of waste-management options such as landfilling, incineration, recycling and composting.

LCA may be divided into three separate, but interrelated, components:

- *Inventory analysis* — the identification and quantification of energy and resource use and environmental releases to air, water and land. This component involves quantifying energy and raw-material requirements, atmospheric emissions, waterborne emissions, solid wastes and other releases for the entire life cycle of a product, process, material or activity.
- *Impact analysis* — the qualitative and quantitative characterization and assessment of the environmental consequences. This component involves characterization and assessment of the effects of the resource requirements and environmental loadings (atmospheric and waterborne emissions and solid waste) identified in the inventory stage.
- *Improvement analysis* — the evaluation and implementation of opportunities to reduce environmental burdens. This involves evaluation of the needs and opportunities to reduce the environmental burden associated with energy and raw-materials use and waste emissions.

The quality of a life-cycle inventory depends on an accurate description of the system to be analysed. The necessary data collection and interpretation are contingent on a proper understanding of where each stage of a life cycle begins and ends. It is likely that the attention given to the life-cycle environmental impact of developments (from initial planning considerations through to materials production, construction and eventual operation and demolition) will continue to increase.

2.2. Environmental assessment components

Once the decision has been made that an EA has to be made there are, apart from the actual quantification of the impacts, three key elements within the assessment process:

- focusing on the key impacts (scoping) and outlining possible alternative approaches to the project
- establishing baseline environmental conditions in the likely impacted area
- compilation of the ES (the documentation used for assessing likely impacts).

A number of techniques are available that provide for an objective and comprehensive identification of environmental factors and impacts, and assist in understanding the interrelation between sources and impacts. These techniques offer means for classifying and presenting material for impact analysis or for aiding the presentation of results. Often they can be modified to assist in the identification of impact magnitude and significance.

The potential for the use of simulation and mathematical modelling in EA has been increasing with advancements in software packages suitable for use on microcomputers (Petts & Eduljee, 1994). Models provide a predictive framework within which the temporal and spatial effects of a release to the environment can be studied. In principle, therefore, simulation models may be used to assist with both network and cause–effect analysis.

2.2.1. Scoping

For any project there will be a large number of potential issues that may have a bearing on the siting, design and operation of the facility under consideration. However, the number of these issues which are of such importance as to influence the eventual decision is usually much smaller. The purpose of scoping is to provide a focus for the EIA by identifying the key issues of concern and ensuring that they are subject to assessment at a level of detail appropriate to the scale of the project in question. The focus must not be solely upon environmental impacts, but must also encompass all the significant policy, legal, technical, economic and social implications of the project. Refining the focus on the most significant impacts continues throughout the EIA process.

The importance of the scoping stage cannot be overemphasized. Not only can poor scoping be wasteful of resources, but a failure to scope can also mean that fundamental deficiencies in project design are not identified or that the chosen site and environment cannot accommodate the facility with safety. An essential procedure in undertaking scoping is consultation and discussion, both between the project team and relevant experts, and with external interested parties, such as the planning authority, other statutory authorities, conservation groups and the public.

The first requirement of scoping is a familiarity with the project (having an understanding of the activities throughout its life cycle) and the study area to identify potentially sensitive land uses and locally important issues. A site visit, preferably as an assessment team, is essential not only to gain familiarization, but also to ensure a more effective review of extant environmental data.

Impact identification brings together project characteristics and baseline environmental characteristics, with the aim of ensuring that all potentially significant environmental impacts (adverse or favourable) are identified and taken into account in the assessment process. The basic risks are physical, chemical and biological in nature.

A wide range of methods (each with its own strengths and weaknesses) has been developed and the approaches can be divided into check-lists, matrices, networks and overlay maps.

Check-lists. Most of these are based on a list of special biophysical, environmental, social and economic factors that may be affected by a development. Check-lists may be general (relating to all types of project or environment) or generic (relating to a particular class of project or environment). The simple check-list (Figure 2.4) can only help to identify impacts and ensure that impacts are not overlooked, although quantitative methods attempt to compare the relative importance of all impacts by weighting, standardizing and aggregating impacts to produce a composite index. However, check-lists do not usually include direct cause–effect links to project activities. Nevertheless, they have the advantage of being easy to use.

Matrices. These combine check-lists into a diagrammatic presentation that allows areas of cause–effect relationships, or interactions between the proposed action and the existing situation, to be identified and recorded in a qualitative manner in the cells of the matrix. The recording in the cells can vary from a simple 'yes' or 'no' to a ranking of potential significance. There are no fixed rules as to what should be listed on the axes of the matrix, and the best approach to completion of the cells is to 'brainstorm', involving people with an awareness of the proposal, possible issues of concern and environmental effects. Simple

Data required	Information sources, predictive techniques
Nuisance	
Change in occurrence of odour, smoke, haze, etc., and number of people affected	Expected industrial processes and traffic volumes, population surveys.
Water quality	
For each body of water, changes in water uses, and number of people affected	Current water quality, current and expected effluent
Noise	
Change in noise levels, frequency of occurrence, and number of people bothered	Current noise levels, changes in traffic or other noise sources, changes in noise mitigation measures, noise propagation model, population surveys
Economy	
Local and non-local employment, characteristics of employment, wage levels	Local chamber of commerce, job advertisements, local job and employment agencies, council statements and statistics, housing surveys and property values

Figure 2.4. An example of a scoping checklist

matrices are merely two-dimensional charts showing environmental components on one axis and development actions on the other (Figure 2.5). The aim at the early stages is to achieve a comprehensive coverage of the interactions between the two components under study; matrices can be refined as the process advances and more information becomes available. Matrices are the most commonly used method of impact identification in EIA.

Networks. These methods explicitly recognize that environmental systems consist of a complex web of relationships. They try to reproduce that web by extending the matrix concept to include the environmental subsystems or pathways along which the environmental effect can be traced. This provides an understanding of direct and indirect effects and of required linkages in the system. Impact identification using networks involves following the effects of development through changes in the environmental parameters in the model. Water is one of the six environmental components, the others being climate, geophysical conditions, biota, access conditions and aesthetics.

Overlay maps. In this technique a base map is prepared, showing the general area within which the project may be located. Successive transparent overlay maps are then prepared for the key restraining factors of a development, such as engineering factors, existing areas of ecological or landscape sensitivity, and population centres. The degree of importance of each factor is shown by the intensity of shading, with darker shading representing a greater impact. The composite impact of the project is found by superimposing the overlay maps and noting the relative intensity of the total shading. Unshaded areas are those where a development project would not have a significant impact. Overlay maps are an excellent way of showing the spatial distribution of impacts. They are easy to use and understand, are popular in practice and have been used in environmental planning since the 1960s. The development of computer mapping, and in particular geographical information systems, is allowing this technique to be developed on a quantitative basis with different importance weightings being assigned to the impacts.

In considering the choice of scoping techniques, the following requirements must be satisfied:

- all issues and impacts that are perceived to be potentially important by the interested parties are considered
- all potential impacts are identified
- all parameters for assessment, and relevant data sources, are identified

Figure 2.5 — scoping matrix. Legend: H, high impact; L, low impact; M, medium impact.

Characteristics of existing environment (columns), grouped:
- **Biological:** Flora and fauna · Food chain
- **Chemical:** Air quality · Water quality · Soil · Sewage
- **Physical:** Groundwater · Landscape · Natural resources · Climate · Hydrogeology
- **Human:** Population · Housing · Local economy · Employment · Health and safety · Land use · Risk · Traffic · Leisure and amenity · Infrastructure · Energy requirements · Cultural heritage · Industrialization of the area

Construction phase

Activity	Flora & fauna	Food chain	Air quality	Water quality	Soil	Sewage	Groundwater	Landscape	Natural resources	Climate	Hydrogeology	Population	Housing	Local economy	Employment	Health & safety	Land use	Risk	Traffic	Leisure & amenity	Infrastructure	Energy requirements	Cultural heritage	Industrialization
Site preparation	H		L	L	L		L	H	M	M	M	L	L	L	L		M		L		M	L		M/L/H
Ground works	L	L	L	M	M	M	L	H	M	M	M	L	L	L	L		M	L	L		L	L		M/L/H
Building	L	L	L	L			L	H	M			L	L	L	L		M	L			L			M
Ancillary works	L	L	L	L			L	M	M			L	L						L					
Raw materials demand	M	L	L	L		L	L	M	M	L		L	L						L		L			
Man-made materials demand	M	L	L	L				L				L	L				L	L						
Site deliveries	L	L	M		L							L	L				L	M	L					M
Noise	M	H						L	M	L		L	M	L		H	M	L						M
Vibrations	M	M						L	M	L														
Dust and mud	L	L	L	L	L							L							L					L
Gaseous emissions	L	L	L	L	L							L							L					L
Liquid discharges	L			L													L							L
Solid waste disposal	L						L	L			M	L					L		L					
Immigration												L	L	L		L			L		L			
Employment												L	L	L	L	L			L		L	L		
Local expenditure												L	L	L	L	L			L		L			
Accidents/hazards	L	L		L	L	L										L								

Operational phase

Activity	Flora & fauna	Food chain	Air quality	Water quality	Soil	Sewage	Groundwater	Landscape	Natural resources	Climate	Hydrogeology	Population	Housing	Local economy	Employment	Health & safety	Land use	Risk	Traffic	Leisure & amenity	Infrastructure	Energy requirements	Cultural heritage	Industrialization
Physical changes	M	H	L	L	L	L	M	H	M	L	M	L	L									L		M/L
Earthworks	M	H	L	M	M	M	M	H	M	L	M	L			L		L				L			M/L
Raw materials demand	L	L	L			L	M	L	L	L					L			L			L	L		L
Traffic	L	H	L	M		M		L	M			L	M	M	L	L	M	L	M	H	M		L/H	M
Site noise	L	H	M	M				L	M	L		L	M	M	L		L	M	L	L	L			L/H/M
Vibration	M	M	M									L	M	L			L				L			
Dust and mud	L	L	L	L								L	M								L			
Gaseous emissions	L	L	L	L				L	L				M						L	L	L	M	L/M	L
Liquid discharges	L	H	L	H	M	M					L		L	L	L		L		L	L	L			
Solid waste disposal	L	M	L	M				L				L		L	L	L		L						
Employment		L										L	L	M	L		L		L	L	L			
Immigration												L	L	L		L			L	L	L			
Local expenditure								L	L	L		L	L					L	L		L		L	
Accidents/hazards	M	L		L	L	L										L								
Site control	L							L													L		M	M
Odours	M		M									L	M			L		L	M				M	M
Energy generation/conservation	L		L					L	L			L	L		L		L		L	L	L			
Land reclamation	H			M			L	M	M		M	L					M		M		M			L

Post-operation

Activity	Flora & fauna	Food chain	Air quality	Water quality	Soil	Sewage	Groundwater	Landscape	Natural resources	Climate	Hydrogeology	Population	Housing	Local economy	Employment	Health & safety	Land use	Risk	Traffic	Leisure & amenity	Infrastructure	Energy requirements	Cultural heritage	Industrialization
Accidents/hazards	L	L		L	M	L														L	L			
Land reclamation	M						M													L	L			L
Energy generation			L					L	L			L	L	L	L		L		L		L			
Odours			L					L				L												L
Solid waste disposal		L		M	L			L			L						L	L	L	L				
Liquid discharge	L	L		M	M	L													L	L	L			
Gaseous emissions	L			L				L									L	L	L	L				
Noise	L																		L	L				
Traffic	L			L																L				
Raw materials demand	L	L						L	L		M						L				L			
Physical changes	H	H		M		L	L	M	M	L		L						L		L	L	L		L
Cessation of activity	H	H		M	L			L	M	M	L	L	L	L	L	L	L	L	M	M	L	L	L	M

Figure 2.5. A scoping matrix for a waste treatment and disposal facility. H, high impact; L, low impact; M, medium impact

- temporal and spatial elements of the potential project impacts are identified
- impacts resulting from the project are differentiated from those produced by other causes
- the chosen technique should be appropriate to the skills of the user.

2.2.2. Baseline studies

The baseline survey should evolve from the scoping exercise in that it should aim to measure and describe those components of the specific environment identified as being significantly affected. The general objective of a baseline survey is to collect information and data so as to characterize the conditions (physical, ecological, social, etc.) which are currently existing, and which would be likely to exist in the future if the project was not undertaken. This prediction takes into account changes resulting from natural events and from other human activities which are independent of the proposed project. With this background knowledge three objectives can be achieved:

- potential impacts can be predicted and evaluated
- operational and emission standards can be set and mitigation measures devised
- any environmental changes resulting from the project once operational can be audited and detected.

Initial baseline studies may be wide ranging, but comprehensive overviews can be wasteful of resources. The studies should focus as quickly as possible on those aspects of the environment that may be significantly affected by the project, either directly or indirectly. The rationale for the choice of focus should be explained with reference to the nature of the project and to initial scoping and consultation exercises. There are numerous potential sources of baseline data; some examples are given in Table 2.6. However, the environment is a complex arrangement of interrelated systems which are very difficult to define and model, owing to their inherent variability. For a major activity there are numerous environmental characteristics (Table 2.7) and not all of them can be identified explicitly or quantified precisely.

2.2.3. The environmental impact statement

The Environmental Impact Statement (EIS) provides documentation of the information and estimates of impacts derived from the various steps in the assessment process. It is not a substitute for decision-making, but it does help to clarify some of the trade-offs associated with a proposed development action, which should lead to more rational and structured decision-making. The statement is normally wider in scope and less quantitative than other evaluation techniques, such as cost–benefit analysis. Matters to be considered for inclusion in an EIS are summarized in Table 2.8.

An EIS must include a description of the proposed development, the data necessary to identify and assess the main effects of the development, a description of the likely effects on the environment, and a statement of the measures foreseen in order to avoid significant adverse effects. The statement may also include:

- information on the physical characteristics of the proposed development
- the main characteristics of the main production processes proposed
- an estimate of the type and quantity of residues and emissions
- the main alternatives that can be foreseen
- the effects on the environment resulting from the use of natural resources or the emission of pollution

Table 2.6. Examples of sources of baseline data (after Fortlage, 1990)

Source	Baseline data
Rates roll and estate agents	Value of property, ownership of land
Mailing lists (local authorities sometimes sell lists of ratepayers)	Postal questionnaires
Aerial surveys which may have been carried out for other purposes	Updating of maps and land uses
Local history and conservation societies	Unrecorded information on rights of way, monuments, etc.
Residents' associations and amenity groups	Gauging public opinion
Railway companies, bus companies, local transport-user groups	Use of public transport
Schools, parents' groups	School journeys
Chambers of Commerce, local residents	Shopping patterns
Local residents, commuters	Private transport problems
Land-use classification maps or local planning authority	Land uses
Soil survey of England and Wales	Agricultural land classification
Geological survey maps	Geological and hydrological data
Mineral planning authority	Mineral resources
Water authority	Catchment areas, sewage and water supplies
River quality survey	Condition of streams and rivers
Local planning authority and English Heritage	Listed buildings and historic monuments, conservation areas
Railway Heritage Trust	Conservation of railway structures
Royal Society for Protection of Birds	Bird populations and habitats
Forestry Commission	Tree surveys
British Coal	Construction in coal-mining areas
DETR	Road policies and programme Aircraft noise, preferential routes
Gas companies	Gas mains and supplies
Electricity generating companies	Electricity supplies and transmission lines
HM Inspectorate of Pollution (HM Industrial Pollution Inspectorate in Scotland)	Limits of emissions, air pollution, radiochemical pollution, hazardous waste
Marine Pollution Control Unit	Offshore pollution control
Central Unit on the Environment, Department of the Environment	Policies on environmental protection
Ministry of Agriculture, Food and Fisheries (MAFF)	Levels of pollutants in water and farmland
National Radiological Protection Board (NRPB)	Natural radon levels
Nuclear Industry Radiation Waste Executive (NIREX)	Disposal of nuclear waste
Institute of Waste Management	Waste disposal techniques
Hazardous Waste Inspectorate (in Scotland)	Waste disposal

Table 2.7. Baseline data requirements for a waste management project

Characteristic	Appropriate data
Flora and fauna*†	Habitat type/diversity, current population, protected species, sensitivity to soil chemistry/water quality/air quality, migratory/overwintering animals
Food chain‡	Soil chemistry, vegetation, land use, type of agriculture
Air quality*	Odours, chemistry, dust or other mobile particles, seasonal pollen levels
Water quality*	Utilization, conductivity, colour/clarity
Soil‡	Type, structure, chemistry
Sewage*	Metals content, disposal outlets, industrial contamination, treatment works
Groundwater*†	Quality, utilization, extraction points, water collection/storage systems
Landscape†	Topography, land use, vegetation cover, areas of natural beauty
Natural resources	Lining materials source (local geology), construction materials
Climate*	Air quality, wind regime, rainfall
Hydrogeology*	Drainage, surface run-off, topography, site geology, water extraction, surface vegetation
Population‡	Density, demographic features
Housing*	Existing stock (for workers), housing quality/price range adjacent to sites
Local economy‡	Employment statistics, current employers
Employment‡	Available skills, employment statistics
Health and safety*	Medical statistics, traffic-accident statistics
Land use†	Derelict/blighted land, development plans, regional aid, agricultural land values
Risk	Mineral workings, site accessibility, centres of population, sensitive structures/locations
Traffic*	Statistics, road network, railways, noise/vibration surveys, congestion points, property surveys, parking
Leisure and amenity*†	Footpaths, watersports, air quality, traffic, tourism
Infrastructure	Road system, traffic statistics
Energy requirements*	District heating schemes, nature of local industry
Cultural heritage†	Tourist attractions, cultural assets, local history
Industrialization*	Property values, traffic composition

* Survey/assessment may be necessary.
† Important sites are generally already well documented.
‡ Not usually of major relevance to waste management works.

Table 2.8. Matters for inclusion in an environmental statement

Topic	Indicative content
Project overview	Purpose and physical characteristics of the project, including details of proposed access and transport arrangements
Land	Land-use requirements and other physical features of the project at all stages from during construction to after decommissioning
Raw materials	Types and quantities of raw materials used in the production processes; energy and other resources consumed
Pollution	Residues and emissions (to air/water, noise, vibrations, deposits, etc.) from the production process by type, quantity, composition and strength
Site selection	Main alternative sites and processes considered, where appropriate, and reasons for final choice
Site details	Information describing the site and its environment (topography, soil, water, flora, fauna, etc.)
Impacts	Assessment of effects on human beings, buildings, flora, fauna, geology, land, water, air and climate; indirect and secondary effects (transport, materials, consumption, etc.)
Impact alleviation	Mitigating measures (planning, technical measures, effectiveness, etc.)

- the methods of forecasting
- any difficulties in compiling the information.

A non-technical summary of the information supplied should always be included in order to improve communication between the various parties involved. Information acquired through an environmental statement will therefore help authorities to identify and take account of the various factors when considering whether to give planning permission for particular developments.

An EIA revealing many significant unavoidable adverse impacts would provide valuable information that could contribute to the abandonment or substantial modification of a proposed development action. If adverse impacts can be successfully reduced through mitigating measures, there may be a different decision.

The environmental effects of a development during its construction and commissioning phases should be considered separately from the effects arising when it is operational. Where the operational life of a development is expected to be limited, the effects of decommissioning or reinstating the land should also be considered separately.

Some environmental impacts will be avoidable and some will require specialized mitigation measures. Where significant adverse effects are identified, a description of the measures to be taken to avoid, reduce or remedy those effects should be provided. Assessment of the likely effectiveness of the mitigating measures should also be undertaken and appropriate means of monitoring their effectiveness should be considered. It should be remembered that environmental enhancement can sometimes be achieved with little or no additional cost if considered at an early stage. Different types of preventive/mitigation measures include:

- avoidance of the action/operation by selecting an alternative location
- modification of working practices during construction

- designing the development to minimize operational impacts
- minimization of impacts off-site
- limiting the magnitude of the action
- rectification of impacts by rehabilitation or restoration
- impact compensation (e.g. replacing/substituting resources or environments).

3 Basics of soil materials

3.1. Introduction

Ever since properly engineered construction work has been undertaken using geotechnical materials it has been necessary to have methods of characterizing materials and determining their suitability for use. Geotechnical materials are inherently variable and they are used in large quantities in various forms of construction. Furthermore, the practical methods of working with soil materials are relatively unsophisticated. The foregoing factors have fostered the development of index tests that can be undertaken relatively easily to classify soil materials and to provide benchmarks for assessing their suitability.

One of the primary elements of a water-retaining embankment is its very low permeability core, which traditionally has been formed by puddling clay. 'Puddling' is a process which destroys the natural structure of the clay by remoulding and working in extra water. This produces a very wet clay fill which will only allow very slow passage of water through itself (i.e. it has low permeability). Most embankment dams in the UK (around 70%) were built before the 20th century and, while the clay used for puddling had to be obtained from the vicinity of the dam, the material brought onto site still needed to be suitable. Hence simple index classification methods were developed for rapid characterization and classification of soil. An 1859 specification required puddle clay to be formed of 'the best and most tenacious clay' that could be found within the area of the reservoir. Before being used the clay was to be cut, wetted and worked, and to have any stones exceeding ½ lb in weight (equivalent to a stone diameter of approximately 50 mm) taken out. A much more recent specification (1987) for alterations at an existing dam called for the determination of basic parameters such as moisture content, index properties and particle-size distribution (Johnston *et al.*, 1990). The specification also required that, when at placement moisture content, the clay should pass the following empirical index tests:

- *Pinch test*. The clay is kneaded manually to form a ball about 75 mm in diameter; it should not be fissured at this stage. The ball is then squeezed flat to form a 25 mm thick disc; if no cracks form, the clay has passed the test.
- *Tenacity test*. A cylinder of clay (300 mm long, 25 mm diameter), is rolled. This cylinder is then held vertically from one end so that 200 mm of it is unsupported for a period of 15 s. If the clay supports its own weight, it has passed the test.
- *Elongation test*. A cylinder (formed as for the tenacity test) is gripped firmly at each end in the hands so as to leave 100 mm unsupported. With the cylinder held horizontally it is stretched by pulling on both ends so that it 'necks' and finally breaks. The longer the neck at failure the more suitable the clay.

3.1.1. Geotechnical materials

Rock is the hard, rigid deposits forming the part of the Earth's crust that underlies the soil. Rocks can be divided into three basic types based on their mode of origin: *igneous* (formed by the solidification of molten magma ejected from deep in the Earth's mantle), *sedimentary* (formed when deposits of gravel, sand, silt and clay, which are formed by weathering, are compressed and cemented together to form a solid material) and *metamorphic* (this involves changing of the composition and texture of rocks, without melting, by heat and pressure).

Soil is the uncemented aggregation of different-sized mineral grains and decayed organic matter, with liquid and gas in the spaces between the solid particles, overlying the rock. In the ground soil is not of one unique type and a typical soil profile consists of a succession of strata or horizons that have been formed by interaction of rock material with organic material and with the atmosphere. The development of a soil profile is determined by climate, vegetation, parent material, topography and time. The two groups of naturally occurring soils are those formed in situ and those transported to their present location. There are two different types of soil formed in situ: weathered rocks and peat. Transported soils are moved by the principal agents of water, wind and ice, although they can also be formed by volcanic activity and gravity.

The stress history of a soil (i.e. the loading it has undergone in its lifetime), has a major effect on its engineering behaviour. During deposition stresses in the soil will increase as more soil particles are placed. During erosion the overburden stress (that resulting from the weight of material overlying a particular point) will decrease as soil is removed. *Normally consolidated soil* has never been subjected to pressure greater than that imposed by its existing overburden. *Overconsolidated soil* has, at some time in its history, been subjected to pressures in excess of those due to the existing overburden. Overconsolidation is normally caused by the removal of overlying sediments, but can be caused in other ways (e.g. by being loaded with ice which subsequently melts, desiccation through evaporation).

Overall the ground profile can be divided into three broad bands:

- a top stratum that has been altered by natural processes such as leaching and washing away of material
- a stratum of accumulation or deposition containing the material that is leached or washed down from the upper layer
- a layer containing soil that is unaltered by weathering subsequent to deposition or formation.

3.1.2. Formation processes

The mineral grains that form the solid phase of a soil aggregate are the product of rock weathering. Weathering is the process of breaking down rocks by mechanical and chemical processes into smaller pieces. The forces associated with weathering and erosion constitute the chief agencies for production of soil from rock. Agents of change include:

- glaciation
- running water
- wind erosion
- expansion and contraction due to alternate freezing and thawing
- carbonic acid (water and carbon dioxide from the atmosphere)
- dissolution of soluble salts.

Water (e.g. wave action, erosion by running water, water expansion due to freezing) has by far the greatest effect. Other contributory factors are temperature, pressure from glaciers or other sources, wind, bacteria and human activity. The most erosive locations are in the highland or mountainous regions and upper reaches of rivers (especially during flood conditions) and along the coastline (particularly at high tides and during storms). The products of the erosive forces are subsequently deposited in lower regions such as deltas, valleys and plains. Weak cementing can occur due to carbonates or oxides being precipitated between the particles or due to the presence of organic matter.

With physical degradation processes the resultant soil particles usually retain the same composition as that of the parent rock. Particles of this type are approximately equi-dimensional, although the actual particles may be described as angular, subangular or rounded. The particles occur in a wide range of sizes, from boulders down to the fine rock flour formed by the grinding action of glaciers. Depending on the way in which the particles are packed together, the soil structure may be categorized as loose, medium dense or dense. Chemical weathering results in the formation of groups of crystalline particles of colloidal size (<0.002 mm) known as the 'clay minerals'. Most clay mineral particles are of plate-like form and thus have a high specific surface (i.e. a high surface area/mass ratio), with the result that their properties are influenced significantly by surface forces.

If the products of weathering remain at their original location they constitute a residual soil. If the products are transported and deposited in a different location they constitute transported soil, the agents of transportation being gravity, wind, water and glaciers. During transportation the size and shape of particles can change and the particles can be sorted into size ranges.

3.1.3. Soil constituents

Most types of soil consist of a mixture of gravel, sand, silt and clay particles (some soils also contain organic material). The terms 'clay', 'silt' and 'sand' are used in geotechnics to define the sizes of particles between specified limits as indicated in Table 3.1. The coarse constituents comprise sand, gravel, cobbles or boulders while the fine soil fraction consists of silt and clay particles (Table 3.1).

Gravels are pieces of rock with occasional particles of quartz, feldspar and other minerals. *Cobbles* and *boulders* are particles larger than 200 and 600 mm in diameter, respectively, and are classed as very coarse soils. Sand particles are composed primarily of quartz and feldspar, with the parent rock likely to have been granite.

Silts are usually the result of mechanical weathering and are similar in shape and composition to the coarse fraction particles (i.e. very fine quartz grains and some flake-shaped fragments of micaceous minerals). The grains in a deposit of silt are often rounded with smooth outlines. The particle shape and size influence the degree of packing of the particles and are also responsible for the volume increase (dilatancy) that a dense mass of silt or sand undergoes when it is sheared.

Table 3.1. Particle-size ranges

Designation	Subdivision	Particle size (mm)	Particle size (μm)	Category
Cobbles	—	>60	—	Coarse materials
Gravel	Coarse	20–60	—	
	Medium	6–20	—	
	Fine	2–6	—	
Sand	Coarse	0.6–2	—	
	Medium	0.2–0.6	200–600	
	Fine	0.06–0.2	60–200	
Silt	Coarse	0.02–0.06	20–60	Fines
	Medium	0.006–0.02	6–20	
	Fine	0.002–0.006	2–6	
Clay	—	<0.002	<2	

Clays are the finest soil particles (< 0.002 mm) and generally result from chemical weathering of rock fragments, although not all clay-sized particles are composed of clay minerals. Clays are mostly flake-shaped microscopic and submicroscopic particles of mica, clay minerals and other minerals. There are three major clay minerals:

- Montmorillonite is the most active mineral and exhibits marked volume changes upon drying or wetting
- Kaolinite is one of the last clay minerals to be formed in the weathering process and is very stable
- Illite is a very commonly occurring clay mineral and its structure results in it having a significant activity (intermediate between Kaolinite and Montmorillonite).

The active clay soils are more plastic than the inactive silt soils.

Peat is an accumulation of partially decomposed and disintegrated plant remains that have been fossilized under conditions of incomplete aeration and high water content. Peat can be divided into three basic groups: amorphous granular (with a high colloidal fraction), coarse and fine fibrous (both composed of distinct woody fibres). Moor and bog peats tend to be brown, fibrous and lightly decomposed, while fen peats are darker, less fibrous and more highly decomposed.

3.1.4. Soil characteristics

In addition to signifying specific sizes of particle, the terms 'clay, 'silt' and 'sand' are also used to denote 'types' of soil which exhibit certain behavioural characteristics.

In general a clay soil is one which exhibits apparent cohesion (it appears to have some inherent shear strength), plasticity (it can be flexed/distorted and will still stay integral) and a slow-draining nature. For instance, if a soil contains more than about 25% (by weight) of clay-sized particles, these fine particles begin to dominate the behaviour of the soil and overall it would be classified as a clay.

Sandy soils are cohesionless (they will fall apart when dry or saturated), frictional, non-plastic and fast-draining. For a granular soil (coarse silt, sands, gravel) the soil structure and resultant properties will depend primarily on the range of particle sizes present, the denseness of packing of the particles and the presence of fines and organic matter.

The general behaviour of silts is intermediate between that of sands and clays. They exhibit apparent cohesion but are clearly frictional, they can be non-plastic or plastic, and they drain rapidly by comparison to clays but compared to sands they have low permeability. When a significant amount of peat material is present in any of the preceding soils they are described as organic, and typically they will have enhanced compressibility and reduced permeability.

It is the physical properties that largely determine the engineering uses of a particular soil. The properties of prime importance in geotechnics are strength, compressibility, permeability, volume-change behaviour, compactability and frost susceptibility. As indicated briefly in the preceding paragraphs, soils can be divided into several general categories each of which exhibits certain overall behavioural characteristics:

The principal minerals in a clay deposit have a significant influence on its engineering behaviour. The ability of a clay soil to deform plastically is influenced by the size of its clay fraction and the type of clay minerals present, since the latter greatly influence the amount of water held in a soil. The shear strength is related to the amount and type of clay minerals present in a clay deposit, together with the presence of cementing agents. Geological age also has an influence on the engineering behaviour of a clay deposit. The porosity, water

Figure 3.1. Macro-structure (fabric) in clay soils

content and plastic nature normally decrease in value with increasing depth, whereas the strength and stiffness (load needed to produce a given deformation) increase. The macro-structure of a clay soil comprises the structure which can be seen with the naked eye and generally consists of features produced during deposition (such as inclusions, partings, laminations, varves) and features produced after deposition (e.g. fissures, joints, shrinkage cracks, root holes) (Figure 3.1).

Densely packed sands are almost incompressible, whereas loosely packed deposits are relatively compressible but otherwise stable (when located above the water table). Moist sands tend to give a misleading picture, as they will often stand almost vertically for a short time, but eventually the face either breaks up or slumps — just like a sand castle. Sands and gravels are highly permeable, and water flows into excavations below the water table will cause erosion at the foot of the slope (Figure 3.2), resulting in collapse. A coarse soil generally described as 'sand and gravel' may contain up to 35% silt and clay particles and will still behave as a free-draining granular material (BSI, 1981). However, if these fines are predominantly clay then they will probably dictate the behaviour of the soil, and so it may be better described as a clay.

Silty sands, when dry, will stand almost vertically. However, when wet, silts are extremely troublesome and are the most problematic soil material in practice. Excavation of saturated silt is difficult, as the soil flows into the excavation almost as quickly as it is removed, thus producing a very flat shape (Figure 3.3).

Figure 3.2. Wash-out of a sand pocket within a clay due to groundwater

Figure 3.3. Waterlogged silt

It is difficult to remove the water from silty soils and, in general, open excavations in such material are impracticable.

In coarse fibrous peat a fine mesh exists within the large interstices of the overall network of fibres, while in fine fibrous peat the interstices are very small and contain colloidal matter (Figure 3.4). The mineral content significantly

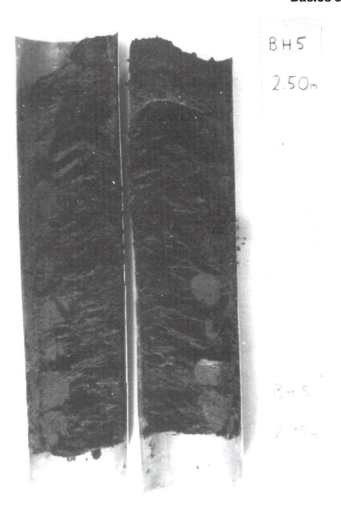

BH.5
2.50m

Figure 3.4. A split sample of peat

affects the engineering properties and usually consists of quartz sand and silt and often it increases with depth. Peats have extremely high water contents and most of the peculiarities of peat are attributable to this. This water may be in macropores (as in the fibrous peats) or in micropores (as in the more humified amorphous peats), and the high water content ensures that compression under load is large and rapid (10–75% of the original volume) and the peat has very low strength.

As the glaciers melted and retreated during the Ice Ages they left many large lakes wherein fine particles of clay and silts were deposited (glacio-lacustrine deposits) producing laminated and varved clays (Figure 3.5). Close to the glaciers the coarser particles (boulders, gravels, sands) were deposited as moraines. Glacial drift (glacial till, boulder clay) is composed of broken-up fragments of rock deposited through ice. It may be predominantly fine grained, or it may be composed of gravel, cobble- and boulder-size lumps of rock embedded in a fine clay matrix. Lodgement till was formed at the base of the glaciers and is often described as boulder clay. It was highly compressed beneath the moving glacier and, because of the weight of the overlying ice, such deposits are overconsolidated. The proportion of silt- and clay-size material is relatively high in lodgement till (typically the clay fraction is 15–40%). The fine fraction consists mostly of rock flour (finely ground rock) and the proportion of clay minerals is low. Lodgement till is commonly stiff, dense, relatively incompressible and practically impermeable (Figure 3.6).

Some clay soils exhibit significant sensitivity (i.e. loss of strength when distorted), the classic examples being the quick clays of Scandinavia. These

Figure 3.5. Laminated clay

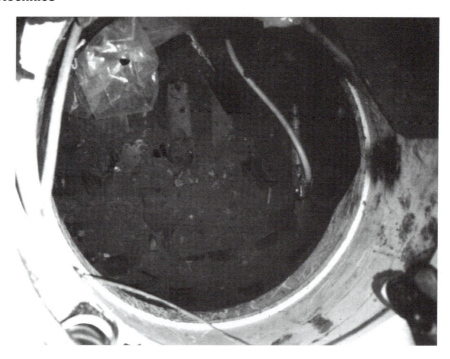

Figure 3.6. Mining through boulder clay

marine clays were laid down in a loose state under saline conditions. Subsequent leaching removed the bonding agency, leaving the particles in a very open, metastable configuration (like a 'house of cards') that needs only a slight stimulus to induce collapse. The soil then behaves like a dense fluid and can flow for long distances.

Chalk may be categorized as either a hard, relatively impermeable soil, or as a soft, earthy limestone rock. It is composed of calcium carbonate in a relatively pure state, formed from the shells and skeletons of creatures that lived in the sea. Intact chalk can exhibit high strength, but it can be reduced to the consistency of a putty by remoulding it at high water content.

Soils and rocks have been created by many processes out of a wide variety of materials. Geotechnical materials are notoriously variable and often have properties that are undesirable from a construction point of view. Unfortunately, the decision to develop a particular site is not usually made on the basis of its complete suitability from the engineering viewpoint, and hence geotechnical solutions are needed. To achieve these solutions definition and quantification of relevant parameters is required.

3.2. The soil model

In general, soils have solid, liquid and gaseous components. The solid phase is represented by mineral and some organic particles, varying in size from submicroscopic clay to visually discrete coarse gravel. The liquid phase is essentially water. The gaseous phase is very similar to the composition of air, but with a high content of water vapour. The variation in any, or all, of these three phases provides for non-uniformity in soil composition (and associated characteristics) and properties.

3.2.1. State parameters

Basic soil properties may be defined by visualizing the three component phases in the soil (solid, liquid and gaseous), as separate entities, as shown in Figure 3.7. The presence of voids between the soil grains and the medium occupying these voids are very important, and hence various geotechnical parameters have been defined to express and quantify the internal state of a soil:

SOIL MODEL	VOLUME	MASS	WEIGHT
Air	V_a	$V_a\,\rho_a \equiv V_a\,G_a\,\rho_w$	$V_a\,G_a\,\rho_w\,g \equiv V_a\,G_a\,\gamma_w \equiv W_a$
Water	V_w	$V_w\,\rho_w$	$V_w\,\rho_w\,g \equiv V_w\,\gamma_w \equiv W_w$
Soil grains	V_s	$V_s\,\rho_s \equiv V_s\,G_s\,\rho_w$	$V_s\,G_s\,\rho_w\,g \equiv V_s\,G_s\,\gamma_w \equiv W_s$
TOTALS	$V_a + V_w + V_s$	$\rho_w\,(V_aG_a + V_w + V_sG_s)$	$\gamma_w\,(V_aG_a + V_w + V_sG_s)$
	$V_v + V_s$	$\rho_w\,(V_w + V_sG_s)$	$\gamma_w\,(V_w + V_sG_s)$

Figure 3.7. The schematic soil model. For the definition of parameters, see text

The **voids ratio e** is defined as the ratio of the volume of voids V_v to the volume of solids V_s:

$$e = \frac{V_v}{V_s} \tag{3.1}$$

For regularly packed, identical spheres the densest packing gives a voids ratio of 0.35, while the loosest packing gives a value of 0.91. The voids ratio cannot be measured directly and it is calculated from other parameters which can be measured.

The *porosity n* is defined as the percentage of the total volume V that consists of voids:

$$n = \frac{V_v}{V} \tag{3.2}$$

The voids ratio and porosity are related as follows:

$$V = V_v + V_s = V_s + eV_s = (1 + e)V_s \tag{3.3}$$

Hence

$$n = \frac{V_v}{V_s(1 + e)} = \frac{e}{(1 + e)} \tag{3.4}$$

The *moisture content w* is the ratio of the weight of water W_w to the weight of solids W_s:

$$w = \frac{W_w}{W_s} \tag{3.5}$$

The moisture content (or water content) is determined by simply drying a soil sample (at 105°C) and measuring its weight before and after drying — w is then quoted as a percentage.

Many geotechnical engineering situations require the calculation of forces and stresses within the ground. These values can be obtained directly if the bulk unit weight of the soil is known. This bulk unit weight is also an indicator of the state of denseness of the soil mass. The unit weight γ is directly related to the density ρ: $\gamma = \rho g$.

The *bulk unit weight of a soil* γ is simply the total weight per unit volume:

$$\gamma - \frac{W}{V} = \frac{(W_s + W_w + W_a)}{(V_s + V_w + V_a)} \tag{3.6}$$

The preferred units of unit weight are kN/m^3 while densities should be quoted in terms of Mg/m^3.

The *specific gravity* G is defined as the ratio of the unit weight of a given material to the unit weight of water. Thus for the soil solids $\gamma_s = \gamma_w G_s$ (for most soils G_s lies in the range 2.6–2.8 and a commonly assumed average value is 2.65) and $W_s = G_s \gamma_w V_s$. (Note: It is common practice to denote the specific gravity of the soil grains simply as G).

Note that

$$W_a \cong 0, \quad V_v = V_w + V_a, \quad W_w = \gamma_w V_w, \quad e = \frac{V_v}{V_s} \tag{3.7}$$

and thus

$$\gamma = \frac{(G_s \gamma_w V_s + \gamma_w V_w)}{(V_s + V_v)} = \frac{\gamma_w \left(G_s + \dfrac{V_w}{V_s}\right)}{(1+e)} = \frac{\gamma_w (G_s + m)}{(1+e)} \tag{3.8}$$

wherein m is the *volumetric moisture content* and is expressed as:

$$m = \frac{V_w}{V_s} = \frac{\left(\dfrac{W_w}{\gamma_w}\right)}{\left(\dfrac{W_w}{G_s \gamma_w}\right)} = wG_s \tag{3.9}$$

While m is a very useful indicator of the moisture content, it is not as convenient to measure as w because of the need to determine G_s.

The *degree of saturation* S_r is defined as the proportion (expressed as a percentage) of the total voids volume that is occupied by water:

$$S_r = \frac{V_w}{V_v} \tag{3.10}$$

For a fully saturated soil all the voids are completely filled with water so that the volume of voids is equal to the volume of water and $S_r = 100\%$.

From the previous definition (equation 3.9)

$$m = wG_s = \frac{V_w}{V_s}$$

but

$$\frac{V_w}{V_s} = \left(\frac{V_w}{V_v}\right)\left(\frac{V_v}{V_s}\right) = S_r e$$

so that

$$m = wG_s = S_r e \tag{3.11}$$

Thus the expression for bulk unit weight (equation 3.8) can be written as

$$\gamma = \frac{\gamma_w(G_s + S_r e)}{(1+e)} = \frac{\gamma_w(G_s + wG_s)}{(1+e)} = \frac{G_s \gamma_w(1+w)}{(1+e)} \tag{3.12}$$

The preceding equations can be inverted to provide expressions for determining the voids ratio and degree of saturation:

$$e = \frac{G_s \gamma_w(1+w)}{\gamma} - 1 \tag{3.13}$$

$$S_r = wG_s \left[\frac{G_s \gamma_w(1+w)}{\gamma} - 1 \right] \tag{3.14}$$

To determine the bulk density of soil it is necessary to measure the volume of a sample and weigh it. The simplest procedure is to test samples of regular shapes, such as a right cylinder, so that the volume can be obtained by linear measurement. This approach is only suitable for soils of a cohesive nature and which are little affected by sample preparation. The volume of an irregular lump of soil can be found by coating it with a layer of molten paraffin wax, weighing it, then immersing it in water contained in a measuring cylinder and recording the volume of water displaced. Soils which do not exhibit apparent cohesion cannot be sampled in an undisturbed manner.

The extreme density states of a granular mass (i.e. its densest and loosest conditions), can be measured by placing a measured weight in a measuring cylinder and measuring its maximum and minimum volumes after shaking. The *relative density* (RD) ranges from 0% (loosest) to 100% (densest) and is defined as:

$$\text{RD} = \frac{(\text{density} - \text{minimum density})}{(\text{maximum density} - \text{minimum density})} \tag{3.15}$$

The *dry unit weight* γ_d is the weight of dry solids per unit volume, so it is the value of the bulk unit weight γ for $S_r = 0$ and $w = 0$:

$$\gamma_d = \frac{G_s \gamma_w(1+0)}{(1+e)} = \frac{G_s \gamma_w}{(1+e)} \tag{3.16}$$

or

$$\gamma_d = \frac{\gamma}{(1+w)} \tag{3.17}$$

Typical values of some of the foregoing parameters, for natural soils, are given in Table 3.2.

3.2.2. Principle of effective stress

The pore spaces between the soil particles are not discrete but consist of small irregular passages. It is possible for water to be stored in the voids and also for it to move through the soil via the void passages, although this movement may in some cases be extremely slow. In general, any external total stress applied to the soil element will be carried partly by the pore fluid and partly by the soil solids.

Table 3.2. Typical ranges of state parameters

	e	w (%)	γ(kN/m^3)
Sand	0.45–0.90	6–20	15–18
Silt	0.40–1.10	10–25	18–20
Clay	0.40–1.30	11–30	16–22
Till	0.30–0.60	8–15	18–24
Peat	1–20	200–2000	10–12

The *pore pressure u* is the pressure induced in the fluid (water, or vapour and water) filling the pores. Pore fluid is able to transmit normal stress, but not shear stress (a fluid simply flows when it is sheared), and it is therefore ineffective in providing shearing resistance within soil. For this reason, the pore pressure is sometimes referred to as *neutral pressure.*

Consider the section through the soil shown in Figure 3.8. It is possible to draw a section line (such as XX) which passes through the points of contact between its solid particles but which primarily lies within the pore fluid. If force equilibrium is considered in any direction then:

Total boundary force = force transmitted between grains + force transmitted by pore fluid

or

$$P = P' + Uw \tag{3.18}$$

If A is the cross-sectional area under consideration, then

$$P = \sigma A \tag{3.19}$$

where σ is the average total normal stress acting on the boundary of the element.

The area of physical contact between solid particles is actually very small because the grains have curved surfaces and do not fit together perfectly. The contact area is also very small by comparison to the total cross-section. Thus the area over which the pore fluid acts is approximately equal to the total cross-sectional area of the soil element. Hence the pore fluid force is equal to the pore fluid pressure multiplied by the cross-sectional area uA:

$$\sigma A = P' + uA \tag{3.20}$$

or

$$\sigma = \frac{P'}{A} + u \tag{3.21}$$

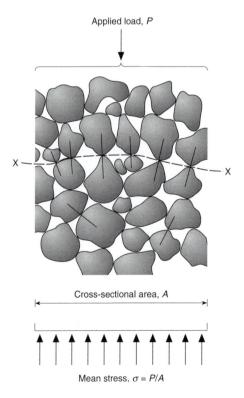

Applied load, P

X — — — X

Cross-sectional area, A

Figure 3.8. A section through a soil element

Mean stress, $\sigma = P/A$

P'/A is the notional mean stress transmitted through the solid particles (i.e. through the soil skeleton), and it is directly related to the contact stress between particles. Hence this notional mean stress will be indicative of skeleton compression or deformation and shearing resistance at intergranular contact points (Terzaghi, 1966). Thus P'/A is a prime indicator of the geotechnical behaviour of a soil; it is denoted by σ' and is called the *effective stress*. Terzaghi formulated the *principle of effective stress* and demonstrated that, for a saturated soil, effective stress may be defined quantitatively as the difference between the total stress and the pore water pressure:

$$\sigma' = \sigma - u \tag{3.22}$$

Thus

$$\sigma = \sigma' + u \tag{3.23}$$

The concept of effective stress is very important because water has negligible resistance to shearing, it is relatively incompressible and its properties are unaffected by an increase in pressure. In general, the pore fluid is a mixture of gas (usually air) and water, and the pressures in these two phases will not be the same (due to surface tension). However, for soils with a high degree of saturation (90% or more) the pore pressure can be taken as equal to the water pressure (in the UK most slow-draining engineering soils exhibit high degrees of saturation).

In the case of partially saturated soils, the pore fluid will consist of liquid water, which is virtually incompressible, and air/water vapour, which is highly compressible. There are thus two components of pore pressure: the pore water pressure u_w and the pore air pressure u_a. Due to surface tension, the presence of air reduces the pore pressure:

$$u = u_a - \chi(u_a - u_w) \tag{3.24}$$

The parameter χ is dependent mainly on the degree of saturation and, to a lesser extent, on the soil fabric structure. It is possible to determine χ experimentally and it seems to vary almost linearly from zero for dry soil to unity for saturated soil. For most practical situations in the UK, engineering soils have degrees of saturation of 90% or more, so that χ is very near to unity. In such cases, the small amount of air present will be in the form of occluded bubbles, which affect the compressibility of the pore fluid without significantly lowering the pore pressure. A reasonable approximation for effective stresses in the partially saturated zone can be obtained from:

$$\sigma' = \sigma + \gamma_w d S_r \tag{3.25}$$

where d is the elevation above the water table and S_r is the degree of saturation.

3.3. Soil classification

Soil has not been made under carefully controlled conditions like some engineering materials (e.g. steel, cement), and even within a 'uniform stratum' basic properties (e.g. density, moisture content) may vary by 10% or more within a volume equal to that of a cupboard. However, soils may be categorized, or classified, according to their overall engineering characteristics. The object of soil classification is to divide soils into groups such that all the soils in a particular group have similar characteristics, by which they may be identified, and exhibit similar behaviour in given engineering situations. Soil classification, however, should be regarded as the first step only in the evaluation of a soil, as the classification tests use samples of the soil in a disturbed form and the properties of the soil in its in situ condition may not be accurately represented.

There are many methods available for the classification of soils and the choice of method depends on the specific use intended for the soil. Geological classifications of soils differ from those used in soil engineering and from those used in agriculture. The soil classification system used in geotechnical engineering divides soils into a number of groups, each denoted by a group symbol. A soil is identified and allocated to the appropriate general group on the basis of particle-size distribution and its consistency characteristics. These properties are determined either by standard tests in the laboratory or by simple, and less accurate, visual and manual procedures in the field. Designations such as 'gravel', 'sand', 'silt' or 'clay' are used according to the predominant size of particles within the soil. The British Standard particle size classification gives full guidance on how to designate a soil (BSI, 1981). In addition to the appropriate group symbol (Table 3.3), a general description of the soil and its in situ condition should be given. This description should include the colour of the soil, details of the predominant particle shape and, if possible, the mineral composition of the particles.

3.3.1. Particle-size distribution

Most natural soils are composites, that is mixtures of particles of different sizes. The distribution of these sizes gives very useful information about the engineering behaviour of the soil. Particles sizes vary considerably, from those measured in microns (clays) to those measured in metres (boulders). The larger particles (i.e. sands and gravels) form the skeleton of the soil and determine many of its mechanical properties. The finest particles (i.e. the clay) have a large surface area and determine most of the chemical and physico-chemical properties of the soil. Particle-size contribution influences water-holding capacity, strength and compressibility. It is common practice to recognize two general categories of soil particles: granular (gravel, sand, silt) and colloidal (clay). For any particular soil the 'fines' are those particles smaller than $60 \, \mu$m (i.e. the silt and clay particles). The general physical characteristics of generic soil 'types' are outlined in Table 3.4.

The particle-size analysis of a soil sample involves determining the percentage by weight of particles within the different size ranges. There are two general methods for finding the particle-size distribution of soil:

- sieve analysis (for particles $>63 \, \mu$m diameter)
- sedimentation/hydrometer analysis (for particles $<63 \, \mu$m diameter).

Sieve analysis consists of passing the soil sample through a series of sieves that have progressively smaller openings. Very coarse materials (e.g. cobbles) are not usually included in a particle-size distribution test by sieving. First, the soil is oven dried and all lumps are broken into small particles before being passed through the sieves. To ensure separation of particles, wet sieving is preferred, where the soil particles are soaked and washed through the sieves in order to remove fine particles that otherwise might adhere to larger particles of the material. If the soil fines are cohesive a dispersant should be added to the wash

Table 3.3. Soil-group symbols

Primary letter		Secondary letter	
G	gravel	W	well-graded
S	sand	P	poorly-graded
M	silt	M	non-plastic fines
C	clay	C	plastic fines
O	organic	L	low plasticity
Pt	peat	H	high plasticity

Table 3.4. Physical characteristics of soil 'types'

Sand	Silt	Clay
Single particles visible	Some particles visible	No particles visible
Crumbles very easily, falls off hands when dry	Easy to crumble, can be dusted off hands when dry	Hard to crumble, sticks to hands when dry
Gritty between fingers	Rough between fingers	Smooth between fingers
No plasticity	Some plasticity	Plasticity
A small lump of damp sand placed in water disintegrates to form a conical heap immediately	A small lump of damp silt placed in water will disintegrate to form a low conical heap after some minutes	A small lump of clay placed in water will disintegrate to form a very flat conical heap (but only after some hours or even days)

water to separate the fine grains. The dry weight of soil retained on each sieve is determined and the cumulative percentage (by weight) passing each sieve is calculated. Grain shape may also be assessed visually.

To determine the fines content a sedimentation method is used with either a pipette or a hydrometer. It is essential that the soil particles are separated from each other before sedimentation so a pretreatment dispersing procedure is carried out. When a soil specimen is dispersed in water, the individual particles settle at different velocities depending on their shape, size and weight and the viscosity of the water. If the soil particles are assumed to be spheres then the settling velocity of the particles is proportional to the square of their diameter (Stokes' law). Hence we can estimate the time that it will take for particles of a particular size to settle by a given distance. All particles larger than this size will settle more quickly, or more will settle in the same time, while finer particles will settle less. Hence by measuring the weight of soil in suspension at a particular depth at different times from the start of settling the proportion of particles in different size bands can be estimated.

The pipette method involves taking a small sample (about 10 ml) of the soil suspension at a depth of 100 mm below the fluid surface at particular times, corresponding to specific particle diameters. In total, three successive samples of the suspension are taken. Typical sampling times (after shaking) are 4 min 5 s, 46 min and 6 h 54 min. These periods, which have been derived on the assumption that the specific gravity of the soil particles is 2.65, correspond to the times for particles of coarse silt, medium silt and fine silt, respectively, to settle by 100 mm.

The alternative hydrometer method is based on measuring the density of the fluid and soil suspension as it reduces with time due to the particles settling. More readings can be taken with this test than the pipette method, but it can be more prone to errors since difficulties can be experienced in reading the meniscus around the hydrometer and calibration of the hydrometer scales is required.

The results of a particle-size analysis are presented as a semi-logarithmic plot (the percentage of material passing against the logarithm of the particle size) to give the particle size distribution (PSD) curve (Figure 3.9). The flatter the distribution curve the larger the range of particle sizes in the soil, the steeper the curve the smaller the size range. A soil is described as well-graded if there is no excess of particles in any size range and if no intermediate sizes are lacking. In general, a well-graded soil is represented by a smooth, concave distribution curve. A soil is described as poorly-graded if a high proportion of the particles have sizes within narrow limits (a uniform soil) or if particles of both large and

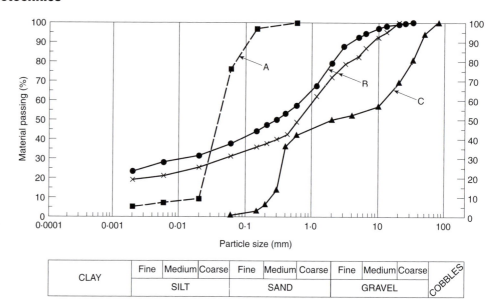

Figure 3.9. Particle-size distribution curves. A, poorly graded sandy silt (single-size material); B, well-graded boulder clay (two samples from adjacent sampling points); C, gap-graded sandy gravel

small size are present, but with a relatively low proportion of particles of intermediate size (a gap-graded soil). There are two primary coefficients for indicating the shape of the grading curve, the coefficient of uniformity (CU) and the coefficient of curvature (CZ):

$$CU = \frac{d_{60}}{d_{10}} \tag{3.26}$$

$$CZ = \frac{d_{30}^2}{d_{60}d_{10}} \tag{3.27}$$

where d_{10}, d_{30} and d_{60} are the particle sizes corresponding to the 10%, 30% and 60% 'finer than' (% material passing) points, respectively, on the PSD curve.

The higher the value of the CU the larger is the range of particle sizes in the soil. For example, if the soil is well-graded the CU will be greater than 4 for gravels and 6 for sands. A well-graded soil has a CZ of 1–3; very high or very low values indicate that the PSD curve is irregular.

3.3.2. Consistency limits

With a free-draining soil the strength and compressibility are only slightly affected by a change in water content, and silts and sands are not particularly susceptible to shrinkage. However, the consistency of fine soil varies drastically according to the amount of water present — when completely dry the soil may be hard/solid while at high water contents it may be almost a slurry/liquid. This behaviour results from the nature of the water near the surface of a clay particle, which tends to exhibit excess negative charges on its flat sides and positive charges at the edges. Because of these surface electrical charges the water in the vicinity of the clay particle is effectively immobilized (it is adsorbed) and may be considered essentially solid. With distance from the particle surface this effect decreases until it disappears and the water has the properties of free water. With an abundance of water the soil particles are separated by free water and the mixture is more or less fluid; as the amount of water decreases the particles are separated by increasingly stiffer water and the mixture becomes like a solid.

If a soil starts with a high moisture content (i.e. it is a viscous liquid), it gradually shrinks as it loses water (Figure 3.10) until a state of equilibrium is reached at which further loss of moisture will result in no further volume change

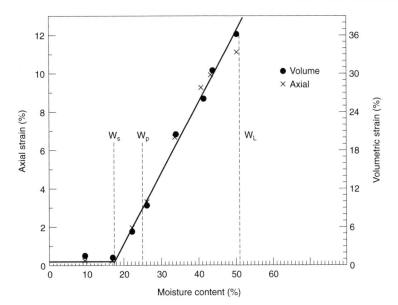

Figure 3.10. The effect of moisture content changes on the volume of a clay soil

(i.e. it is effectively a solid). In field situations shrinkage may occur naturally due to moisture movement near the ground surface (as moisture evaporates in dry weather) and by transpiration of soil moisture from greater depths (via the roots of vegetation). A classification of the behaviour of a soil into four discrete, basic states, i.e. solid, semi-solid, plastic and liquid, may be made on the basis of its moisture content (Figure 3.11). In reality the transition between the foregoing states is gradual rather than abrupt. The boundaries between these phases are called the *consistency limits* and their definition is rather arbitrary:

- *Shrinkage limit* (w_S) — the water content at which a further decrease in moisture content does not cause a decrease in volume of the soil.
- *Plastic limit* (w_P) — the minimum moisture content at which a thin thread of soil can be rolled by hand without breaking up.
- *Liquid limit* (w_L) — the minimum content at which the soil is assumed to flow under its own weight.

The difference between the liquid and plastic limits is the range of water contents over which a soil is plastic, and it is designated as the *plasticity index* (I_p). Liquid and plastic limits are determined by relatively simple laboratory tests, which have been used extensively for the correlation of various physical parameters as well as for soil identification. The liquid limit is determined using the cone penetrometer (Figure 3.12). Soil passing the 425 μm sieve is mixed to a paste and placed in a cup and trimmed level with the top. A cylindrical cone, with a smooth, polished surface, is lowered to just touch the soil surface and its initial position is recorded. The cone is then released to sink freely into the soil paste for

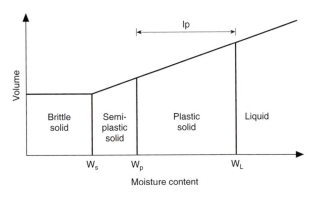

Figure 3.11. The consistency states of soil

Figure 3.12. Cone penetrometer apparatus

5 s, after which the new position is noted so that the height through which the cone has fallen is known. The moisture content of the soil is then increased, the paste is thoroughly mixed and the cone test repeated, at least five times. The data pairs (moisture content and associated cone penetration) are plotted and the best straight line fitted to them. The liquid limit is the moisture content corresponding to a penetration of 20 mm (Figure 3.13).

In the plastic limit test the soil is initially dried (preferably from its natural state) by moulding it into a ball and rolling it between the hands. When the soil is near its plastic limit a thread about 6 mm in diameter is rolled back and forth over the surface of a glass plate with just enough rolling pressure to reduce the thread to a diameter of 3 mm. If the soil does not shear at this stage then it is dried further and the test is repeated until the thread crumbles or shears both

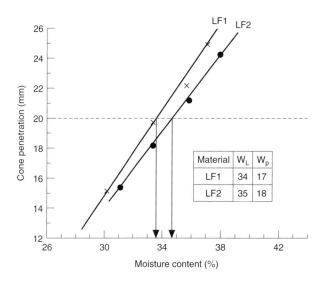

Figure 3.13. The determination of the liquid limit

Material	W_L	W_p
LF1	34	17
LF2	35	18

longitudinally and transversely. The soil is now considered to be at its plastic limit and its moisture content is determined.

Linear shrinkage is determined by mixing soil with water to form a paste, with a moisture content approximating to the liquid limit. The sample is placed in a brass mould shaped like a semi-circular trough. The sample is dried out completely with the *linear shrinkage* (LS) being equal to (1 − final length/initial length). Alternatively, a cylindrical sample may be prepared and dried in stages and measurements made (length, average diameter, weight) to determine relationships between volumetric/linear shrinkage and moisture content (Figure 3.10) and to identify the shrinkage limit. The amount of shrinkage will depend on the clay content and its mineralogy, but it will also be significantly affected by the structural arrangement of the particles. For this reason a test may be carried out on an undisturbed cylindrical specimen. The determination of the liquid and plastic limits may be difficult in the case of soils having low percentages of clay mineral particles, in which case the plasticity index may be estimated from the linear shrinkage (expressed as a percentage):

$$I_p = 2.13 \times LS \tag{3.28}$$

3.3.3. Soil plasticity

Different soils may be distinguished by their plasticity characteristics because these vary with the surface activity of the constituent particles. The results obtained from the liquid and plastic limit tests can be employed to classify a fine-grained soil by using the plasticity chart proposed by Casagrande (Figure 3.14) in which the plasticity index is plotted against the liquid limit. The chart separates fine-grained soils into two basic groups by means of the empirical A line. Over most of its length the A line has the equation $I_p = 0.73 (w_L − 20)$ although it is horizontal at its lower end (when $w_L < 28$). This line divides the inorganic clays (above the line) from the inorganic silts and organic soil (below the line). There is

Silt (*M*) plots below the *A* line
Clay (*C*) plots above the *A* line } *M* and *C* may be combined as *fine soil* (*F*)

The letter *O* is added to the symbol of any material containing a significant proportion of organic matter.

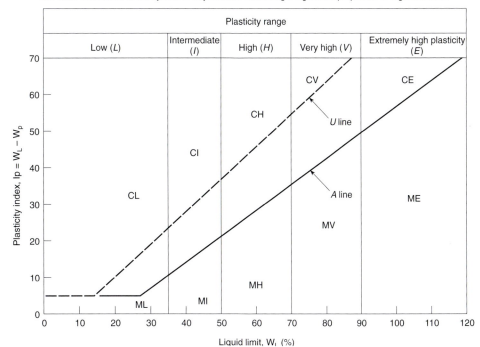

Figure 3.14. The plasticity chart

another line above the A line, which is the approximate upper limit of combinations of plasticity index and liquid limit found for any soil. This is the U line, which has the equation $I_p = 0.9(w_L - 8)$ (Figure 3.14). The Casagrande chart is subdivided into zones of different plasticity (according to liquid limit values) from low ($w_L < 35$) to extremely high ($w_L > 90$), with symbols for each different type of soil (e.g. CH, clay with a high upper plasticity range).

4 Ground investigation

4.1. Introduction

Site investigation in its widest sense is the process by which relevant site information (e.g. geotechnical, environmental, chemical), which might affect the construction or performance of a civil engineering or building project, is acquired (Clayton *et al.*, 1982). On the other hand, the term 'ground investigation' is used to cover the detailed work undertaken (excavation, boring, drilling, sampling, testing) to identify the geological/geotechnical characteristics of the site. Ground investigation involves exploring the characteristics of a site as laid down by nature and modified by man, and representing the results in a form that can be rapidly understood and assimilated.

The ruling in the case of Rylands vs Fletcher (1868) was for many years a keystone in the law of tort. The case itself is a prime example of the importance of a thorough ground investigation. The plaintiff in the case was the occupier of a mine and works under a plot of land. The defendants were the owners of a nearby mill who decided to construct a water reservoir adjacent to their mill, some distance from Rylands' land. Beneath the defendants' plot of land, on which they proposed to construct their reservoir, there were old and disused mining passages and works comprising five vertical shafts and some horizontal shafts communicating with them. Apparently the vertical shafts had been infilled with soil and waste materials and it does not appear that anybody was aware of the existence of either the vertical shafts or the horizontal works communicating with them. The reservoir was subsequently constructed under the supervision of an engineer. As the reservoir was being filled the water broke into the underlying disused mine workings thereby flooding the neighbouring mine workings and causing considerable damage. According to the ruling Law Lord, the principles on which the case was determined in law were extremely simple in that the defendants had used their plot of land for 'a non-natural use'; that is, they had introduced into the area something which did not naturally belong there. Consequently, if, as a result of any imperfection in the mode of retaining this item on their land, it escaped and caused damage then the defendants were liable. This was exactly the position in this case, and the defendants had to pay substantial damages (All England Law Reports, 1873). After the shafts had been properly sealed the reservoir was successfully filled (Figure 4.1).

Unrealistically cheap ground investigations usually fail to present an accurate picture of ground or groundwater conditions, and it is not surprising that the geotechnical works designed for the site are often not suited to the actual ground conditions. Apart from meeting unforeseen ground, other consequences of poor ground investigation have included:

- differential settlements leading to foundation problems
- encountering unrecorded subterranean voids (chambers, tunnels, shafts)
- intercepting unknown springs, resulting in site flooding
- underpinning required for adjacent buildings
- generation of excessive ground vibrations.

Figure 4.1. Completed reservoir at the centre of the Rylands vs Fletcher case

In such circumstances the costs of remedying wrongly-designed works, or of mobilizing alternative construction methods, are usually far in excess of the money saved on an inadequately-designed site investigation.

Various reports over the past 25 years have shown that in civil engineering and building projects the largest element of technical and financial risk usually lies in the ground. Claims for unforeseen ground conditions can easily amount to 10% or more of the contract value. However, expenditure on site investigation as a percentage of the total project cost is low and typically ranges from a mere 0.1% to 0.3% for building projects (Littlejohn *et al.*, 1994). There seems to have been a general, adverse, decrease in site investigation expenditure over the past 50 years. Rowe (1972) cited a typical mean investigation expenditure of 0.7–1.5% of the total cost of the works (Table 4.1), whereas in the 1940s Harding (according to Clayton *et al.*, 1982) assessed the cost of site investigations to be usually 1–2% of the total cost of the works.

4.1.1. Objectives of investigation

It is necessary to collect and collate information and make assessments of the likely ground conditions so that construction projects can be designed, methods of construction can be determined and the overall costs and risks can be estimated. Thus various parties (i.e. the designer, the contractor, the client), will require information and its interpretation, but their needs, responsibilities and concerns will differ. However, the general objectives of ground investigation can be summarized as the provision of geotechnical data for the following processes:

- Site selection: the construction of major projects is dependent on the availability of a suitable site.

Table 4.1. Typical site investigation costs

Rowe (1972)		Littlejohn *et al.* (1994)	
Type of works	% of total cost	Consumer	% of total cost
Earth dams	0.9–3.3	Government authorities	0.29
Embankments	0.1–0.2	Civil engineering contractors	0.23
Docks	0.2–0.5	Developers/builders	0.11
Buildings	0.1–0.2	Civil engineering consultants	0.29
Roads	0.2–1.6	Structural engineers	0.16
Overall mean	0.7	Average	0.21

- Permanent works design: to allow a safe and economical design to be prepared.
- Temporary works design: the actual process of construction may impose greater stress on the ground than the final structure.
- Investigation of existing construction: to permit assessment of the current situation.
- Remedial works: to obtain parameters for design.
- Assessment of the effects of the proposed project on its environment: structural distress to neighbouring structures, ground pollution, etc.

The foregoing objectives are achieved by deriving a three-dimensional model for the ground at the site. Exploratory holes such as boreholes and trial pits establish the ground conditions at various points and cross-sections across the site are completed by interpolation. For uniform homogeneous conditions and simple geology this could be achieved fairly confidently with widely-spaced exploratory holes, but for variable conditions an accurate impression can only be gained from closely-spaced boreholes. Boreholes only provide a view of the ground at their specific locations and their interpretation, and the inference of ground between holes, is a matter of judgement based on geotechnical knowledge and experience.

4.1.2. General methodology

At one extreme an investigation may consist of a brief site walk-over and consultation with records, supplemented possibly by the excavation of a few trial pits. At the other extreme, for major developments, a comprehensive programme of borings, laboratory and field testing and analysis may be required. Ideally, ground investigation should always be carried out by geotechnical specialists and there should be field supervision of drilling by experienced engineers who are conversant with the aims of the proposed works. Major ground investigations

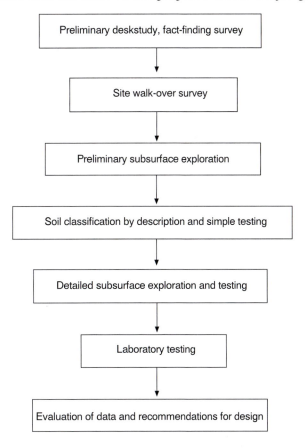

Figure 4.2. Ground investigation activities

involve a considerable number of activities, with the general sequence of activities being as illustrated in Figure 4.2.

In the early stages of any projected development, a desk study should be made of all sources of information already available. These would normally include old and current Ordnance Survey maps and geological maps and records. Information may also be available from past development of the site or of adjacent property. Potential sources of information are outlined in Box 4.1. Aerial photographs of the location may be available. Examination of these photographs of the site will provide an appreciation of the local topography and may reveal anomalies or items not recorded on maps. The desk study should identify the general expectations of the site — the superficial and solid geological formations and associated soil and rock types, the likelihood of structural discontinuities or subsurface topographical variations, etc.

A walk-over inspection of a site is essential in all ground investigations, particularly since topographical and geological features of an area are frequently indicative of subsurface conditions. The presence of fill, compressible soils, shallow rocks, high water-table, etc., may often be deduced from observations of excavations or natural escarpments on, or in, the area around the site. An appropriate physical investigative technique may subsequently be selected to provide information on the anticipated variations in ground conditions. There is a considerable variety of methods of ground investigation, and normally a combination of methods is employed to cover the technical requirements and the range of ground conditions that are encountered (BSI, 1981). As a result of the information obtained from the desk study and site reconnaissance, together with the nature and scale of the project, the extent of ground investigation required can be decided. In planning these investigations attention should be paid to the safety of personnel involved in site or laboratory work, particularly where the site has a history of industrial usage.

For most sites ground investigation will involve a pattern of trial pits (shallow excavations) and/or boreholes (deep holes). These may be set out in relation to the proposed position of construction works. Alternatively, a grid layout may be satisfactory, unless the site topography and features determine the locations. Trial pits are usually dug relatively close to one another. For many building purposes a borehole spacing of 20–25 m is satisfactory, but considerably wider spacing may be appropriate for an initial overall appraisal of a large site. At least three boreholes are usually required for anything but very small developments. It is essential that the location of every excavation, boring or probing is recorded accurately during the execution of the fieldwork. It is equally vital that the ground levels of these locations be established and recorded (unrecorded infilling and waste disposal can have a significant, local effect on levels).

The depth of ground boring is generally determined by the depth of excavation associated with the site development and the possible depth of influence of foundations. It is usual to take boreholes in thick compressible strata to a depth where the changes in stress applied by the works are minimal, and for shallow loaded areas this depth may be taken as three times the width of the likely foundation. Typically, the recommended borehole depth for retaining walls is up to twice the wall height below the base of the bottom of the wall (Clayton & Mililitsky, 1986). Where piled or other deep foundations are envisaged as a possibility, investigation should extend some 5 m below the possible toe level. The required depth of boreholes rarely exceeds 30 m on most industrial sites, but old mine workings and the presence of subterranean passages/voids may warrant special criteria.

Topographical maps
- The 1 : 50 000 series is useful for general planning and location purposes, but contains too little detail for most site investigations.
- The 1 : 25 000 scale map is useful for preliminary assessment of a site. The use of colour facilitates identification of important features such as water-course, springs, woodlands, and public rights of way.
- Large-scale maps will be needed for the later stages of site investigation: 1 : 2500 (cover the whole of Great Britain), 1 : 1250 (cover most towns and urban areas).

Geological maps
- 1 : 50 000 geological maps are generally available in drift editions (shows the distribution of surface deposits) or solid editions (shows the distribution of formations that would be exposed at the ground surface if the drift was not present).
- 1 : 10 000 sheets should be examined for detailed geological information on a site. They often have descriptive notes printed on them and the positions of boreholes with brief data.

Geological records
- A regional picture of geology within England, Wales and Scotland can be obtained from the *Handbooks on the Regional Geology of Great Britain* published by the Institute of Geological Sciences (IGS).
- Detailed local information on the composition of geological strata, exposures, well logs and groundwater can be obtained from *One-Inch Sheet Memoirs* (from the IGS).
- The water supply papers and memoirs of the IGS contain information on groundwater.

Land use
- Soil survey records consider deposits to a depth of about 1.5 m below ground level and are intended to provide information primarily on agricultural soil conditions. However, the properties of this upper soil are related to the underlying materials. The most useful cover is provided by maps with scales of around 1 : 25 000.
- Land-use maps have 13 main categories (e.g. residential and commercial, industry, transport, derelict land), which are further subdivided. The maps are produced at a scale of 1 : 25 000.

Mineral extraction records
- Records of coal-mining prior to 1946 are often difficult to find. Recent information (position and extent of known workings and their depths as measured at shafts) could be obtained from the Plans Record Offices at British Coal.
- The Mining Record Office of the Health and Safety Executive keeps plans and records of abandoned mines and their associated tips.
- Information on quarries may be found in the Directory of Quarries and Pits.

Local archival material
- Local authority records and informal discussions: planning department, building control section, environmental health division.
- Aerial photographs.
- Records of the relevant Waste Regulation Authority and the regional office of the Environment Agency.
- Historical books produced by local authors about the area in general and also about specific industries and works of construction. Back copies of local papers.

4.2. Physical investigation

There are two basic methods that allow direct classification of soil by visual examination and laboratory testing: examination in situ, and boring and drilling. Examination in situ is the only completely satisfactory method of obtaining a continuous record of soil conditions with depth but it is only practicable for shallow excavations. A large number of methods are available for obtaining samples or details of strata. The principal methods in use are:

- trial pits
- hand auger methods
- probing
- light percussion drilling
- rotary core drilling
- power auger.

The provision of relevant groundwater information is a prime requirement of reliable site investigation work. The flow of water into a hole or excavation below the water table is dependent on factors such as the dimensions of the excavation and the permeability of the deposits. All groundwater strikes should be recorded, in any excavation below ground level, together with water levels on completion of work in the evening and at the start of work on the next morning.

4.2.1. Shallow investigations

The simplest, and for many purposes still the most satisfactory, method of subsurface investigation is by the excavation of trial pits. These permit the in situ condition of the ground to be examined, both laterally and vertically, so that detailed information on groundwater conditions, soil strength, stratification and discontinuities can be obtained. Trial pits provide access for carrying out manual in situ tests and for taking samples. Tube samplers can be driven into the floor of the pit, while high-quality block samples can be taken only from trial pits. However, it must be remembered that every year a number of people are killed during the collapse of unsupported trenches, and personnel should not enter trenches or pits more than 1.2 m deep without either supporting the sides or battering back the sides. Unfortunately, pits do not provide any indication of conditions at depths below the base of the excavation, which must therefore be assessed by geological inference or by boring.

Trial pits may be excavated by hand or by machine. Usually the excavation is done by a back-hoe excavator which forms a trench about 3–5 m long with a width equal to the back-acter bucket (0.6–0.9 m wide). The depth is determined by the reach of the hydraulic arms and is usually 3–6 m, depending on the type of machine. Shallow trial pits provide a cheap method of examining near-surface deposits in situ, but the cost increases dramatically with depth, particularly when the ground needs to be supported and manual excavation is performed.

The hand auger is a light, portable apparatus for sampling soft to stiff soils near the ground surface. The principal types of auger head are posthole (or Iwan), Dutch and gravel (Figure 4.3). The most commonly used auger for ground

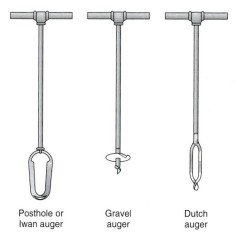

Figure 4.3. Types of hand auger head

Posthole or Iwan auger   Gravel auger   Dutch auger

Figure 4.4. An assembled hand auger

investigation is the Iwan auger (typically with a diameter in the range 100–200 mm) (Figure 4.4). The hand auger system consists of a crossbar (above ground level) attached by rods to an auger below ground level, and usually the auger is operated by two men to form a borehole. The men press down on the crossbar as they rotate it, thus scraping away soil at the bottom of the hole. When the auger has collected sufficient material it is lifted out of the hole and the soil is removed. 'Undisturbed' samples can be obtained from the bottom of an auger hole by using a sampling tube (usually about 40 mm diameter) fastened to the bottom of the drill rods and driven into the soil. The auger hole can also be used for groundwater observations.

This method of investigation is cheap but, because of the simplicity of the equipment, it does suffer from several disadvantages:

- The depth of boring is generally limited to about 5 m.
- In stiff or very stiff clays augering progress is very slow/difficult. When such clays contain gravel, cobbles or boulders it is impossible to advance the hole at all.
- In uncemented sands or gravels it is impossible to advance the hole below the water table. In addition, it is impossible to carry out in situ tests to determine the relative density of granular deposits.
- The samples obtained from the auger hole are fairly small.

The Mackintosh prospecting tool consists of rods (1.2 m long, 12 mm diameter) that can be threaded together. The rods are normally fitted with a streamlined driving point at their base and a light hand-operated driving hammer at their top. The hammer has a total weight of about 4 kg. It is raised manually and then released to drive the rods into the ground. This tool provides a very economical method of determining the thickness of soft deposits such as peat. The device may also be used to provide an indication of the strength profile with depth by driving the point and rods into the ground (utilizing the full drop height available for the hammer) and recording the number of blows needed for each 150 mm of penetration.

4.2.2. Drilling

In the UK the formation of investigation boreholes in soil deposits is most commonly undertaken by *light percussion drilling* using a shell-and-auger rig (although the auger is rarely used with this method) (Figure 4.5). A rope from a winch passes over a pulley at the top of an A frame and is used to raise and drop a series of weighted tools onto the soil being drilled (Figure 4.6).

In clays a borehole is formed by dropping a weighted steel tube (known as a clay cutter) onto the soil so that soil is forced into the tube. The tube is then

Figure 4.5. A shell-and-auger drilling rig

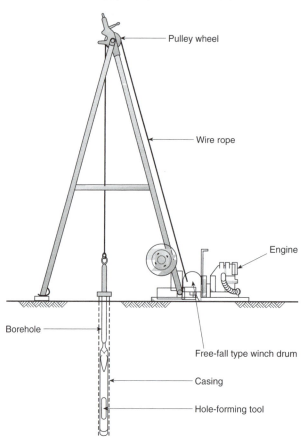

Figure 4.6. Light percussion drilling

pulled out of the borehole so that the soil wedged inside the clay cutter can be removed. To penetrate granular materials, such as sands or gravels, a 'shell' is used. The shell is a tube containing a simple non-return (clack) valve at its lower end. The bottom of the borehole is flooded with water and the shell is then surged up and down. This surging loosens the soil at the base of the hole, and as the shell is dropped to the bottom of the hole a mixture of soil and water flows past the non-return valve and into the tube. As the shell is raised, the clack closes and retains the soil. By repeatedly surging the shell up and down at the base of the hole, soil can be collected and removed from the hole.

Casing should either be allowed to follow the hole down (if it is loose) or should be driven so that it is just above the base of the hole, otherwise large cavities will be formed on the outside of the casing or the soil will be loosened for a considerable distance around the hole. The casing may be omitted when drilling in stiff clay without boulders. A nominal minimum casing diameter of 150 mm is used to enable 100 mm diameter samples to be taken. When considerable depths are to be bored it is normal practice to commence a hole at a diameter of 200 mm or larger to permit the reduction of friction by using successively smaller diameter casings inserted telescopically.

Rotary drilling is used for the exploration of hard soils or rocks when light percussion methods cannot penetrate. The technique is generally much easier with intact hard rock than with soft highly-weathered rock or soil. The drill string (the hollow drill rods and the bit) is rotated and pushed downwards (at the same time) to grind away the material in which the hole is being made. Usually flushing fluid is pumped down through the drill rods so that it passes outwards over the bit and travels upwards in the annular space between the drill rods and the outside of the hole. The purpose of this flushing is to remove the cuttings, to cool the drilling bit and to help retain an open hole without the use of casing, wherever possible. A large number of different types of flush fluid are in use, but they generally fit into four classes: water based, oil based, air and stable foam. The most common flush fluid in use in the UK is water, with air being used when water causes serious softening of the formation being drilled. The use of drilling mud (a thin mixture of water and bentonite) has various advantages over water. Firstly, it is more viscous and can therefore lift cuttings adequately at a lower velocity. Secondly, it will cake the edges of the borehole and will largely eliminate the seepage of water out of the borehole. The mud cake formed on the outside of the borehole also improves the stability of the borehole considerably.

The most common use of rotary drilling is to obtain intact samples of the rock being drilled, at the same time as advancing the borehole. To do this a core barrel fitted with an annular core-bit at its lower end is rotated and grinds away an annulus of rock. The stick or rock in the centre of the annulus passes up into the core barrel, and is subsequently removed from the borehole when the core barrel is full. Typical cores are shown in Figure 4.7.

The method known as *open-holing* involves the formation of a hole in the subsoil without taking intact samples. Sampling during open-holing is usually limited to collecting the material abraded away at the bottom of the borehole as it emerges mixed with flush fluid at the top of the hole. Open-hole rotary drilling techniques have the disadvantage that the structure of the deposits is totally destroyed by the boring.

4.2.3. Sampling

Samples are usually obtained for laboratory analysis from both trial pits and boreholes, but the latter only allow examination of a small percentage of the soil in its undisturbed state. The sampling procedure chosen depends on the quality of sample required and the character of the ground, in particular the extent to which it will be disturbed by the sampling process. In choosing a sampling method, it should be decided whether it is the mass properties or the intact material properties of the ground that are to be determined. The behaviour of the ground is often dictated by the presence of weaknesses and discontinuities and it is quite feasible to retrieve high-quality soil samples that are unrepresentative of the mass. Representative samples should be taken from every stratum encountered in a trial pit or a borehole and at intervals within an apparently uniform stratum. The

Figure 4.7. Typical cores from rotary drilling

frequency of taking undisturbed samples or carrying out field tests depends on the nature of the investigation but a common requirement is to sample, or do an in situ test, at intervals of about 1–2 m initially and then to increase the interval at depth. The samples should be labelled immediately with information about the site, the date, borehole or pit number, depth, etc., in order to have a ready reference to the deposits.

Samples recovered from shell-and-auger boreholes fall into the general categories indicated in Table 4.2. Bulk samples are acceptable for soil identification and classification, but undisturbed samples of clay and other cohesive soils are required for laboratory testing of shear strength, consolidation/ drainage and permeability. Samples taken for routine purposes are normally of 100 mm nominal diameter (giving rise to the description 'U100 sample') and about 450 mm long. It should be appreciated that the term 'undisturbed' is a relative one, as the action of taking a sample inevitably disturbs the soil to a greater or lesser degree. The sampler can be driven into the ground by dynamic means using a drop weight or sliding hammer, or by a continuous static thrust using a hydraulic jack. The fundamental requirement of a sampling tool is that it should cause as little remoulding and disturbance as possible when forced into the ground.

Open-tube samplers consist essentially of a tube that is open at one end and fitted at the other end with means for attachment to the drill rods (Figure 4.8). A

Table 4.2. Classes of geotechnical samples

Sample category	Material	Container
Bulk	All types of soil	Jar with air-tight lid, sealed polythene bag
Undisturbed	Primarily cohesive soils	Sample tube sealed with end caps
From in situ test apparatus	Usually granular	Core box or longitudinally-split tube
Groundwater	Liquid	Sealed uncontaminmated flask/bottle

Figure 4.8. A sample tube attached to drill rods

non-return valve permits the escape of air or water as the sample enters the tube, and assists in retaining the sample when the tool is withdrawn from the ground. Thin-walled samplers, which are used for soils that are particularly sensitive to sampling disturbance, have a lower end shaped to form a cutting edge with a small inside clearance. These samplers are suitable only for fine soils, which have a firm consistency at most, and which are free from large particles. The piston

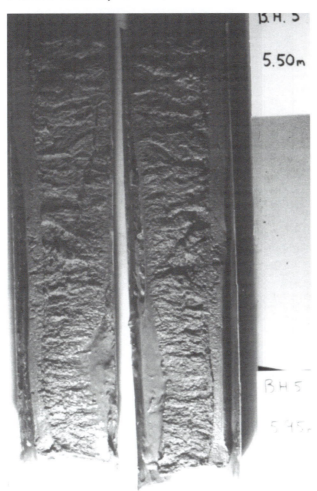

Figure 4.9. A split U100 sample

sampler contains a close-fitting sliding piston which is pushed back as the soil enters the sample tube. The suction between the piston and the top of the sample helps to retain the soil when the sample tube is extracted from the ground.

The structure within a cohesive sample can be seen by splitting a core, which thus reveals a section. The bedding planes are sometimes made more distinct by partial or complete drying. The core is split longitudinally by using a knife to cut half-way through and then forcing the two halves outwards, by hand (Figure 4.9). It is possible to obtain undisturbed samples of sand from above the water-table (Figure 4.10). However, the recovery of tube samples of sand from below the water-table is virtually impossible because the sand tends to fall out of any sample tube. Furthermore, the virtually insurmountable problems of stress relief and fabric changes following sampling from the ground make the study of sand in situ very difficult. Emphasis has therefore been placed on the use of in situ testing techniques to assess the state of sand strata.

4.3. In situ testing

In situ soil tests can be divided into classification and identification tests, and those intended to provide an estimation of the strength in situ.

4.3.1. Standard Penetration Test

The Standard Penetration Test (SPT) employs a 50 mm diameter split tube (referred to as a 'spoon') which is driven 450 mm into the base of a borehole by a

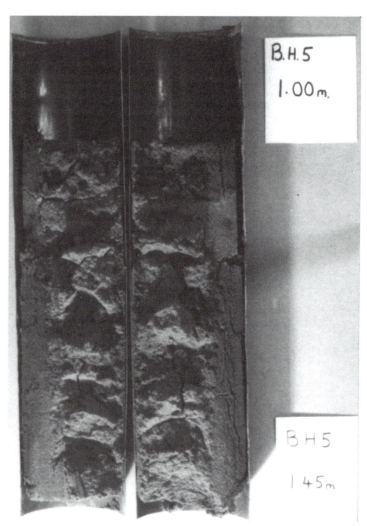

Figure 4.10. An undisturbed sand sample

Figure 4.11. Standard Penetration Test apparatus attached to drill rods

65 kg hammer which has a standard free fall of 760 mm (Figure 4.11). The split tube is hollow and takes a core of about 35 mm diameter (shown in the foreground in Figure 4.8). After an initial 150 mm seating drive, the number of blows taken to produce four successive penetration increments of 75 mm are recorded and summed to give the SPT N value. If resistance is very high, the second stage of the process is discontinued at 50 blows, even if the total penetration has not reached 450 mm. When SPT tests are carried out in sands and silts below the standing water level the results should be treated with caution, as disturbance caused by water flows into the borehole may cause an underestimate of the compactness of such deposits. When the granular soil encountered is a gravel or gravely sand, the test is carried out using a solid cone in place of the split barrel sampler.

The SPT results can be correlated in a general way with soil density state (Table 4.3) and physical properties. Relationships between SPT values and properties of cohesive soils have been produced (Table 4.3), but it is generally accepted that such relationships are unreliable because the results are seriously affected by the sensitivity and compressibility of such soil. The most common use of SPT values is to estimate bearing pressures and the settlement of foundations.

4.3.2. Cone Penetration Test

The Cone Penetration Test (CPT) was originally developed for use in the relatively uniform geological conditions of The Netherlands. Nowadays it is carried out using highly sophisticated electronic equipment and may be very useful as a second-stage investigation, particularly where deposits of sand or silt are present. A solid cone is pushed into the soil at a steady rate and the resistance to

Table 4.3. Soil characterization from the Standard Penetration Test N value

N value	Sand		Clay	
	Relative density	Friction angle (°)	Consistency	Unconfined compressive strength (kN/m^2)
<2	Extremely loose	27	Very soft	<25
2–4	Very loose	28	Soft	25–50
4–8	Loose	29	Medium	50–100
8–15	Loose to medium	29–31	Stiff	100–200
15–30	Medium to dense	31–36	Very stiff	200–400
30–50	Dense	39–41	Hard	>400
>50	Very dense	41–45	—	—

penetration is measured. Most cone penetrometers also have friction sleeves to provide measurement of the resistance to penetration due to the contact between the soil and the shaft attached to the cone. The test produces a continuous trace of the resistance to penetration of the cone and the shearing resistance on the sleeve with respect to depth. As no soil samples can be recovered for identification the cone resistance and friction ratio (shaft friction/cone resistance) values are used to differentiate between various soil types. Clean sands generally exhibit low friction ratios, and the value of the ratio increases with increasing clay content. A soil-classification chart based on the standard friction cone is shown in Figure 4.12.

One of the major advantages of the CPT is that it is not necessary to make boreholes to conduct the test. However, unlike the SPT, with the CPT no soil samples can be recovered for visual observation and laboratory tests.

4.3.3. Vane apparatus

The vane shear apparatus is used for the in situ determination of the undrained strength of non-fissured, fully-saturated clays. The test is not suitable for other

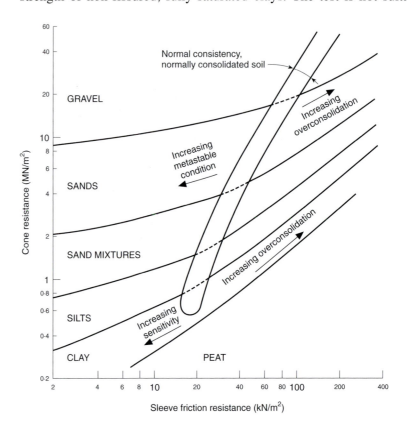

Figure 4.12. Soil identification from the Cone Penetration Test

types of soil. The vane test is particularly suited to deep soft deposits of alluvium, such as silt and soft clay, in the saturated condition. In such cases the site sampling process causes significant sample disturbance, as does the eventual extrusion and sample preparation procedure in the laboratory. However, the test may not give reliable results if the clay contains sand or silt laminations.

The equipment consists of a stainless steel vane (composed of four thin rectangular blades) mounted on the end of a high tensile steel rod. The length of the vane is equal to its overall width, typical blade dimensions being 150 mm ×

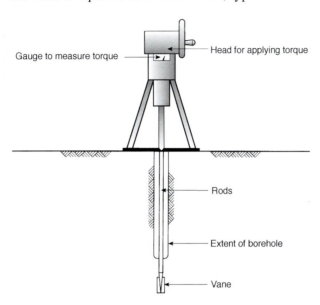

Gauge to measure torque

Head for applying torque

Rods

Extent of borehole

Vane

Figure 4.13. Shear vane apparatus

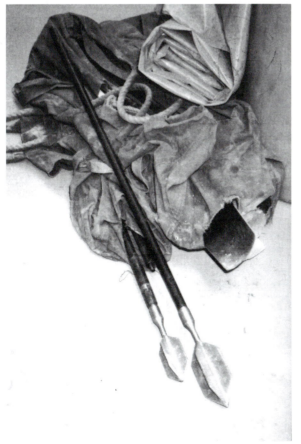

Figure 4.14. Shear vanes

75 mm and 100 mm × 50 mm (Figures 4.13 and 4.14). The vane is pushed into the soil at the bottom of a borehole to a depth of at least three times the borehole diameter. The vane is then rotated at a constant rate by manually applying a torque to the head of the vertical connecting rod. The apparatus which is used to rotate the rod incorporates a means of measuring the maximum applied torque and the amount of vane rotation. When the vane rotates it carries with it a cylindrical plug of soil, and hence shearing of the soil occurs on the sides and ends of this cylinder. The undrained shear strength S_U of the soil is then calculated from the peak value of applied torque (see Chapter 6, equations 6.24 and 6.25).

A small hand vane (approximately 13 mm overall diameter) is used for rapid in situ testing of soils in trial pits and other open excavations.

4.3.4. The pressuremeter

The pressuremeter was developed in the 1950s for in situ measurement of the stress–strain modulus of soil. The basic test involves expanding a pressure probe inside a borehole and measuring the increase in volume (Figure 4.15). Pressuremeter probes are grouped according to their method of installation (prebored, self-bored, pushed in) and method of measuring displacement (volume, radial). The deflated probe diameter is 40–85 mm and the total length is 1–2 m. The length/diameter ratio of the expanding section usually exceeds 5. The Menard pressuremeter is a particular volume displacement type pressuremeter which contains three expanding sections and is installed in a prebored hole. The expansion of the central test section is monitored, and the guard cells are inflated to ensure that the length of the test section remains constant.

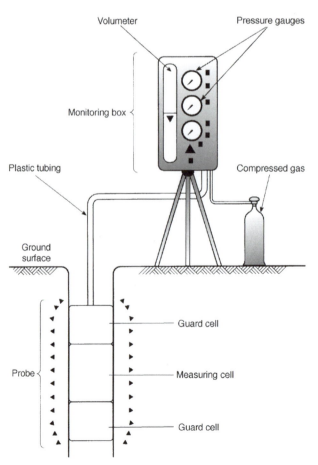

Figure 4.15. The pressuremeter test

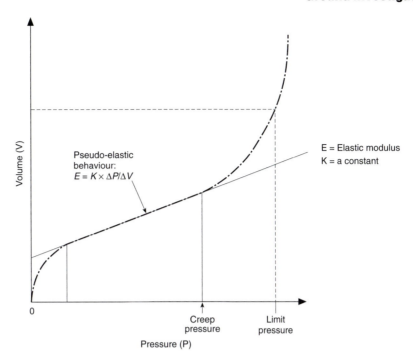

Figure 4.16. A pressuremeter curve

A plot of applied pressure against membrane displacement is known as a pressuremeter curve or ground response curve (Figure 4.16). From this information it is possible to obtain directly the values, in situ, of horizontal stress, stiffness and strength.

4.3.5. Plate bearing tests

These are direct tests that determine the bearing capacity and settlement by jacking small plates into the ground and measuring the response. Plate bearing tests are rarely conducted because of their high cost in relation to the quality of information derived from them. In particular, it is unrealistic to relate the behaviour of a small plate (0.3–0.8 m square) to large loaded areas which, because of their greater width, stress soil layers over a much greater depth. Plate tests are carried out occasionally on uniform granular soils, and a modern form of the test involves using a loaded refuse skip to transmit pressure through a base plinth of suitable dimensions.

4.3.6. Geophysical ground investigation

Geophysical surveying can be used to make a preliminary and rapid assessment of ground conditions. The technique indicates ground variations and anomalies, which may be correlated with geological or man-made features. Geophysical survey data can be used to interpolate the ground conditions between boreholes, and to indicate locations where further boreholes are needed so that the significance of a geophysical anomaly can be investigated.

Resistivity is a commonly used technique for investigating relatively simple geological problems. A current is passed through the ground and the potential difference is measured between a pair of electrodes. With suitable deployment of the electrodes the system may be used to provide information on the geoelectrical stratification of the ground (depth probes), lateral changes in resistivity (constant separation traversing) or local anomalous areas (equipotential survey). The unsuspected presence of electrical conductors (e.g. pipes, cables), under the site will, of course, render the results unreliable. The interpretation of the results obtained by this method does not always provide a definite solution, particularly as the number of layers increases, because it involves curve matching with idealized conditions.

Local changes in the Earth's magnetic field are associated with changes in rock types. In suitable circumstances, magnetic surveys may locate boundaries between rocks that display magnetic contrasts (e.g. faults, dykes). In civil engineering the main use of magnetic surveys is the location of buried metallic man-made objects, such as cables or pipelines. However, the technique can also be used to locate old mine shafts and areas of fill.

The seismic technique (reflection or refraction) may be used to locate boundaries within the ground between materials having different dynamic moduli. The refraction method is the one most frequently used, primarily for the determination of rockhead level. Seismic waves are produced (either from a small explosive charge or a heavy hammer blow), and the time taken for them to travel from the point of origin to vibration detectors (geophones), set at various distances, is measured.

4.4. Reporting

The ground investigation report is the culmination of the investigation, exploration and testing programme. There are essentially two types of report; factual and interpretative. The comprehensive factual soils report is prepared after site operations and laboratory tests are complete. Only factual information is presented, although the results may be summarized, analysed or compared in different ways. The interpretative report is usually prepared separately from, and after, the factual report and it contains detailed comments and recommendations. It should be written by a geotechnical specialist who has up-to-date information about the project. The recommendations for design must be based on the facts stated in the factual report (i.e. on the borehole records and test data), and not on conjecture.

Although individual reports will vary according to the particular brief given and conditions encountered, a typical report will contain the items indicated in

Box 4.2. The contents of a ground investigation report

Introduction
- Brief summary of proposed works, investigations conducted, site location, time-scale of the work, key personnel.

Site description
- Topography, main surface features, details of access, history of the site, presence of existing works/underground openings, site location map.

Site geology
- Overall/regional geology, geology of the specific area, main soil and rock formations/structures.

Soil conditions
- Detailed account of the conditions encountered (in relation to the proposed works), description of all strata, results of laboratory and in situ tests, details of groundwater conditions.

Discussion
- Comments on the validity and reliability of the information presented, further work (if required). Definition of appropriate design parameters and methods of both design and construction (interpretative report).

Appendices
- Borehole logs, laboratory test results, in situ tests, geophysical survey records, references, literature extracts.

Box 4.2. One of the key items of information in deriving a ground profile from the investigation is the driller's borehole log. At the end of each day's drilling, the drilling foreman prepares a report incorporating the following information:

- depth at the start and end of the shift
- depth of each stratum change
- groundwater records
- brief description of each soil type
- locations where water was added to the boring
- depths when chiselling was required.

A typical representation of the foregoing information is shown in Figure 4.17.

						Geotechnical Borehole Log	
Method	Shell & Auger		**Date**				**Bore-hole**
Diameter (mm)	150 mm Cased to 9.10 m		**Coords**				
			Ground level	0.35 = 00		**Client**	**Sheet**
Soil samples/test		**SPT value (U100 blws)**	**Field records**	**Level (x00)**	**Depth (m)**	**Strata description (thickness in metres)**	**Legend**
Depth	Type						
0.20	D1			0.25	0.10	MADE GROUND (tarmacadam) (0.01)	
0.50	D2					MADE GROUND (stiff black, dark brown and dark grey silty sandy ashy clay fill with pockets of silty clay and fragments of brick, roadstone, ash and concrete, and occasional gravel) (2.10)	
1.00–1.45	C1	N = 7	7/3/2/1/1				
1.00–1.50	B1		Seepage at 1.30 m sealed at 3.00m				
1.80	D3						
2.00–2.45	C2	N = 4	5/1/1/1/1				
2.00–3.00	B2			–1.85	2.20	Soft to very soft grey silty to very silty CLAY with many brown organic fibres at top of stratum. Becomes organic in places (2.80)	
3.00–3.45	S3	N = 0	1////				
3.00	D4						
3.70	D5						
4.10–4.55	U1	(5)					
4.60	D6						
5.00–5.45	U2	(6)		–4.65	5.00	Very soft dark brown, with pockets of grey, silty sandy CLAY with occasional flint gravel (0.70)	
5.50	D7						
5.70	W1		Slow inflow at 5.70 m rose to 4.9 m not sealed	–5.35	5.70	Medium dense fine to coarse sub rounded to sub angular flint GRAVEL with brown fine to medium sand and pockets of firm brown silty sandy clay (0.70)	
6.00–6.50	B3						
6.00–6.45	C4	N = 11	7/2/3/4/2	–6.05	6.40	Soft brown silty very sandy CLAY with some fine to coarse flint gravel at top of stratum, and occasional pockets of clayey sand (2.10)	
7.00	D8						
7.50–7.95	S5	N = 17	2/3/3/4/7				
7.50	D9						
8.50–8.95	C6	N = 15	10/4/5/2/4	–8.15	8.50	Strong grey fine to coarse gravel grade fragments of claystone with tan brown silty slightly sandy clay (0.60)	
8.50–9.00	B4						
9.10–9.10	C7	50	50 for 0 mm seating 24.03.93/3.90	–8.75	9.10	Hard grey ROCK*	
9.10	D10						

Remarks 1. * Drillers description.
            2. Groundwater strike at 5.70 m. Water rose to 5.5 m in 5 mins; 5.3 m in 10 mins; 5.10 m in 15 mins; 4.90 m in 20 mins.
            3. Hard grey rock encountered at 9.10 m. Chiselled for 1 hour with no progress being made. Drilling ceased.
            4. Piezometer tip installed at 6.30 m.

Note: for explanations of symbols and abbreviations see accompanying notes	Logged by	Scale (approx.) 1 : 50	Figure number

Figure 4.17. A borehole log

5 Compaction and earthworks

5.1. Introduction

Compaction is, in general, a process of reducing the voids ratio of a soil through removal of air by applying mechanical kneading and shearing of the soil. The purpose of imparting mechanical energy to the soil is to force the particles closer together and thereby increase the density of the soil.

Compacted clay liners (low-permeability layers) have been used extensively beneath landfilled refuse to prevent pollution of the underlying ground and groundwater. An early UK example of this type of construction was the conversion of a disused water reservoir near Bury (Sarsby, 1987a). At the commencement of this project there was sufficient suitable clay on site to form the required liner, provided that it was placed and compacted to form a coherent layer. The site is located in an area of relatively high annual rainfall and this severely restricts the times when earth-moving and compaction can be undertaken. Even during the drier parts of the year the numerous groundwater issues at this site meant that there was always a lot of standing water on the site. Unfortunately the initial attempt to construct a liner was undertaken by the owner of the site (who had no geotechnical expertise or experience of earthworks) without any technical guidance. He tried to place the clay regardless of the weather and the presence of standing water and he had no specification for the compaction. Hence a lot of clay was transformed into a slurry and was wasted. After ownership of the site changed hands a geotechnical engineer was employed to guide the liner construction. Clay samples were taken from various parts of the site and were subjected to compaction tests to provide both a specification for placement of the clay and a rapid means of checking that the completed liner would have the requisite low permeability. Quality control was undertaken by specifying that the clay should be compacted in accordance with the relevant Department of Transport specification regarding the weight of the roller, the number of times the loose soil should be rolled, etc. Under this regime the liner was successfully constructed (Figure 5.1) with much of the compacted clay having greater bulk density and lower permeability than was achieved in the laboratory tests. The site has now been filled with domestic and commercial waste (Figure 5.2) and the low-permeability clay liner has been effective in preventing pollution of the underlying groundwater.

5.1.1. Earthworks

Earthworks or earth structures are usually constructed where it is required to alter the existing topography (e.g. in the construction of embankments, cuttings or roads). After excavation and placement the soil is normally loose and bulked and must be compacted to prevent post-placement distortion, settlement and softening. For conventional earthworks the objective of soil compaction is to economically and efficiently reduce the volume of the soil by mechanical means, thereby reducing the compressibility and increasing the shear strength — the permeability may also be reduced at the same time (although this is not usually a concern for conventional earthworks).

Figure 5.1. Clay compacted to form a liner for a landfill site

Figure 5.2. The completed landfill site near Bury

In concept, compaction is a simple process whereby energy in some form is applied to loose material to bring about densification to produce a soil mass with controlled engineering properties. Mechanical compaction is used to form coherent fills and embankments. The soil is laid in thin layers and then subjected to momentary application of load (via rolling, tamping or vibration). The expulsion of air from the soil voids usually occurs without significant change in moisture content. In highway and airfield construction, granular materials are compacted (in unbound or bound states) to provide strong layers that can spread load and provide good drainage characteristics.

Engineered fill comprises material that has been selected, placed and compacted to an appropriate specification, thereby producing a material that has engineering properties which are known and which are considered adequate for the purpose for which the fill has been placed. Non-engineered fills generally arise due to intermittent infilling of voids and levelling of ground surfaces (using on-site materials or wastes) or simple disposal of waste materials. Placement of these latter materials has usually been undertaken without specification or control, and consequently it is difficult to characterize the engineering properties of the material and to predict its behaviour.

5.1.2. Compaction

The mechanical effort involved in compaction results in a change in soil structure in terms of both the aggregation of the soil lumps and the arrangement of particles within individual lumps. When a clay soil is relatively dry it achieves very high shear strength upon compaction due to the interlocking action of strong lumps of soil. If water is added to this clay, the soil lumps become weaker and softer and, during compaction, the soil particles slip over each other and move closer together. Thus compaction creates a denser soil mass and the effect of compaction can be described quantitatively in terms of the resultant dry density.

If a soil is very dry then its compacted dry density is raised if the moisture content before compaction is increased (Figure 5.3). However, beyond a certain moisture content (the optimum value) an increase in moisture content tends to reduce the dry density because the water occupies an increasing proportion of the void spaces. Hence it is impossible to force the soil particles closer together because a clay soil cannot drain instantaneously and the fluid in the voids is essentially incompressible. The volume of any further water added exceeds that

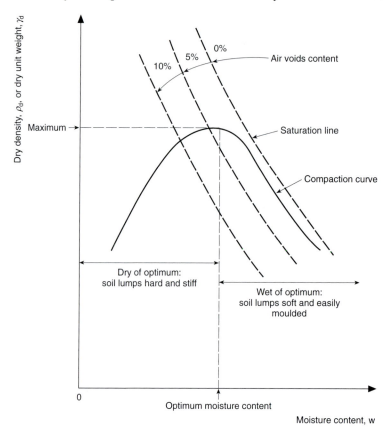

Figure 5.3. A compaction curve for a fine-grained soil

of the air expelled during compaction, so the degree of compaction obtainable falls. Eventually all the voids are filled with water (i.e. the soil is saturated), and the resultant compacted soil is very soft. Typically the relationship between the dry density and the moisture content on the 'wet side' of optimum is more or less parallel with the theoretical relationship for zero air voids (i.e. saturation) — see Figure 5.3.

The compaction behaviour of free-draining granular materials is different from that of cohesive soils because of the inability of the former to retain large quantities of water and because of the physical size of the particles. Hence compaction of granular materials produces the maximum dry density at low moisture contents. Idealized plots of moisture content versus dry density for a clean sand (Figure 5.4) are typically S-shaped. That is, for a given compactive energy the highest density states are attained when the water content is either very low (essentially zero) or relatively high (approaching its liquid retention capacity) — optimum conditions. The dry density at zero moisture content is usually slightly higher than that at optimum moisture content. The characteristic S-shape of the compaction curve is due to surface tension in the soil moisture, which provides the soil with shear strength to resist the compaction forces.

On-site compaction is carried out by rolling, tamping or vibrating. The attainable compaction depends very much upon the soil type and moisture content, and while heavy rolling will significantly reduce the voids in granular soils, cohesive materials will consolidate only very slowly under a sustained pressure. In order to obtain satisfactory compaction of the vast range of soils to be met in practice, several different types of compaction plant have been developed, and these are generally classified into those imparting energy by static weight, impact, vibration or kneading.

The effect of compaction can be assessed both in the laboratory and in the field. Laboratory tests determine the degree of compaction achievable when using a standard amount of compactive effort with a view to defining the probable density obtainable in the field. Measurement of the compacted state (moisture content and dry density) in a field situation is used to assess whether the completed earthworks satisfy the design requirements and/or specification.

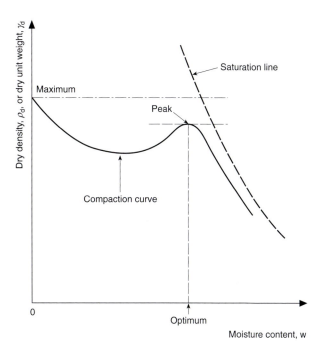

Figure 5.4. A compaction curve for coarse-grained soil

5.2. Laboratory testing

The main value of laboratory tests lies in the classification and selection of soils for use in fills and embankments and for providing a specification for the works. The dry density to be achieved after field compaction is often expressed as a percentage of the maximum dry density from a particular laboratory test. Alternatively, the laboratory tests may be used to define a suitable moisture content range for field compaction. However, it must be remembered that laboratory compaction tests do not correspond directly to the field situation because of the different compactive energies imparted in the laboratory and in the field. Furthermore, most laboratory tests are only conducted on material finer than 20 mm (BSI, 1990). Nevertheless, laboratory test results provide a guide to the water content at which maximum dry density is achievable in the field.

5.2.1. The compaction curve

In laboratory tests on cohesive soils a standard, dynamic, compactive effort is used to compact soil at several different moisture contents (ranging from fairly dry to wet) so that the moisture content which produces the highest dry density (i.e. the optimum moisture content) may be determined. In the UK there are two standard compaction tests for cohesive soils, both employing a metal rammer (a solid cylinder) falling freely onto soil contained in a mould. The basic difference between these two tests is the total amount of energy imparted to the soil during the compaction process. The British Standard light compaction test (the British equivalent of the Proctor test) consists of compacting soil in a mould (approximately 100 mm diameter, 120 mm high), using a rammer weighing 2.5 kg dropped through a height of 300 mm. The soil is compacted in three equal layers, each layer being given 27 blows uniformly distributed over the surface of the soil. The work on which the Proctor test is based was undertaken in the 1930s. To account for the development in compaction techniques, a modified Proctor test has been created that is representative of the performance of today's heavier, more efficient, earth-moving plant. The British Standard heavy compaction test (the British equivalent of the modified AASHO compaction) utilizes a heavier hammer (4.5 kg) falling through a greater height (450 mm). Furthermore, the soil is compacted in five equal layers, with each layer receiving 27 uniformly distributed blows. Figure 5.5 shows the manual apparatus for undertaking light and heavy compaction in the laboratory.

The density of the final compacted mass is described in terms of the weight, or mass, of solid particles per unit volume (i.e. dry unit weight or dry density). This is usually determined from measurement of the bulk unit weight or density and moisture content according to equations (3.13) and (3.14). Thus dry unit weight is given by

$$\gamma_d = \frac{G_s \gamma_w}{(1+e)} = \frac{\gamma}{(1+w)} \tag{5.1}$$

and dry density by

$$\rho_d = \frac{G_s \rho_w}{(1+e)} = \frac{\rho}{(1+w)} \tag{5.2}$$

Percentage air voids contours (typically for 0, 5% and 10% air content) should be included on the dry density versus moisture content plot to assist interpretation of compaction curves. These contours represent the variation in the theoretical upper limit of the dry density with sample moisture content and percentage air V_a (the percentage of the total sample volume occupied by air). These upper-limit values of the dry density are obtained from the equation

$$\rho_d = \frac{G_s \rho_w (1 - V_A)}{(1 + w G_s)} \tag{5.3}$$

Figure 5.5. Laboratory compaction apparatus

It has been observed that the 5% air voids contour passes, more or less, through the optimum moisture content values obtained for different compactive energies. Furthermore, soil that is wet of optimum typically has 2–3% air voids, although for very wet samples the compacted soil is very close to full saturation. In practice, under the influence of mechanical compaction it is not possible to achieve saturated soil conditions, and typically a soil mass compacted at moisture contents higher than the optimum moisture content will contain about 2% air voids.

Changing the compactive effort per unit volume of soil results in different compaction curves (Figure 5.6). As the compaction effort is increased, the maximum dry unit weight increases and the optimum moisture content decreases. Typical values are given in Table 5.1. Thus the compaction curve (dry density versus moisture content) obtained for a soil in a laboratory test is not a fundamental property of a soil, but rather is a function of the amount of energy imparted, the means of doing this and the soil characteristics. However, at high moulding water contents the influence of compactive energy decreases significantly.

For granular soil, compaction using a vibrating hammer is more efficient than impactive means as it excites individual particles and causes them to move so that interparticle voids are reduced in size (this behaviour also occurs in field compaction as confirmed by practical experience with vibrating rollers). In the vibrating hammer test the soil is compacted in three equal layers (in a 150 mm diameter mould) to give a depth of around 125 mm. The hammer vibrates with a frequency of 25–45 cycles/s and is pressed down on the soil during compaction. For cohesionless granular soils greater energy input results in greater densification (as with cohesive soils). Generally granular materials have higher dry densities and lower optimum moisture contents, for a given compactive effort, than do cohesive materials. Some particles, especially weak rocks such as shale and chalk, are prone to break down during compaction, and so this material

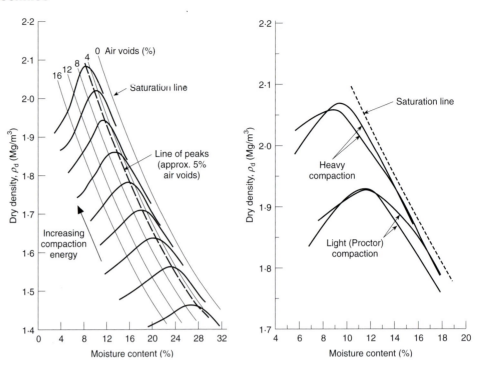

Figure 5.6. The effect of compactive effort

Table 5.1. Typical compaction densities and optimum moisture contents*

Soil type	Typical value		Illustrative effect of compaction energy			
			Standard		Heavy	
	$\rho_{d\ max}$ (Mg/m^3)	w_{opt} (%)	$\rho_{d\ max}$ (Mg/m^3)	w_{opt} (%)	$\rho_{d\ max}$ (Mg/m^3)	w_{opt} (%)
Heavy clay	1.44–1.69	20–30	1.47	26	1.87	18
Silty clay	1.60–1.85	15–25	1.57	21	1.94	12
Sandy clay	1.76–2.17	8–15	1.87	13	2.05	11
Sand	1.80–1.96	6–13	1.94	11	2.09	9
Sandy gravel[†]	1.85–2.15	6–14	2.06	8	2.15	8
Coal-mining waste	1.46–1.91	4–11	1.57	11	1.87	7

* Sources: Carter & Bentley (1991), Horner (1981), O'Flaherty (1976) and Sarsby (own records).
† Vibrating hammer; $\rho_{d\ max}$ = 2.25 Mg/m^3, w_{opt} = 6%.

must be discarded after each compaction test and fresh materials must be used each time the moisture content is adjusted.

As the different laboratory tests yield different compaction curves for the same soil, it is only to be expected that different items of compaction plant will perform differently on the same soil. Also, one would not expect there necessarily to be any direct relationship between the laboratory compaction curves and the relationships obtained from compaction plant, particularly because the actual remoulding processes are different. Furthermore, research work (Servais & York, 1993) has shown that the precision of the laboratory compaction test is poor — in practical terms, the reproducibility of the maximum dry density is about ±3% for the modified compaction test and ±4% for the standard test. There is also uncertainty over the precision with which the field density test is performed. Consequently, there is a significant likelihood that the compliance of an earthworks with the specification may be incorrectly judged on the basis of a simple specified limit, because the true quality of the work lies within a wide band.

5.2.2. Compaction effects

Compaction induces variations in the structure of cohesive soils, and the results of these structural variations include changes in strength, compressibility and permeability. The strength of compacted clay soils generally decreases with increasing moulding moisture content, and at wetter-than-optimum water content there is a significant loss of strength (Figure 5.7). The effect of negative pore pressure on the characteristics of compacted clay soil is very significant. When a suction exists in a soil it will readily absorb water so that the effective stress decreases and it swells. The higher the initial suction after compaction the greater will be the subsequent reduction in the effective stress and the amount of swelling. The relationship between the initial suction and the moulding moisture content is illustrated in Figure 5.8.

Increasing the moulding water content results in a significant decrease in permeability on the dry side of optimum moisture content, but causes only a slight increase in permeability on the wet side of optimum (Figure 5.9). It is generally accepted that the influence of moulding water content on the permeability of compacted clay soils is explained by the 'clod' (clusters of clay particles) theory. Most of the flow of water in compacted clay occurs in the pore spaces between the clods rather than between the clay particles in the clods. Soft wet clods of soil are easier to remould and knead together than are hard dry clods. The greater the applied compactive effort the lower the permeability of the soil due to enhanced remoulding of the soil (see Figure 5.9). When a soil is compacted wet of optimum, the clay particles tend towards a dispersed structure so that the void spaces are small and distributed throughout the soil mass — hence the development of the clay puddling technique mentioned in Chapter 3 (Section 3.1).

The soil type (particle size distribution, grain shape, clay minerals present, etc) has a great influence on the maximum dry unit weight and optimum moisture content. It can be seen from Figure 5.10, which shows typical compaction curves for different soil types, that different soils subjected to the same amount and type of compactive effort achieve quite different degrees of compaction. A flat curve denotes a uniformly-graded soil, while a curve with a pronounced peak denotes a well-graded soil. Very plastic materials do not achieve high densities when compacted, the density being fairly insensitive to changes in moisture content, and so a relatively flat compaction curve is obtained.

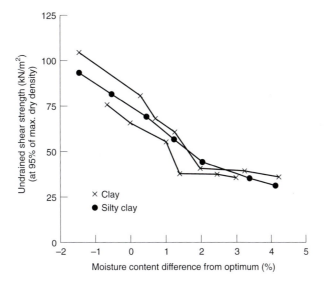

Figure 5.7. The effect of moulding moisture content on soil strength (after Ervin, 1993)

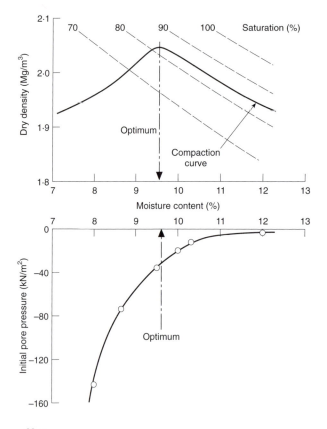

Figure 5.8. The initial suction in compacted clay (after Bishop & Bjerrum, 1960)

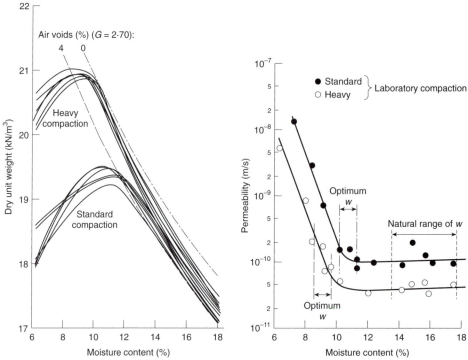

Figure 5.9. The effect of moulding moisture content on permeability

5.2.3. Moisture condition value

For a given sample of soil, at any given value of moisture content there is a unique compactive effort at which the density of the sample ceases to increase. The higher the moisture content, the lower the compactive effort beyond which no further increase in density occurs. The compactive effort (in terms of the number

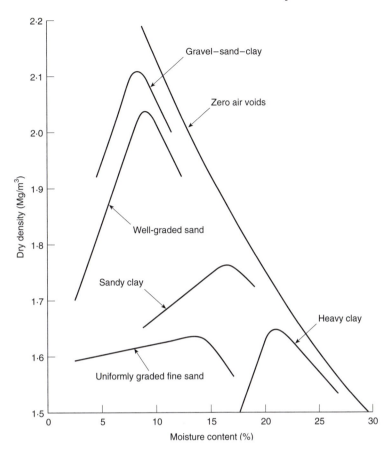

Figure 5.10. The effect of soil type on the compaction curve (after Parsons, 1992)

of blows of a rammer) required to compact a sample of soil to the foregoing density state determines the Moisture Condition Value (MCV) of that sample.

The MCV apparatus consists of a mould (100 mm diameter) into which fits a free-falling rammer (7 kg), which is raised and released to maintain a constant drop height (250 mm) onto the surface of a soil specimen (Figure 5.11). The progressive change in the density of the soil is inferred by measuring the length of the rammer protruding from the top of the mould at various stages. Soil is placed loosely in the mould and successive blows of the rammer are applied. The protrusion of the rammer at any given number of blows is compared with the protrusion for four times as many blows and the difference in penetration of the rammer (a measure of the change in density) is determined. This change in penetration is plotted against the cumulative number of blows (on a logarithmic scale) — Figure 5.12. The descending part of the curve is normally a straight line. To define the fully compacted state a change in penetration of 5 mm has been arbitrarily selected as indicating the point beyond which no significant change in density occurs. The MCV is defined as ten times the logarithm of the number of blows B corresponding to a change in penetration of 5 mm, i.e.

$$\text{MCV} = 10 \log_{10} B \qquad (5.4)$$

If the soil is in a fully remoulded condition the relationship between the MCV and the moisture content is a straight line over a substantial range of moisture contents. With overconsolidated clays, however, a concave-upwards curve can be produced on a similar plot. The curvature is probably caused by the different degrees by which the overconsolidated clay structure (which results from its stress history) is destroyed. A tentative relationship between soil strength (determined by the hand vane) and the MCV has been derived by Parsons & Boden (1979).

Figure 5.11. Apparatus for determining the Moisture Condition Value (MCV)

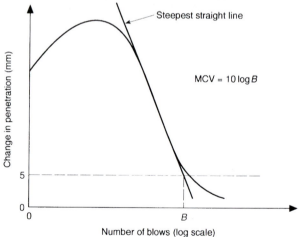

Figure 5.12. The determination of the moisture condition value (MCV)

5.3. Field compaction

The aim of earthworks specifications is to provide a means whereby both the engineer and the contractor are able to agree upon a satisfactory quality of placed and compacted fill. In this respect the specification should not require a contractor to wait days for approval to continue work, nor should it be uneconomical to implement (in terms of either construction or testing/

verification). The role of the specification in the formation of acceptable engineered earthworks is two-fold: the definition of acceptable materials and the identification of satisfactory methods of placement.

5.3.1. Specification

Definition of the requirements of the materials to be used for earthworks, and in particular any characteristics that may influence or limit their performance, is essential. Relevant factors are; particle breakage under compaction, degradation with time, swelling and shrinkage characteristics, and loss of strength on saturation. Earthworks materials fall into two general categories:

- *Acceptable materials* meet the requirements of the specification. In particular 'selected materials' are materials that are essentially devoid of the detrimental characteristics of compressibility and high variability. These materials are used in a range of fill situations by being spread in thin layers and then being well compacted. This gives an end-product that has high shear strength and low compressibility, thereby providing stability and ensuring that subsequent settlements are small. Typical acceptable earthworks materials are defined in Table 5.2.
- *Unacceptable materials* should not be used in permanent works. Unsuitability can arise by virtue of non-compliance with permitted constituents, unstable composition, hazardous chemical or physical properties or high compressibility.

*Table 5.2. Acceptable earthworks materials**

Class	Description	Typical properties	Usage
General granular fill	Well graded Uniformly graded Coarse granular	<125 mm, <15% fines, CU >10 <125 mm, <15% fines, CU <10 <500 mm, <15% fines, CU >5	General fill
General cohesive fill	Wet cohesive Dry cohesive Stony cohesive Silty cohesive	<125 mm, >15% fines, $w > w_p - 4$ <125 mm, >15% fines, $w < w_p - 4$ <125 mm, 15% >2 mm <125 mm, >80% fines	General fill
Selected granular fill	Well graded	<500 mm, <5% fines, CU >10	Below water
	Coarse granular Uniformly graded	<500 mm, 90% >125 mm, CU >5 <125 mm, 90% >2 mm, CU <10	Starter layer
	Fine grading Coarse grading	<75 mm (OMC −2) < w < OMC <125 mm (OMC −2) < w < OMC	Capping
	Well graded	<75 mm, CU >10	Fill to structures
Selected cohesive fill	Cohesive	15–100% fines, $w_L < 45$, $I_p > 25$	Fill to structures
	Wet cohesive, stony cohesive	Minimum strength and chemical stability	Reinforced earth fill
	Silty cohesive	>15% fines, 80% <2 mm	Capping
Stabilized materials	Cement-stabilized silty cohesive Lime-stabilized cohesive	>15% fines, 80% <2 mm, 2% cement minimum, MCV <12 >15% fines, $I_p > 10$, organics <2%, 2.5% lime minimum	Capping

CU, coefficient of uniformity; OMC, optimum moisture content; I_p, plasticity index; w_p, plastic limit; w_L, liquid limit.
* Based on the Highways Agency specification (1998).

Factors to be taken into account when choosing suitability criteria can be obtained from consideration of stability, settlement characteristics, trafficability and economics. In general, the parameters used to define the suitability of cohesive materials are the moisture content, the liquid and plastic limits, and the strength and compaction characteristics. However, the most common requirement is that the fill must have a moisture content no greater than a given value. This basis is attractive because the shear strength increases as the moisture content decreases and the measurement of moisture content is easy to perform. A frequently used way of selecting a maximum allowable moisture content w_{max} is as a function of the plastic limit (w_p). In the UK typically the range of acceptable moisture contents lies from ($w_p - 4$) to $1.2 \times w_p$ (although a factor of 1.3 has been adopted for wet clay fill).

However, on site the plasticity limits and the grading characteristics will vary, so for one value of moisture content a wide range of shear strengths can be obtained. Furthermore natural soils are frequently wet of optimum and the low permeability of clay soils makes it impractical and uneconomic to dry them out. On the occasions when soils are dry of optimum it is possible to wet them up using a water spray. Thus an alternative control is to determine the soil strength directly. Such a result can be obtained quickly, and for highly cohesive stone-free soil quite simply, using a hand vane apparatus (see Chapter 4, Section 4.3.3) or the unconfined compression apparatus (see Chapter 6, Section 6.3.6), either on-site or in a laboratory. If MCV is used as an acceptability criterion then a minimum value of 8 approximates the limits of strength for trafficability purposes as well as for the stability of an earth structure. The suitability of granular soils is usually determined on the basis of maximum moisture contents, which are related to the optimum moisture content (to give high density) and the particle-size distribution (to ensure a free-draining nature).

The specifications for field compaction work, which are intended to ensure that the finished job will be satisfactory, are basically of two types:

- *Method specification*: this stipulates the compaction procedures to be followed, giving the type of equipment to be used together with the number of passes to be made over each layer.
- *End-result specification*: this indicates the end result required, usually in terms of the in-place dry density of the compacted soil.

In some instances a combination of both methods is specified.

Use of a method specification reduces the amount of quality control testing to a minimum by making use of a standard list of compaction methods. For instance, the Series 600 Earthworks section of the Specification for Highway Works (Highways Agency, 1998) contains tables defining the minimum number of passes for a maximum depth of compacted layer, for different categories of compaction plant and their applicability to different soil types. A sample of the data is given in Table 5.3. Plant, materials and methodologies not included in the table can only be used providing that the contractor demonstrates in site trials that the state of compaction achieved by the alternative method is equivalent to, or better than, that obtained using the specified method. Drawbacks of method specification are:

- constant supervision of the compaction is required
- a significant amount of testing is required to ensure a reasonable indication of the quality of the fill placed.

Typical end-result specifications require the fill to be compacted to a dry density of 90–95% of the maximum dry density determined from a standard compaction test (there is a trend towards adopting the modified compaction test as the basis for control testing). The disadvantages of this form of specification are:

*Table 5.3. A method of compaction for earthworks**

Compaction plant	Roll mass (t) per metre width	Cohesive		Well graded granular and dry cohesive		Uniformly graded		Stabilized cohesive	
		D (mm)	N	D (mm)	N	D (mm)	N	$N^†$	$N^‡$
Smooth roller	2.1–2.7	125	8	125	10	125	10	U	U
	2.7–5.4	126	6	125	8	125	8	U	U
	> 5.4	150	4	150	8	U	U	12	U
Grid roller	2.7–5.4	150	10	U	U	150	10	U	U
	5.4–8.0	150	8	125	12	U	U	16	U
	> 8.0	150	4	150	12	U	U	8	U
Pneumatic-tyred roller	1.0–1.5	125	6	U	U	150	10	U	U
	1.5–2.0	150	5	U	U	U	U	12	U
	2.0–2.5	175	4	125	12	U	U	6	U
	2.5–4.0	225	4	125	10	U	U	5	U
	4.0–6.0	300	4	125	10	U	U	4	16
	6.0–8.0	350	4	150	8	U	U	U	8
	8.0–12.0	400	4	150	8	U	U	U	4
	> 12.0	450	4	175	6	U	U	U	4
Vibratory roller	0.27–0.45	U	U	75	16	150	16	U	U
	0.45–0.7	U	U	75	12	150	12	U	U
	0.7–1.3	100	12	125	12	150	6	U	U
	1.3–1.8	125	8	150	8	200	10T	U	U
	1.8–2.3	150	4	150	4	225	12T	12	U
	2.3–2.9	175	4	175	4	250	10T	10	U
	2.9–3.6	200	4	200	4	275	8T	10	U
	3.6–4.3	225	4	225	4	300	8T	8	U
	4.3–5.0	250	4	250	4	300	6T	8	U
	> 5.0	275	4	275	4	300	4T	6	12

D, maximum depth of the compacted layer; N, minimum number of passes; T, roller must be towed by a track-laying unit; U, unsuitable.

* From the Highways Agency (1998) specification.
† Number of passes for a 150 mm layer.
‡ Number of passes for a 250 mm layer.

- laboratory tests can only be performed on material finer than 37.5 mm (in fact, in most laboratory compaction tests all particles larger than 20 mm are removed), and thus the results may need to be adjusted for the material actually used on site
- the laboratory compactive effort may bear little resemblance to the field compactive effort applied by the compaction plant
- the laboratory and field dry densities are dependent on the grading of the soil, and thus changes in grading would require re-testing.

An alternative end-result specification is to define the maximum air voids allowed for the compacted material. In this case the upper limit of the moisture content has to be carefully observed otherwise, for wet materials, the specification could be achieved with little compactive effort. The method also requires accurate determination of the specific gravity of the soil in order to calculate the air voids content.

It is important to recognize that any in situ test only provides information on the condition of the fill at the test location. It is therefore critical that the uniformity of the compaction process is also monitored as the earthworks are carried out. Variables such as layer thickness, number of passes of the compaction equipment, uniformity of moisture conditioning and the consistency of the materials being compacted all need to be controlled.

5.3.2. Field measurement

Field compaction control tests usually involve the determination of the in-place bulk density and the water content of the fill. The field density can be determined directly (core cutter, sand replacement and balloon methods) or indirectly (nuclear density and moisture content gauge method).

In the *core cutter method* a steel tube (approximately 100 mm diameter, 130 mm long) is driven into the ground until it is filled with soil. It is then dug out, the ends are trimmed and the filled tube is weighed to find the bulk density. The soil plug is extruded from the tube and dried to determine its moisture content. By its very nature the method is applicable only to soils that are relatively free of stones. It is imperative that the complete depth of layer being compacted is included in the sample or that the least compacted part of the layer (usually the bottom where the compaction stresses are lowest) is sampled.

In the *sand replacement method* a small hole (approximately 100 mm diameter, 150 mm deep) is dug (Figure 5.13) and the mass and moisture content of the excavated material is carefully determined. The volume of the hole thus formed is obtained by pouring a dry, uniform sand into it from a special container and measuring the weight of the sand in the hole. The bulk density attained by the sand when it fills the hole is determined separately (by filling a container of approximately the same dimensions as the excavated hole) so that the weight of sand in the hole can be converted to a volume.

The *balloon method* also involves excavating a cylindrical hole in the ground and weighing the soil removed. The volume of the hole is determined by measuring the quantity of fluid that has to be pumped into a balloon to fill the hole. Alternatively, a flexible plastic sheet is used to line the excavated hole, which is then filled with water. The diameter of the hole should be at least four times the size of the largest particle encountered in the soil.

A *nuclear density and moisture (NDM) gauge* is an instrument that can measure both the bulk density of a soil and its water content. Nuclear density gauges consist of two parts: a source, which emits radiation; and a counter–

Figure 5.13. In situ density determination by the sand replacement method

Figure 5.14. A nuclear density and moisture (NDM) gauge

receiver, which records the amount of radiation received. In the direct-transmission method the source is inserted to a known depth in the soil while the counter remains on the surface. In the surface back-scatter method both the radioactive source and the detector are placed on the surface of the soil (Figure 5.14). The intensity of emissions sensed by the counter varies with the density and water content of the soil. A calibration curve, which relates the detected radiation intensity to the density, is established for each meter using blocks of known density. The blocks need to be such that their ranges of densities are similar to those expected for the soils to be tested. Similarly, a relationship between radiation count and apparent moisture content can be obtained using standard blocks of known moisture equivalence. Nuclear meters have gained wide acceptance in recent years because of the rapidity with which a test can be performed, the non-destructive nature of the test and acceptance that the calibration techniques are reliable.

The use of nuclear equipment requires a licence (from the Department of the Environment or equivalent legislative authority). There are established monitoring procedures for the operators, including medical examinations and the issue of film badges to ensure that the radiation exposure is within acceptable limits.

5.3.3. Compaction plant

Compaction plant may be classified (by mode of operation) as:

- construction traffic
- dead-weight rollers
- vibrating rollers
- rammers.

Two types of construction traffic are often considered to give some degree of effective compaction: rubber-tyred plant (scrapers and dump trucks) and tracked plant (bulldozers). However, the tyres and tracks do not give complete coverage with each pass and drivers tend to follow existing tracks, and so this type of compaction can only be considered incidental, as it is not engineered in any way.

In selecting plant for engineered compaction the following points should be considered:

- the weight of the compactor and the ability to adjust its weight
- the contact area of the rolls/wheels/protrusions with the ground and the adjustment capability
- the drive mechanism (i.e. self-propelled or tractor-towed)
- manoeuvrability (the effect of turning and irregular compaction at the edges of the site)
- the suitability of plant for compacting a soil type and the number of passes required.

The most common types of smooth-wheeled rollers are tandems or large diameter, single-axle, drums (Figure 5.15). The towed single-axle roller consists of a frame and smooth steel cylinder ballasted with sand or water to increase self-weight. Smooth-wheeled rollers perform best on granular materials, but they also produce high states of compaction with nearly all soils (except perhaps wet clays) provided that thin layers are used. It is common practice to use smooth-wheeled rollers to shape and seal clay fill layers during construction. While this may encourage run-off of precipitation it does not help bonding of one soil layer with the next, and in fact it builds in a potential slip surface.

Drum rollers, which have projections attached to the drum or which have drums made of heavy-duty mesh, are most effective in compacting clayey soils. The optimum moisture content for the roller approximates to that of the modified,

Figure 5.15. A smooth-wheeled roller

heavy compaction test. The compactors are usually unsuitable for granular soils as local shear failure and surface loosening is induced. The sheepsfoot roller consists essentially of a cylindrical steel drum, with projecting feet (club-shaped or tapered) mounted on the surface of the drum which can be ballasted (Figure 5.16). The small surface area of each foot transmits high pressure, thus kneading the soil particles together. With grid rollers the surface of the roll is composed of heavy steel mesh of a square pattern. The roller is suitable for crushing and compacting cohesive soil dry of optimum, soft rocks, slag or soils having a high gravel content because of the high localized pressures applied. It is an all-purpose roller (except for wet clays, when it becomes clogged) which will produce a high-density state, combined with a low air voids content in most soils.

Pneumatic-tyred rollers basically comprise a ballast box mounted between two axles, the rear axle having one wheel more than the front, and the wheels are mounted such that the rear wheels travel over the area not covered by the front wheels. The ballast box may be loaded to the desired weight and the tyre inflation pressures adjusted to vary the contact area with the ground. These rollers achieve compaction by a combination of pressure and kneading action, and can be used for sandy and clayey soils. The rubber-tyred roller produces a smooth compacted surface, which consequently does not provide significant bonding between successive layers. Furthermore, the relatively smooth surface of a rubber tyre can neither aerate a wet soil nor mix water into a dry soil. When used to compact cohesive soils, rubber-tyred rollers give the best performance when the soil is about 2–4% below the plastic limit.

Figure 5.16. A sheepsfoot roller

Figure 5.17. A smooth roller with a vibratory unit

By setting the rim of a roller into oscillation (vibratory units can be attached to any type of roller) vibrations are transmitted to the soil particles and improve compaction. Vibratory rollers (Figure 5.17) are very efficient in compacting granular soils — a much heavier static weight roller would be needed to achieve equivalent results. The frequency and deadweight of vibratory rollers must be suited to the material being compacted:

- heavyweight rollers with high-frequency vibrations for gravel
- light- to medium-weight rollers with high-frequency vibrations for sands
- heavyweight rollers with low-frequency vibrations for clays.

Vibrating plate compactors have a large surface area and are generally light. They do not satisfactorily compact heavy clay and even well-graded sand requires many passes for a minimum of 10% air voids to be achieved. Hence they are only suitable for compacting shallow fills. Vibrotampers are usually operated in confined spaces for the reinstatement of trenches or compacting close to structures. Tampers are particularly effective in compacting cohesive soils.

6

Shear strength

6.1. Introduction

In geotechnical engineering there are two common limiting states: shear failure of the soil and excessive displacement. Stability analyses in geotechnical engineering, whether they relate to cut slopes, earth dams, retaining walls or foundations, involve calculation of the shear stress that can be applied to a soil before it fails in shear (i.e. its shear strength). The Factor of Safety for many practical situations is then essentially the ratio of this shear strength to the applied shear stress.

Kettleman Hills Hazardous Waste Repository consists essentially of a very large, oval-shaped bowl excavated in the ground to a depth of about 30 m, into which the waste fill was dumped. The bowl has a nearly horizontal base with side slopes of 1:2 to 1:3. To prevent the escape of hazardous materials into the underlying and surrounding ground and the groundwater below, the base and sides of the excavation were lined with a multilayer system of impermeable membranes, compacted clay layers and drainage layers.

In March 1988 a slope failure occurred at the site, with lateral displacements of up to 11 m and vertical settlements of up to 4 m being measured. The failure developed over a period of a few hours, from early morning to early afternoon. Surface cracking was clearly visible, as were cracks and displacement on the exposed portions of the containment system (the liner) beneath the refuse. It was evident from field observations, photographic and survey records, and stability analysis, that the failure developed by sliding along interfaces within the composite, multilayered containment system beneath the waste fill (Mitchell *et al.*, 1990b). The multilayer nature of the protective liner meant that it contained contact surfaces between sheets of various synthetic materials and these materials and compacted clay.

Laboratory tests were performed to evaluate the interface shear strength characteristics of the various components of the multilayer liner system and the shear strength characteristics of the compacted clay. It was found that the maximum interface shearing resistance (between the man-made material and the clay) was developed after very small displacements (0.5–1.5 mm). Further displacement caused a drop in the shearing resistance, with a minimum value (about 80–95% of the maximum) generally being attained after movements of less than 2.5 mm. The interfaces between the synthetic materials were characterized by very low shearing resistance (approximately 50% of the peak strength of the compacted clay, upon which slope stability analyses of the refuse had been based). The frictional resistance was significantly affected by small variations in geosynthetic material structure, surface texture, surface cleanliness, and sample orientation.

A properly constructed composite double-liner system of the type used at Kettleman Hills provides protection against transport of leachate out of the containment system. At the same time, however, the system contains a number of interfaces between different materials and various layers, and these interfaces may act as potential surfaces of sliding whenever elevation differentials exist. Furthermore, a sliding surface will develop in a uniform geotechnical material if

Figure 6.1. An exposed shear plane

its shear strength is exceeded (Figure 6.1). Hence it is vital to have means of determining the shearing resistance along potential sliding surfaces.

6.1.1. Shearing behaviour of soils

If the load on a soil mass is gradually increased then the soil will eventually fail, either through the individual soil particles sliding relative to one another (shear failure), or by fracture of the particles (i.e. crushing — since in geotechnical engineering most direct stresses are compressive). In practical situations the shear failure mode is more critical than crushing because the soil is not confined (so the particles can move) and load changes invariably increase the applied stress differential (i.e. the shear stress). The maximum stress that can be resisted by a soil is termed its *shear strength*. The value of the shear strength depends on the loading and drainage conditions being applied.

Typical shear stress–strain curves for soil are shown in Figure 6.2. At the start of the shearing process the distortion is approximately proportional to the stress

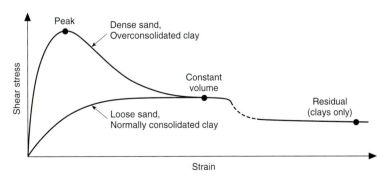

Figure 6.2. Typical shear stress–strain curves for soils

applied. If the applied stress is removed, much of the distortion is recovered and the soil behaves in an approximately elastic manner. As the shear stress is increased the soil structure increasingly deforms in a plastic manner (i.e. removal of the stress results in little recovery of deformations). Since soil is an uncemented collection of grains, mobilization of its internal shearing resistance requires significant deformation (by comparison to other 'solid' construction materials such as steel and concrete). Eventually a peak shear stress (i.e. the maximum shear stress which can be sustained by the soil) will be reached. Generally speaking the larger the grain size, the greater is the strength, and deposits consisting of a mixture of different-sized particles are usually stronger than those that are uniformly graded. If the soil is normally consolidated or loose the peak stress produces continuous plastic deformation. However, for an overconsolidated or dense soil the shear stress that can be sustained by the soil in a post-peak state reduces with further strain (i.e. the soil work-softens). After a considerable amount of shear strain a soil achieves a constant-volume state (after the soil structure has expanded or contracted) and it continues to shear without change in volume or voids ratio so that all energy input to the sample is used to overcome shearing resistance. A reduction in strength beyond a peak value is usually accompanied by the formation of shearing surfaces in the soil on which strains are concentrated.

If a clay soil is subjected to very large shear displacements along a single slip zone the platey particles either side of this surface rearrange to a parallel orientation. This produces a reduction in shearing resistance to a lower limit (below which it is impossible to go without changing the stress regime), which is termed the *residual strength*.

Usually the shear strength of a soil is taken as the maximum shear stress that can be sustained (i.e. the peak value). However, it may be dangerous to rely on this value for some brittle soils, due to the rapid loss of strength if the soil is strained beyond this peak. On the other hand, for loose sands and soft clays very large strains may be needed to develop the maximum resistance, and so a maximum deformation limit (e.g. 20% axial strain) may be imposed.

To obtain the shear strength of a clay soil as it exists in situ its structure must not be altered before it is sheared in the laboratory apparatus. To achieve this it is necessary to:

- take samples in a manner that produces the least disturbance
- avoid moisture content changes after sampling
- use a test procedure that controls water leaving or entering the specimen.

In addition to drainage conditions, a prime factor influencing the shear strength characteristics of a clay is its stress history. Many soils have been subjected to overconsolidation due to high stresses being applied at some stage in their deposition history. Overconsolidated soils inevitably have lower porosities than do their normally consolidated counterparts. This leads to their having stiffer behaviour with regard to deformations under an applied load and more dilatant behaviour under shear because of the denser packing of the particles.

6.1.2. The Coulomb failure envelope
The shearing resistance of a soil is 'traditionally' defined as arising from two components (by analogy with the shearing interaction between solid materials), namely cohesion and friction. In the geotechnical sense cohesion is strictly some inherent strength of the soil that is determined by the nature of the soil and its formation history. Cohesion is analogous to the shearing resistance developed when two pieces of metal become rusted together, in other words, the prevailing strength of the bond depends on some previous event and is more or less independent of the stress pressing the two pieces of metal together. Furthermore,

when one piece of metal is forced to move relative to the other, this bond strength rapidly disappears — likewise true cohesion in solids.

The friction arises because of the direct contact between solid materials (the grains), so that for sliding to occur frictional resistance must be overcome. When two solid bodies are in contact with each other the frictional resistance at the contact, F, is proportional to the normal force N between the two bodies. The constant of proportionality is the coefficient of friction (often denoted by μ). The value of the coefficient depends on the nature and condition of the surfaces in contact and is essentially independent of the applied forces:

$$F = N\mu \tag{6.1}$$

If the shearing resistance in a soil T is equal to the sum of the cohesion resistance and the frictional resistance, then

$$T \equiv \text{cohesion} + \text{normal force} \times \mu \tag{6.2}$$

The shear strength τ of a soil is equivalent to the maximum shear stress that it can withstand, so that

$$\tau \equiv T \text{ per unit area}$$
$$= \text{cohesion per unit area} + \text{normal force per unit area} \times \mu \tag{6.3}$$

Coulomb is credited with proposing the original form of the shear strength law for soil:

$$\tau = c + \sigma_n \tan\phi \tag{6.4}$$

where c is the cohesion, σ_n is the normal stress acting on the sliding surface and ϕ is the friction angle ($\tan\phi$ is thus equivalent to the coefficient of friction μ).

The foregoing simple law for the shear strength of soils (equation 6.4) has been used successfully in geotechnical engineering for many years. However, it must be remembered that soils are an uncemented collection of various sizes of grains that interact with one another as they slide over each other. Hence a sliding surface (Figure 6.3) is not a flat plane and relative sliding of the soil grains on either side of the sliding surface must, in general, be accompanied by volume changes, or attempted volume changes, in the region of sliding. Furthermore, the sliding surface must pass between grains and through the fluid in the voids (pores) between particles.

The basis of calculating the shearing resistance (E) based on a physical model for the sliding behaviour of a soil mass should therefore be

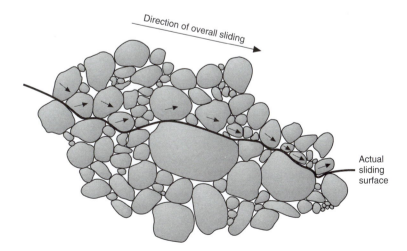

Figure 6.3. A failure surface through a soil

Shearing resistance $\equiv$ energy needed to cause sliding failure

$$E = E_C' + E_F' + E_V' + E_{PF} \qquad (6.5)$$

Where E_C', E_F' and E_{PF} are the energies required to overcome the cohesion between particles, to overcome interparticle friction and to overcome the shearing resistance of the pore fluid, respectively. E_V' is the energy needed to cause volume changes in the soil skeleton. The components E_C', E_F' and E_V' are due to the interaction of the soil grains alone (i.e. there is no contribution from the pore fluid).

Energy is the product of force and displacement. However, the primary concern in deriving a shear strength law for soils is to have a means of predicting the ultimate shearing resistance (regardless of deformations at this stage). Thus:

$E \rightarrow T$ the force to overcome the shearing resistance of the soil mass

$E_C' \rightarrow T_C'$ force to overcome cohesion between grains only

$E_F' \rightarrow T_F'$ force to overcome friction between grains only

$E_V' \rightarrow T_V'$ force needed to cause volume change of the soil mass

$E_{PF} \rightarrow T_{PF}$ force to overcome shearing resistance of the pore fluid.

At a sliding surface the frictional resistance is proportional to the effective force (i.e. that transmitted between the solid grains — see Chapter 3, Section 3.2.2) acting perpendicularly (normally) to the sliding surface. Because a soil is composed of irregularly shaped grains the local direction of sliding and frictional resistance must vary along the failure surface. Clearly the normal stress across sliding contacts will not in general correspond to the mean stress normal to the overall direction of the failure surface. However, for a given arrangement of particles within a soil both the frictional and volume change resistances can be expected to be proportional to the mean effective force normal to the overall failure surface (i.e. to the effective force 'clamping' the soil grains together). Thus

$$T_F' + T_V' \equiv \text{constant} \times N' \qquad (6.6)$$

and so

$$\tau = \frac{T}{\text{area}}$$

$$= \frac{T_C'}{\text{area}} + \text{constant} \left(\frac{N'}{\text{area}} \right) + \frac{T_{PF}}{\text{area}}$$

$$\equiv c' + \text{constant} \left(\sigma_n' \right) + \tau_{PF} \qquad (6.7)$$

Since the pore fluid is typically water and/or air, its shearing resistance is negligible by comparison to the other components of shear strength and $\tau_{PF} \approx 0$. Therefore,

$$\tau = c' + \sigma_n' \text{ (constant)} \qquad (6.8)$$

To be in keeping with the form of the shear strength law proposed by Coulomb, the constant can be defined as $\tan \phi'$. However, it should be noted that $\tan \phi'$ does not just represent the interparticle friction of a soil but also includes effects due to volume changes. Practical separation of the resistance components due to interparticle friction and volume change is problematical (for further reading see the analyses given by Taylor (1966) and Rowe (1962)). Hence, when undertaking laboratory shear strength tests, it is vital that the soil samples have the same internal structure as exists in the ground. Furthermore, for large stress ranges the shear strength envelope defined by equation (6.8) is likely to be curved due to the effects of suppression of volume change at high pressures and the change from an overconsolidated to a normally consolidated state. Notwithstanding these comments, a suitable law for representing the shear strength of a soil, for most practical situations, is the Coulomb law written in effective stress form:

$$\tau = c' + \sigma'_n \tan \phi' \tag{6.9}$$

where c' is the effective cohesion, σ'_n is the effective normal stress and ϕ' is the effective frictional angle.

It is sometimes convenient to define the shear strength of a soil in terms of cohesion and friction components that relate to the whole soil mass (i.e. both the soil grains and the pore fluid, even though the pore fluid itself has negligible shear strength). If the soil mass is treated as an entity then the shear strength components (cohesion and friction) are the total values (because they automatically account for both the soil and the water). However, the values of the components are dependent on the actual composition of the soil, and hence they are only applicable to the soil in this state (i.e. if nothing enters or leaves the soil — it is kept in an undrained state). Hence for this simplified representation of the shear strength law for soils (i.e. for soil in an undrained state):

$$\tau = \text{total cohesion} + \text{total friction}$$

$$= \text{total undrained cohesion} + \text{total undrained friction}$$

$$= c_u + \sigma_n \tan \phi_u \tag{6.10}$$

6.1.3. The Mohr–Coulomb failure criterion

A typical element of soil in the ground is represented by a cube and its stress state will be defined by the stresses acting on each of the three orthogonal faces of the cube. In general, the stresses acting on each face consist of a direct (or normal) stress σ_n acting perpendicularly to the chosen plane, and two orthogonal shear stresses that act tangentially to the plane. However, even though geotechnical situations are three-dimensional, if instability or failure occurs the movement of the soil will essentially be in a particular plane because of the geometry of real situations. For instance, consider failure of a long retaining wall. Movement of the soil behind the wall will be essentially downwards and directly outwards (at right-angles to the face of the wall), because the length of the wall precludes sideways movement, and thus failure will take place in a vertical plane normal to the length of the wall. A similar condition applies to cuttings, embankments, strip foundations, etc. Figure 6.4 shows a view along the top of an earth dam showing the outcrop of a slip surface, which is clearly parallel to the crest of the embankment. Hence satisfactory stability solutions may be obtained for most practical geotechnical situations by two-dimensional analysis. For a two-dimensional element of material (Figure 6.5), in general there must be a shear stress and a normal stress acting on each face.

Since the failure/slip surface within a soil mass can occur at various inclinations it is necessary to have some means of relating the stresses acting on different planes that pass through a common point. The Mohr's circle of stress, which is a graphical representation of all of the combinations of normal and shear stress acting on different planes within a material, is one such means (Figure 6.6). The Mohr's circle of stress cuts the normal stress axis at two points (A and B in Figure 6.6). This means that within the two-dimensional soil element there are two planes (defined by points A and B) on which the shear stress is zero; these are termed the principal planes. The normal stresses acting on these principal planes are known as principal stresses, and the larger is termed the major principal stress and the smaller is called the minor principal stress, with the corresponding principal planes being designated in the same manner. The directions of these

*Figure 6.4. The upper edge of
a failure surface in an
embankment*

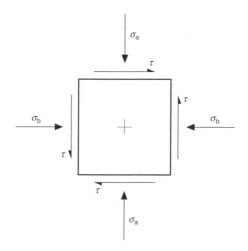

*Figure 6.5. Stresses on a
two-dimensional soil element*

σ_a and σ_b: direct stresses
τ : shear stress

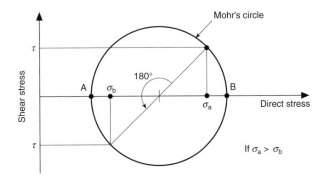

Figure 6.6. Mohr's circle of stress

planes do not automatically correspond to vertical and horizontal although for many geotechnical situations such a correspondence is either more or less true or represents a conservative simplifying assumption for a very complex problem. Because the real situation is three-dimensional there also exists a principal stress that is orthogonal to the major and minor stresses and which is intermediate in value to the other stresses. Hence this third stress is termed the intermediate principal stress and the complete principal stress system is designated as:

$$\sigma_1 \text{ (major principal stress)} \geq \sigma_2 \text{ (intermediate principal stress)} \geq \sigma_3 \text{ (minor principal stress)}$$

The principal stresses have associated major, intermediate and minor principal planes. For practical purposes the controlling factor in shear strength analysis is the relationship between the largest and the smallest principal stresses.

If the Coulomb failure criterion is plotted as a graph of shear stress against normal stress (i.e. using the same axes as for the Mohr's circle) then a straight line results. When a soil element fails in shear its state of stress must correspond to a point on the strength envelope. That is, its Mohr's circle of stress must just touch (be tangential to) the Coulomb line (Figure 6.7). This situation is true whether the stresses are expressed in total or effective terms. From triangle ABC in Figure 6.7,

$$\sin \phi = \frac{\text{Opposite}}{\text{Hypotenuse}} = \frac{BC}{(AD + DC)} = \frac{\left(\dfrac{\sigma_1 - \sigma_3}{2}\right)}{\left(\dfrac{\sigma_1 + \sigma_3}{2}\right) + \left(\dfrac{c}{\tan \phi}\right)} \tag{6.11}$$

Rearranging gives

$$(\sigma_1 - \sigma_3) = (\sigma_1 + \sigma_3) \sin \phi + 2c \cos \phi \tag{6.12}$$

or

$$\sigma_1 = \sigma_3 \frac{(1 + \sin \phi)}{(1 - \sin \phi)} + 2c \frac{\cos \phi}{(1 - \sin \phi)} \tag{6.13}$$

If $(1 + \sin \phi)/(1 - \sin \phi)$ is written as K, then

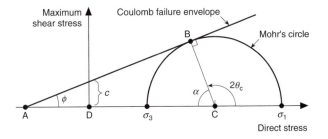

Figure 6.7. The Mohr–Coulomb failure diagram

$$\frac{c \cos \phi}{(1 - \sin \phi)} = \sqrt{K}$$

Therefore

$$\sigma_1 = \sigma_3 K + 2c\sqrt{K} \qquad (6.14)$$

Equation (6.14) has been derived in an illustrative way, i.e. using c and ϕ to represent actual shear strength parameters (c_u, c', ϕ_u, ϕ', etc.), and usable forms of this equation are:

$$\sigma_1' = \sigma_3' K' + 2c'\sqrt{K'} \quad \text{with } K' = \frac{(1 + \sin \phi')}{(1 - \sin \phi')} \qquad (6.15)$$

for effective stresses and

$$\sigma_1 = \sigma_3 K + 2c_u \sqrt{K} \quad \text{with } K = \frac{(1 + \sin \phi_u)}{(1 - \sin \phi_u)} \qquad (6.16)$$

for total stress.

The Mohr–Coulomb failure diagram also contains information on the direction of the failure plane. This plane is represented by point B in Figure 6.7. A property of the Mohr's circle diagram is that an angular rotation of θ in reality is represented in the diagram by a rotation of 2θ. Thus if θ_c is the angle between the failure plane and the plane on which the major principal stress acts, then (from the Mohr–Coulomb diagram):

$$\alpha + 2\theta_c = 180° \quad \text{and} \quad \phi' + \alpha + 90° = 180° \qquad (6.17)$$

From which

$$\theta_c = 45° + \frac{\phi'}{2} \qquad (6.18)$$

The Mohr–Coulomb criterion implies that the effective intermediate principal stress σ_2' has no influence on the shear strength of the soil. While this failure criterion is widely used in practice (because of its simplicity), it is by no means the only model for the shear strength behaviour of soils (see Chapter 6, Section 6.6.5).

6.1.4. Volume changes in granular soils

For an appreciation of the volume change behaviour of granular soils, consider an assemblage of rotund particles arranged in two extreme regular packings, i.e. the loosest and densest states. In two-dimensions, for stable conditions, these packings would correspond to the configurations shown in Figure 6.8, with the corresponding external boundaries. The typical elements of the structures are a square (for the loosest state) and a rhombus (for the densest state). The diagram represents a section (of unit width) through the granular medium so that the plan area indicates a volume.

If these two particle arrangements are subjected to an all round pressure increase then each particle will become slightly smaller, but the particles will not change shape and there will be no distortion of the particle arrangement, i.e. the soil skeleton (represented by the lines joining the centres of touching particles as shown in Figure 6.8) will shrink but will not distort. Thus the external boundaries will attempt to move inwards to produce an overall decrease in the sample volume.

If the soil element is free-draining, then pore fluid will be expelled as the volume of voids is decreased (i.e. the soil remains in a drained state). It is important to realize that in geotechnical engineering the term 'drained' is used to denote the absence of excess porewater pressure — 'drained' does not mean that

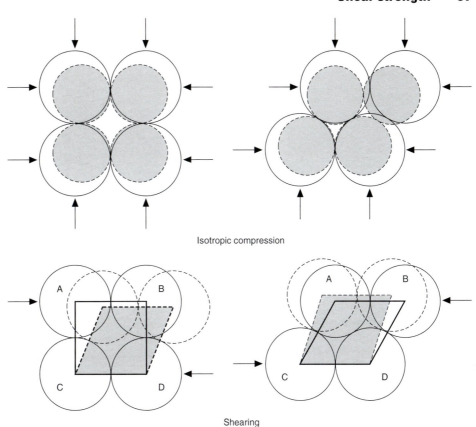

Isotropic compression

Shearing

Figure 6.8. Idealized loose and dense packings of particles

the soil has zero water content, nor does it automatically mean that the soil contains zero porewater pressure. Thus, in general, an increment in applied all round pressure will cause a decrease in the volume of a free-draining soil.

If a shear stress is now applied to the two particle arrangements, so that grains such as A and B slide over grains C and D, then distortion of the typical elements for each particle packing will result (the particles must remain in contact during shearing because all the applied stresses are compressive). For material in a loose state the square element distorts to a rhombus with the same base width as before but with less height (i.e. the element undergoes a decrease in volume). Since the volume of solid grains in the typical element has not changed there is an attempted decrease in the volume of the voids. Hence if the soil is free-draining there will be an overall volume change as pore fluid is expelled.

If a shear stress is applied to the dense packing so that particles A and B slide over particles C and D the typical element would change from a rhombus to a square. The base width stays the same but the height of the element increases (i.e. the typical element is attempting to increase in volume). In common with the loose packing, the volume of solid particles in the typical element remains constant, and thus the volume change being induced must occur within the pore voids. If the soil is free-draining then fluid can flow freely into the voids and the overall soil element volume increases (i.e. dilation of the particle mass occurs). This dilation of a dense particulate medium when all of the surrounding stresses are compressive is one of the special phenomena of geotechnics, and it is also the reason for the appearance of a 'dry' patch beneath the foot when one walks across damp sand on a beach. Placement of the foot induces localized shearing of the densely-packed soil. This causes a localized volume increase of the soil skeleton and water lying on the surface of the soil is instantaneously drawn inside the soil skeleton (it is not squeezed out as is commonly believed) and thus the sand surface suddenly becomes dry.

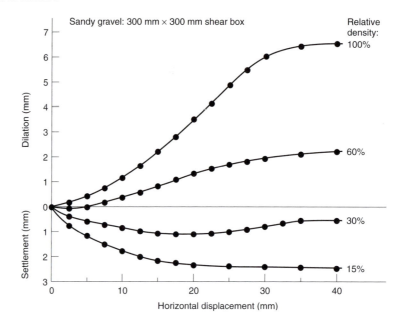

Figure 6.9. Volume changes during shear of free-draining soils

Thus a change in the shear stress applied to a free-draining soil causes a volume change. The magnitude and size (increase or decrease) of the change depends on the internal structure and hence the stress history of the soil. Figure 6.9 gives volume change data from actual soils subjected to shearing.

6.1.5. Pore pressure behaviour

In Section 6.1.4 it was explained how changes in applied stress cause changes in the volume of free-draining, particulate media. If, on the other hand, a soil is slow-draining, or if it is maintained in an undrained state (so that pore fluid can neither exit or enter the soil), an attempted change in the pore volume will be resisted by the pore fluid occupying the voids. Consequently, from Figure 6.8, an increase in the all round (ambient) stress applied to a slow-draining soil will cause an attempted decrease in the pore volume and hence an increase in the pore pressure. If the soil is dry (i.e. the voids are full of air), the rise in pore pressure will be negligible because the air is highly compressible. If, on the other hand, the pore spaces are completely filled with water, the pore fluid will be much less compressible than the soil skeleton and hence the all round pressure increase will be carried, for all practical purposes, entirely by the pore water. If the soil is partially saturated, then the relative compressibility of the pore fluid and the soil skeleton will depend on the degree of saturation of the soil, the stiffness of the skeleton and the compressibility of the soil particles themselves.

The pore pressure change resulting from the application of a shear stress depends on the density state of the soil. An approximate correlation of the internal structure of soils is that loose sands and normally consolidated clays can be thought of as behaving like loose packings of particles, whereas dense sands and overconsolidated clays have densely packed arrangements of particles. Thus undrained shearing of a normally consolidated clay will cause the pore water pressure to increase. On the other hand, an overconsolidated clay will undergo pore pressure reduction during shearing as the attempted volume increase is resisted. The pore pressure changes will be greatest if the soil is fully saturated, and essentially zero if the soil is dry.

Thus the overall change in pore pressure Δu, in soil maintained in an undrained state, resulting from a change in the surrounding stress system, is equal to the change due to the all round pressure plus the change due to the shear stress. By treating soil as an elastic medium, Skempton (1954) derived an equation that expresses these individual components of pore pressure change in terms of the boundary stresses:

$$\Delta u = B\,\Delta\sigma_3 + AB\,(\Delta\sigma_1 - \Delta\sigma_3) \tag{6.19}$$

where σ_3 and σ_1 are the minor and major principal stresses, respectively, and A and B are pore pressure coefficients (originally derived using elastic theory).

Since, in reality, soil is not an elastic isotropic medium, Skempton's expressions for A and B cannot predict individual values of the pore pressure coefficients, which have to be determined by laboratory testing. However, the expression for the parameter B $(= 1/(1 + nC_f/C_s))$ does indicate the limiting values of this parameter. If a soil is dry the pore fluid (air) is much more compressible than the soil skeleton so that $C_f \gg C_s$ and thus $B \to 0$. If the soil is fully saturated, because water is virtually incompressible $C_f \ll C_s$ and thus $B \to 1$. For intermediate states of saturation and the extremes of the soil skeleton stiffness, B varies generally as indicated in Figure 6.10. For a soft clay soil, because of the compressibility of the soil skeleton, the value of B is close to unity if the degree of saturation is greater than approximately 90% — this is a common situation for clay soils in the UK.

Skempton's value for the pore pressure parameter A was $\frac{1}{3}$. However, the magnitude of the pore pressure change due to shear depends significantly on the internal structure of the soil (as indicated previously), which itself depends upon the previous history of the material. Soil in a very dense or highly overconsolidated state will have a negative A value. On the other hand, soils in loose or normally consolidated states will exhibit high positive values of A (up to and even exceeding unity). A typical variation in A with degree of overconsolidation is presented in Figure 6.11 and the general range of values is indicated in Table 6.1. As indicated previously, the pore pressure change caused by shear stress also depends on the magnitude of the strain developed, and this is also true of the coefficient A. The value of A at the failure condition is usually regarded as the most important for practical purposes.

The Skempton pore pressure expression can be rearranged (Bishop & Bjerrum, 1960) as

$$\Delta u = B\left\{\frac{\Delta\sigma_3}{\Delta\sigma_1} + A\left(1 - \frac{\Delta\sigma_3}{\Delta\sigma_1}\right)\right\}\Delta\sigma_1 \equiv \bar{B}\Delta\sigma_1 \tag{6.20}$$

It can be seen that, provided the stress ratio $\Delta\sigma_3/\Delta\sigma_1$ remains constant, the pore pressure response can be calculated merely from the change in the major principal total stress. In many practical situations the foregoing stress ratio will be approximately constant. Furthermore, the major principal stress can often be

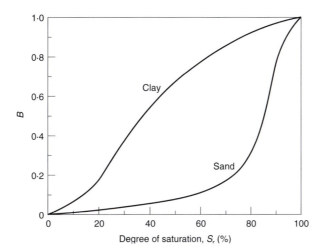

Figure 6.10. The effect of the degree of saturation on B

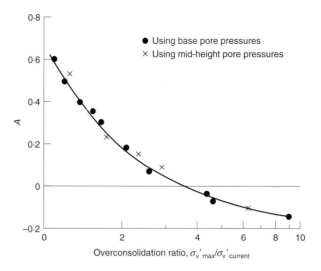

Figure 6.11. The effect of overconsolidation on A

Table 6.1. Typical values of the pore pressure coefficient A

A	Soil category
> 1.0	Sensitive soils (strength decrease during shearing)
1.0 to 0.5	Normally consolidated soils
0.5 to 0.0	Lightly overconsolidated soils
0.0 to -0.5	Heavily overconsolidated soils

represented by the vertical stress (which can be calculated relatively easily). Hence

$$\Delta u \approx \bar{B}\,\Delta\sigma_{\mathrm{v}} \tag{6.21}$$

For a clay soil with a high degree of saturation $\bar{B}$ is usually in the range 0.9–1.0.

6.2. The direct shear box

In the direct shear test (Figure 6.12) the soil sample is contained in a square box (typically 60 mm × 60 mm × 40 mm high, or 100 mm × 100 mm × 30 mm high), which is split horizontally into two equal halves. For coarse-grained materials (e.g. gravel, crushed demolition rubble), a large shear box (300 mm × 300 mm × 180 mm high) (Figure 6.13) may be used. A vertical load is applied to the soil via a loading platen placed on top of the sample. The split box is seated in a rectangular container (the carriage), which holds the bottom half of the box rigidly in place. This carriage is filled with water (which can pass freely through the base of the split box) to saturate the soil and hence eliminate surface tension effects within the soil voids. After the requisite vertical stress has been applied to the upper surface of a sample the lower half of the split box is displaced horizontally, while the top half is restrained, so that shear stress is induced in the

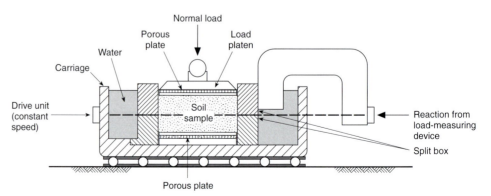

Figure 6.12. The direct shear box

Figure 6.13. The large direct shear box apparatus

soil along a plane that coincides with the horizontal division of the box. The bottom half of the box is driven further and further until the soil fails in shear (i.e. the top half of the sample slides relative to the bottom half). During the testing process the horizontal force applied and the resultant sample deformations (both horizontally and vertically) are measured periodically. Hence the stress–strain and volume change characteristics of the soil may be determined. Typical curves are presented in Figure 6.14. Failure of the soil is usually taken as the maximum shear stress that the soil can withstand, but testing continues until either the limit of travel of the apparatus is reached or the shearing resistance remains constant with displacement.

The foregoing procedure is repeated for at least two further samples of soil under different vertical loads (i.e. for different normal stresses on the sliding plane). A graph is then drawn of maximum shear stress against applied normal stress, from which the shear strength parameters are obtained (Figure 6.15). The slope of the failure envelope is the friction angle, and the intercept (the value of the shear stress at zero normal stress) is the cohesion.

In this apparatus the test sample is not sealed within the shear box and hence the test is not really suitable for clay (slow-draining) soils, for the following reasons:

- Drainage conditions cannot be controlled.
- Pore pressures cannot be measured, and thus effective stresses can only be determined if the pore pressure is zero (i.e. in a drained test). This would take a very long time to perform.
- The failure plane is predetermined and may not be the weakest plane in the soil.
- The internal stress conditions are not uniform and the cross-sectional area under shear is not constant.

The most serious limitations are the lack of control of drainage of the sample and the inability to measure the pore pressure within the soil failure zone. However, these limitations are not relevant to the behaviour of free-draining soils and so the shear box is eminently suitable for testing sands and gravels. It is virtually impossible to obtain undisturbed samples of free-draining granular soils and so test samples must be re-formed within the split box. To ensure accurate results the

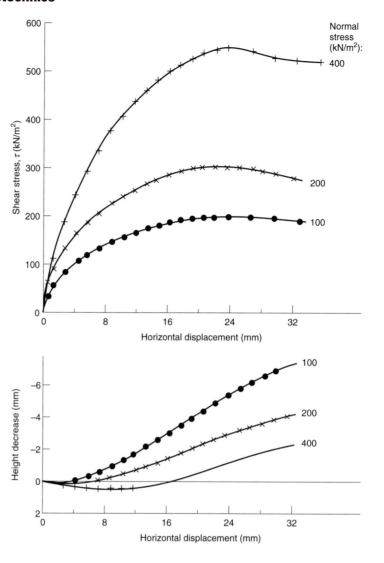

Figure 6.14. Typical stress–strain curves from the direct shear box (300 mm × 300 mm)

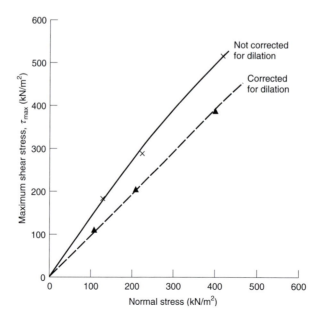

Figure 6.15. A shear strength envelope from the direct shear box

samples should be formed by compacting the soil in layers (3–5) to obtain a value of dry density that corresponds to the field situation (see Section 6.1.2).

6.3. The triaxial test

The triaxial compression test apparatus (Figure 6.16) overcomes the limitations of the direct shear box for the testing of clay soils. Since the sample setting-up procedure requires the formation of a cylinder of soil which is self-supporting for some time, the apparatus is not suitable for routine testing of free-draining granular soils. Tests specimens are typically 38 or 40 mm diameter and 76 or 80 mm high, but larger sizes are also tested (e.g. 100 mm diameter and a height of up to 200 mm).

6.3.1. The triaxial apparatus

This apparatus has two primary components, a base plate and a cell body (see Figure 6.16). The soil sample sits on a base pedestal, which may contain a sensing element to detect the pore pressure and transmit it to an external measuring device. A loading platen, which may incorporate a drain to allow water in or out of the sample, is placed on top of the sample and the whole specimen is sealed within an impermeable flexible membrane. Hence liquid flows into or out of the soil sample can be controlled as required. The test specimen is enclosed within the cell body, which is clamped to the base plate and the space between the sample and the cell body is filled with pressurized water (Figure 6.17). Soil samples are usually brought to failure by progressively adding an additional stress (the deviator stress) in the vertical direction (i.e. to the top of the sample). Readings of the deviator force (measured by a calibrated proving ring or load cell), axial shortening and pore pressure (if measured) are taken at regular intervals. The test continues until the strength of the soil has been exceeded and the shearing resistance has become constant or until a set large axial strain (typically 20%) has been achieved. For each determination of the shear strength parameters, three or more samples are tested under different cell pressures (the cell pressure range applied should be similar to that likely to be incurred in the field situation).

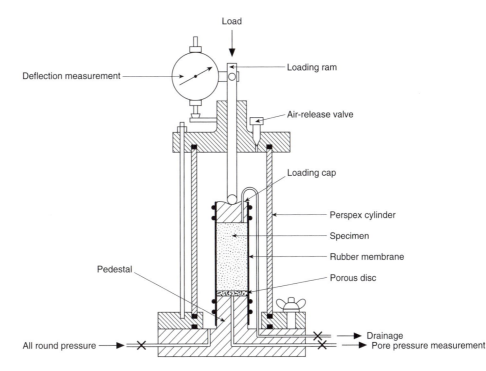

Figure 6.16. The triaxial compression apparatus

Figure 6.17. The triaxial compression apparatus

Barrelling (bulging of a sample at mid-height) of a triaxial sample may occur due to friction at the ends of the specimen (where it is in contact with the loading platens). Thus the common geometry of triaxial samples is a height/breadth ratio of 2, so that the central vertical portion of the soil undergoes more or less uniform deformation. The friction between soil and loading plates may be reduced to a minimal level by smearing the surface of the plates with high vacuum grease and placing a disc of latex membrane on top of the grease; the soil sample then sits on top of the latex disc (Rowe & Barden, 1964).

For stiff, stoney clays it may not be possible to prepare small-diameter specimens from an undisturbed soil sample (typically 100 mm diameter). Hence a multistage triaxial test has been developed in which one sample (100 mm diameter) is tested under several different cell pressures. Under the first confining stress the deviator stress is slowly increased until the stress–strain curve starts to flatten. At this point the cell pressure is increased and shearing is continued for the second stage. This procedure is repeated for a third stage. The stress–strain curves are then extrapolated to a common axial strain value (such as 20%) and the resultant deviator stresses are used in the Mohr's circle analysis. It is also possible to conduct multistage tests wherein drainage is allowed after each cell pressure rise.

6.3.2. Analysis of test data

The triaxial stress system is defined as being composed of three orthogonal principal stresses. These act in the vertical and horizontal directions (because there is no shear stress on the periphery of the soil since it is loaded by water) and they are denoted by σ_1, σ_2 and σ_3. By definition, $\sigma_1 \geq \sigma_2 \geq \sigma_3$ (Section 6.1.3). For the conventional triaxial compression test $\sigma_2 = \sigma_3$.

Loading of the soil in the triaxial test can be thought of as being in two distinct stages (Figure 6.18):

- Ambient stage: the soil is subjected to an equal, all round pressure σ_3 due to pressurization of the water in the triaxial cell.
- Deviatoric stage: additional stress (i.e. the deviator stress D) is applied to cause the sample to fail in shear.

Thus, by considering the stress in the vertical direction,

$$\sigma_1 = \sigma_3 + D \tag{6.22}$$

or

$$D = \sigma_1 - \sigma_3 = (\sigma_1' + u) - (\sigma_3' + u) = \sigma_1' - \sigma_3' \tag{6.23}$$

Failure occurs on a plane that is inclined at some unknown angle to the horizontal, and so the Mohr's circle plot is used to represent all the combinations of shear stress and normal stress on the different planes within the soil. At failure the values of σ_1 and σ_3 are known and so the relevant Mohr's circle may be drawn. If a series of triaxial tests is conducted, then at failure all circles must touch the Coulomb envelope because within each sample there must be a plane on which the failure conditions are satisfied. Hence the failure envelope is the best-fit tangent to all circles (see Figure 6.7) and this line defines the cohesion (the intercept on the shear stress axis) and the friction angle (the gradient of the line). This same form of construction applies whether the principal stresses are plotted as total or effective stress values, although the designation of the resultant shear strength parameters will be different. There are two different drainage conditions (undrained or drained) that can be applied to the ambient and deviatoric stages, according to the shear strength parameters to be measured. The names of the resultant tests and the shear strength parameters obtained from them are outlined in Table 6.2.

The pore pressure coefficients A and B are determined in the triaxial test where pore pressure measurement is undertaken. The coefficient B is measured by applying different all round pressures (in the ambient stage of the triaxial test) and measuring the pore pressure response. The coefficient A is derived from the pore pressure measurements taken during the shearing stage, when the all round cell pressure is kept constant and the soil is maintained in an undrained state.

6.3.3. The undrained test
Each sample is surrounded by a membrane and the drain is closed so that no water goes in or out of the specimen during the shearing stage. Shearing is

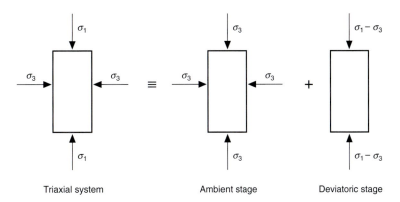

Figure 6.18. The triaxial stress system

Triaxial system   Ambient stage   Deviatoric stage

Table 6.2. Drainage conditions in triaxial tests and the associated strength parameters

Ambient stage	Deviatoric stage	Test name	Mohr's circle analysis	Strength parameters obtained
Undrained	Undrained	Undrained	Total	Total: c_u, ϕ_u
Drained	Undrained	Consolidated–undrained	Effective Total	Effective: c', ϕ' Total: c_{cu}, ϕ_{cu}
Drained	Drained	Drained	Effective	Effective: c'_d, ϕ'_d

typically applied at a rate of 1% vertical strain per minute, so that a test takes between 10 and 20 minutes. Even if the drain were not closed there would be insufficient time for a clay soil to drain to any significant degree.

If the soil is fully saturated it will be found that all samples have essentially the same shear strength (maximum deviator stress). Thus, when the Mohr's circles are plotted in total stress terms the best-fit tangent (i.e. the shear strength envelope), will be essentially horizontal (Figure 6.19). Since, in total stress terms, the equation of the shear strength envelope is $\tau = c_u + \sigma_n \tan \phi_u$ (see Section 6.1.2) then, by definition, ϕ_u will be equal to zero and c_u will be half the deviator stress at failure. This result ($\phi_u = 0$) is often taken by people as meaning that the soil has no friction angle and that it is a purely cohesive material. This latter perception is incorrect and, in fact, the apparent undrained cohesion results principally from the frictional resistance within the soil (this is demonstrated in Section 6.6.2).

If a clay soil is partially saturated, then at low ambient stress levels the strength is increased by increasing the cell pressure, because B is less than unity and an increase in the cell pressure will enhance the effective stress. However, under high cell pressures the strength becomes constant because the resultant high pore pressures cause the air in the voids to dissolve in the pore water, thereby causing the soil to become saturated.

Pore pressure can be measured during an undrained test, but the effective shear strength parameters c' and ϕ' cannot be determined reliably. If the soil samples have a high degree of saturation (the most usual situation) the effective stress circles are either identical or very close together and the common tangent can be fitted at a range of gradients. If the soil has a low degree of saturation these measurements of pore pressure are unreliable because of the presence of the highly compressible air. Consequently, a form of triaxial test, the consolidated–undrained test, was devised to enable the effective shear strength parameters to be determined accurately. It should be noted that this latter test was not devised to replicate the field loading sequence.

6.3.4. The consolidated–undrained test

In this test the sample is allowed to drain via the top platen after the all round cell pressure has been applied. Samples are usually allowed to drain overnight and then sheared the next day. During the deviatoric stage the drain is kept closed so

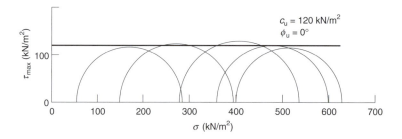

Figure 6.19. An undrained failure envelope for saturated clay

the soil is sheared to failure under undrained conditions, but at a slower rate than in the conventional undrained test (a rate of approximately 5% axial strain per hour is typical). This slow rate of strain encourages uniform pore pressure distribution throughout the sample.

For accurate determination of the pore pressure and to prevent volume changes, the soil sample should either have a high degree of saturation or should be saturated. Saturation is achieved during the ambient loading stage by raising the cell pressure and ensuring that the pore water pressure does not fall below a certain value so that air in the voids dissolves in the pore water (a pore water pressure of about $100 \, kN/m^2$ is sufficient). To do this the drain outlet from the sample is connected to a constant-pressure source (rather than simply venting to atmosphere) so that the soil is 'back-pressurized'. The connection to the back-pressure source is closed at the end of the consolidation stage, before application of the deviator stress is commenced.

The pore pressure u is measured during the shearing stage so that at failure the effective stresses σ_1' and σ_3' can be calculated and the Mohr's circles drawn. The best-fit tangent to the failure circles (typically three in number) defines the shear strength parameters c' and ϕ' (Figure 6.20). The failure circles may be also plotted using total stresses (i.e. the extremes of the diameter are σ_1 and σ_3), and the best-fit tangent then defines the total stress shear strength parameters (from a consolidated–undrained test), i.e. c_{cu} and ϕ_{cu}.

Instead of drawing circles and tangents the results may be analysed by using the Mohr–Coulomb diagram in a different (i.e. equational) form. The Mohr–Coulomb failure diagram can be represented by equation 6.12 in effective stress form, i.e.

$$\sigma_1' - \sigma_3' = (\sigma_1' - \sigma_3') \sin \phi' + 2c' \cos \phi'$$

Thus if $\frac{1}{2}(\sigma_1' - \sigma_3')$ is plotted against $\frac{1}{2}(\sigma_1' + \sigma_3')$, then at failure the data will lie on a straight line with a gradient of $\sin \phi'$ and the intercept on the τ-axis will be $c' \cos \phi'$ (Figure 6.21). If all the test data are plotted, as in the figure, then a stress-path is obtained which touches, or becomes tangential to, the failure envelope. This plot usually resolves the problem of normally consolidated plastic clays, where the deviator stress often continues to rise throughout the test (even up to 20% axial strain). When plotted by this method the behaviour of the material is clarified in that the stress-path travels along the failure envelope (and hence upwards) after some stage.

6.3.5. The drained test
In a drained test the drainage connection remains open throughout the test, drainage taking place against a back-pressure. This back-pressure is then the datum for excess pore water pressure. This test gives effective stress parameters, with the effective cohesion usually being zero (for all soils except heavily overconsolidated clays). A consolidated–undrained test with porewater pressure measurements gives a similar effective stress strength envelope (but not identical, because the drained test involves volume changes of the samples).

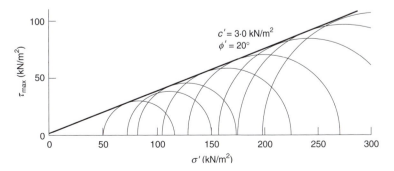

Figure 6.20. The Mohr–Coulomb diagram from consolidated–undrained triaxial tests

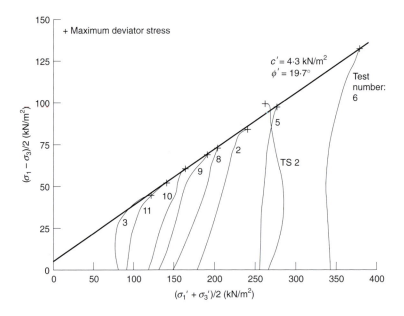

Figure 6.21. Stress-path plots

6.3.6. The unconfined compression test

Sometimes a very rapid estimate is required of the shear strength of a slow-draining soil, but it is necessary to test a reasonably sized soil element (e.g. on an earthworks scheme a minimum shear strength of around $40 \, kN/m^2$ might be required for earth-moving traffic). If the soil is slow-draining then an undrained triaxial test can be undertaken without the use of a surrounding membrane or confining pressure if the loading is applied rapidly. Such an apparatus is the unconfined compression test (usually for testing 38 mm diameter samples), which is commonly carried out in the field using portable apparatus.

The sample is placed between two platens, one of which is fixed and the other moveable (Figures 6.22 and 6.23). A vertical load is applied by winding a screw that pulls a calibrated spring connected to the moving platen. Springs of different

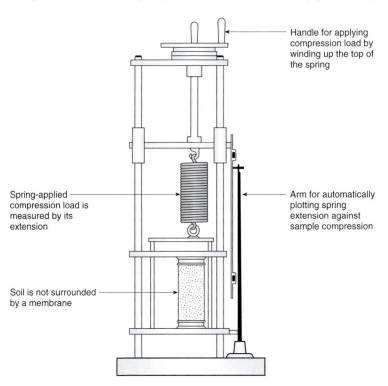

Figure 6.22. The unconfined compression test

Figure 6.23. The unconfined compression apparatus

stiffness can be used according to the strength of the specimen. A pivoted scribe arm is attached to the lower platen, and this produces a plot of spring extension against sample deformation as the test proceeds. The force applied to cause shearing of the soil is obtained by multiplying the extension of the spring by its stiffness, so that the compressive strength of the soil (the deviator stress to cause failure) can be calculated.

The test is fast, simple, compact and inexpensive. However, fissures, gravel particles or other defects may induce premature failure and the drainage conditions are not controlled (so sample preparation and testing should be undertaken rapidly). Since the test data only permit one Mohr's circle of stress to be drawn it is not possible to determine the undrained friction angle for the soil. The undrained cohesion of the clay is typically taken as half the compressive strength (i.e. ϕ_u is assumed to be zero).

6.4. The ring shear apparatus

Skempton (1954) observed that when shearing of a clay is continued beyond the peak strength the resistance decreases to a limiting value, which is termed the residual strength. Individual slip surfaces are quite thin, less than a millimetre thick, and when separated present a polished, fluted or striated, slicken-sided appearance. This degree of polish is due to the reorientation of individual clay particles or platelets in the direction of shear. Under a given effective pressure, the residual strength of a clay is essentially independent of whether the soil was originally normally consolidated or overconsolidated. The formation of a shear surface destroys all earlier fabric in the soil, by remoulding, and it is unnecessary to acquire undisturbed soil specimens for the determination of residual strength — remoulded specimens are perfectly adequate. The peak shear stress data from a set of residual shear strength tests usually define a failure envelope (with parameters c' and ϕ'), with the residual strengths for the same material being represented by a lower line and the parameters c_r' and ϕ_r' (Figure 6.24). In moving from peak to the residual strength, effective cohesion usually falls to zero and the effective angle of shearing resistance may be reduced to a few degrees (as little as 10° for some clays). Soils with high clay contents exhibit the largest fall in shear strength from the peak to the residual state. In general, if the clay content exceeds about 40% or if the Plasticity Index is 30–40% or more then ϕ_r' can be expected to be lower than 15°. Previous shear surfaces, in plastic clays, may be reactivated at low residual friction angles. Furthermore, fissures exist in most overconsolidated clays and produce planes of weakness. The strength of these fissures can be much lower than the intact strength of the clay between the fissures — in fact it may approach the residual value. These fissures may dictate the strength en masse of the clay, so that the mass strength or design strength should be close to the fissure strength.

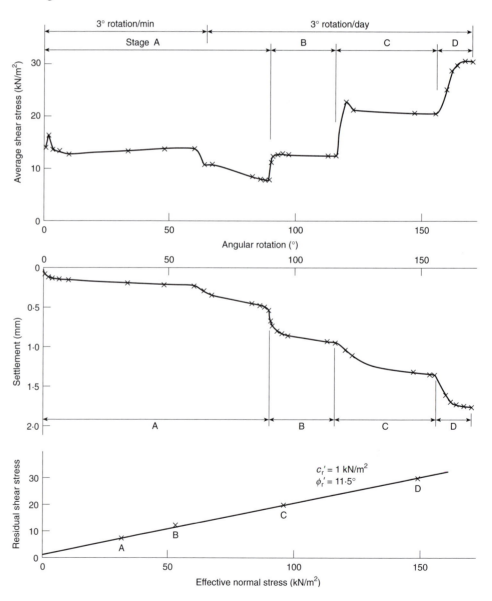

Figure 6.24. Residual shear strength data

Figure 6.25. The ring shear apparatus

The residual shear strength can be determined using a ring shear apparatus (Bishop *et al.*, 1971), in which a thin annular sample of remoulded soil is sheared in a direct shear manner, by rotating the upper half of the sample over the lower half (Figure 6.25), until a polished slip surface is formed. The ring shear device allows almost infinite shearing displacement to be applied along a sliding surface (while still keeping a uniform sample), so that it is relatively easy to produce real residual conditions in the laboratory, starting from an initially unsheared specimen. Also, since a test will eventually be fully drained, regardless of the strain rate used, the problem of measuring pore pressures on the sliding surface does not exist.

6.5. The vane test

The laboratory vane (approximately 13 mm diameter, 13 mm high) is a small version of the field vane apparatus (see Chapter 4, Section 4.3.3) and is used to determine the undrained shear strength of soft clays within a sampling tube or a compaction mould. The test has the advantage of being fast and straightforward. The cruciform vane is pushed vertically into the soil and a torque T applied to the shaft to cause failure to occur due to rotation of a cylindrical plug of soil, thereby overcoming the shearing resistance S_u on the sides and base of the soil plug. The applied torque is equal to the resistance moments generated by these two components of soil shearing, and if the cylinder has a diameter d and height h then

$$T = S_u \pi d h \frac{d}{2} + S_u 2 \frac{\pi d^2}{4} \frac{d}{3} \tag{6.24}$$

If the soil is saturated, then $\phi_u = 0$ and $S_u = C_u$, and

$$c_u = \frac{T}{(\pi d^2 h/2 + \pi d^3/6)}$$

$$= \frac{T}{\pi d^2 (h/2 + d/6)} \tag{6.25}$$

Vane test results can be severely affected by macro-fabric and the undrained conditions assumed may not exist if the soil contains permeable layers.

Furthermore, if a fast rate of rotation is applied to ensure undrained conditions the strength may be overestimated because of the high rate of strain.

The field vane is particularly useful for obtaining an in situ undrained shear strength profile with depth for soils that are difficult to sample, such as soft clays and sensitive clays (see Chapter 3, Section 3.1.4). The sensitivity S_t of a clay soil is defined as the ratio of the undisturbed undrained strength to the remoulded undrained strength:

$$S_t = \frac{S_{U(\text{undisturbed})}}{S_{U(\text{remoulded})}} \tag{6.26}$$

It is commonly found that the remoulded (or residual) strength of a natural clay is much less than its undisturbed counterpart. Typically the value of sensitivity ranges from about 5 to 10, but there are some clay deposits where the sensitivity exceeds 100.

6.6. Applications of shear strength theory

Pore water pressure has a major influence on the shearing strength of soil by virtue of its influence on the magnitude of the effective stress. With coarse-grained soil, drainage is normally so good that the effective stress changes simultaneously with a change in total stress (i.e. the soil is free-draining). With fine-grained soils the rate of water movement is so slow that a considerable time may have to elapse after a total stress change before the effective stress changes.

6.6.1. Influence of pore suctions

For a fundamental appreciation of the shearing behaviour of soils the strength law must always be remembered in its effective stress format. It is crucial to appreciate that the prime factor determining the shearing resistance of a given soil is the effective stress state.

The 'simple' sand castle demonstrates the importance of appreciating the fundamental influence of effective stress on shear strength. A dry sand is clearly not cohesive — it certainly does not possess any true intrinsic cohesion ($c' = 0$). If a mass of dry sand is slowly tipped out of a container then it forms a conical pyramid (i.e. it will not stand up vertically). If, however, a small quantity of water is mixed with the sand then the familiar sand castle, with nearly vertical faces, can be made (Figure 6.26). Thus the damp sand could be described as a cohesive material (as per the common usage of the term 'cohesive'), but the sand does not possess effective cohesion or inherent latent strength. The sand castle can be made with vertical faces because the small quantity of water present in the damp sand partially fills the voids between the sand particles and surface tension forces in the water pull the particles together (the pressure in the interparticle water is negative). The effective stress σ' is equal to $(\sigma - u)$, and so a negative pore pressure enhances the effective stress, i.e. it increases the contact stress between the particles, which in turn increases the frictional resistance to sliding at the contacts ($\sigma'_n \tan \phi'$). Since the frictional resistance has been increased, the overall shear strength of the soil mass has been increased (from its dry state) and this increase is sufficient to enable the sand particles to stand vertically. Hence the damp sand mass is simply exhibiting the fact that it is a frictional material, although it might be described as being cohesive if the internal behaviour is not appreciated.

To test whether a soil has true inherent cohesion (as opposed to apparent cohesion due to negative pore pressure effects), leave it submerged in water for several days and observe whether it remains as an integral mass.

(a)

Figure 6.26. Materials that appear to be cohesive due to pore suctions: (a) immediately after formation; (b) after drying

(b)

6.6.2. Undrained shear strength

Confusion often occurs with regard to whether soils are cohesive or not, because of association of the common usage of the term 'cohesion' with the technical meaning within soil mechanics. Consider the behaviour of a soil subjected to an undrained triaxial test. The deviator stress at failure D_f is given by

$$D_f = \sigma_{1f} - \sigma_{3f} = \sigma'_{1f} - \sigma'_{3f} \tag{6.27}$$

The fundamental definition of the shear strength of a soil is in terms of effective stresses. Thus, from equation (6.15)

$$D_f = \sigma_3'(K' - 1) + 2c' \sqrt{K'} \tag{6.28}$$

Assume that the soil sample had a pore pressure of u_0 prior to being installed in the test apparatus (when $\sigma_3 = 0$ and $D = 0$). At failure the cell pressure is σ_3 and the deviator stress is D_f. Hence from Skempton's equation the pore pressure change up to failure is given by

$$\Delta u = B\,\Delta\sigma_3 + AB\,\Delta D = B\sigma_3 + ABD_f \tag{6.29}$$

Thus the pore pressure at failure u_f is equal to $u_0 + \Delta u$:

$$u_f = u_0 + B\sigma_3 + ABD_f \tag{6.30}$$

Hence, at failure,

$$\sigma'_{3f} = \sigma_3 - u_f = \sigma_3 - u_0 - B\sigma_3 - ABD_f = \sigma_3(1 - B) - u_0 - ABD_f \tag{6.31}$$

Substitution into equation (6.28) gives

$$D_f = [\sigma_3(1 - B) - u_0 - ABD_f](K' - 1) + 2c' \sqrt{K'} \tag{6.32}$$

or

$$D_f[1 - AB + ABK'] = \sigma_3(1 - B)(K' - 1) - u_0(K' - 1) + 2c' \sqrt{K'} \tag{6.33}$$

Therefore,

$$D_f = \frac{\sigma_3(1 - B)(K' - 1) - u_0(K' - 1) + 2c' \sqrt{K'}}{(1 - AB + ABK')} \tag{6.34}$$

If three identical fully saturated soil samples are tested in the undrained triaxial test they will all have a B value of unity and the same values of c', ϕ' (and hence K'), u_0 and A (they all have the same stress history and internal structure). Consequently, the equation for their deviator stress at failure will reduce to

$$D_f = \frac{-u_0(K' - 1) + 2c' \sqrt{K'}}{(1 - A + AK')} = \text{constant} \tag{6.35}$$

Hence all three samples will fail under the same deviator stress. If the Mohr–Coulomb diagram is plotted in total stress terms the failure envelope will be horizontal since all the circles are of the same size and, by definition, ϕ_u will be equal to zero (see Figure 6.19). However, it must be strongly emphasized that this does not mean that the soil has no friction. Rather it means that the total stress changes during testing are balanced by the changes in pore pressure, so that the effective stresses (and hence the shear strength) remain constant. If ϕ_u is equal to zero, then

$$c_u = \frac{D_f}{2} = \frac{[-u_0(K' - 1) + 2c' \sqrt{K'}]}{2(1 - A + AK')} = \text{function } (\phi') \tag{6.36}$$

Even if the soil is truly cohesionless (i.e. $c' = 0$), c_u still has a value provided there exists a suction within the pore fluid (u_0 is negative) — compare with the behaviour of the sand castle; see Section 6.6.1.

6.6.3. Parameters

For granular materials (gravels, sands, silts) and normally consolidated clays the effective cohesion is zero. Overconsolidated clays may exhibit effective cohesion due to particles having been forced into very intimate contact with one another at some time in the history of the soil. While significant values of c' may be reported from laboratory shear tests, caution should be exercised in using these values in practice. Poor quality laboratory testing can lead to overestimation of c', and progressive movement in field situations may mean that the effective cohesion is destroyed.

Typical values of the effective friction angle are given in Table 6.3. The value of ϕ' is generally found to increase with:

- increasing particle size
- increasing particle angularity
- increasing uniformity coefficient (well-graded soils will produce better interlock and have more interparticle contacts)
- particle strength (weak particles are prone to crushing, especially at higher confining stresses).

Typical ranges of undrained shear strength are illustrated in Table 6.4, with ϕ_u being essentially zero for fully saturated soils.

The test methods for determining the friction angle apply to triaxial compression conditions, whereas in reality many geotechnical constructions (e.g. retaining

Table 6.3. Typical effective
friction angles for soils

Soil type	ϕ' (°)
Uniform sand, rounded particles	28 (loose) to 36 (dense)
Well-graded sand, angular particles	33 (loose) to 45 (dense)
Sandy gravel	38 (loose) to 55 (dense)
Silt	27 (loose) to 36 (dense)
Sandy clay	25 (NC) to 29 (OC)
Silty clay	20 (NC) to 25 (OC)
'Fatty' clay	15 (NC) to 22 (OC)

Table 6.3. Typical effective friction angles for soils

Soil consistency	Undrained strength (kN/m²)
Very stiff or hard	> 150
Stiff	100–50
Firm to stiff	75–100
Firm	50–70
Soft to firm	40–50
Soft	20–40
Very soft	< 20

Table 6.4. Typical shear strengths of slow-draining soils

walls), exist under a plane strain situation. The value of ϕ' in plane strain is only slightly higher than that obtained from triaxial compression (typically being higher by between 0.5–4°). Hence, in theory, the use of the triaxial values in stability analyses provides a marginally enhanced Factor of Safety. However, the value is not grossly overconservative, because real failure surfaces are much longer than those in laboratory tests and there will be progressive development of shearing resistance so that not all soil elements will reach failure simultaneously.

6.6.4. Practical situations

If ϕ_u is equal to zero it means that the undrained shear strength is independent of the confining stress. Hence representation of a uniform clay deposit by one set of parameters such as c_u, $\phi_u = 0$ would mean that the strength of the soil is defined as being constant with depth. While such an approximation may be acceptable for limited depths, it must be remembered that it is a significant simplification of reality. For a uniform soil the effective stress will automatically increase with depth (because the bulk unit weight of soil is much greater than the unit weight of water) and hence the shear strength will increase naturally with depth.

Skempton found that the undrained shear strength S_u of a normally consolidated clay increased approximately linearly with effective stress p'_0. He proposed the following equation:

$$c_u = \frac{c_u}{p'_0} = 0.11 + 0.0037 I_p \tag{6.37}$$

where I_p is the Plasticity Index. For overconsolidated soils the following relationship has been proposed (Ladd *et al.*, 1977):

$$\left(\frac{S_u}{\sigma'_v}\right)_{OC} = \left(\frac{S_u}{\sigma'_v}\right)_{NC} (OCR)^{0.8} \tag{6.38}$$

where OCR is the overconsolidation ratio.

When the total stress regime within a zone in the ground is changed then there is an associated pore pressure change (see Section 6.1.4). The effective stress (which controls the shear strength) is given by equation (3.18), i.e. $\sigma' = \sigma - u$. Thus, if the total stress and pore pressure change by $\Delta\sigma$, and Δu respectively, the change in effective stress $\Delta\sigma'$ is given by

$$\Delta\sigma' = \Delta\sigma - \Delta u \qquad\qquad (6.39)$$

With free-draining soils local pore pressure excess (whether above or below the steady-state value) will dissipate immediately so that the total stress change produces the same effective stress change (i.e. $\Delta\sigma' = \Delta\sigma$). Thus there is an instantaneous change in the shear strength. Hence undrained shear strength is not applicable to coarse, granular soils (sands, gravels, etc.). Because of the free-draining nature of such soils the parameters c' and ϕ' are always used in strength analyses, together with the relevant effective stress.

For slow-draining soils (clays and, to a certain extent, silts) total stress changes produce pore pressure changes that dissipate slowly. For very low permeability clays it may take decades for full dissipation to occur. In general, excavations (cut slopes, trenches, etc.) will induce negative pore pressure changes, while ground-loading and fill situations (embankment construction, erection and building of retaining walls, etc.) will cause pore pressures to rise (Figure 6.27).

If the total stress change is relatively rapid, the change in pore pressure may be essentially the same as the total stress change (if the soil has a high degree of saturation). Hence there will be little or no immediate change in effective stress or shear strength. Thus the shear strength immediately after construction would be the same as it was before any works, and so an undrained, total stress analysis (using c_u and ϕ_u) would provide a suitable estimate of the shear strength. However, in such an analysis it is vital to be sure that the soil does not contain any free-draining layers, such as sand or silt partings, which would permit drainage and thereby lead to a reduction in undrained shear strength. With time, and certainly in the long term (for permanent works), the pore pressure excess will dissipate; for positive values of Δu the pore pressure will fall, and hence the effective stress and shear strength will increase. However where Δu is negative

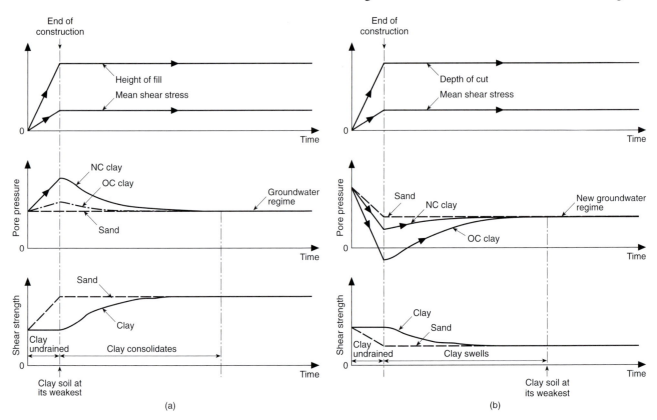

Figure 6.27. The shear strength behaviour of cuts and embankments (after Bishop & Bjerrum, 1960): (a) loading situations (embankments, fills behind retaining walls); (b) unloading situations (cut slopes, temporary excavations)

the pore pressure will increase with time, thereby leading to a reduction in effective stress and shear strength. Thus permanent excavations in slow-draining soil will induce a reduction in shear strength which will have to be predicted in stability analyses using c' and ϕ' and appropriate effective stresses. On the other hand, for loading and fill situations the shear strength will increase with time, and the most critical period for stability is usually the end of construction (for which the parameters c_u and ϕ_u give a safe, and not overconservative, estimation of shear strength).

The applicability of the various shear strength parameters is summarized in Table 6.5.

6.6.5. Alternative shear strength models

The Critical State model for soil behaviour was developed at Cambridge University and provides a unified model for the geotechnical behaviour of soil (Schofield & Wroth, 1968). The soil is treated as an amorphous, elastoplastic continuum, and the variables that define the state and behaviour of the soil are generalized to account for the three-dimensional nature of the world. These variables are:

- the mean effective stress p' (the generalized equivalent of the normal effective stress σ'_n)
- the shear stress stress q (the generalized equivalent of the shear stress τ)
- the specific volume v or voids ratio e (indicative of the density state of the material).

The foregoing variables define an overall state boundary that separates impossible states (above the state boundary) and possible states (on and below the state boundary). In essence, plastic deformation of a soil can only occur when its state coincides with the state boundary (below the state boundary surface is the elastic domain). When a soil reaches the Critical State it undergoes continuous plastic deformation through homogeneous shearing (i.e. all the material is undergoing the same shear strain). At this stage the soil is completely remoulded and so any effects of its previous stress history are eradicated. Thus at the Critical State the energy put into causing shearing is only dissipated in overcoming internal resistance; none is used to cause volume changes (see Section 6.1.2).

Table 6.5. Use of shear strength parameters

Field situation	Critical time	Appropriate shear strength parameters
Small embankments and foundations on saturated clay Temporary cuts in intact, non-fissured clay Support for retaining walls installed in clay	End of construction	c_u and ϕ_u, often with the undrained friction angle being taken as zero
Permanent cuts Temporary cuts in fissured and/or very heavily overconsolidated clay Pressures acting on the rear face of retaining walls	Long-term	c' and ϕ' or c'_d and ϕ'_d. For very heavily overconsolidated, fissured clays it may be necessary to use the residual parameters c'_r (typically zero) and ϕ'_r
Large earth embankments (involving construction over more than one earth-moving season).	End of construction	c' and ϕ' (with estimated pore pressures) to account for pore pressure dissipation and consequential shear strength increase during construction

Thus the Critical State strength can be used to define an ultimate strength that is a lower bound, and so can be used with a very low Factor of Safety.

The Critical State line separates the overall state boundary into two parts: the Hvorslev and Roscoe surfaces (Figure 6.28). The Roscoe surface can be thought of as relating to normally consolidated and lightly overconsolidated soil, while the Hvorslev surface is the part of the state boundary surface that represents the behaviour of heavily overconsolidated soils.

Critical state theory is often thought of as being an entirely theoretical model for soil. However, it must be remembered that the basic relationships employed correspond to shearing and volume change relationships that are observed, and defined, in conventional soil mechanics. In the Critical State model the behaviour is expressed in terms of generalized parameters. For routine geotechnical analyses the benefit of the Critical State model is that it provides information on both stresses and strains. The disadvantage is that correspondence has to be established between the principal directions and the relevant directions in the real situation (typically the horizontal and vertical planes).

Stress–dilatancy theory, on the other hand, was derived by treating soil as a particulate medium (Rowe, 1962). The shear strength of a granular soil results from two components: friction between the grains (because of contact of solid materials), and interlocking of the particles (which causes volume changes during shearing) (see Section 6.1.2). The first strength component is defined by the true interparticle friction angle ϕ_μ, which depends upon the mineral type (e.g. for mica it is less than 15°). This component can be considered as the lowest possible shear strength of a soil (if effects due to specific particle orientation and stress history are excluded). This behaviour is comparable to the energy dissipation that is assumed at the critical state.

In order to shear a dense particulate mass it is necessary to move the particles up and over each other because of their physical interaction. This requires a shear force to be applied to the soil (in addition to that required to overcome interparticle friction). This additional shear force increases with increasing degree of interlock. Shearing a dense sand entails a volume increase, and therefore a dense sand exhibits greater shear strength than does a loose assemblage of particles. However, during shearing of a dense mass, as the peak strength is approached the particles have moved so far apart that the degree of interlock is significantly reduced. After the peak resistance has been attained the shear strength of an initially dense mass decreases (strain-softening) because of the loss

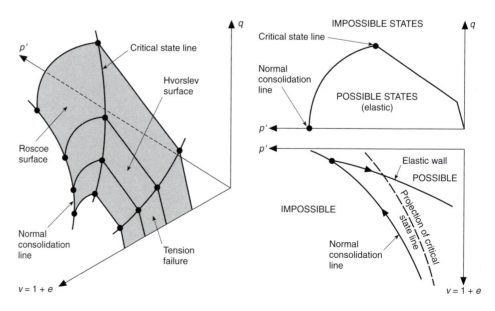

Figure 6.28. The Critical State model

of interlock. A loose sand will contain groups of particles that can move closer together during shear, so a volume decrease results (and energy does not have to be input to produce a volume change), but as the particles move closer together there is a tendency for greater interlock and the shear stress increases continually (strain-hardening).

7 Groundwater and permeability

7.1. Introduction

Soils have interconnected voids through which water can flow from points of high potential to points of low potential. The ease with which water can pass through a soil is indicated by its permeability — sands and gravels are free-draining materials, while clays generally have low permeability and can be practically impermeable.

In 1992 the collapse of an abandoned mineshaft in Gunnislake resulted in an entire back garden disappearing down a hole approximately 10 m wide and 8 m deep (*New Civil Engineer*, 1992b). The collapse had occurred at the site of the original engine shaft of the Old Gunnislake Mine. During the centuries of mining activity mine waste had been dumped over the bedrock and at the location of this shaft the waste was 20–35 m thick. A reinforced concrete cap had been placed over the shaft in 1972, 3 years before a council housing development on the site. Unfortunately, the cap was seated on the deep sandy mine waste rather than on bedrock. There is increasing groundwater recharge in the area, due to mine closure and the cessation of water pumping, and a combination of a high water-table (as water backed up an adit) and the deep drain provided by the shaft itself led to the mine waste being washed away from under the cap. As a consequence a large cavity (with an estimated volume of about 600 m^3) formed just above bedrock level. The cap then disappeared, probably whole, down the shaft and out of sight. Due to the high and fluctuating water-table and deep superficial cover it was not economically viable to physically excavate to bedrock level to undertake remedial capping of the shaft. Hence the top of the shaft and the overlying waste were stabilized by injecting cement into the ground (grouting).

Analytical treatment of the flow of water is essential for excavations below the water-table and for water-retaining dams and barriers. It is necessary to estimate the quantity of water entering an excavation (for pumping and drainage purposes) and also the effect of the water pressure on the stability of the excavation. Water flows through or beneath dams and barrages reduce the efficiency of the containment and may detrimentally affect stability by erosion and wash-out of soil materials.

7.1.1. Groundwater

Because of rainfall and infiltration it is inevitable that in the ground the voids between soil particles contain water. At shallow depths the voids will also contain gas, usually air (i.e. the soil will be partially-saturated), but at some depth the voids will be completely filled with water and below this depth the ground will be saturated. At the top of the zone of saturation the pore pressure will be negative due to surface tension and capillary effects within the fine passages in the soil. The voids in the soil form an intricate network of continuous channels, which can be imagined as fine capillary tubes. The height to which water will rise in a capillary tube due to surface tension effects increases as the diameter of the capillary tube decreases, so this capillary zone will be greatest for finer grained soils. With increasing depth below the saturation line the pore pressure will increase and the level at which the porewater pressure is equal to that of the

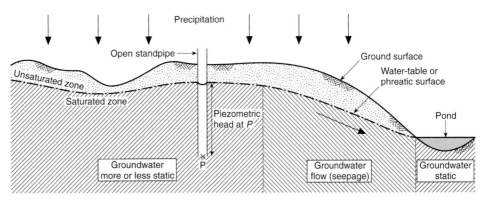

Figure 7.1. The groundwater regime

atmosphere is termed the *water-table* or *phreatic surface* (Figure 7.1). The water-table generally follows the shape of the ground surface topography, but in a subdued manner. A sloping water-table is an indication of the flow of groundwater or seepage in the direction of the fall. The piezometric head or standing water level at any particular point is the level to which the water will ultimately rise in a borehole or pit given sufficient time for this to be established. For bodies of water (e.g. at a river or the side of an excavation below groundwater level), the water table corresponds to the free water surface. Water-tables change with varying rates of infiltration, so that in winter they can be expected at high levels and at lower levels in summer.

The voids of permeable deposits such as sand provide natural storage of water and allow water to flow out easily, so they are called *aquifers* (water-bearing layers). The void spaces in a clay will also contain water, but these void spaces are so small that flow of water is significantly impeded, making a clay practically impermeable. Thus clay deposits impede the movement of water and act as *aquicludes* (water-confining layers). The location and state of groundwater in soil deposits is often determined by the stratification of sand–clay or permeable–impermeable sequences. A perched water-table may arise due to rainfall seeping down from the ground surface and resting on top of a low-permeability layer below which there is more permeable material (Figure 7.2). If an aquifer is confined above and below by strata of low permeability an actual water-table will not be visible as such but standpipes or wells sunk into the aquifer will indicate the phreatic surface level. *Artesian* conditions exist in a stratum when its phreatic surface lies above ground level (see Figure 7.2) — i.e. if a standpipe is sunk into the layer, the water in the pipe rises above ground level. The porewater pressure in a confined aquifer is governed by the conditions at the place where the layer is

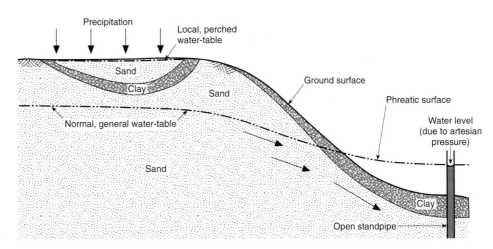

Figure 7.2. Perched water-table and artesian conditions

unconfined, so when the water-table changes here the artesian pressure will change, although there may be a time lag between the two events.

7.1.2. Darcy's law and permeability

Since soils are made up of numerous discrete particles with interconnected pore spaces, water may pass from zones of high pressure to zones of low pressure. In most soils the flow through the void spaces can be considered laminar (i.e. not turbulent, but flowing in a steady 'streamlined' way). The hydraulic gradient i between two points is the water-head difference Δh between the points, divided by the distance between the points l:

$$i = \frac{\Delta h}{l} \tag{7.1}$$

Darcy published a simple equation for the flow of water through a porous medium, which states that the velocity of flow through a stable medium is proportional to the hydraulic gradient:

$$v = ki \tag{7.2}$$

where k is the coefficient of permeability.

The foregoing equation was based primarily on observations of the flow of water through clean sands. In the Darcy equation the term v is the superficial discharge velocity, i.e. the quantity of water flowing in unit time through a unit gross cross-sectional area of the soil at right angles to the direction of flow. This is not the same as the velocity of the water percolating through the voids of the soil, which is known as the seepage velocity v_s, whereby $v_s = v/n$ (where n is the porosity).

Typical values of the coefficient of permeability can be ascribed to the generic soil categories, as illustrated in Figure 7.3. In addition, several relationships have been proposed to predict the coefficient of permeability of different soils (Table 7.1). Most of these relationships are empirical and relate to uniform granular materials. However, the presence of a small quantity of fines (silt and clay) can change the permeability by orders of magnitude. Such relationships therefore have very limited application to most natural soil deposits because of the wide range of particle sizes within a stratum and the major influence that the stress history and non-homogeneity of material have on permeability.

k (m/s)	1	10^{-1}	10^{-2}	10^{-3}	10^{-4}	10^{-5}	10^{-6}	10^{-7}	10^{-8}	10^{-9}	10^{-10}	10^{-11}
Generic soil type		Gravel		Sand			Silt			Clay		
Actual soils		Clean gravels		Clean sands, sand and gravel mixtures			Very fine sands, silts, sand/ silt–clay mixtures			Unweathered, unfissured, homogeneous natural and compacted clay layers		
							Stratified clay deposits					
						Clays modified by; weathering, vegetation, fissuring, desiccation						
Effect on seepage		Aquifers					Aquicludes					
Uses		For embankment construction, drainage material					General fill			Clay cores for dams, barriers for waste disposal sites and contaminated land containment		

Figure 7.3. Typical values of the coefficient of permeability

Soil type	Relationship
Table 7.1. Equations for the prediction of the coefficient of permeability k	
Clean sands (Hazen, 1892)	$k = C(d_{10})^2$ mm/s d_{10}, the 10% particle size (in mm); C, a factor varying between 10 (dense) and 15 (loose state) approximately
Sandy soils (Carmen, 1956 — Korzeny-Carmen equation)	$k = C_1\left(\dfrac{e^3}{1+e}\right)$ mm/s C_1, factor of the order of 1 (fine sand) to 100 (coarse sand); e, voids ratio
Soils (general) (Taylor, 1966)	$k = d^2\left(\dfrac{\gamma_w}{\mu}\right)\left(\dfrac{e^3}{1+e}\right)C$ d, equivalent particle size; μ, viscosity of the permeant; C, factor accounting for the shape of the cross-section though which flow occurs; γ_w, unit weight of water
Medium to fine sand (Shahabi *et al.*, 1984)	$k = 1.2\, C^{0.735}\, d_{10}^{0.88}\left(\dfrac{e^3}{1+e}\right)$ C, shape factor; d_{10}, finest 10% of the particles; e, voids ratio
Remoulded soft clay (Carrier & Beckman, 1984)	$k = \left(\dfrac{1}{1+e}\right)\left(\dfrac{LI + 0.242}{95.21}\right)^{4.29}$ m/s LI, liquidity index $= (w - w_p)/(w_L - w_p)$
Clay soils for low-permeability barriers in waste disposal sites (Sarsby & Williams, 1995)	$k_{opt} = \dfrac{25(1+e)\,100^{\log d_{10}}}{CU}\left(\dfrac{d_0}{d_{10}}\right)^2$ mm/s CU, coefficient of uniformity; d_0, particle size for zero percentage on the grading curve; d_{10}, the 10% particle size; e, voids ratio

The coefficient of permeability is particularly affected by the:

- pore size distribution/voids ratio
- grain size distribution (particularly the fines content)
- degree of saturation
- fluid viscosity.

Consequently, the actual permeability of a real soil containing a variety of particle sizes can only be determined by testing.

7.2. Laboratory measurement of permeability

The value of the coefficient of permeability of a real soil is sensitive to many factors and, regardless of the accuracy or precision of the testing regime, for natural or compacted materials 'identical' samples are likely to exhibit a range of different permeability values, by up to a factor of 2 or 3 (Beeuwsaert & Sarsby, 2000).

7.2.1. Constant head permeameter

Constant head tests (Figure 7.4) are suitable for coarse-grained soils that have relatively high coefficients of permeability so that measurable quantities of water flow through the soil in reasonably short times. Typically this means soils having a coefficient of permeability of around 10^{-2} to 10^{-5} m/s (i.e. clean sand and sand–gravel mixtures with less than 10% fines). However, it must be remembered that it is very difficult to obtain 'undisturbed' samples of coarse-grained soils, so the main use of the test is for assessing or checking the suitability of fill for drainage purposes.

The soil is contained within a cylinder (approximately 80 mm diameter, 150 mm long) (Figure 7.5), and is fully saturated. A steady supply of water is

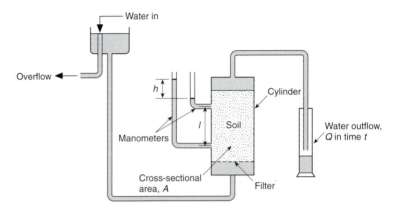

Figure 7.4. Constant head permeameter

Figure 7.5. Constant head permeameter

introduced to the top of the soil column and the head between two points within the soil sample and the quantity of water flowing in a selected time are measured. By adjusting a control valve it is possible to vary the flow rate q through the sample and the hydraulic head loss h between the manometer points, and these values are measured for steady flow conditions. The test is repeated for different hydraulic gradients by adjusting the control valve.

If the quantity of water passing through the soil in time t is Q, then the observed flow rate q is Q/t. The flow rate is also equal to the superficial velocity of flow v multiplied by the cross-sectional area A and if this velocity is given by Darcy's law then

$$\frac{Q}{t} = q = Av = \frac{Akh}{l} \tag{7.3}$$

Thus

$$k = \frac{Ql}{Aht} = \frac{ql}{Ah} \tag{7.4}$$

7.2.2. Falling head permeameter

This test was developed for fine-grained soil with relatively low permeability, and essentially the apparatus (Figure 7.6) consists of a soil sample within a cylindrical mould to which a water head is applied via a vertical standpipe (Figure 7.7). The apparatus was specifically designed to enable small flow rates to be measured accurately. The volume of water that flows through a sample originates from the standpipe, which has a much smaller bore than the diameter of the soil sample. Thus a small flow through the soil corresponds to a large, easily measured change in the level of water in the standpipe (a vertical scale is mounted alongside the standpipe), but it also means that the head applied to a sample varies during a test.

At any time t after the start of a test the flow rate through the soil q is given by

$$q = Av = Aki = \frac{Akh}{l} \tag{7.5}$$

The head applied to the sample h increases vertically upwards, so that a fall in the water level (occurring within a very small time interval dt) has to be represented mathematically as a head change of $-dh$. Equating the rate of water flow from the standpipe with the flow rate through the sample gives

$$q = \frac{Akh}{l} = -a\frac{dh}{dt} \tag{7.6}$$

where a is the cross-sectional area of the standpipe and A is the cross-sectional area of sample. Thus

$$dt = -\frac{al}{Ak}\frac{dh}{h} \tag{7.7}$$

Integration and application of limits gives

$$(t_2 - t_1) = \frac{al}{Ak} \ln\left(\frac{h_1}{h_2}\right) \tag{7.8}$$

where h_1 and h_2 are the water heads (relative to the outflow level in the tray which holds the apparatus) at times t_1 and t_2 from the start of measurement. Thus

$$k = \frac{al}{A}\frac{1}{(t_2 - t_1)} \ln\left(\frac{h_1}{h_2}\right) = 2.3\frac{al}{A}\frac{1}{(t_2 - t_1)} \log\left(\frac{h_1}{h_2}\right) \tag{7.9}$$

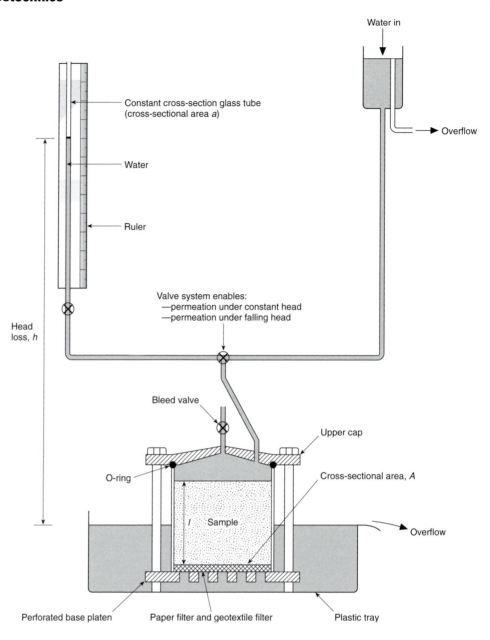

Figure 7.6. Falling head permeameter

The coefficient of permeability can be calculated by inserting data at different times (i.e. t_1 and t_2), from the start of the test into equation (7.9). However, an alternative method is recommended wherein a series of values of h, for different elapsed times from the start of a test, is taken so that $\log(h_0/h)$ can then be plotted against time from the start of the test (h_0 is the head when time t is zero). When steady-state conditions are attained, a straight-line plot is obtained (Figure 7.8), with the slope of the line being proportional to the coefficient of permeability. Initial curvature of the plot indicates either the presence of a significant amount of air in the voids or that the sample is swelling or contracting. The straight-line portion which is obtained at large time values is essentially parallel to the steady state value (see Figure 7.8).

Since this test is intended for clay soils a sample is obtained by: compacting soil directly into the cell body, or driving the cell body (with a cutting shoe attached) into the ground, or jacking soil from a sample tube and into the cell body. Potential limitations of the conventional falling head test are:

Figure 7.7. Falling head permeameter

Test	Initial k (m/s)	Final k (m/s)
1	0.11×10^{-9}	0.02×10^{-9}
2	0.03×10^{-9}	0.01×10^{-9}
3	0.01×10^{-9}	0.01×10^{-9}

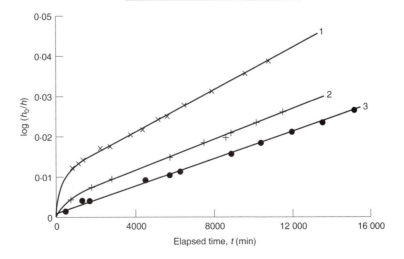

Figure 7.8. Falling head test data

- It is not possible to saturate a test specimen by pressurization of the pore fluid (since the sample is not 'sealed' within the apparatus) and achievement of complete saturation by flushing out the air from the voids is unlikely to work with fine-grained soil. However, for most real situations, where it is necessary to determine the coefficient of permeability of a clay soil, the undistorted or compacted material will usually have a high degree of saturation.
- The effective stress acting on the sample changes as the head changes, so that consolidation and swelling of the soil will occur. At the top of the sample the effective stress is zero, and for fine-grained soils the surface material swells significantly and becomes very soft.
- Preferential flow paths can form between the soil and the relatively smooth inside of the rigid permeameter body, particularly if the material being tested is an undisturbed sample which has been obtained by driving the cell body (or a sampling tube) into stiff ground. However, when permeation is commenced swelling of the soil usually establishes intimate contact between the soil and the cell body.

Notwithstanding the foregoing comments it must be appreciated that the falling head apparatus is straightforward and easy to set up, and for clay soils accurate; reproducible values of the coefficient of permeability are obtained with this system (Beeuwsaert & Sarsby, 2000).

7.2.3. Triaxial cell permeameter

As devices for measuring small flow rates accurately have developed, so has the trend to apply constant head tests to measure the permeability of fine-grained soils. This has facilitated the development of permeability tests using the triaxial apparatus. The sample is set up as for a drained triaxial test (see Chapter 6, Section 6.3.5) with drains connected to the top and bottom of the specimen. The sample is subjected to an all round pressure, but no deviator stress. A constant water head is applied, using the drainage connections, across the sample and the resultant steady-state flow rate is measured (see Section 7.2.1). The permeability is then estimated from equation (7.4).

The confining ambient pressure ensures intimate contact between the sides of the confining walls (the membrane) and the soil, even if it has a significantly uneven surface, provided a sufficient pressure differential (at least 200 kPa) exists between the cell pressure and the porewater pressure (Sarsby *et al.*, 1995). Because the sample is sealed inside a membrane the pressure in the porewater can be raised to dissolve air/gas in the pores and hence produce full saturation — this is the prime reason for development of this test arrangement. Hence the apparatus is suitable for natural or compacted samples of cohesive soil. Furthermore, different effective stress levels can be applied (by changing the cell pressure) to reflect the stress range applicable to the field situation. However, the equal all round stress state does not correspond to the field situation, and consolidation in the test may result in inappropriate permeability values being obtained. For compacted clay samples the triaxial and falling head tests have been found to give very similar values of the coefficient of permeability (Sarsby *et al.*, 1995).

7.2.4. Consolidation cell permeameter

The Casagrande oedometer (see Chapter 8, Section 8.3.1) is sometimes used to obtain an indirect estimation of the coefficient of permeability. However, this method is not recommended because the value obtained depends greatly on the relative magnitude value of small quantities, each of which can vary significantly according to the structure of the soil.

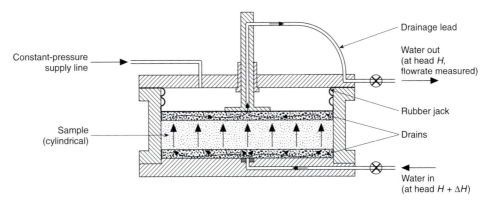

Constant-pressure
supply line

Drainage lead

Water out
(at head H,
flowrate measured)

Rubber jack

Sample
(cylindrical)

Drains

Water in
(at head H + ΔH)

*Figure 7.9. Permeability
measurement in the Rowe
cell*

On the other hand, the Rowe cell (see Chapter 8, Section 8.3.2) can be used to measure the coefficient of permeability directly in both the vertical and horizontal directions (Figure 7.9). Hence the apparatus is used to test natural soils containing 'fabric' that can significantly affect the permeability (e.g. decayed vertical rootlets, horizontal laminations). The relevant vertical stress level can be applied, thereby inducing the appropriate lateral stress, as opposed to subjecting a sample to an equal all round pressure. A constant head test is conducted by applying a head to the bottom drain and allowing water to exit from the top drain. The apparatus can also be used to test compacted material, in which case the soil is compacted directly into the cell body so that there is good contact between the soil and the inside of the rigid body.

7.3. Field permeability tests

In the field the overall coefficient of permeability of a soil deposit can be determined from pumping tests, which involve pumping water in or out of the ground. The tests may be performed by using wells that extend for a significant depth (such that they measure an average permeability of the soil) or by using cylindrical or 'point' sources isolated within the ground so as to measure the permeability of a specific stratum.

7.3.1. Well tests

This type of pumping test is conducted using a well that has a perforated casing installed in an aquifer (i.e. a layer with relatively high permeability). During the test, water is pumped out or in at a constant, measured rate from the bottom of the well, using suction pumps (for depths less than about 5 m) or submersible pumps. Pumping out will lower the groundwater table in the vicinity of the well and a cone of depression will form. This cone should be symmetrical about the well once steady-state conditions have been attained (i.e. when the rate of pumping water out of the well equates to the rate of inwards water flow within the aquifer). It may take several days to achieve a steady state for relatively low permeability soils. Determination of the coefficient of permeability requires information on the shape of the cone of depression, and so observation wells are installed at various radial distances. Typically four observation wells are installed in the aquifer, in two rows at right angles to each other, and water levels are observed until a steady state has been achieved or the rate of change of drawdown is small. Readings are taken (pumping rate and levels in observation wells) when steady conditions are attained.

There are two main cases to consider: an unconfined aquifer underlain by a relatively impermeable layer, and an aquifer confined between layers of relatively low permeability (Figure 7.10). In each case it should be noted that it is the relative permeability of the adjacent strata that matters. A layer only needs to have a permeability that is significantly less (by an order of magnitude or more) than that of its neighbour for it to qualify as essentially impermeable.

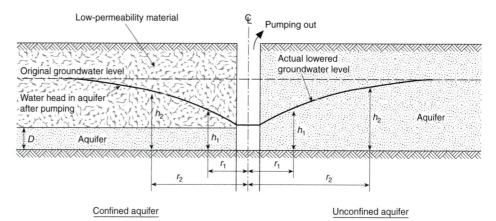

Figure 7.10. Unconfined and confined aquifers

The *unconfined aquifer case* involves the following assumptions:

- the groundwater surface (phreatic surface) represents the variation in the hydraulic gradient with distance from the well
- the phreatic surface intercepts the well at the same level as the water level in the well
- the diameter of the wall is small (by comparison to the zone of influence of the well).

At a radius r from the centre of the well, the rate of increase in h with distance (i.e. the gradient of the phreatic surface), is taken as the hydraulic gradient i. Thus

$$\mathrm{d}h/\mathrm{d}r = i \tag{7.10}$$

The water in the ground is assumed to flow radially at a velocity given by Darcy's law:

$$v = ki \equiv \frac{k\mathrm{d}h}{\mathrm{d}r} \tag{7.11}$$

The surface area across which the foregoing flow occurs (see Figure 7.10) is equal to the surface area of a cylinder of radius r and height h (corresponding to the depth of water in the aquifer). Hence the area for flow is equal to $2\pi rh$ and the flow rate q is given by:

$$q = 2\pi rhk\frac{\mathrm{d}h}{\mathrm{d}r} \tag{7.12}$$

This previous expression can be rearranged to give:

$$\frac{\mathrm{d}r}{r} = \frac{2k\pi}{q}h\,\mathrm{d}h \tag{7.13}$$

For a constant rate of pumping and constant k, the equation can be integrated and, if the heights of water in the observation wells are h_1 and h_2 (at distances r_1 and r_2 respectively), then

$$\ln\left(\frac{r_2}{r_2}\right) = \frac{\pi k}{q}(h_2^2 - h_1^2) \tag{7.14}$$

Hence

$$k = \frac{q}{\pi(h_2^2 - h_1^2)}\ln\left(\frac{r_2}{r_1}\right) \equiv \frac{2.3q}{\pi(h_2^2 - h_1^2)}\log\left(\frac{r_2}{r_1}\right) \tag{7.15}$$

The *confined aquifer case* involves the same components as for the unconfined case. However, the area through which flow occurs is now a cylinder of radius r

and height D (the thickness of the aquifer) (see Figure 7.10). Thus the surface area for flow at radius r from the centre of the well is equal to $2\pi rD$. Hence

$$q = 2\pi rDk\left(\frac{\mathrm{d}h}{\mathrm{d}r}\right) \tag{7.16}$$

giving

$$\frac{\mathrm{d}r}{r} = \left(\frac{2\pi k}{q}\right)D\,\mathrm{d}h \tag{7.17}$$

Integration and insertion of the conditions at the observations wells leads to:

$$k = \frac{q}{2\pi D(h_2 - h_1)}\ln\left(\frac{r_2}{r_1}\right) \equiv \frac{2.3q}{2\pi Dk(h_2 - h_1)}\log\left(\frac{r_2}{r_1}\right) \tag{7.18}$$

7.3.2. Borehole tests

Both falling and rising head tests may be conducted in a section of borehole or by using a sealed-in piezometer (detailed information on in situ permeability testing is given in Somerville (1986)). In borehole tests a length of unlined hole is created by filling the borehole with gravel for the required depth and then pulling the casing back. The whole test section must be below ground water level.

In falling head tests water is added to the borehole to raise the water level as high as possible. Readings are then taken of the falling water level at frequent time intervals. The water pressure used in the test should be less than that which will rupture the ground by hydraulic fracturing. Serious errors will be introduced if excessive water pressures are used and, in general, the total increase in pressure should not exceed one-half the effective overburden pressure. With high-permeability soils (greater than about 10^{-3} m/s) flow rates are likely to be large and head losses at entry or exit and in the borehole may be high.

In the simple pumping-out test (rising head) the preparations are more or less the same as for the falling head test, except that water is baled out to just above the bottom of the casing and frequent readings are taken of the rising water level at various time intervals.

The calculation of the coefficient of permeability is similar for falling and rising head tests. When the groundwater level is known accurately, values of $\log(H/H_0)$ are plotted against time t, where H is the measured head at time t and H_0 is the initial head (Figure 7.11). The best straight line is then drawn through the experimental points. If the initial part of the graph is curved, a straight line (parallel to the final straight part of the graph) is drawn through the origin. From the straight line the value of t for which $H/H_0 = 0.37$ is read off — this value of t is termed the basic time lag T. The coefficient of permeability is then obtained from the following equation:

$$k = \frac{a}{FT} \tag{7.19}$$

where a is the cross-sectional area of the standpipe and F is the shape factor (the appropriate value is obtained by comparing the geometry of the test with those arrangements having analytical solutions) (Figure 7.12).

If the groundwater level is only known approximately then an initial $\log(H/H_0)$ versus time plot is produced using the assumed groundwater level. If all of the resultant graph is curved, then a revised groundwater level is used to re-calculate the $\log(H/H_0)$ values until the graph is essentially a straight line (the initial part may still be curved) (see Figure 7.11). The remainder of the analysis is then as before.

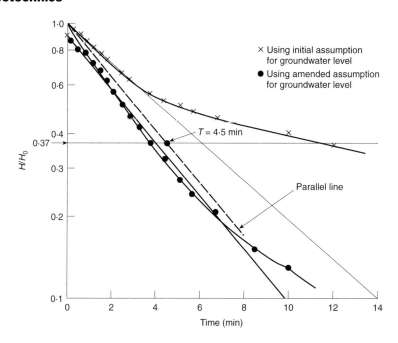

Figure 7.11. Data from borehole falling head test

The permeability of particular strata or horizons can be measured by using 'sockets', which are sealed, temporarily, in the ground at particular levels. This type of measurement is conducted primarily in unlined boreholes in rock and is usually called a 'packer test' after the packers that are used to close off the borehole above and below the socket. In essence a packer is a rubber bag which is inflated against the sides of the borehole to isolate a section of the hole. Water is then pumped in (the most common approach) or out of the socket and the permeability deduced from the flow rate and applied head. The applied head can simply be the gravity head of water in the tube above the water-table, although it is more common to apply a pressure to increase the water head. Falling head tests are often used.

Care should be exercised in the interpretation of the results, for several reasons:

- the socket may not be properly isolated from other strata
- leakage past the packers may occur
- enhanced water pressures may cause fracturing of the ground

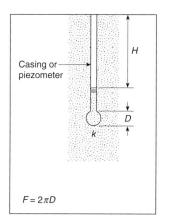

$F = 2\pi D$

Spherical intake or well point in uniform soil

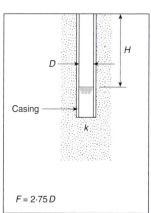

$F = 2.75 D$

Soil flush with bottom in uniform soil

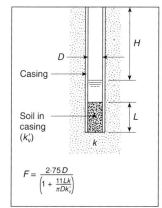

$$F = \frac{2.75 D}{\left(1 + \frac{11Lk}{\pi Dk'_v}\right)}$$

Soil in casing with bottom in uniform soil

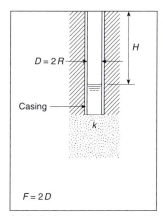

$F = 2D$

Soil flush with bottom at impermeable boundary

Figure 7.12. Shape factors for borehole permeability test (from Somerville, 1986)

- if fines are not washed completely from the hole they may get washed into open fissures and joints in the rock.

A piezometer (a porous element buried in the ground to measure pore pressures; see Chapter 11, Section 11.4) can be used in a similar way to the packer system to measure permeability of a limited zone. The head is applied to the sand filter surrounding the piezometer, so the dimensions of this zone must be known with reasonable accuracy. Furthermore, the permeability of the piezometer tip material itself must be at least ten times more permeable than the surrounding soil, otherwise the test will merely measure the permeability of the ceramic tip. For this reason the test is only suitable for soils of low permeability. Further details are given within BS 5930 (BSI, 1981).

7.4. Seepage

Seepage is the flow of water through a soil as a result of a hydraulic head or gradient. It can give rise to a variety of effects:

- water flowing into an excavation below groundwater level
- water passing through and/or beneath a water-retaining barrage
- the development of quicksand conditions and instability
- influence on effective stress and hence soil strength.

'Poor soil' conditions are usually attributable to an excess of groundwater with consequential lack of strength and associated excessive compressibility. The so-called 'running sand' is not a material that exists naturally in the ground. These 'runs' are in fact man-made by allowing water to flow out of the sides of an excavation, carrying with it soil particles (see Chapter 3, Figure 3.2). This results in the sides of the excavation slumping, the soil above being undermined, and leads to collapse of the excavation. In Nature, groundwater may be flowing through the ground, but this flow will not normally be large enough to cause instability, so the ground is stable or in equilibrium. Groundwater flow becomes more significant as the permeability of the soil increases. However, while the groundwater flow is much slower in silts and clays than in sands and gravels, it does not mean that there are no problems associated with groundwater in silts and clays.

For seepage situations an estimate of the flows or resulting water pressures can be made using appropriate analyses, but high precision should not be ascribed to these values because of the limited accuracy of mean permeability values for field conditions. For routine problems it is not feasible to account for all ground characteristics without the complexity of the situation becoming excessive.

7.4.1. Stratified deposits

Depending on the nature and past history of a soil deposit, the coefficient of permeability may vary with the direction of flow. For clearly stratified deposits an equivalent, overall permeability may be derived according to the direction of flow. To derive suitable relationships, consider a simple system of two parallel layers, of width w, with thicknesses and coefficients of permeability as shown in Figure 7.13.

For *flow parallel to the soil strata*, each layer is subjected to the same head loss H over a given length L, but each layer has a different flow rate (i.e. q_1 and q_2). Applying Darcy's law to each layer:

$$q_1 = wd_1 \frac{H}{L} k_1 \quad \text{and} \quad q_2 = wd_2 \frac{H}{L} k_2 \qquad (7.20)$$

The total flow rate q is equal to $(q_1 + q_2)$ but it is also given by $[k_m w(d_1 + d_2)H]/L$, where k_m is the overall coefficient of permeability of the two-layered system. Equating the two expressions for total flow rate gives

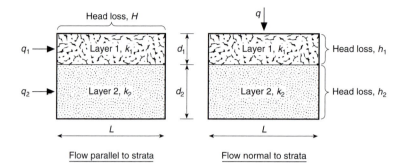

Figure 7.13. Flow parallel and normal to stratification

$$q = wd_1 \frac{H}{L} k_1 + wd_2 \frac{H}{L} k_2 = w(d_1 + d_2)\frac{H}{L} k_m \qquad (7.21)$$

and rearranging gives

$$k_m = \frac{(d_1 k_1 + d_2 k_2)}{(d_1 + d_2)} \qquad (7.22)$$

If the *flow is at right angles to the soil strata*, then the flow rate q through each layer must be the same, although there will be different head losses (h_1 and h_2) across each layer (see Figure 7.13). The total head loss is ($h_1 + h_2$) and if the two-layer system is treated as a single material with overall permeability k_m, then from Darcy's law

$$q = \frac{wbk_m(h_1 + h_2)}{(d_1 + d_2)} \qquad (7.23)$$

but

$$q = \frac{wbk_1 h_1}{d_1} = \frac{wbk_2 h_2}{d_2} \qquad (7.24)$$

Substituting for h_1 and h_2 gives

$$q = \frac{wbk_m[qd_1/wbk_1 + qd_2/wbk_2]}{d_1 + d_2} \qquad (7.25)$$

Hence

$$k_m = \frac{d_1 + d_2}{(d_1/k_1 + d_2/k_2)} \qquad (7.26)$$

7.4.2. Steady-state seepage

Consider a cube of soil with water passing through each of the faces as shown in Figure 7.14. The x, y and z axes represent the three orthogonal directions in a three-dimensional situation. If the water is incompressible and the soil mass does not change in volume, then the total flow into the soil must equal the total flow outwards. For each direction x, y and z the water enters with a certain velocity (v_x, v_y and v_z, respectively) and, on passing through the soil, the velocity is assumed to change by a small amount given by the rate of change in velocity with respect to distance multiplied by the distance travelled. For instance, in the x direction the entry velocity is v_x and, in travelling the distance to the exit face, this becomes

$$v_x + \left(\frac{\partial v_x}{\partial x}\right) dx \qquad (7.27)$$

Similar expressions apply to the y and z directions. (Note that the mathematical term ($\partial v_x/\partial x$) is short-hand notation for 'the amount of change of the velocity in the x direction that results *only* from a small increment of movement in the x

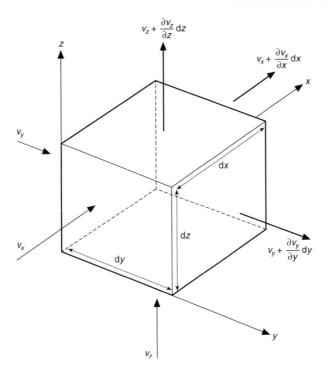

$v_z + \dfrac{\partial v_z}{\partial z}\,dz$

z

$v_x + \dfrac{\partial v_x}{\partial x}\,dx$

x

dx

v_y

dz

v_x

$v_y + \dfrac{\partial v_y}{\partial y}\,dy$

dy

y

v_z

Figure 7.14. An element of soil subject to seepage

direction'. If the differential term was written as (dv_x/dx) it would mean 'the *total* amount of change in the velocity in the x direction resulting from a small increment of movement in the x direction'.)

For each direction the flow rate is the product of the velocity of flow and the area through which it flows (i.e. the area at right angles to the direction of flow). The total inflow rate is the sum of the inflow rates of the three orthogonal directions, thus:

$$q\,(\text{in}) = v_x\,dy\,dz + v_y\,dx\,dz + v_z\,dx\,dy \tag{7.28}$$

$$q\,(\text{out}) = \left(v_x + \frac{\partial v_x}{\partial x}\,dx\right)dy\,dz + \left(v_y + \frac{\partial v_y}{\partial y}\,dy\right)dx\,dz$$

$$+ \left(v_z + \frac{\partial v_z}{\partial z}\,dz\right)dx\,dy \tag{7.29}$$

If

$$q(\text{out}) = q(\text{in})$$

then

$$q(\text{out}) - q(\text{in}) = 0 \tag{7.30}$$

and

$$\frac{\partial v_x}{\partial x}\,dx\,dy\,dz + \frac{\partial v_y}{\partial y}\,dx\,dy\,dz + \frac{\partial v_z}{\partial z}\,dx\,dy\,dz = 0 \tag{7.31}$$

The term $(dx\,dy\,dz)$ can be eliminated, because it is not zero, to give the *three-dimensional continuity equation*:

$$\frac{\partial v_x}{\partial x} + \frac{\partial v_y}{\partial y} + \frac{\partial v_z}{\partial z} = 0 \tag{7.32}$$

If Darcy's law is applied to each direction, and h is the water head, then

$$v_x = k_x\, i_x = k_x \frac{\partial h}{\partial x}, \quad v_y = k_y\, i_y = k_y \frac{\partial h}{\partial y}, \quad v_z = k_z\, i_z = k_z \frac{\partial h}{\partial z} \tag{7.33}$$

Substituting for v_x, v_y and v_z in the continuity equation, the following equation is obtained:

$$k_x \frac{\partial^2 h}{\partial x^2} + k_y \frac{\partial^2 h}{\partial y^2} + k_z \frac{\partial^2 h}{\partial z^2} = 0 \tag{7.34}$$

If the soil is isotropic with respect to permeability, then $k_x = k_y = k_z = k \neq 0$ and the three-dimensional *Laplace equation* is obtained:

$$\frac{\partial^2 h}{\partial x^2} + \frac{\partial^2 h}{\partial y^2} + \frac{\partial^2 h}{\partial z^2} = 0 \tag{7.35}$$

7.4.3. Flownets

Solution of the Laplace equation (equation 7.35) is very difficult for three-dimensional situations with complicated boundary conditions, although the power of modern computing has made this much more feasible. However, for many practical situations it is sufficiently accurate to consider the problem as being two-dimensional, and a simple graphical technique of solving the two-dimensional form of the Laplace equation has been derived. The solution is represented by two orthogonal families of curves (flow lines and equipotentials), that intersect each other at right angles to form a pattern of curved squares with the sides in a fixed ratio.

A *flow line*, or streamline, is an imaginary line along which a water particle travels from a zone of high potential (upstream) to a zone of low potential (downstream). There are an infinite number of flow lines, and if the flow is not turbulent the flow lines cannot intersect or merge. Each flow path (the space between adjacent flow lines) is unique and the flow rate is the same in all paths in the set. Buried impermeable boundaries (e.g. the surface of a clay layer, the vertical surface of sheet piling, the underside of a concrete dam) are flow lines.

An *equipotential* is an imaginary line along which the potential head is constant (in other words if an open standpipe is moved with its lower tip following an equipotential the elevation of the water in the standpipe remains constant). As the water flows through the pore spaces its energy is dissipated by friction and the equipotential lines act like contours to show how the energy is lost. Each equipotential is associated with a different head/elevation and, obviously, equipotentials cannot intersect or merge. The intervals between adjacent equipotentials represent a constant difference in total head loss and the total head lost is shared equally among equipotential drops. Where a permeable soil boundary is in contact with open water as on the upstream face of an earth dam, this boundary is an equipotential (i.e. the total head is constant on this boundary).

Flownets are drawn by trial and error (Figure 7.15). The first step is to identify the extreme upstream and downstream equipotentials (usually free surfaces) and all assumed or actual impermeable boundaries (extreme flow lines) (Figure 7.13(a)). Several intermediate flow lines are then sketched in (Figure 7.13(b)). Identification of the extreme flow lines is particularly helpful in this initial guess as the intermediate flow lines will reflect the gradual transition in shape from one extreme to the other. Most solutions will be sufficiently accurate if four or five flow paths are used; drawing too many flow lines will complicate the result unnecessarily. Intermediate equipotentials reflect the transition from the extreme upstream equipotential to the extreme downstream one. After sketching in the initial guess for the flow lines, sufficient intermediate equipotentials are inserted to produce a pattern that has some semblance of being composed of curved

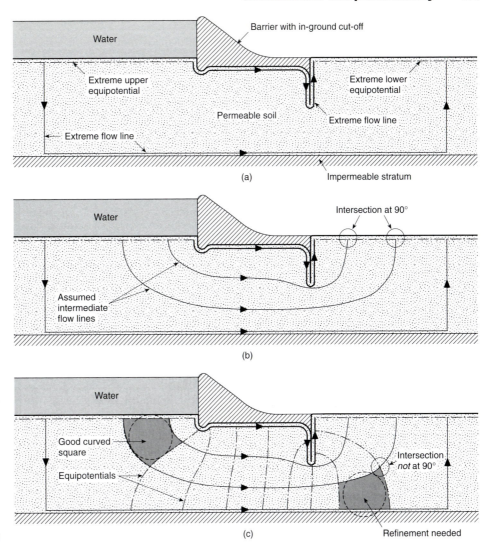

Figure 7.15. Drawing flownets

squares (strictly, a circle can be inscribed within the curved square) (Figure 7.13(c)). Unless the situation has very simple boundary conditions, the outer edges of the flownet will always have some shapes that are not curved squares (increased accuracy can be obtained by subdividing areas on a local basis so that the part of the net not represented by curved squares is reduced). The shape and position of the flow lines and equipotential are then gradually adjusted until the final pattern (flownet) of curved squares is obtained throughout the section. Figure 7.16 shows examples of flownets for different practical situations.

Although there is a unique solution of the Laplace equation in terms of flow lines and equipotentials, the positions of these lines in a sketched flownet must be regarded as approximate. To obtain a repeatable, high degree of accuracy, a large diagram with many flow lines and equipotentials must be used, the dimensions of the curved squares must be carefully measured, and each flow line–equipotential intersection must be measured to ensure that it is a right-angle. Hence water uplift distributions or discrete water head values, which are estimated from the pattern of equipotentials in a conventional flownet, must be treated with caution.

However, the flownet analysis rapidly produces an accurate estimate of seepage and flow rates, because these values are related to the ratio that the total number of flow paths holds to the total number of equipotentials, and this ratio is fairly insensitive to the accuracy of the flownet.

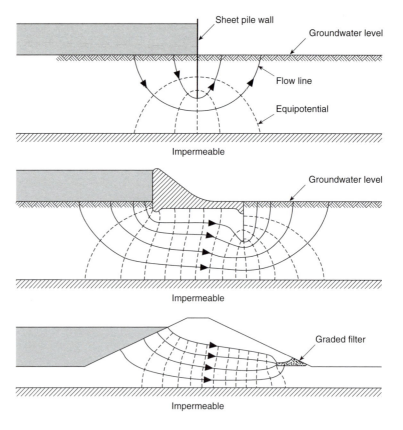

Figure 7.16. Examples of flownets

For each curved square (Figure 7.17) the inflow and outflow rates Δq are equal. For a square of unit width (into the plane of Figure 7.17),

$$\Delta q = \text{area} \times \text{velocity} = av \qquad (7.36)$$

If there are N_q equipotentials the head loss across each square ΔH is the total head H divided by N_q, so that

$$\Delta H = H/N_q \qquad (7.37)$$

From Darcy's law, the flow velocity within a flowpath is equal to $(\Delta H/b)$, and thus

$$\Delta q = a\Delta H/b \qquad (7.38)$$

There are N_f flowpaths, so that the total flow rate q is equal to $N_f\,\Delta q$. Hence

$$q = kH \frac{N_f}{N_q} \frac{a}{b} \qquad (7.39)$$

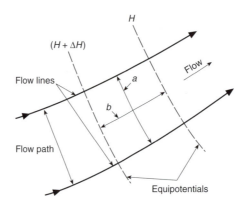

Figure 7.17. A typical element of a flownet

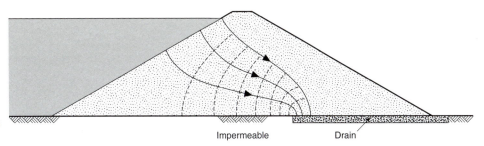

Figure 7.18. A flownet for earth dams (Casagrande solution)

Impermeable   Drain

For curved squares $a/b = c = 1.0$, so that the seepage flow per unit width of the flownet is

$$q = kH \frac{N_f}{N_q} \tag{7.40}$$

With flow through an earth embankment the upper flow line is subject to atmospheric pressure. Hence the boundary conditions are not completely defined and it is difficult to sketch a flownet until the upper flow line has been located. There are three factors that help the location process:

- The top flow line is at atmospheric pressure so the only type of head that can exist along it is elevational. Hence between successive points where equipotentials cut the upper flow line there must be equal drops in elevation.
- The upstream face of the dam is an equipotential, so the upper flow line must start at right angles to it.
- The downstream end of a flow line will tend to follow the direction of gravity and the upper flow line either exits at a tangent to the downstream face of the dam or takes up a vertical direction as it exits into a filter.

The flownet for the middle and downstream portions of a dam consists of two sets of parabolas. The Casagrande method involves drawing an actual parabola and then correcting its upstream end (Figure 7.18). This solution is only applicable to a dam that has its toe resting on a permeable material. In the absence of such a situation the phreatic surface cuts the downstream slope at some distance up the slope from the toe (Figure 7.19).

7.4.4. Flownets for anisotropic soil

The flownet sketching technique works well for a medium that is isotropic and homogeneous with respect to permeability. Unfortunately, many soil deposits are not like this. While there are techniques for dealing with a degree of anisotropy, these extensions reduce the attractiveness of the flownet technique (i.e. its simplicity), and are of limited application.

The depositional history of an actual soil deposit will frequently result in stratification and anisotropy with regard to permeability, the horizontal permeability often being several orders of magnitude greater than the value in the vertical direction. If the soil is anisotropic but uniform, then the two-

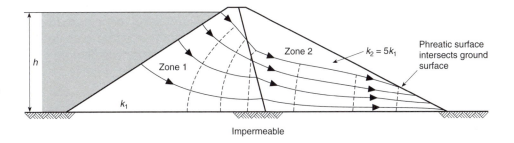

Figure 7.19. A flownet for an earth dam over an impermeable base (after Cedergren, 1967)

h

Zone 1

k_1

Zone 2 $k_2 = 5k_1$

Phreatic surface intersects ground surface

Impermeable

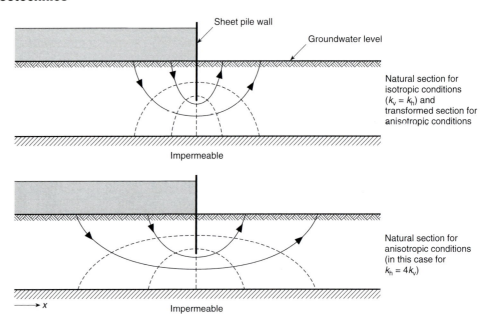

Figure 7.20. A flownet for an anisotropic soil

dimensional form of the Laplace equation may be modified by substituting x_m for x (where $x_m = x\sqrt{k_z/k_x}$). Hence the initial equation

$$k_x \frac{\partial^2 h}{\partial x^2} + k_z \frac{\partial^2 h}{\partial z^2} = 0 \qquad (7.41)$$

is transformed to

$$k_x \frac{\partial^2 h}{\partial x_m^2} \frac{k_z}{k_x} + k_z \frac{\partial^2 h}{\partial z^2} = 0 \equiv \frac{\partial^2 h}{\partial x_m^2} + \frac{\partial^2 h}{\partial z^2} \qquad (7.42)$$

Thus a flownet which is drawn with vertical and horizontal axes represented by z and x_m will once again form a pattern of curved squares. So, in the graphical analysis, the actual situation is first redrawn in a distorted way by adopting a horizontal (x_m) scale that is equal to the vertical scale multiplied by $\sqrt{k_z/k_x}$. The flownet is then drawn in the usual way. The seepage rate is obtained from

$$q = H \frac{N_f}{N_q} \sqrt{k_x \, k_z} \qquad (7.43)$$

To obtain the actual distribution of flow lines and equipotential the correct scale must be reintroduced. This will distort the flownet, so that the pattern of curved squares is lost (as shown in Figure 7.20).

When flow lines cross a boundary between two different permeability zones there is a change in the direction of these lines, unless they are normal to the boundary. When water flows from a soil of high permeability to one of low permeability, both the flow area and hydraulic gradient must increase (to maintain continuity of flow) and this can be represented in a flownet by increasing the flow channel width or by decreasing the distance between equipotentials (see Figures 7.19 and 7.21). The opposite is true when water flows from a low-permeability zone to one of high permeability. By considering continuity of flow across the boundary it can be shown that the directions (see Figure 7.21) of the flow lines before and after the interface are related by the expression: $k_1/k_2 = \alpha_1/\alpha_2$.

7.4.5. Quicksand conditions
The behaviour of an unconfined mass of granular material when water flows upwards through it is dependent on the flow rate. When the head differential

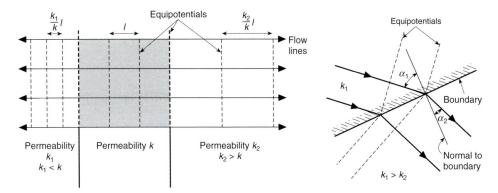

Head loss between adjacent equipotentials is constant for any flow path

Figure 7.21. The effect of change in permeability on flownets

applied to the soil is small the water will flow through the soil at a velocity directly related to the hydraulic gradient in accordance with Darcy's law. As the head differential is increased, the velocity of flow increases. However, as the hydraulic gradient is increased the upward force exerted on the soil particles by the seeping water is also increased. Eventually this seepage force will be sufficiently large to lift the soil particles and cause heaving. Further increase in the hydraulic gradient will result in localized piping or 'boiling' of the soil (i.e. the formation of distinct flow conduits within the soil deposit). This is often observed as small 'volcanoes' of soil. This results in separation of the particles, increased permeability and progressive increase in seepage rate. Once piping has developed in a soil then most of the water flows through channels or pipes within the soil.

When the soil is on the point of 'boiling', the downward force (from the weight of the saturated soil) equals the total upward force from the water (buoyancy and seepage), (Figure 7.22):

$$\text{Downward force} = \gamma_s A l \tag{7.44}$$

$$\text{Upward force} = \gamma_w A l + h \gamma_w A \tag{7.45}$$

Thus

$$h \gamma_w A = \gamma_s A l - \gamma_w A l \tag{7.46}$$

Hence

$$\frac{h}{l} = i_{\text{crit}} = \left(\frac{\gamma_s - \gamma_w}{\gamma_w} \right) = \frac{\gamma_w}{\gamma_w} \left(\frac{G_s + S_r}{1 + e} \right) - 1 = \left(\frac{G_s + S_r}{1 + e} \right) - 1 \tag{7.47}$$

Since the soil is fully saturated, $S_r = 1$ and

$$i_{\text{crit}} = \frac{G_s - 1}{1 + e} \tag{7.48}$$

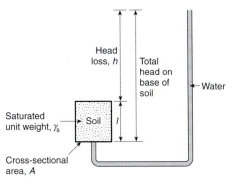

Figure 7.22. Quicksand conditions

Equation (7.48) predicts the critical hydraulic gradient (for the onset of quicksand conditions) and, with the values of G_s and e typically encountered, i_{crit} is close to unity for most soils. This gradient makes the soil into a quicksand (i.e. it is a fluid soil), without effective stress and therefore without strength, and with a density intermediate to that of water and a soil. In general, quicksand (which is not a type of soil but a flow condition) occurs with fine sands, but coarser soils can be made 'quick' if a sufficiently large water flow rate is provided.

The erosive force produced by water passing out of an earth structure can be large, resulting in the formation of 'pipes' that increase in size and flow capacity, leading to progressive erosion, undermining and eventual instability. When heaving or piping occurs in front of a sheet-piling retaining wall there is progressive and rapid loss of resistance in front of the piling, followed by complete collapse of the retaining structure. Certain soils (non-cohesive, fine soils) have a low resistance to erosion and should not be used in earth dams, flood banks or canal linings unless they are protected by other resistant materials. Quick conditions and piping are associated with silts, and ground heave with expansive clay soils. The groundwater and the pressure within it can be confined beneath a relatively impermeable stratum (clay or silt) producing an uplift pressure. If the downward pressure from the clay at the base of an excavation is insufficient to balance the uplift pressure then the clay will be lifted, causing it to crack and allowing water to flow through.

Piping is easier to prevent than it is to cure because, once it has started, the quicksand condition causes the soil to lose its strength, which results in instability. Preventive measures include increasing the path length of the water flow, and enhancing the downward weight of the soil mass by loading it with free-draining material.

7.5. Groundwater control

In order to carry out construction work below groundwater level it is normally necessary for the working area to be reasonably free of standing water. Ground engineering works, particularly excavations, will disturb the groundwater equilibrium and alter the pattern of flow. Treatment methods are therefore aimed at preventing ingress of groundwater to, or removing it from, the site in question on the one hand, or improving soil strength on the other. The type of soil, the height of the water-table, the depth of the excavation and its shape all influence the choice of method of dealing with water flow in and around the excavation works, and various methods are available. Soil-treatment techniques may be either temporary or permanent.

7.5.1. Groundwater barriers
Methods of forming barriers include (Bell, 1993):

- sheet piling
- contiguous bored and secant pile walls
- bentonite cut-off walls
- geomembrane barriers
- concrete diaphragm walls
- grout curtains and panels
- ice walls.

The economy of providing a barrier to exclude groundwater depends on the existence of an impermeable stratum beneath the excavation to form an effective natural cut-off for the barrier. If such a stratum does not exist or if it lies at too great a depth to be used as a cut-off, then upward seepage may occur, which in turn may give rise to instability at the base of the excavation. Artesian conditions, in particular, can cause serious trouble in excavations and, therefore, if such conditions are expected it is essential that groundwater pressures are determined before work commences. Excavations that extend close to strata under artesian pressure may be severely damaged due to heave or blow-out (see Section 7.4.5).

Sheet-piling may be composed of steel, timber or concrete piles, each pile being linked to the next to form a continuous wall. Steel sheet-piles are most widely used for forming a water barrier (and for supporting the ground at the same time) in temporary structures such as strutted excavations and cofferdams, and in permanent structures such as retaining walls and bulkheads. Sheet-pile walls are generally sufficiently watertight for most practical purposes but if piles come out of interlock during driving this means that the effectiveness of the cut-off will be severely impaired (Figure 7.23).

In some soils, notably fine sands and silts beneath the water-table, there is a danger of quick conditions developing if the critical gradient is exceeded as ground is removed. From model tests Terzaghi (1966) concluded that heaving generally occurs within a horizontal distance of $0.5 D$ from the sheet piles (where D is the depth of embedment of sheet piles into the permeable layer). Therefore, we need to investigate the stability of soil in a zone measuring $D \times 0.5 D$ in cross-section. If h_m is the average uplift head on the base of this zone, then from equation 7.48, for stability,

$$\frac{h_m}{D} < \left(\frac{G-1}{1+e}\right) \cong 1 \tag{7.49}$$

Slurry trenches are used extensively as a means of groundwater cut-off, for example to control seepage beneath dams and to contain groundwater pollution (notably from landfills and hazardous waste impoundments). Slurry trenches can achieve permeabilities of the order of 10^{-9} m/s. They are formed by excavating a narrow, deep trench and filling it with a low-permeability, flexible medium such as soil–bentonite or cement–bentonite. To ensure stability of the trench during excavation it is filled with a dense bentonite slurry through which the excavation is undertaken (Figure 7.24).

Figure 7.23. Water inflow to the base of a sheet-piled excavation

*Figure 7.24. Slurry trench
formation*

If a trench is filled with a dense fluid then the fluid pressure acts as shoring, supporting the trench walls to prevent cave-ins and slumping during excavation. A suitable dense fluid can be obtained by mixing bentonite with water to form a thixotropic slurry. Left undisturbed this slurry 'sets' to form a weak gel, but subsequent agitation as a result of excavation through the slurry causes the gel to revert to its liquid state. As a general rule, in most soil conditions a net slurry head of about 1.5 m above the highest piezometric pressure to be encountered should maintain trench stability. The weight of the slurry also forces bentonite into the soil matrix at the trench faces and base to form a filter cake, which restricts water inflow. The fundamental requirements for an excavation slurry are thus that:

- it exerts sufficient hydrostatic pressure to stabilize the excavation
- it controls loss of slurry to the ground
- it resists groundwater inflow
- it should not bleed (i.e. solids should not settle out of the water).

The foregoing requirements are met by bentonite–clay, bentonite–cement and polymer slurries.

Soil–bentonite backfill consists of excavated soil, with the addition of material from a borrow pit, mixed with bentonite mud, to form a low-permeability, highly plastic cut-off wall. In the cement–bentonite cut-off process, cement is added to a fully hydrated bentonite slurry. The addition of the cement causes the slurry to harden, giving a strength comparable to that of stiff clay. Diaphragm walls may be constructed by replacing the bentonite slurry with concrete. These walls compare favourably with cut-offs formed of steel sheet-piling, precast piles or cast in situ piles in terms of watertightness, stiffness and mechanical strength. Diaphragm walls may also be used as load-bearing and retaining walls, and in such cases they are reinforced by incorporating a steel cage.

Ground freezing involves the artificial lowering of ground temperatures so that porewater is converted into ice. A series of pipes is inserted into the ground and cold fluid, either a liquid such as brine or liquid nitrogen, is circulated through the pipes. If the original ground formation was saturated it will become impermeable after the ice forms. Furthermore, the ice bonds adjacent particles together, adding shear strength to the formation — this increase may be significant for particles of sand size and smaller. In this way an ice wall can be formed around an excavation until a permanent structure has been constructed. Frozen walls are frequently used in shaft sinking.

Freezing may be accomplished in any formation, regardless of structure, grain size or permeability. Furthermore, there are no depth limitations to the method, except for the possible inability at great depths to place the cooling pipes with sufficient accuracy. However, the initial cost of a freezing plant and brine circulation system is high, and the freezing process is slow (weeks or months), precluding economical use on small projects. The use of liquid nitrogen as a coolant may overcome these limitations. The problems associated with the freezing method occur primarily during the thawing period and they are mainly related to the volume decrease that occurs when ice changes into water.

7.5.2. Dewatering

Pumping methods are most appropriate for the majority of granular soils. Silty soils often cannot be effectively dewatered by pumping, and they are particularly troublesome. For these materials electro-osmosis, ground freezing or grouting (see Chapter 14, Section 14.1.2) should be considered.

If an electric potential is applied to a saturated porous material, electrolyte will flow from the anode to the cathode — this is the principle of electro-osmosis. Electro-osmosis was originally developed by Casagrande as a method for dewatering low-permeability active clay soils. It has been used with varying degrees of success. The fundamental drawback with the process is that when it is used solely for dewatering it is a decelerating process, becoming progressively less efficient as the soil water content is reduced. Where electro-osmosis has been successful, it has usually been as a result of using the process to introduce a chemical into the soil, either through anode solution or by direct electrolyte replacement. This improves soil stability either by chemical change (ion replacement) in the clay mineral or by partial cementation of the pore space.

Dewatering is a method of improving the soil properties by reducing the water content and/or porewater pressure. When an excavation is made below the water-table in slow-draining soil, the quantity of water entering the excavation will be so small that it can be collected in a sump within the site and then pumped away. However, for permeable soils it is not acceptable to have water flowing into an excavation because of the high inflow rate and wash-out of particles. Under these circumstances it is better to lower the water-table temporarily before excavating,

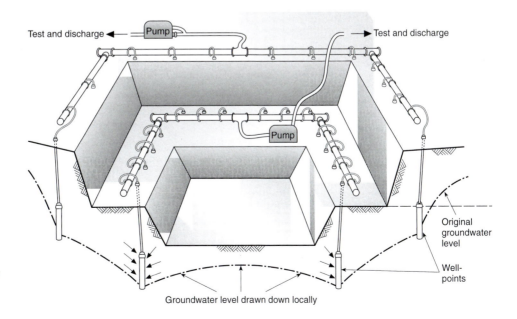

Figure 7.25. Dewatering by well-pointing (after Harris, 1994)

(a) Well-points alongside excavation

Figure 7.26. Groundwater removal by well-pointing

(b) Collector pipes taking extracted water to discharge

so that the excavation work can be undertaken in dry conditions. This may be achieved by using dewatering techniques such as pumping from well-points (vertical, perforated tubes) inserted below the water table. Pumping from wells positioned outside the excavation boundary is usually the preferred technique (Figures 7.25 and 7.26). Lowering of the water-table reduces buoyancy of the soil and may cause significant settlement of weak, compressible strata. Where settlement must be kept to a minimum, recharge water (i.e. water introduced into the ground between the wells and adjacent structures) should be provided to maintain the groundwater level beneath the structures.

Dewatering systems usually consist of small wells installed so that their cones of drawdown overlap. The drawdown at any point on the composite cone of depression can be assumed to be equal to the sum of the individual drawdowns. In the case of well-points fully penetrating a permeable deposit overlying a relatively impermeable stratum a simple method of system design has been proposed by Harris (1994). The wells are assumed to be arranged in a circle (of radius x) about the excavation (of length l and width b), wherein

$$x = \sqrt{\frac{bl}{\pi}} \tag{7.50}$$

The purpose of the dewatering is to lower the height of the water-table above the impermeable stratum from H_0 to H. If this is achieved by using n wells, of radius r, each of which extracts water at a constant rate q, then from the analysis of a single well (Harris, 1994),

$$H_0^2 - H^2 = \frac{nq}{\pi k} (\ln R - \ln x) \tag{7.51}$$

where k is the permeability of the deposit being dewatered and R is the radius of the outer extent of the drawdown. For open ground:

$$R \cong 3000(H_0 - H)\sqrt{k} \tag{7.52}$$

For a temporary well of radius r_0 and effective height h_e (which can be assumed to be $0.4\ H_0$) the well capacity is given by

$$q = 2\pi r_0\, h_e\, \frac{\sqrt{k}}{15} \tag{7.53}$$

8 Consolidation and settlement

8.1. Introduction

The application of load to a soil creates a state of excess pore pressure (as explained in Chapter 6, Section 6.1.5). These excess pore pressures dissipate by the gradual movement of water through the voids of the soil, and this results in a volume change that is time dependent. A soil experiencing such a volume change is said to be *consolidating* and the vertical component of the volume change is called *consolidation settlement*. These consolidation volume change processes must be distinguished from compaction, which is the expulsion of air from a soil by applying compaction energy.

The major problem with the construction of the Great Yarmouth bypass was the ground — in places it consisted of waterlogged bog to a depth of more than 20 m. A 2 m high railway embankment, which was built in 1901 along the same line, suffered a ground punching failure (probably during construction). Even when the last section of the railway line was closed over 70 years later, the embankment was still settling and having to be topped up (in places it had sunk by as much as 6 m). The engineering solution derived for the new bypass was a combination of vertical drains, a fabric-enclosed granular mattress and a continuously reinforced concrete road base (*New Civil Engineer*, 1984). The vertical wick drains discharge upwards into a 500 mm thick mattress of granular material (enclosed within a geogrid), which carries the groundwater to side ditches. The mattress also restrains longitudinal and lateral spreading as the embankment settles, so preventing movement of granular material in the embankment and reducing further the risk of a slip in the silts below.

The works were heavily instrumented, with more than 200 piezometers (to measure pore pressures), 200 rod settlement monitors, 60 surface profile monitors, 35 inclinometers (to measure lateral displacements) and 35 magnet settlement gauges and numerous studs for direct measurement of movements — descriptions of the principles and applications of such instrumentation are contained in Chapter 11. The total settlement under a typical 2.5 m high embankment was around 1 m and the greatest rate of settlement was about 10 mm/week.

To minimize the imposed loading on the compressible ground adjacent to a railway bridge a short section of embankment incorporates polystyrene as an ultra-lightweight fill. As polystyrene weighs only 1 kN/m³ (as compared to about 19 kN/m³ for granular fill or 15 kN/m³ for pulverized fuel ash) the weight saving is very large. Blocks about 2 m × 1 m × 100 mm thick were laid in bond and sealed in a polyethylene sheet as protection against petrol spillage. The top of the polystyrene mass was covered with sand and granular fill, and the sides are protected by a covering of earth.

In the natural process of deposition of fine-grained soils they undergo a process of consolidation in which the water between the particles is gradually squeezed out by the weight of layers deposited above. After a period of time, which may be a considerable number of years, a state of equilibrium is achieved and compression ceases. The soil is then said to be *fully consolidated*. If the soil is sustaining the maximum effective stress of its life it is *normally consolidated*, but if the current effective stress level is less than that applied during the history of the soil then it is said to be *overconsolidated*.

8.1.1. Consolidation and settlement

All soils consolidate when compressed, but for free-draining soils (i.e. sands and gravels), the compression is small and rapid. It is the consolidation of the fine-grained clay soils that requires particular attention, because it takes a significant time period for the process to be completed and the resultant compressions are much larger. The amount of volume change or settlement resulting from an applied pressure is defined by the compressibility of the material.

In addition to the direct effect of structural loading, clay consolidation can be induced by imposed drainage. On a local scale, seasonal variations in water content in clay soils cause shrinkage and heave, creating annual surface oscillations commonly up to 30 mm and locally up to 50 mm in the UK (Waltham). Vegetation plays a major role in this, with Poplar and Willow trees in particular abstracting large quantities of soil water. Figure 8.1 shows the corner of a house damaged due to ground water extraction by adjacent Poplar trees. Removal of the vegetation causes gradual swelling and heaving of the surface.

In consolidation problems, it is often found that the porewater pressure distribution re-equilibrates to that which existed before the load was placed. This is inevitable where the groundwater conditions are controlled externally (e.g. by a river) or when the construction is above ground level. Any permanent excavation such as a cut slope, however, is almost certain to change the external hydraulic boundary conditions, by changing their shape if not their type. Hence it is unusual to find a permanent excavation re-equilibrating to its original pore pressure state.

The complex subject of soil–structure interaction, which involves the effects of ground movements on structures, is being given increasing recognition by practising engineers and researchers, but its relative importance can be assessed only by considering such displacements in the general context of building movements. While settlement of structures will produce problems, the damage effects are much greater when the settlement are differential (i.e. when different

Figure 8.1. Settlement-induced cracking due to groundwater extraction by trees

Figure 8.2. Settlement-induced sliding along the damp-proof course

parts of the structure settle by different amounts). Figure 8.2 shows the relative sliding of brick courses along a dampproof layer. Natural soils are rarely uniform enough to allow perfectly even settlement (Figure 8.3). The classic case of differential settlement is the bell tower at Pisa, which stands 58 m high and now leans about 4 m out of true. If the tower had been built in a single phase it would have failed before completion, but the soil strength was increased by the unplanned staged consolidation. The settlement of the tower has been mainly due to compression of normally consolidated clays, and the largely consolidated clays beneath the tower can now safely bear the imposed load, but drained creep is continuing and the lean is increasing.

Clay consolidation involves a major primary phase of water expulsion and a lesser secondary phase of internal restructuring. The settlement behaviour can be separated into three components:

- *Immediate undrained settlement*: this is due primarily to lateral displacement of the soil and is high for some very plastic or organic clays.
- *Consolidation settlement*: this is due to the squeezing out of porewater (primary consolidation) as the pore pressure falls, and it accounts for the major part of most ground movements with clay soils.

Figure 8.3. Tilting of buildings due to non-uniform settlement

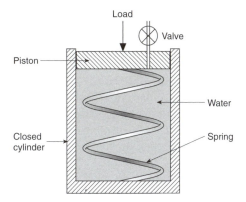

Figure 8.4. Terzaghi's piston and spring analogy

- *Drained creep* (or *secondary compression*): this results from the long-term creep of the soil skeleton under uniform stress, and its rate declines progressively with time.

8.1.2. Piston and spring analogy

Terzaghi (1966) proposed a piston and spring analogy for the process of consolidation (Figure 8.4). The cylinder is filled with water (which corresponds to the porewater) and the spring represents the soil skeleton (i.e. the spatial arrangement of the soil particles). The valve opening in the piston represents the permeability of the soil mass. If a load is applied to the piston with the valve closed, the length of the spring cannot change because the water is incompressible. Hence the increase in pore pressure must balance the total stress applied on top of the piston. Consequently, there is no change in the force or stress in the spring (analogous to effective stress), and of course there is no reduction in the volume of this soil mass. This behaviour is comparable to that occurring during the ambient stage of an undrained triaxial test on fully saturated clay samples, wherein the pore pressure B is equal to 1.0 and the effective stress is unaffected by changes in all round cell pressure (see Chapter 6, Section 6.1.5). When the valve is opened, water starts to flow out of the cylinder, the pressure in the water falls and the piston sinks as the spring is compressed. The rate at which the pore pressure dissipates depends on the size of the valve opening, which is thus analogous to the soil permeability. When the excess pore pressure has fallen to zero, the piston ceases to move. At this stage the effective stress (as indicated by the force in the spring) has increased by an amount equal to the applied total stress, and the amount of piston settlement has been determined by this effective stress increase.

Terzaghi's model is a useful analogy for the consolidation process, although the water pressure conditions (i.e. uniform throughout) do not conform to reality. The pore pressure in the soil immediately adjacent to the drain will fall to zero as soon as the drain is opened, and it will increase with distance from the drain.

8.2. Consolidation theory

For consolidation or settlement behaviour, reference is usually made back to Terzaghi's (1966) consolidation model and theory of one-dimensional consolidation. Two- and three-dimensional analyses, incorporating lateral drainage, are more difficult to apply, and comparisons of results show that the benefits are often limited on a practical scale. One difficulty in accurate settlement prediction arises from the critical importance of the preconsolidation stress, which is not easily determined in the laboratory due to sampling disturbance. Thus in situ tests can be of major benefit.

8.2.1. One-dimensional consolidation (Terzaghi)

Initial assumptions are that consolidation (i.e. the flow of water in the soil and the resultant settlement) is one-dimensional and occurs in the direction of loading (usually assumed to be vertical so that there is no lateral deformation of a soil). This is the situation that will exist for a soil layer that is thin by comparison to the width of the loaded area, particularly at the centre of that area. If a deposit of clay is subjected to an imposed load then the ground surface represents a drain, since it is either a free surface or a drainage layer is placed over the original ground before the load is applied. The soil is also assumed to be homogenous and fully saturated.

The behaviour of a typical element during consolidation is illustrated in Figure 8.5. Water enters the element with a certain pore pressure u and at velocity v and it exits with an increased velocity $(v + \mathrm{d}v)$, while the pore pressure fills to $(u - \mathrm{d}u)$. If the thickness of the element in the direction of flow is $\mathrm{d}z$ then the hydraulic gradient across the element is equal to the change in head divided by the element thickness. This part of the analysis is concerned only with water flow due to head differential. Thus the total change in head $\mathrm{d}h$ over a small increment of vertical distance $\mathrm{d}z$ is equal to the rate of change of head, with respect to distance z only, multiplied by the actual increment of distance:

$$i = \frac{\mathrm{d}h}{\mathrm{d}z} = \left(\frac{\partial h}{\partial z}\right)\frac{\mathrm{d}z}{\mathrm{d}z} = \frac{\partial h}{\partial z} = \frac{1}{\gamma_\mathrm{w}}\left(\frac{\partial u}{\partial z}\right) \tag{8.1}$$

The velocity of water flow within a soil mass is assumed to be governed by Darcy's law (see Chapter 7 Section 7.1.2):

$$v = ki = \frac{k}{\gamma_\mathrm{w}}\frac{\partial u}{\partial z} \tag{8.2}$$

If the cross-sectional area of the soil element is taken as A, then in time $\mathrm{d}t$ the rate of volume decrease (consolidation) for an element of thickness $\mathrm{d}z$ is $\mathrm{d}V/\mathrm{d}t$, where

$$\mathrm{d}V/\mathrm{d}t = A(v + \mathrm{d}v) - A(v) = A\,\mathrm{d}v \tag{8.3}$$

From Section 7.4.2 (for flow in the z direction only) and equation (8.2),

$$A\,\mathrm{d}v = A\,\frac{\partial v}{\partial z}\,\mathrm{d}z = A\,\frac{k}{\gamma_\mathrm{w}}\frac{\partial^2 u}{\partial z^2}\,\mathrm{d}z \tag{8.4}$$

The volume decrease of the element (equation 8.4) can only come about through a change in the volume of voids, since the soil grains are essentially incompressible. The initial volume of voids is equal to the total volume multiplied by the porosity, i.e. $A\,\mathrm{d}z\,[e/(1+e)]$. The term $\mathrm{d}z/(1+e)$ is the

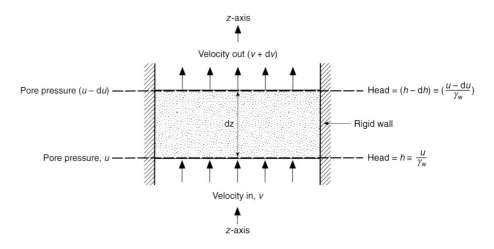

Figure 8.5. Consolidation of a soil element

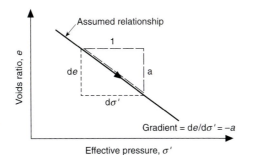

Figure 8.6. The idealized relationship between the voids ratio and the consolidation pressure

equivalent height of the soil skeleton, and if strains are small it can be taken as constant and equal to $dz/(1 + e_0)$ — this is the small strain approximation. Thus the rate of change of the element volume (dV/dt) is given by

$$\frac{dV}{dt} = A\,\frac{dz}{1 + e_0}\,\frac{de}{dt} \tag{8.5}$$

A further assumption is that there is a linear relationship between the voids ratio and the effective pressure σ' (Figure 8.6). This assumption is in-keeping with the spring analogy and, although it is not strictly true for all soil conditions (a relationship between e and $\ln p'$ is more true, so that the value of a may change with the stress level), for small pressure changes it is an acceptable approximation. Hence

$$de = -a\,d\sigma' = -a\,d(\sigma - u) = -a\,d\sigma + a\,du \tag{8.6}$$

If the total stress is constant (with time and position within the soil) during consolidation (which is the usual situation) then $d\sigma = 0$. Since it is only the variation in u with time that is being considered, then $du = (\partial u/\partial t)dt$. Thus

$$\frac{dV}{dt} = A\,\frac{dz}{1 + e_0}\left(a\,\frac{\partial u}{\partial t}\right) \tag{8.7}$$

Equating the two expressions for rate of volume change (equations 8.3 and 8.5), i.e. that due to outward flow of water and that resulting from change of voids ratio, gives

$$A\,\frac{k}{\gamma_w}\,\frac{\partial^2 u}{\partial z^2}\,dz = A\left(\frac{dz\,a}{1 + e_0}\right)\frac{\partial u}{\partial t} \tag{8.8}$$

The term $a/(1 + e_0)$ is defined as m_v (the coefficient of volume compressibility), so that

$$\left(\frac{k}{\gamma_w}\right)\left(\frac{\partial^2 u}{\partial z^2}\right) = m_v\left(\frac{\partial u}{\partial t}\right) \tag{8.9}$$

Furthermore, $k/(m_v\gamma_w)$ is defined as C_v (the coefficient of consolidation), and hence *Terzaghi's equation for one-dimensional consolidation* is obtained:

$$C_v\left(\frac{\partial^2 u}{\partial z^2}\right) = \frac{\partial u}{\partial t} \tag{8.10}$$

8.2.2. Solution of the consolidation equation

What is important from an engineering point of view is a solution of the preceding one-dimensional consolidation equation to express how pore pressure changes with time, because this indicates how shear strength and settlement will develop. For a soil deposit that has a drainage path length H, solution of equation (8.10) gives the value of the pore pressure excess u_z at distance z (from the drain) and at time t (from the start of consolidation):

$$u_z = \sum_{M=\pi/2}^{M=\infty} \frac{2u_i}{M} \left(\sin \frac{Mz}{H} \right) e^{-M^2 T_v} \tag{8.11}$$

where u_i is the initial pore pressure excess (uniform over the whole depth of the soil deposit), M ranges from $\pi/2$ to infinity, T_v is the time factor ($= C_v t/H^2$) and e is the exponential. It should be noted that the drainage path length H is the maximum distance a particle of water has to travel to reach a drain. This distance is not necessarily equal to the thickness of the soil layer; it depends on the boundary drainage conditions.

At any particular level within the soil (expressed by $Z = z/H$), the proportion of the pore pressure excess that has dissipated is the degree of consolidation U_z, which is given by

$$U_z = \frac{u_i - u_t}{u_i} \equiv 1 - \sum_{M=\pi/2}^{M=\infty} \frac{2}{M} (\sin MZ) e^{-M^2 T_v} \tag{8.12}$$

Equation (8.12) can be conveniently represented as a graph of U_z versus Z for different T_v values (Figure 8.7). The resultant curved lines refer to constant values of time (or constant T_v) and so are called *isochrones*. The figure relates to an initially uniform distribution of pore pressure excess that has been set up within the soil. Other initial distributions are possible, and these would produce differently shaped isochrones. The instantaneous loading state ($t = 0$) is represented by the curve in Figure 8.7 with T_v and U_z zero, since consolidation has not yet commenced. Very soon after commencement of the consolidation process the soil adjacent to the permeable boundaries will have consolidated fully, but at the middle of the soil deposit consolidation will not have started. With time (increasing T_v) the isochrones become flatter, and eventually they reach the 100% consolidation state.

Since equation (8.12) defines the degree of consolidation for every individual level within the soil deposit, the average degree of consolidation U_v for the whole deposit at any particular value of T_v can be obtained. The consolidation achieved for any value of T_v is the total possible consolidation (i.e. the initial pressure distribution less the area under the relevant isochrone). The average degree of consolidation is then the preceding difference divided by the sample thickness:

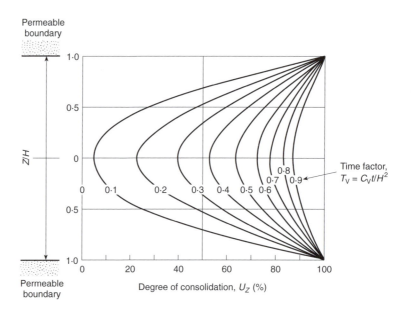

Figure 8.7. Consolidation isochrones

$$U_{\mathrm{v}} = \frac{(Hu_{\mathrm{i}} - \text{area under isochrone})}{Hu_{\mathrm{i}}} \equiv 1 - \sum_{M=\pi/2}^{M=\infty} \frac{e^{-M^2 T_{\mathrm{v}}}}{M^2} \qquad (8.13)$$

Equation (8.13) is plotted in Figure 8.8 for three likely initial distributions of pore pressure excess within a soil layer and the possible permeability of the soil boundaries. The three distributions are:

1. Uniform pore pressure with depth: this corresponds to an area of applied pressure that is wide or extensive by comparison to the thickness of the soil layer, such as would occur with an embankment or general filling over a large area.
2. Triangular pore pressure distribution (maximum at base of layer and zero at the surface): this corresponds to placement of hydraulic fill or the self-weight effect of soil forming an embankment.
3. Triangular pore pressure distribution (from a maximum at ground surface to zero at the base of the layer): this corresponds to situations where pressure is applied over a small area only, such as loading from a foundation.

For curve 1 the theoretical relationship between U_{v} and T_{v} can be approximated as follows:

For $U_{\mathrm{v}} < 0.6$: $T_{\mathrm{v}} = \dfrac{\pi U_{\mathrm{v}}^2}{4}$ \qquad (8.14)

For $U_{\mathrm{v}} > 0.6$: $T_{\mathrm{v}} = [-0.933 \log(1 - U_{\mathrm{v}})] - 0.0851$ \qquad (8.15)

Individual values of U_{v} against T_{v} are given in Table 8.1, and typical values of C_{v} are given in Table 8.2.

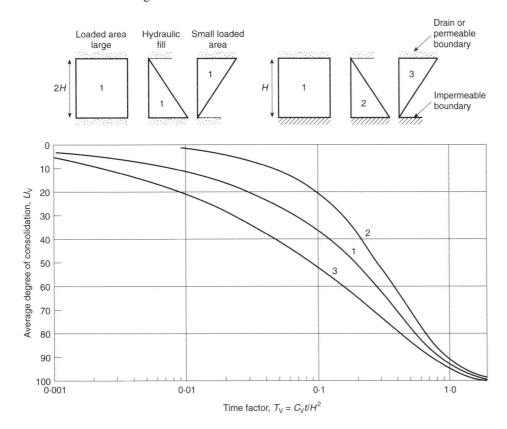

Figure 8.8. Solutions of the consolidation equation

Table 8.1. Values of the time factor T_v for the cases identified in Figure 8.8

	Average degree of consolidation, U_v (%)										
	0	10	20	30	40	50	60	70	80	90	100
Curve 1	0	0	0.008	0.071	0.126	0.197	0.287	0.403	0.567	0.848	∞ (≈ 2.0)
Curve 2	0	0.003	0.010	0.024	0.049	0.091	0.160	0.270	0.440	0.730	∞ (≈ 3.0)
Curve 3	0	0.047	0.100	0.158	0.221	0.294	0.383	0.500	0.665	0.940	∞ (≈ 2.0)

*Table 8.2. Typical values of the consolidation parameters**

Parameter	Typical values
m_v	Heavily overconsolidated boulder clays: $<0.05\,\mathrm{m^2/MN}$ Boulder clays with low compressibility: 0.05–$0.10\,\mathrm{m^2/MN}$ Weathered clays with medium compressibility: 0.10–$0.30\,\mathrm{m^2/MN}$ Normally consolidated alluvial clays: 0.3–$1.5\,\mathrm{m^2/MN}$ Highly organic alluvial clays and peats: $>1.5\,\mathrm{m^2/MN}$
C_v	Clay with high proportion of clay-sized particles: 0.1–$0.5\,\mathrm{m^2/year}$ Silty clay: 0.5–$5\,\mathrm{m^2/year}$
C_c	Peats: $0.01 \times$ moisture content Undisturbed clays: $0.009\,(W_L - 10)$ Remoulded clays: $0.007\,(W_L - 10)$, $0.234\,W_L/G_s$ Normally consolidated medium sensitive clays: 0.2–0.5 Organic silt and clayey silts: 1.5–4.0 Organic clays: >4.0
C_α	Organic silts: 0.09–0.12 Organic clays: 0.12–0.20 Peats: 0.15–0.30

* Sources: Terzaghi & Peck (1967), MacFarlane (1969), Nagaraj & Murty (1985), Carter & Bentley (1991).

Pore pressure changes will produce volume changes in the soil and hence settlement of the surface of the material. If the soil is fully saturated there will be direct correspondence between average degree of consolidation and the settlement of loaded areas on the surface of a soil layer. Thus if ρ_t is the settlement at time t and ρ_c is the final consolidation settlement,

$$U_v \equiv \frac{\rho_t}{\rho_c} \tag{8.16}$$

8.2.3. Volume change due to consolidation

A soil may be considered as a flexible skeleton of mineral particles (which are themselves incompressible) so that the decrease in volume ΔV that results from loading and consolidation reflects a change in the voids ratio. Hence, from equation (8.5),

$$\Delta V = \frac{V_0\,\Delta e}{1 + e_0} \tag{8.17}$$

The shape of the voids ratio versus the effective stress curve is dependent on the stress history of a soil (as indicated in Figure 8.9). When a normally consolidated

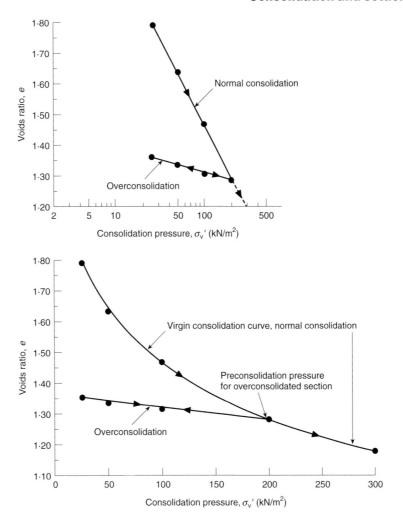

Figure 8.9. The relationship between the voids ratio and the consolidation pressure

soil (i.e. one which has never been subjected to a higher effective stress than at the present time) is subjected to a significant increase in load and allowed to consolidate, the relationship between e and σ'_v is non-linear. The actual relationship is well represented by an equation of the form

$$\Delta e = -C_c\,\Delta(\log\,\sigma'_v) \tag{8.18}$$

where C_c is the compressibility or compression index (see Table 8.2). Thus, after consolidation,

$$\Delta V = -\left(\frac{V_0}{1+e_0}\right)C_c\,\Delta(\log\,\sigma'_v)$$

$$\equiv -\left(\frac{V_0}{1+e_0}\right)C_c\,\log\left(\frac{\sigma'_{vf}}{\sigma'_{vi}}\right) \tag{8.19}$$

where σ'_{vf} and σ'_{vi} are the final and initial vertical effective stresses, respectively.

If a soil is subjected to normal consolidation and then the stress is reduced so that the soil swells, the relationship between e and σ'_v obtained during unloading is approximately linear (i.e. it does not follow the origin consolidation curve; see Figure 8.9). In this unloading situation the soil is overconsolidated and the gradient of the e versus σ'_v curve is much smaller than that for a normally consolidated material. If the soil is then reloaded, the recompression curve is similar to the swelling curve, with the change in the voids ratio being proportional to the change in the effective stress. Once the previous maximum effective stress level (the preconsolidation pressure) has been attained the soil becomes normally consolidated again and the virgin consolidation curve is followed (i.e. there is a noticeable increase in the compressibility of the soil). The ratio of the preconsolidation pressure to existing effective overburden stress is the overconsolidation ratio. This ratio is unity for normally consolidated clays and greater than unity for overconsolidated clays.

In the UK, natural processes of deposition and the Ice Ages have produced many overconsolidated clays. During the deposition process the virgin consolidation curve was followed, but subsequent erosion or melting of the ice removed some of the pressure. As the effective stress was reduced from the previous maximum value to its present level, the soil became overconsolidated.

For overconsolidated soils the linear relationship between the voids ratio and the effective stress assumed by Terzaghi (1966) (see Section 8.2.1) is an accurate representation of the real situation for small changes in stress. This relationship is represented by equation (8.6) and hence, after consolidation,

$$\Delta V = -\left(\frac{V_0}{1 + e_0}\right) a\, \Delta \sigma'_v \equiv -V_0\, m_v\, \Delta \sigma'_v \tag{8.20}$$

The parameter m_v is the coefficient of volume compressibility (see Section 8.2.1). Values of m_v (see Table 8.2) can also be defined for normally consolidated soils, but then its value will decrease as the effective stress level increases (to reflect the relationship between e and $\log \sigma'_v$ as the soil becomes stiffer and the voids ratio decreases).

8.3. Consolidation testing

The purpose of consolidation testing is to determine the relevant parameters: C_v (for the rate of change of pore pressures), and C_c and m_v (for the estimation of the amount of settlement). The first apparatus developed to measure these parameters was the Casagrande oedometer, which tests a small, thin sample by dead loading. Subsequently, the hydraulic consolidation cell (Rowe cell), which uses a flexible hydraulic jack to apply the consolidation load, was developed for testing much larger samples. In both types of apparatus the soil sample is contained within a rigid cylinder which sits on a base plate and a vertical load is applied to the upper surface of the sample. There is thus no lateral deformation of the soil, and all compression (or settlement) is in a vertical direction.

8.3.1. Casagrande oedometer

In the Casagrande oedometer (Figures 8.10 and 8.11) the base plate is covered by a saturated porous disc that acts as a drain and connects to a passage through the

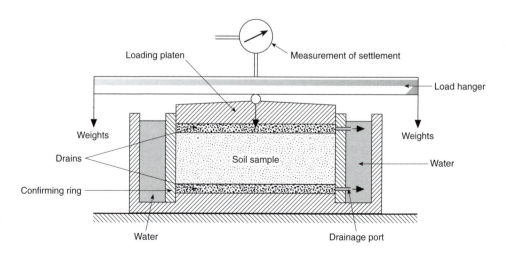

Figure 8.10. The Casagrande oedometer

Figure 8.11. The Casagrande oedometer

cylinder. A sample is obtained by driving a cutting ring (approximately 76 mm diameter, 19 mm deep) into undisturbed soil or by extruding compacted soil into the ring. Surplus soil is trimmed off from around the cutting ring to form a disc of soil (contained within the ring) with smooth top and bottom surfaces. The cutting ring and included sample are then placed on top of the drain and clamped inside the oedometer. The space between the cutting ring and the outer body is filled with water and a top drain is seated on top of the sample. On top of this drain is a rigid loading platen. Using a series of lever arms, load hangers and dead weights the requisite vertical load is applied to the top of the loading platen. A dial gauge (or displacement transducer) is installed to measure vertical movement of the top loading platen and, after a sample has reached equilibrium, a test is started by increasing the vertical stress and taking readings of the vertical dial gauge at selected times after the start of loading. After full consolidation under this load, the vertical stress is again increased (typically by 50% or 100% of the existing pressure) and the variation in settlement with time is again obtained. The moisture content, unit weight, thickness, etc., of the soil sample should be measured at the start and end of a test.

Although for the vast majority of routine practical investigations the Casagrande oedometer is perfectly adequate and provides sufficiently accurate parameters, it does have several limitations:

- The pore pressure cannot be measured accurately because of a lack of control of drainage, and hence the degree of consolidation has to be inferred (by measuring settlements).
- Drainage is entirely vertical. As some soils are strongly anisotropic, their properties, particularly drainage, are very different in horizontal and vertical directions.

- Drainage starts as soon as the load is applied. A uniform pore pressure may not be developed throughout a sample, and the initial undrained compression cannot be measured directly.
- Samples cannot be saturated. Compression of gas or air in the soil produces a false initial undrained compression.
- The apparatus uses a small sample that does not necessarily incorporate the structural 'fabric' of the soil (which has resulted from its history).

In a number of practical situations Rowe (1968) found that calculated rates of settlement of clay soils (based on laboratory C_v values obtained using the Casagrande oedometer) were significantly smaller than those back-calculated from the actual observed rates of settlement (Table 8.3). These differences were due to the influence of the fabric found in most clay soils, but may not be represented in the small-scale oedometer test. The effect depends on the orientation of the fabric or internal structure, which may be horizontal (laminations, layers of silt and fine sand), vertical (rootlets) or inclined (fissures, etc.). Such fabric can drastically reduce the time required for consolidation by decreasing the drainage path lengths and increasing the mass permeability of the clay. To overcome this problem, larger, more representative samples must be tested. The Rowe cell was developed for this purpose.

8.3.2. Rowe cell (hydraulic oedometer)

This apparatus (Figures 8.12 and 8.13) was developed for consolidation testing of soils where the Casagrande oedometer would be inadequate and the differences in design of the two apparatuses reflect the limitations to be eliminated. In the Rowe cell, samples are typically 250 mm diameter by 100 mm thick, and this means that samples can only be obtained economically from shallow deposits. The apparatus should only be used where it is acknowledged that the conventional Casagrande oedometer is likely to give unreliable and uneconomical data. Soil can be compacted directly inside the apparatus to provide thoroughly remoulded

Table 8.3. The effect of soil fabric on the consolidation parameters (after Rowe (1968))

Stratum description	σ'_v (kN/m^2)	C_v (m^2/year)			Ratio*
		CO	RC	In situ	
Multifissured weathered shale	96–575	5.6 (h)	2973	6132	1100
Boulder clay with occasional thin silt or sand layers	48–192	1.8	—	1858	1050
Lake clay coarsely layered with well-defined silt and sand	240–479	1.1 (v) 2.6 (h)	—	836	320
Estuarine clay with vertical roots	24	9.3 (v)	186–1858 (v)	929	100
Lake clay coarsely layered with fine sand and silt	240–480	2.6 (h)	—	121	47
Estuarine clay, very silty	28	18.6 (v)	465	483	26
Lake clay laminated with occasional silt dustings and a few silt partings	240–480	1.4 (h)	—	11.1	8
Estuarine clay with vertical rootlets	288	0.7 (v)	2.8 (v)	—	4
Uniform boulder clay	—	1.7 (h)	—	5.6	3

CO, Casagrande oedometer (76 mm diameter); RC, Rowe cell (254 mm diameter); in situ, from an in situ permeability test; v, vertical drainage; h, horizontal drainage.
* In situ C_v/Casagrande oedometer C_v.

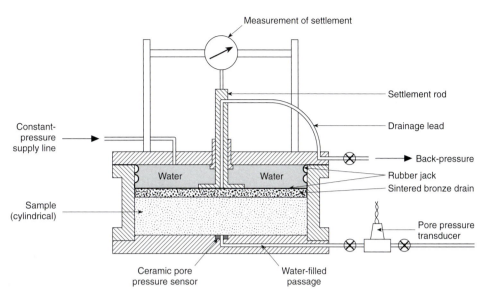

*Figure 8.12. The Rowe cell
(hydraulic oedometer)*

Figure 8.13. The Rowe cell

samples. The ability to achieve full saturation of samples (by applying a back-pressure) is particularly advantageous.

In the Rowe cell the specimen is enclosed within a cell base, a cell body and a convoluted rubber jack, which sits on top of the soil. A hollow vertical spindle is attached to the rubber jack and it passes through the cell top (so that porewater can pass in or out of the sample via the spindle). A total stress increment is applied to the rubber jack by a constant-pressure supply, and the resultant consolidation settlement is recorded by a dial gauge or a displacement transducer which follows the movement of the drainage spindle. Consolidation can be monitored directly by measuring the porewater pressure on the underside of the sample, if drainage is only from the upper surface.

Various drainage conditions can be applied: vertical only (one-way or two-way) or radial only (inwards to a central sand drain or outwards to a peripheral porous lining) (Figure 8.14). In addition to measurement of coefficients of consolidation in both the vertical and horizontal directions the Rowe cell can be used to determine the coefficients of horizontal and vertical permeability directly and under the correct effective stress level (Section 7.2.4).

8.3.3. Analysis of results

When the oedometer was first developed, the apparatus and technology did not permit accurate measurement of pore pressure, which is the true parameter for consolidation. However, it was feasible to measure settlement accurately, and this is a measure of the average degree of consolidation of a sample (equation 8.16). Unfortunately, this produces a complication in the analysis of the oedometer data, since to define the degree of consolidation the ultimate settlement of the soil (i.e. 100% consolidation) has to be known. This would require the test to be continued for a very long period of time (because the time factor for 100% consolidation approaches infinity) and it would be necessary to wait until negligible further settlement was observed. In addition, many soils exhibit some ongoing settlement (or creep) for a very long time after pore pressures have become zero. Hence analytical methods were derived to compare theoretical and experimental consolidation curves and to identify common points that do not require the cessation of settlement or the precise identification of 100% consolidation values. If the consolidation theory is truly applicable to the real behaviour of a saturated soil, the time factor for the actual soil at these points of similarity can then be equated to precise theoretical values of the time factor. The two commonly used analytical constructions are: the Taylor method, where the settlement is plotted against the square root of elapsed time; and the Casagrande method, where the settlement is plotted against the logarithm of elapsed time.

Taylor's method utilizes the fact that for up to approximately 60% consolidation the theoretical consolidation equation can be represented as

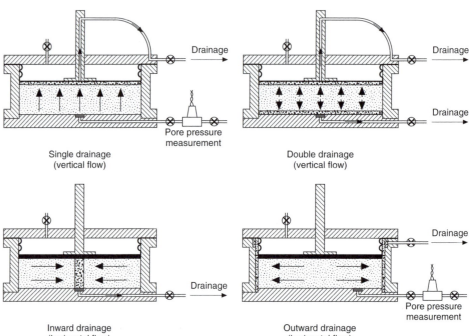

Figure 8.14. Drainage conditions in the Rowe cell

Single drainage
(vertical flow)

Double drainage
(vertical flow)

Inward drainage
(horizontal flow)

Outward drainage
(horizontal flow)

$T_v = \pi U_v^2/4$ (equation 8.14), i.e. the plot of U_v against $\sqrt{T_v}$ for this consolidation range is linear. Hence, when settlement is plotted against the square-root of time, the initial part of the relationship should be a straight line — this correlates with the observed behaviour of clay soils during consolidation. This fact is then used to produce a method for identifying a common point (i.e. 90% consolidation) on both the theoretical and the actual consolidation curves. The 90% consolidation point is a reasonable value to use since it represents a large portion of the consolidation stage, and with the Casagrande oedometer it is usually achieved in a relatively short time (within a working day).

From the theoretical analysis of consolidation the time factor at 90% consolidation is 0.848 (see Table 8.1). If the initial straight-line portion of the consolidation curve was extended to 90% consolidation (i.e. $U_v = 0.90$) the corresponding value of $\sqrt{T_v}$ would be 0.798. The root of the actual time factor divided by this value is 1.15. Hence to define the 90% consolidation point for the actual data the reverse process is used (Figure 8.15). The time factors (theoretical and experimental) are then equated, $T_v = 0.848 = C_v t_{90}/H^2$, and this equation is solved for C_v. To determine the 100% consolidation point (for the determination of the settlement parameters) the 90% settlement value is simply scaled up and a horizontal line is drawn to intersect the settlement curve.

The Casagrande method of analysis takes into account the fact that many slow-draining soils exhibit long-term, on-going compression (and thus settlement) after all the pore pressure has dissipated. This settlement, or creep, is characterized by a logarithmic relationship (i.e. settlement is proportional to the logarithm of elapsed time). Hence settlement is plotted against the logarithm of time to produce a characteristic form of graph, which contains a well-defined point of contraflexure and a linear portion at large values of time (Figure 8.16). The 100% consolidation point is defined as the intersection of the tangent to the consolidation curve through the point of contraflexure and the long-term linear relationship.

The theoretical time factor for 100% consolidation is infinity (in practice a value of 2.0 or 3.0 is an acceptable approximation), and so C_v cannot be determined by equating time factors for theoretical and actual 100% consolidation. Instead, the 50% point is chosen. This is located by dividing the settlement resulting from 100% consolidation in half and finding the corresponding point on the consolidation curve. The theoretical time factor for this point is 0.197, and C_v is thus obtained from the relationship $T_v = 0.197 = C_v t_{50}/H^2$.

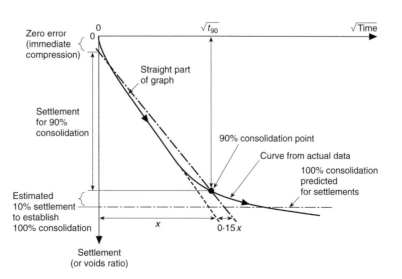

Figure 8.15. Taylor's construction for analysing consolidation data

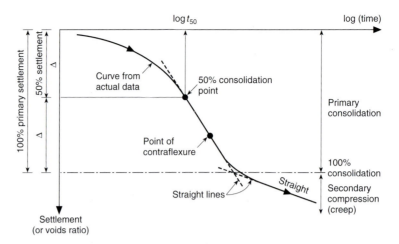

Figure 8.16. Casagrande's construction for analysing consolidation data

As both of the foregoing methods of analysis utilize settlement values, it is essential to correct for any initial compression of the soil. In the Taylor method this is simply done by extrapolating the straight-line portion backwards to time zero and redefining the zero consolidation settlement point. For the Casagrande plot the correction is based on the parabolic form of the initial part of the consolidation curve. Points are selected whereby the consolidation times are in the ratio of 4 : 1 and the difference between the settlement values is taken as the amount of settlement that should have occurred at the first of the two points. Several pairs of values are selected to give an average corrected zero (for settlement).

The coefficient of volume compressibility m_v is calculated for each pressure increment in a consolidation test using equation (8.20). To determine the compression index C_c the results from several pressure increments are combined in plots of voids ratio or settlement against logarithm of pressure (equation 8.18 and 8.19 apply). Typical values of consolidation parameters are given in Table 8.2.

8.4. Compressibility behaviour

In general, when a soil is subjected to an increase in compressive stress (e.g. due to a foundation load), the resulting soil compression consists of immediate consolidation and creep components. The quantities that are of importance are:

- the value of the total anticipated settlement
- the rate at which settlement will occur
- the time required for the majority of the settlement to be completed.

8.4.1. Immediate settlement

The application of load to clay soils causes immediate elastic compression and lateral deformation. The vertical component of the load produces immediate settlement, which for partially saturated soil is due mainly to the expulsion or compression of gases. With saturated, slow-draining soils immediate settlement effects are assumed to occur under undrained conditions, the settlements being due to the lateral displacement of the soil. Equations for estimating the immediate settlement of a loaded area of width B founded on a slow-draining soil are usually of the form

$$\rho_I = \frac{pB(1 - \nu^2)I_F}{E} \tag{8.21}$$

The contact pressure p is usually taken as uniform, whether the loaded area is flexible or rigid. Poisson's ratio for the soil ν is usually taken as 0.5 (i.e. the fully saturated case). The modulus of elasticity E is obtained from the stress–strain graphs from triaxial tests or from empirical relationships (e.g. $E = 400c_u$ to $1500c_u$ (Tomlinson, 1975)). The influence factor I_F accounts for the shape of the loaded area, the thickness of the soil deposit and the dissipation of the vertical load with depth.

For free-draining soils the consolidation is instantaneous, and immediate and consolidation settlements occur contemporaneously. In fact, for cohesionless soils, it is standard practice to use settlements (rather than strength) as a design criterion. Settlements or allowable bearing pressures are defined on the basis of semi-empirical relationships from the standard penetration test, Dutch cone test and plate-bearing test (see Tomlinson, 1975; Burland & Burbridge, 1985).

8.4.2. Consolidation settlement

For one-dimensional consolidation the cross-sectional area of the soil is constant, so that the initial volume V_0 is equal to A_0H_0 and $\Delta V = A\,\Delta H$ where ΔH is the settlement. Hence, from equations (8.19) and (8.20) the consolidation settlement is calculated using either the compression index C_c (for normally consolidated soils) or the coefficient of volume compressibility m_v (for overconsolidated soils or for small pressure increments applied to normally consolidated soils):

$$\Delta H \equiv \left(\frac{H_0}{1 + e_0}\right) C_c\,\Delta\log(\sigma_v') = p_c \tag{8.22}$$

$$\Delta H \equiv H_0 m_v\,\Delta\sigma_v' = p_c \tag{8.23}$$

For deep deposits of soil the ground that undergoes a significant stress increase corresponds to a depth below the load of about three times the breadth of the loaded area, and only compression of material within this depth is calculated (unless there is some obvious highly compressible stratum at greater depth). If the ground comprises discrete layers then the compression of each layer is calculated separately and the values are summed to give the settlement directly below the loaded area. If the ground is uniform it is split into four or five horizontal layers and the compression of each layer determined. The vertical stress change at the centre of a layer is usually calculated using charts (produced by Fadum (1948) from elastic theory), or by assuming that the area over which the load is carried increases in size in some simple way with depth (e.g. each horizontal dimension increases by 1 m for every 2 m of depth increase).

The parameters C_c and m_v are defined in terms of one-dimensional consolidation (this is the deformation assumed in Terzaghi's consolidation theory), which is the actual case in the apparatus used to determine the settlement parameters. However, as pointed out by Skempton & Bjerrum (1957), an element of soil beneath a loaded area of finite extent undergoes lateral deformation as a result of the applied loading. Thus the initial increase in pore pressure, which subsequently dissipates and gives rise to consolidation settlement, is generally less than that which occurs in one-dimensional deformation. For a saturated soil loaded in the oedometer the initial pore pressure increase is equal to the applied overburden, but for saturated soil loaded under 'triaxial conditions' the pore pressure is given by the Skempton pore pressure equation (see Chapter 6, Section 6.1.5). Thus for fully saturated soil

$$\Delta u = \Delta\sigma_v = \Delta\sigma_1 \text{ (oedometer)}$$

$$\Delta u = \Delta\sigma_3 + A(\Delta\sigma_1 - \Delta\sigma_3) \text{ (actual case)} \tag{8.24}$$

The amount of settlement is directly related to the magnitude of the pore pressure that dissipates, so that

$$\Delta H = \text{constant} \times \Delta u \equiv C\,\Delta u \tag{8.25}$$

Thus, the oedometer predicted settlement is

$$\Delta H_{\mathrm{oed}} = C \, \Delta\sigma_1 \qquad (8.26)$$

and the actual field settlement would be

$$\Delta H_{\mathrm{field}} = C(\Delta\sigma_3 + A \, \Delta\sigma_1 - A \, \Delta\sigma_3) \qquad (8.27)$$

For constant $\Delta\sigma_1$,

$$\frac{\Delta H_{\mathrm{field}}}{\Delta H_{\mathrm{oed}}} = A + \frac{(1-A)\Delta\sigma_3}{\Delta\sigma_1} \equiv \mu \qquad (8.28)$$

or

$$\Delta H_{\mathrm{field}} = \mu \, \Delta H_{\mathrm{oed}} \qquad (8.29)$$

The correction factor μ is dependent on the pore pressure parameter A and the relative dimensions of the loaded area and the depth of soil beneath it, as indicated in Figure 8.17.

8.4.3. Secondary compression

Terzaghi's theory assumes that compression is due entirely to a change in effective stress brought about by the dissipation of excess pore pressure. However, experimental data show that for many slow-draining soils compression does not cease when the excess pore pressure has dissipated to zero, but rather it continues at a gradually decreasing rate under constant effective stress (see Figure 8.16). For organic material the rate of this ongoing settlement can be appreciable. The continued compression is termed secondary compression or creep (the consolidation due to dissipation of pore pressure is termed primary consolidation), and it is thought to be due to gradual rearrangement of the soil particles into a more stable configuration (i.e. a decrease in the voids ratio). A further cause is the gradual lateral displacements that take place in thick clay layers subjected to shear stress.

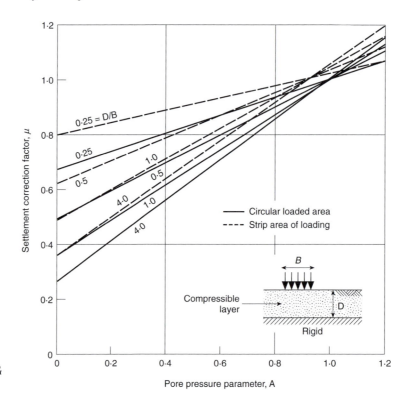

Figure 8.17. Settlement correction factor (Skempton & Bjerrum 1957)

Secondary compression continues over long periods of time. In the consolidation test it becomes apparent towards the end of the primary consolidation stage — the thinness of the sample means that excess pore pressures are soon dissipated and it may appear that secondary compression only occurs after primary compression has been completed. However, in the field situation the large dimensions and long drainage path lengths mean that considerable time is required before the excess pore pressures dissipate. During this time creep is also taking place, so that when primary compression is complete the rate of secondary compression may be relatively small. Hence the terms 'primary' and 'secondary' compression should be seen as rather arbitrary divisions of the total settlement. The magnitude of secondary compression is greater in normally consolidated clays than in overconsolidated soils, and in some clays and peats the secondary compression may completely mask the primary part.

In the oedometer test the rate of secondary compression is defined by the slope C_α of the final part of the consolidation curve when settlement is plotted using the logarithm of time:

$$\Delta H_{\text{creep}} = H C_\alpha \, \Delta \log t = H C_\alpha [\log t_2 - \log t_1] = p_{\text{sc}} \qquad (8.30)$$

Typical values are given in Table 8.2.

8.4.4. Preconsolidation pressure

Preconsolidation pressure is the maximum vertical effective stress under which a soil has been consolidated. It is thus the stress at which the settlement behaviour changes from that of an overconsolidated soil to that of a normally consolidated material, and vice versa (see Figure 8.9). This pressure is a very useful guide to limiting the settlement of overconsolidated strata, since the amount of consolidation compression will be small if the final effective stress is less than the preconsolidation pressure.

Casagrande suggested a graphical method for determining the preconsolidation pressure, as illustrated in Figure 8.18. Point P is located at the maximum curvature between points A and B. Two lines are then drawn through P, one tangential to the compression curve (line PT) and one parallel to the stress axis (line PQ). The bisector (PR) of angle QPT is inserted and its intersection with the back-projection (BD) of the linear part (BC) of the compression curve is the approximate value of preconsolidation pressure.

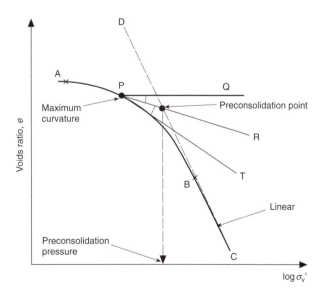

Figure 8.18. Determination of the preconsolidation pressure

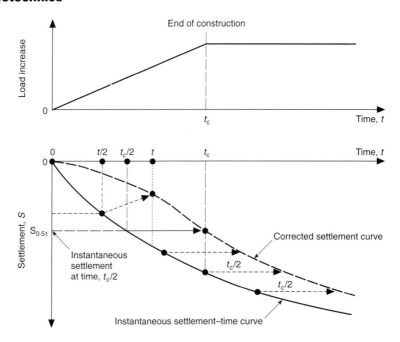

Figure 8.19. Consolidation during construction

8.4.5. Construction aspects

In consolidation theory it is assumed that the load is applied instantaneously. However, in practice most consolidation problems are connected with construction processes extending over several months or even years. Initial excavation works may produce a significant reduction in the applied load, so that the process of consolidation does not commence until the new construction is well advanced. Terzaghi suggested a method (based on his consolidation theory) of correcting the instantaneous curve (Figure 8.19) to account for the gradual application of load.

The total stress increases up to the end of the construction period. The instantaneous settlement–time curve is plotted from where consolidation starts to take place (point O). To obtain the correction it is assumed that up to the end of the construction period (time t_c) the actual amount of consolidation settlement is equal to the amount on the instantaneous curve at $0.5 t_c$. The settlement–time relationship after the construction period is obtained by adding $0.5 t_c$ to all the time values of the instantaneous curve. To obtain the corrected curve during construction, points at different values of time t are chosen and the settlement $S_{0.5 t}$ on the instantaneous settlement curve is determined at each time $0.5 t$. For each point the corrected settlement is calculated as $S_{0.5 t}(t/t_c)$ and the values plotted as the settlements at times t.

8.5. Ground drainage

The time for a given degree of one-dimensional consolidation to take place is proportional to the square of the drainage path length, so that for thick clay layers pore pressure dissipation and settlement can take a very long time to be completed. By inserting vertical drains at fairly close spacings, much shorter drainage paths (horizontal ones) are created, thereby promoting much faster dissipation of pore pressure and accelerated consolidation. Furthermore, natural deposits are often anisotropic, with the horizontal permeability exceeding the vertical value by a significant amount (as demonstrated by Rowe (1968); see Table 8.3). The final amount of consolidation or settlement is unchanged by the presence of the vertical drains, but the rate of consolidation will be enhanced drastically.

8.5.1. Radial consolidation

A three-dimensional form of the consolidation equation may be derived (by a similar process to that presented in Section 8.2.1 for vertical drainage):

$$\frac{\partial u}{\partial t} = C_h \left(\frac{\partial^2 u}{\partial r^2} + \frac{1}{r} \frac{\partial u}{\partial r} \right) + C_v \frac{\partial^2 u}{\partial z^2} \tag{8.31}$$

A solution for purely radial consolidation has been obtained (Barron, 1948) and it is expressed in terms of the degree of radial consolidation U_R and the radial time factor T_R. T_R is equal to $C_h/4R^2$, where C_h is the coefficient of consolidation for horizontal drainage and R is the radius of the equivalent cylindrical block surrounding a drain. Horizontal coefficients of consolidation can be determined from a radial consolidation test in the Rowe cell. The relationship between the radial time factor and the degree of consolidation depends on the ratio n ($= R/r$, where r is the drain radius) (Figure 8.20).

For combined vertical and radial consolidation the overall degree of consolidation is obtained by first calculating the individual amounts of consolidation due to vertical and horizontal water flow (U_V and U_R, respectively) for the given drainage period. These components are then combined using the equation proposed by Carillo (1942):

$$(1 - U) = (1 - U_v)(1 - U_R) \tag{8.32}$$

8.5.2. Vertical drains

In the case of an embankment constructed over a highly compressible clay layer, vertical drains installed in the clay would enable the embankment to be brought into service more quickly (as in the case described in Section 8.1) because the settlement would be completed more rapidly and there would be a more rapid increase in the shear strength of the clay. The original 'sand drains' were formed by sinking boreholes 200–400 mm diameter into the ground and backfilling with a suitable filter sand. However, such drains may lose vertical continuity due to soft clays squeezing into the borehole on removal of the casing or shear of the sand columns as a result of large settlement of the surrounding soil. Prefabricated drains comprising a continuous filter stocking filled with sand (sandwicks) provided a more reliable and cost-effective solution (smaller diameters of about 70 mm could be used). The 'band' drain (consisting of a continuous flat plastic core about 100 mm × 3 mm, corrugated to provide vertical drainage channels and surrounded by a geotextile fabric to act as a filter) is commonly used nowadays. The band drain is installed by attaching one end of the band to the bottom of a

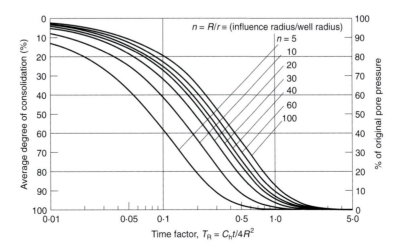

Figure 8.20. Consolidation curves for radial drainage

steel mandrel and by pushing the mandrel into the soil to the depth required. The additional cost of construction is easily offset by the advantages gained.

Since the object of installing vertical drains is to reduce the length of the drainage path, the spacing of the drains is the most important design consideration. Drains are usually arranged in a square or triangular pattern, and for design purposes it is assumed that the zone of influence of each drain extends to the mid-point between columns. The plan area of the zone of influence is equated to a circle having an equivalent radius R. For a square pattern $R = 0.56\,a$ (where a is the centre-to-centre spacing of the drains), while for a triangular pattern $R = 0.525\,a$. The spacing of the drains must obviously be less than the thickness of the clay layers; hence there is no point in using vertical drains in relatively thin layers.

8.5.3. Preloading

If the anticipated settlement of a soil deposit is too high, then preloading of the ground may be used to reduce the amount of post-construction settlement. Preloading consists of subjecting the ground to load (generally in the form of added fill or surcharge) that exceeds that to be applied by the finished construction. The duration of the loading may be several months or years, depending on the desired result and the drainage characteristics of the ground — it is governed by an acceptable defined deformation or rate of settlement of the preloading. Sometimes a system of vertical sand or gravel drains is employed in conjunction with the preloading, in order to decrease the consolidation time and hence permit earlier use of the site. Preloading can be employed for virtually all sites, with some beneficial results, where the ground contains compressible strata, soft silt and clays, or landfill.

Preloading does not increase the rate of drainage of the soil, it works by decreasing the amount of time needed for a certain amount of settlement (Figure 8.21). Assume that the settlement to be completed before construction is undertaken is δ and that under the working load the total anticipated settlement will be S_1. Thus the average degree of consolidation required before construction is δ/S_1 and this will define the time factor T_1 appropriate for this amount of consolidation. However, if the ground is surcharged so that the resultant ultimate consolidation settlement would be S, under these circumstances, a settlement of δ would correspond to an average degree of consolidation δ/S and a requisite time factor T_s. Since $S > S_1$, then $\delta/S < \delta/S_1$, so that $T_s < T_1$. $T = C_v t/H^2$ and C_v and H are constant. Thus $t_s < t_1$.

It is also possible to use preloading to reduce the post-construction rate of observed secondary compression (for further information, see Bjerrum (1972)).

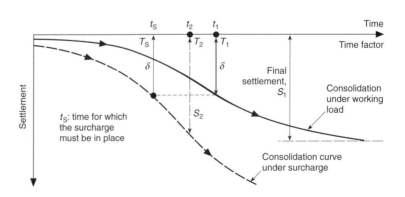

Figure 8.21. The effect of preloading of the ground

9 Slope stability

9.1. Introduction

The first outward sign of slope instability is usually a tension crack in the ground behind the crest of the slope (Figure 9.1), sometimes accompanied by slumping of the soil in front of the crack. The rate of failure is generally relatively slow, developing over a period of hours to days.

The failure of the Carsington Dam in Derbyshire in June 1984 was a classic example of an end-of-construction failure in an earth-fill slope (Skempton & Coats, 1985). The 1.2 km long dam had a maximum height of 35 m and side slopes of 1:3 (upstream) and 1:2.5 (downstream). Highly weathered carboniferous shales were used for the core, with less weathered shales used in transition zones, and unweathered shale used for the shoulders. Finger-type drains were used in both shoulders to control porewater pressures. Construction took place in the three good-weather seasons and was almost complete when a routine inspection showed a tension crack 20–50 mm wide and 220 m long, parallel to, but downstream of, the dam centreline. Within 2 h this had widened, and heaving at the toe of the slope was observed. A slide occurred in the upstream face of the dam over a length of more than 400 m of the crest. Displacements of 10–15 m took place in a matter of days. The Carsington Dam was extensively instrumented with piezometers (see Chapter 11, Section 11.4) in the core, foundation and downstream zones. Instrument readings taken during the sliding showed marked increases in pore pressures, both in the upstream parts of the core and in the foundations.

The slide was brought to a halt with the placement of a berm at the toe of the slide. When the area was stabilized, the main slide came to rest, having moved up

Figure 9.1. A tension crack in an embankment

to 13 m along the slip surface. In places the dam crest had settled 10 m, and elsewhere toe heaves of 2.5 m were recorded. The main embankment, which cost of the order of £15 million, was almost completely destroyed in the slide.

The main slip surface passed through the core of the dam and into the undisturbed natural clay (Figure 9.2), which had a low strength when intact. Unfortunately, the clay had been subjected to shear failure at some time in its history. The resultant shear surfaces were neither highly polished nor striated, but nevertheless there were distinct planes of weakness. The strength along these pre-existing shear surfaces would have been close to the residual value (see Chapter 6, Section 6.4), represented by an effective friction angle of about 12°, and so these surfaces played a significant part in the failure of the dam.

A survey of UK motorway slopes by Perry (1989) revealed a significant incidence of slope failures in side slopes of both cuttings and embankments. Of the 570 km of motorway surveyed, accumulated lengths of over 17 km of embankment slope and over 5.5 km of cutting slope had failed. The type of slope failure observed varied from distinct slab type to shallow circular type, but most slips had a combination of translational and circular movement.

Natural slopes evolve as a result of natural processes such as erosion and movement over a long period of time. Stable slopes result from the soil having sufficient shear strength to resist gravitational forces and the slope acquiring a suitable geometry. On the other hand, man-made slopes are imposed on nature and for stability they have to be designed for a suitable combination of geometry and strength at their most critical/vulnerable period. In general, slope instability results from the in situ shear stresses exceeding the available shearing resistance of the soil, and hence a thorough understanding of the shear strength behaviour of geotechnical materials is essential for determining the stability of slopes.

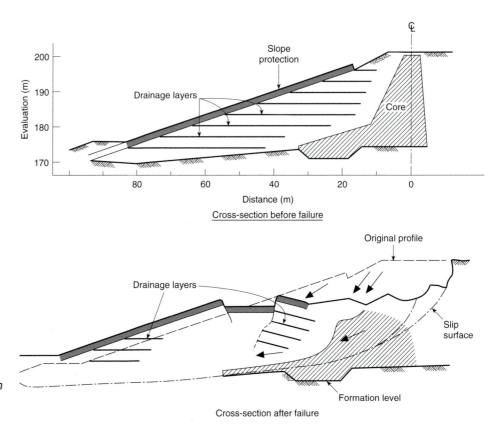

Figure 9.2. A section through the Carsington dam (after Skempton & Coats, 1985)

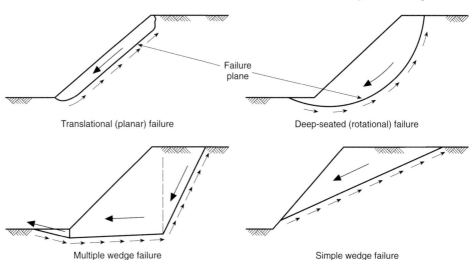

Translational (planar) failure

Deep-seated (rotational) failure

Multiple wedge failure

Simple wedge failure

Figure 9.3. Types of slope failure

9.1.1. Types of slope failure

There are three predominant types of slope instability or failure: translational slide, rotational failure and wedge failure (Figure 9.3). Modes of failure are usually rotational or composite (often with essentially circular slip surfaces); non-circular slip surfaces normally develop because of the influence of ground stratigraphy. Each type of failure has certain general characteristics:

- *A translational slide* is usually relatively shallow (typically 0.5–2 m deep), with the failure surface being more or less parallel to the ground surface. In the survey of motorway slope failures by Perry (1989) it was found that the vertical depth of the failure surface beneath the slope rarely exceeded 1.5 m and ranged from 0.2 m to 2.5 m (Table 9.1). This type of failure is frequently seen on the sides of newly formed cut slopes, where failure of the topsoil cover has occurred. Typically, after failure the slipped material accumulates as a soft heap at the toe of the slope (Figure 9.4).
- *Rotational instability* is usually a deep-seated failure mechanism, with the sliding surface often being more or less circular. There is a large volume of material associated with the movement, and this type of failure is characterized visually by a steep rear scarp slope at the top, an upper plateau which tilts backwards from the slope face and an outwards bulge at the toe (i.e. the bottom of the slope). These features can be seen in Figure 9.5.
- *Wedge failure* represents an intermediate mechanism between the two previous types (Figure 9.6). The failure surface is composed of one or more essentially straight lines, and for a single wedge failure the sliding surface is not parallel to the ground. The shape of the failure surface results from the presence of weak or hard strata orientated in unfavourable directions.

Although the preceding failure types are analysed in different ways, essentially they represent a progression from one extreme to another. For a planar slide the end-effects (i.e. the change in direction of the sliding surface, tension cracks,

Table 9.1. Translational slides in UK motorway slopes*

Depth of failure surface (m)	% of total slip length
0.2–0.4	14
0.5–0.9	35
1.0–1.5	46
1.6–2.0	4
2.1–2.5	1

* From Perry (1989).

Figure 9.4. A shallow translational slide in the face of a new cutting

etc.), have a negligible effect on the overall stability because they are small by comparison to the major part of the planar sliding surface. However, if the depth of the planar sliding surface is increased the end-effects have a significant effect and the complete failure surface moves towards being a compound curve. In the limit, the extent of the planar surface parallel to the ground surface becomes minimal and a curved failure surface results. In uniform soils the sliding surface may be very nearly the arc of a circle in cross-section, but the presence of strata with different shearing characteristics in a stratified deposit causes a slide to adopt a flat-soled shape. The preceding comments relate primarily to a material that is more or less homogeneous in the geotechnical sense (i.e. there is no discrete, continuous formation within the ground that would affect the failure mechanism). The wedge failure is a specific situation in the preceding transition where there is some horizon within the ground that promotes movement of the soil along a straight line (for at least part of the failure surface) (see Figure 9.2).

Figure 9.5. Rotational failure

Figure 9.6. Wedge failure

A category of slope instability that is not covered in this chapter, but which is relevant to the behaviour of slopes of tailings (waste) materials (see Chapter 15), is a flow slide. The term 'flow' is used when the slipping material becomes disaggregated and, although a flow can remain in contact with the surface of the ground it travels over, this is by no means always the case. Earth flows consist of slow movement of softened or weathered debris (such as from the base of a broken up slide mass). Mud flows comprise liquidized clay debris and move very quickly.

9.1.2. Slope instability

Most of the factors that encourage slope failure are fairly obvious and straightforward and can be summarized as:

- Unsuitable geometry: the slope is too steep or too high for the available shear strength of the soil or its geometry changes (by erosion, undercutting, etc).
- Change in the groundwater regime: this results from extreme rainfall and drainage pattern changes, and leads to an increase in the disturbing weight and a decrease in the effective stress (and hence a decrease in the shear strength) in the slope.
- Presence of unforeseen or unrecognized weak planes, bands or layers: in this case the strength of the slope material is overestimated.
- Increase in effective slope height (typically by excavation at the toe).
- Additional imposed loading (deposition of material adjacent to the slope crest or impact loading): this increases the disturbing shear stress.
- Progressive deformation: if the Factor of Safety is too low then localized failures occur, which then cause other areas to be overstressed, and so on.

The shear strength of the soil is the primary stabilizing agent for slopes and the Factor of Safety against instability is often more or less the ratio of the shear strength to the applied shear stress. To assess the stability of a slope it is necessary to identify when, and under what circumstances, it has its lowest Factor of Safety and then select an appropriate analysis to predict the shear strength of the soil at that time. The way in which total stress and pore pressure changes affect the shear strength of soils in the vicinity of slopes was described in Chapter 6 (Section 6.6.4). The influence of the drainage characteristics of soils on slope stability is outlined in Box 9.1.

Both cuts and embankments will increase the shear stress acting on the soil. The Factor of Safety against slope instability is proportional to the ratio of the resistance of the ground to shear stress (i.e. its shear strength), and to the shear stress actually applied. Thus the shear stress and shear strength changes applicable to loading and unloading situations (see Chapter 6, Figure 6.27) can be converted into overall trends of the variation of the Factor of Safety over time (Figure 9.7). Once the critical time for stability has been identified, then the soil shear behaviour illustrated in Figure 9.7 points to the shear strength parameters required to predict, safely and economically, the appropriate shear strength. Appropriate analyses are summarized in Table 9.2. Where necessary, pore pressures may be determined on the basis of steady-state seepage (flownets can be used; see Chapter 7, Section 7.4.3) or by using Skempton's equation (see Chapter 6, Section 6.1.5) with or without consolidation (see Chapter 8, Section 8.2.2). For large embankments (e.g. earth dams), construction will be undertaken over more than one earth-moving season, and so some consolidation and strengthening of slow-draining soils will occur. There will be more than one critical time. The prediction of shear strength changes can be made using the parameters c_{cu}, and ϕ_{cu} (see Chapter 6, Section 6.3.4) or by applying consolidation theory to an effective stress analysis.

An embankment composed of slow-draining material will normally be formed at a moisture content close to optimum. Hence within a compacted layer of slow-draining soil the pore pressure excess will initially be zero (see Chapter 5, Section

Box 9.1. The influence of soil drainage characteristics on slope stability

Slow-draining soils
Slow-draining soils (clays and clay mixtures) can sustain pore pressure excess (both positive or negative) for long periods of time, even decades, provided that they do not include any fabric (such as fissures and laminations) that promotes drainage. If such materials have a high degree of saturation they are likely to remain in an undrained state throughout the construction period, with minimal change in the effective stress (i.e. a negligible change in shear strength from the original state). With time, drainage will occur in the post-construction period, so that the effective stress and shear strength change. Some apparently slow-draining soils may not be so in reality because of the presence of highly permeable layers within the clay matrix (such as laminations or fissures), which open up when the soil is unloaded. This risk is often unappreciated and is one of the most common causes of fatalities in excavations on construction sites.

Free-draining soils
Free-draining soils (sands and gravels) cannot sustain pore pressure excess (either positive or negative), and so the effective stress (and hence the strength) responds immediately to changes in the total stress. If a slope does not fail during construction then it should remain stable in the long-term unless there are subsequent changes in its pore pressure regime or geometry.

Intermediate soils
Silts are not free-draining like sands and gravels, but they are unable to sustain pore pressure excess (either positive or negative) for any significant length of time. Hence, when analysing slopes containing saturated silts it is wise to treat them like sands for unloading situations but like clays for constructions where the total stress is increased.

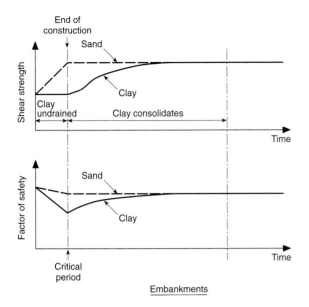

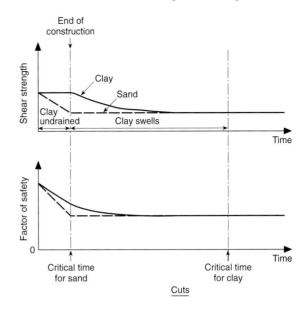

Figure 9.7. Changes in slopes over time

Table 9.2. The selection of shear strength parameters for slope analysis

Slope	Life	Soil type	Critical time	Analysis	Comments
Cut	Permanent	Free-draining	End of construction	c', ϕ' + steady-state seepage	If stable during construction then stable in the long-term
		Slow-draining	Long-term	c', ϕ' + steady-state seepage	May take decades to reach long-term conditions
	Temporary	Free-draining	End of construction	c', ϕ' + steady-state seepage	—
		Slow-draining	End of construction	c_u, ϕ_u	Becomes unsafe if drainage occurs
Fill	Permanent and temporary	Free-draining fill on free-draining ground	During construction	c', ϕ' + original groundwater regime	If stable during construction then stable in the long-term
		Slow-draining fill on slow-draining ground	End of construction	c_u, ϕ_u	Becomes safer if drainage occurs
		Free-draining fill on slow-draining ground	End of construction	c', ϕ' for fill and c_u, ϕ_u for ground beneath	Becomes safer if drainage occurs
		Slow-draining fill on free-draining ground	End of construction	c_u, ϕ_u for fill, c', ϕ' for ground + original groundwater regime	Becomes safer if drainage occurs

5.2.2). As each compacted layer is successively covered, the increase in total stress will cause the pore pressure within the embankment to become positive (Figure 9.8). For the outer edges of the embankment, however, the subsequent loading is minimal, and so on the exterior of the embankment there will be a shallow zone within which the pore pressure excess is initially negative. Thus

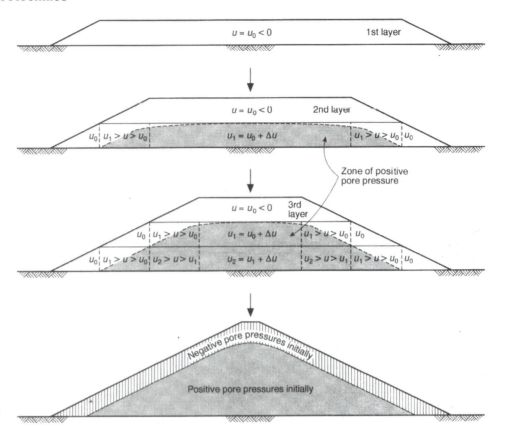

Figure 9.8. Construction pore pressures in an embankment

instability of the surface of such an embankment results from a decrease in the effective stress and a reduction in the shear strength (see Section 9.2).

A natural slope may have reached its present shape by one of several routes and the processes of formation may not be evident from the modified surface features. Where slopes have been affected by instability in the past, the previous mass movements will have produced a first-time slide resulting in considerable strains, usually along a single slip surface. Following large strains, to the residual strength value, this slope will be left with a slip surface along which the lowest possible shear strength exists (the upper surface of the block of clay shown in Figure 9.9 is a pre-existing shear surface). This slope will then be in a metastable condition (only just stable) and liable to undergo progressive creep (incipient failure). Figures 9.10 and 9.11 show the large thickness of road construction at

Figure 9.9. A pre-existing shear surface

Figure 9.10. The slope failure at Mam Tor

Figure 9.11. The road thickness at Mam Tor

Mam Tor (Derbyshire) resulting from long-term, periodic resurfacing of a road located within a zone of incipient slope failure.

The points to be considered when selecting an acceptable Factor of Safety generally fall under two headings: the consequences of failure, and the level of confidence in the quality of the information available. A high Factor of Safety should be chosen where failure poses a risk to life and infrastructure or where slope deformation or movement could affect adjacent structures (even though the slope is stable). For example, lateral movements, toe heave and crest settlements could all affect buried pipes, drains, road surfaces, etc. A low Factor of Safety may be chosen when the risk period is short (as for temporary works) or where only localized, small-scale instability will occur. For cuttings and natural slopes, where a good quality site investigation has been undertaken, a minimum Factor of

Safety of between 1.2–1.5 is normally adequate. Similar values are applicable to embankments, although the overriding concern in the design and construction of these structures may be the need to limit pore pressure increase and minimize settlements.

9.2. Translational slide

This type of failure develops in the face of cuttings and embankments some time after construction of the slope (see Figure 9.4). Initially the pore pressures close to the surface of the slope are low or negative because of either removal of stress (cut slopes) or compaction effects (embankments). With time there is an increase in pore pressure due to infiltration from the surface (primarily rainwater) or a rise in the groundwater towards its steady-state position. This pore pressure change causes the effective stress to reduce and hence the shear strength of the soil decreases.

9.2.1. Analysis

This type of instability has to be analysed using effective stresses, for the shear strength determination, since at failure the shear strength will have reduced significantly from that immediately after construction, and so the original undrained shear strength is not applicable to the analysis. After failure has occurred and the slipped material has been removed, the new face is at the same angle as previously; thus a further failure will occur unless the situation is changed. The translational failure case is sometimes referred to as a 'planar slide' (because of its form) or the 'infinite slope case' (since end-effects are ignored in the analysis).

A typical element within the slope is that indicated in Figure 9.12. The failure surface is essentially parallel to the ground surface. Similarly, the groundwater level is also more or less parallel to the ground surface. Hence for the typical element the normal and shear forces on the vertical faces must be equal and opposite (from symmetry). To assess the stability of the element it is sufficient to compare the disturbing and resisting forces down the slope, since failure would be by sliding along a straight failure surface. Because the normal and shear forces on the vertical faces of the element cancel out, the only disturbing force in the direction of sliding is the component of the weight, $W \sin\beta$. Similarly, the only resisting force in the direction of movement T results from shear strength τ of the soil, and its magnitude is $\tau b \sec \beta$.

The Factor of Safety F for this case is defined as the ratio of resisting force down the slope to the disturbing force down the slope:

$$F = \frac{T}{W \sin \beta} \tag{9.1}$$

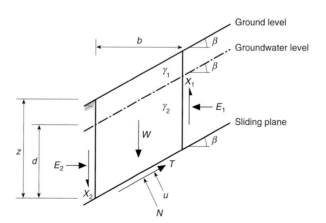

Figure 9.12. A typical element in a translational slide

The shear strength of the soil is predicted using the effective stress form of the shear strength law (see Chapter 6, equation 6.9): $\tau = c' + \sigma'_n \tan \phi'$.

The total normal force N acting on the failure plane is equal to $W \cos\beta$ (because of equilibrium of forces normal to the failure plane). If the mean pore pressure on the sliding surface is u the water uplift force normal to the failure surface is $ub \sec\beta$, so that the effective normal force on the sliding surface is $(W \cos\beta - ub \sec\beta)$. Thus the shearing resistance T can be expressed as

$$T = c'b \sec\beta + (W \cos\beta - ub \sec\beta) \tan\phi' \tag{9.2}$$

The Factor of Safety thus becomes

$$F = \frac{c'b \sec\beta + (W \cos\beta - ub \sec\beta) \tan\phi'}{W \sin\beta} \tag{9.3}$$

This expression apparently depends on the width of the element chosen, and to remove this effect it is convenient to introduce a parameter p, which is equal to the total vertical stress on the failure plane. Thus

$$p = \sum \gamma h = \gamma_{\text{mean}} Z = \frac{W}{b} \tag{9.4}$$

By substituting for W in equation (9.3) an expression is obtained that is independent of the element width:

$$F = \frac{c'}{p \sin\beta \cos\beta} + \frac{\tan\phi'}{\tan\beta} - \frac{u \tan\phi'}{p \sin\beta \cos\beta} \tag{9.5}$$

For translational slides the groundwater level can be approximated to being parallel to the ground surface and also parallel to the failure surface. Hence, because the depth of sliding material is fairly shallow, the groundwater regime in the failing mass can be represented by the flownet shown in Figure 9.13, with flow lines parallel to the ground surface and equipotentials normal to the surface. The pore pressure at any point on the failure surface (e.g. point A) is given by $\gamma_w h_w$. Note that h_w is not the vertical distance of the water-table above point A, but is the hydrostatic head at this point. Since the equipotential through point A intersects the groundwater level at point B, the elevation of this latter point represents the hydrostatic head at point A. From the geometry of the situation an expression for h_w can be obtained:

$$h_w = AB \cos\beta$$

but

$$AB = d \cos\beta$$

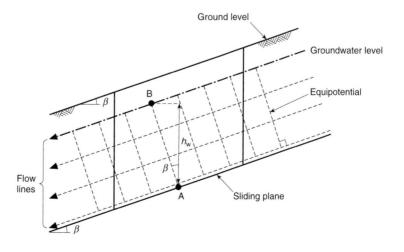

Figure 9.13. A flownet representation of seepage in a translational slide element

so

$$h_w = d \cos^2\beta \tag{9.6}$$

Substituting the preceding expression for pore pressure into equation (9.5) the Factor of Safety can be written in the form

$$F = \frac{c'}{p \sin\beta \cos\beta} + \left[1 - \frac{\gamma_w d}{\gamma_{mean} Z}\right] \frac{\tan\phi'}{\tan\beta} \tag{9.7}$$

9.2.2. Practical aspects

There are several points to observe from equation (9.7):

- The term $c'/p \sin\beta \cos\beta$ is the contribution of the effective cohesion to the Factor of Safety. For typical values of p and β this term is of the order of $0.1c'$ to $0.3c'$ (for c' in kN/m^2). Thus an effective cohesion of only 10 kN/m^2 will give a Factor of Safety greater than unity for the usually encountered slope angles. However, care should be exercised in relying on effective cohesion to provide a suitable Factor of Safety because, even if it has been accurately determined in the laboratory, it can easily be reduced by soil deformation and disturbance. The effective cohesion, where it truly exists, as opposed to being a result of poor-quality soils testing, is a manifestation of the stress history of a soil. Long-term progressive deformation can severely reduce the available effective cohesion. Hence a Factor of Safety in excess of unity should be ensured from the second (frictional) term alone and the contribution of c' should be regarded as an enhancement.
- If the slope is dry or fully drained there is no pore pressure ($d = 0$) and the value of the second term in the equation for the Factor of Safety is equal to $\tan\phi'/\tan\beta$. In other words, it is unrealistic to assume that in the long-term a slope in cohesionless soil will be stable (in terms of a translation slide) if the slope angle is equal to or greater than the effective friction angle of the ground material.
- If the groundwater level is at the ground surface, the second term is $[1 - \gamma_w/\gamma_{mean}]\tan\phi'/\tan\beta$. For typical soils γ_w/γ_{mean} will be approximately 0.5. Thus the effect of full saturation of the slope material is to halve the stabilizing effect of the friction (i.e. the permissible slope angle will be halved approximately). Thus under these conditions a long-term stable slope is not likely to be steeper than 1 vertical to 4 horizontal, approximately. Where the slope angle is small (under residual strength conditions, for example, where ϕ'_r is the residual value), the permissible slope angle will be around $0.5\phi'_r$.
- The preceding point highlights the importance and value of drainage to prevent translational slides in a slope surface and to enable economic slope angles to be utilized. Indeed, an equation may be derived to design the depth of shallow surface drainage systems needed to prevent failure due to planar sliding.

9.3. Rotational failure

In this case the failure surface is curved and penetrates to a significant depth below the ground surface, and hence a considerable volume of material is displaced. This can present problems for structures, objects, etc., located close to the toe or the crest of the slope (Figure 9.14). Most rotational failure surfaces are accurately represented by a circular arc. The rotation means that a steep scarp is formed at the head of the slipped mass and the plateau on top of this mass rotates so that water ponds between the plateau and the scarp (see Figure 9.5). This water

Figure 9.14. Damage to a railway line at the toe of a failed slope

can promote further instability by acting as an additional disturbing force and reducing the shear strength on the sliding surface.

9.3.1. Analysis

Because the motion is rotational, the Factor of Safety F can be defined as the ratio of the total available resisting movement to the total disturbing moment:

$$F = \frac{\text{total resisting moment}}{\text{total disturbing moment}} \equiv \frac{\text{sum of all individual resisting moments}}{\text{sum of all disturbing moments}} \quad (9.8)$$

The resisting moment is provided primarily by the shear strength along the failure surface, while the dead weight of the soil, together with any other extraneous loads, produces the disturbing moment (Figure 9.15). For the failing mass the disturbing moment is the product of the weight of the unstable soil and its lever arm between the point of rotation and the centroid. The problem is that the potential failing mass has an irregular shape, which makes it difficult to calculate its weight and the position of the centroid. If the soil mass is divided into a

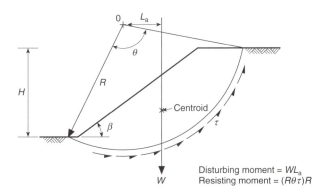

Figure 9.15. Disturbing and resisting moments

Disturbing moment = WL_a
Resisting moment = $(R\theta\tau)R$

collection of regular shapes, each of which has a well-defined centroid, then the moment of each shape can be calculated and all the moments added together. Similarly, if the failure surface passes through a variety of strata or materials of different shear strength characteristics, then the failure circle can be split into a series of arcs, each lying within any one material, such that the shearing resistance of all the material on an arc is defined by one set of shear strength parameters. The resultant shear strength of different portions of the failure surface can be defined in terms of either total stress or effective stress parameters, as appropriate, since either set defines the total available shearing resistance (because the sliding resistance of water in the voids is zero; see Chapter 6, Section 6.1.2). The resisting moment of each arc is calculated and all the moments summed.

To design or analyse slopes a number of different slip surfaces and centres of rotation are assumed and analysed in order to locate the most critical failure system. An indication of the initial location of the critical circle may be provided by the work of previous investigators of slopes, for instance as shown in Figure 9.16. It is often convenient to 'contour' the values of Factor of Safety over the grid of centres. A closed minimum-value contour probably indicates that at least the location of the worst slip circle has been identified. However, there may well be more than one critical slip surface (e.g. an embankment fill could suffer a shallow face failure or a deep-seated slip penetrating into the underlying soil). It is not necessary to find precisely the most critical slip circle, as the accuracy with

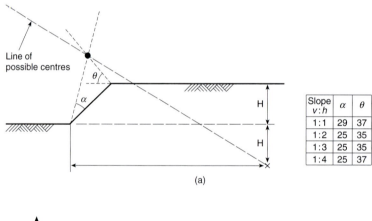

Slope $v{:}h$	α	θ
1:1	29	37
1:2	25	35
1:3	25	35
1:4	25	37

(a)

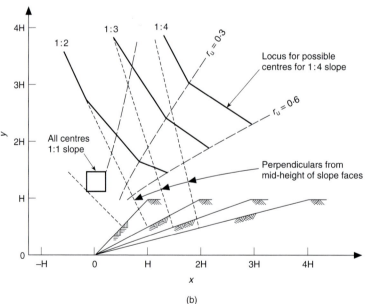

(b)

Figure 9.16. Location of the critical circular failure surface: undrained analysis (after Jumikis, 1962); (b) effective stress analysis (after Mitchell, 1983)

which the Factor of Safety can be expressed is limited (the first place of decimals is reliable, but the validity of quoting the second place of decimals is open to question).

Because of the repetitive nature of slope stability calculations the analysis favours the use of computer programs. In this way it is possible to obtain the Factor of Safety of a trial circle in a matter of seconds, as compared with many minutes when done manually. This speed of calculation is also beneficial in searching for the circle giving the lowest Factor of Safety, where the computer programs will analyse circles over a grid of circle centres and, with the circles passing through chosen points on the slope, over a range of radius values or tangential to particular levels. The programs also permit analysis of complex soil and groundwater conditions, and account can be taken of external loading and seismic effects.

The appearance of a tension crack at the head of a slope (see Figure 9.1) indicates that failure is imminent, and the Factor of Safety is very close to unity. A tension crack reduces the length of the failure surface resisting shear, but it also reduces the disturbing force (because a triangular wedge at the uppermost level of the potentially unstable mass remains stable). Hence it is common practice (since the two foregoing effects more or less cancel out) to ignore tension cracks and shortening of the failure surface in analysis. This approximation is unsafe if the tension crack becomes filled with water, because then an additional horizontal disturbing force will be applied to the unstable mass. While, theoretically, the depth of a tension crack may be in the region of $2c_u/\gamma$ to $4c_u/\gamma$, such a value will be very high, even for the weakest of soils. A more reasonable assumption is that the crack only penetrates to the water-table or at least to the mean seasonal zero pore pressure line. However, it should be remembered that tension cracks are symptomatic of an inadequate Factor of Safety and imminent failure, and they should only be assumed or incorporated in the back-analysis of a slope that is unstable.

9.3.2. The method of slices

The general method of slices involves selecting a failure surface and centre of rotation (for a section of unit thickness). The soil contained within the circular arc is then divided into slices (Figure 9.17) using vertical lines. The slice boundaries are chosen primarily so that the portion of the failure surface at the base of the slice lies entirely within one material (i.e. its shear strength is defined by a unique set of parameters). To maintain accuracy there should be a slice boundary at

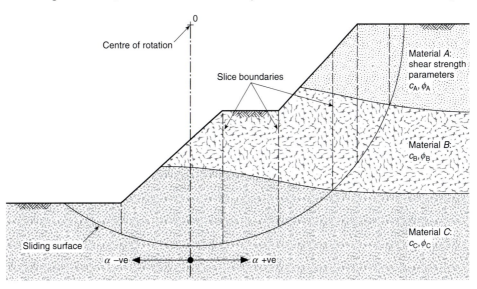

Figure 9.17. Slice boundaries (method of slices)

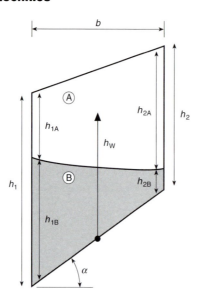

Figure 9.18. A typical slice (method of slices)

every radical change in geometry, and the vertical through the point of rotation should also be used to define a slice boundary as illustrated (this will also ensure that the correct mathematical sign is given to the term α). There is no need to have all slices of equal width, although wide slices should be subdivided to maintain a reasonably uniform distribution of slice size.

After the failure surface has been superimposed on the slope profile, and the slice boundaries drawn in, the relevant parameters for each slice are obtained by scaling off from the diagrams (or by calculation in the case of computer analyses). For each slice the following information is abstracted (Figure 9.18): the angle between the chord at the base of the slice and the horizontal α; the heights of both the vertical faces (or the depth of each stratum within a face); the slice width b; and the mean static porewater head (used to calculate the mean pore pressure u on the slice base). Any external forces present are incorporated into the slice on which they act.

Most slices are approximately trapezoidal, and the centroid of a slice is assumed to lie on the vertical bisector of the slice. Thus the lever arm (of the weight of the slice about the point of rotation) approximates to $(R \sin\alpha)$ (Figure 9.19). Furthermore, the forces that act on the vertical faces of the slices are internal forces, so that their overall sum (and the sum of their moments) is zero and hence they do not appear in the equation for the Factor of Safety. The expression for the Factor of Safety can thus be written as

$$F = \frac{\sum TR}{\sum W \sin\alpha R} \equiv \frac{\sum T}{\sum W \sin\alpha} \tag{9.9}$$

A typical slice will have boundary forces acting on it as shown in Figure 9.20. Unfortunately, the shear and normal forces acting on the slice boundaries cannot be immediately eliminated on the grounds of symmetry. Furthermore, because the failure surface is assumed to be a circular arc at failure there is no relative motion between slices across their vertical boundaries. Consequently the value of the boundary forces on the vertical slice faces are unknown — the situation is statically indeterminate and to obtain working solutions assumptions are made about these boundary forces. The two primary, practical methods of analysis, (Fellenius and Bishop simplified), introduce different assumptions to counter the aforementioned indeterminacy.

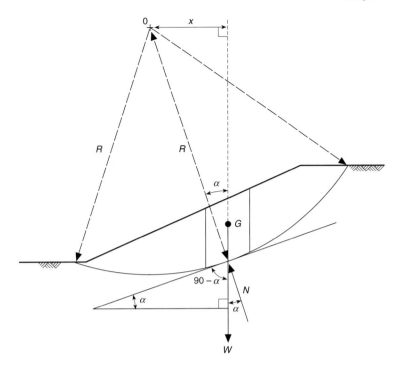

*Figure 9.19. An approximation
for the lever arm of a slice*

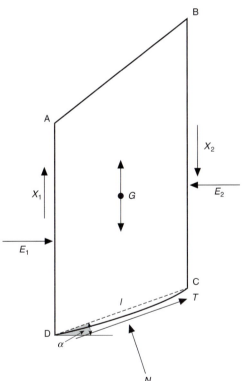

Figure 9.20. Interslice forces

9.3.3. Fellenius method of analysis

In the Fellenius analysis the horizontal and vertical forces on the slice boundaries are assumed to be equal and opposite. This is true if the slice is reduced to the width of a line, but as the width of a slice increases the assumption is patently untrue since the two sides will be very different in size. Thus, if the soil is divided into many slices, as can be done using modern computers, then a reasonably

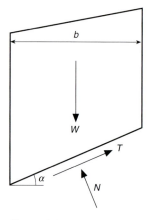

Figure 9.21. The forces acting on a slice: Fellenius analysis

accurate Factor of Safety can be found. However, for manual analysis the number of slices that can realistically be used (i.e. up to 10, approximately) limits the accuracy of the method.

Because the forces on the vertical boundaries are assumed to cancel out (Figure 9.21) the total normal force on the base of the slice is equal to the component of the weight in the direction of the normal force: $N = W \cos\alpha$.

For an *undrained analysis* $\tau = c_u + \sigma_n \tan\phi_u$ and, since the section is of unit thickness,

$$T = \tau(b \sec\alpha) = c_u b \sec\alpha + (\sigma_n b \sec\alpha)\tan\phi_u \equiv c_u + N \tan\phi_u \qquad (9.10)$$

The Factor of Safety is thus given by

$$F = \frac{\sum c_u b \sec\alpha + \sum W \cos\alpha \tan\alpha_u}{\sum W \sin\alpha} \qquad (9.11)$$

The starting point for an *effective stress analysis* is identical to the preceding one, the difference being that the shear strength on the base of a slice is defined in terms of the effective stresses:

$$T = c'b \sec\alpha + N' \tan\phi' \qquad (9.12)$$

All the total forces are as before, but now the effective normal force N' is needed. From the principle of effective stress

$$N' = N - ub \sec\alpha \qquad (9.13)$$

Hence

$$F = \frac{\sum c'b \sec\alpha + \sum(W \cos\alpha - ub \sec\alpha)\tan\phi'}{\sum W \sin\alpha} \qquad (9.14)$$

The failure surface may pass through different materials such that for some sections the parameters c' and ϕ' are applicable while for others the parameters c_u and ϕ_u are appropriate. In this case the expression for the Factor of Safety will be an amalgam of equations (9.11) and (9.14):

$$F = \frac{\left[\sum T\right]_{\text{for slices defined by } c',\phi'} + \left[\sum T\right]_{\text{for slices defined by } c_u, \phi_u}}{\left[\sum W \sin\alpha\right]_{\text{for all slices}}} \qquad (9.15)$$

The Fellenius equation is fairly simple to solve, and yields conservative results (lower than actual Factors of Safety), especially where the slip surface is deep or where the porewater pressures are high. In both these cases the fault lies with the neglect of the interslice forces.

9.3.4. Bishop simplified method of analysis

In this method of analysis the shear forces on the vertical faces of a slice are assumed to be similar in magnitude (but acting in opposite directions) and small by comparison to the other forces, so that they effectively cancel each other out (Figure 9.22). However, the lateral forces E_1 and E_2 still remain as unknowns. Since the Factor of Safety is defined as a ratio of total moments and the lateral forces E_1 and E_2 are internal forces, the latter do not appear in the equation for Factor of Safety. The use of a simple, undrained representation of soil strength is usually inappropriate with a refined method of analysis, and hence the Bishop simplified method is an effective stress analysis.

To derive an expression for the effective normal force, without involving the unknown horizontal forces E_1 and E_2, the vertical equilibrium of the slice is considered — E_1 and E_2 have no component in this direction, but unfortunately the basal shearing resistance does. To obtain a true balance of forces in the vertical direction the actual shearing resistance mobilized on the base of a slice

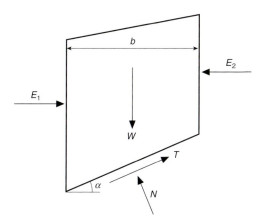

Figure 9.22. The forces acting on a slice: Bishop simplified analysis

(i.e. the ultimate shearing resistance divided by the Factor of Safety), must be utilized. Thus,

$$W = N \cos\alpha + \frac{T}{F} \sin\alpha = (N' + ub \sec\alpha)\cos\alpha + \frac{T}{F} \sin\alpha \qquad (9.16)$$

where $T = c'b \sec\alpha + N' \tan\phi'$ as before (equation 9.12). Hence

$$N' \cos\alpha \left(1 + \frac{\tan\phi' \tan\alpha}{F}\right) = W - ub - \frac{c'b \tan\alpha}{F} \qquad (9.17)$$

When the foregoing relationships are incorporated in the definition of the Factor of Safety and the terms are collated, the final expression contains the unknown Factor of Safety on both sides:

$$F = \frac{\sum m_\alpha [c'b + (W - ub)\tan\phi']}{\sum W \sin\alpha} \qquad (9.18)$$

where $m_\alpha = \sec\alpha / (1 + \tan\alpha \tan\phi'/F)$.

Equation 9.18 has to be solved by iteration, usually started off by substituting $F = 1.0$ in the right-hand side (because the major interest is whether the slope is stable or not) and solving for the new value of F. This value is then substituted in the right-hand side to obtain a more refined value. This iteration converges very rapidly, so little extra computational work is required over the Fellenius method but the accuracy is much greater.

The reason for the accuracy of the results given by the Bishop simplified method lies in the insensitivity of the moment equation to the slope of the interslice forces, as demonstrated by Spencer (1967). It was shown that the accuracy of the Bishop simplified method decreases slightly as the embankment slope, pore pressure and ϕ' increase and as the parameter $(c'/\gamma H)$ decreases. However, the error was usually less than 1%, and the worst combination of these factors resulted in a maximum error in the Factor of Safety of 4%. From his general analysis Bishop (1955) noted that, if the assumption of vertical shear forces cancelling out was not made, the complexity of the analysis was increased drastically, for surprisingly little refinement in the Factor of Safety. Feasible solutions were obtainable with markedly different distributions of vertical shear forces among the slices, but all seemed to give Factors of Safety within a range of about 1%. Hence the Bishop simplified analysis is recommended as the best, most accurate, most effective hand calculation method for calculating the Factor of Safety in non-uniform slope situations.

9.3.5. Non-circular failure surfaces
Where non-homogeneous soil profiles exist, such as with layered strata, a non-circular (but still rotational) slip may develop. Janbu (1973) used the method of

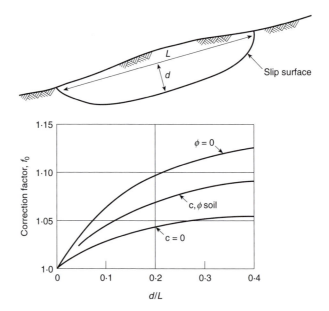

Figure 9.23. Janbu's correction factor for non-circular failure surfaces

slices, together with overall horizontal equilibrium as the stability criterion, to obtain the following expression for the average Factor of Safety:

$$F = \frac{\sum m_\phi [c'b + (W + dX - ub)\tan\phi']}{\sum (W + dX)\tan\alpha} \tag{9.19}$$

The term m_ϕ is equal to $\sec^2\alpha/(1 + \tan\alpha \tan\phi'/F)$ and dX (equal to $X_1 - X_2$) is the resultant vertical interslice force. Janbu suggested a simplified analysis whereby an initial estimate of the Factor of Safety F_0 is obtained using equation (9.19), and assuming that interslice forces can be ignored so that dX is everywhere equal to zero. Hence the unstable mass must be divided into narrow slices. The actual Factor of Safety F, which includes the influence of the interslice forces, is subsequently calculated from the expression

$$F = f_0 F_0 \tag{9.20}$$

where f_0 is a correction factor related to the depth of the slip mass and the soil type (Figure 9.23).

9.4. Stability charts for rotational failure

For a completely uniform slope, wherein the shear strength is defined by one set of shear strength parameters and a simple groundwater regime exists, there will be a unique critical slip surface for which the Factor of Safety can be calculated. If a series of such slopes is analysed, for different values of the relevant parameters, predictive stability charts can be produced, thus eliminating the need for lengthy calculation.

In man-made embankments the material will be more or less uniform, but it is rare for the material of natural slopes and cuts to be completely uniform. However, in many cases the variation in shear strength parameters is limited so that the mildly heterogeneous material can sensibly be ascribed appropriate mean strength parameters. Discontinuities and abrupt changes in the groundwater regime do not exist in most slopes. Furthermore, many real slopes can be represented with sufficient accuracy by a simplified cross-section with a constant face angle and the ground at the toe and at the crest horizontal (Figure 9.24). Consequently, stability charts, derived for entirely uniform conditions, may actually be used to give a very rapid, reasonably accurate estimation of the Factor of Safety of a slope (the process is equivalent to performing an analysis of many

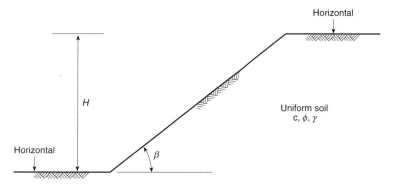

Figure 9.24. The idealized slope

slip circles) and thereby indicate whether further, more precise analysis and detailed investigation is necessary.

9.4.1. Undrained analysis

From inspection of the Fellenius equation (9.11) for undrained analysis of a slope, it can be seen that the parameters determining the Factor of Safety are:

- slope geometry (slope angle and height)
- weight of the failing mass
- shear strength on the failure surface (depends on the shear strength parameters and the total stress level).

Increases in the slope angle β, slope height H and bulk unit weight γ cause a reduction in the Factor of Safety, while increases in the shear strength parameters (c_u and ϕ_u) cause an increase. The Factor of Safety can thus be shown to depend on three non-dimensional terms: $c_u/\gamma H$ (the stability number), $\tan\beta$ and $\tan\phi_u$.

In 1937, Taylor published a slope analysis, for simple homogeneous slopes, based on the friction-circle method. This mathematical solution is precise and thorough (the accuracy of the analysis has been confirmed by subsequent computer analyses of undrained slopes), especially in the determination of locations of critical circles. This method made it relatively easy to analyse simple slopes in homogeneous soil, within which no seepage is occurring. An extensive parametric study (in terms of the previously mentioned non-dimensional groups) was then undertaken and the results were made available in the form of relatively simple charts (tables are also available). The Taylor stability chart (Figure 9.25) is really only applicable to the undrained analysis of slopes having the simple cross-section indicated in Figure 9.24.

The chart represents an equilibrium diagram whereby any point defines the combination of the slope angle, mobilized friction angle and mobilized stability number that gives equilibrium. The Factor of Safety is incorporated in the diagram by using the mobilized cohesion and friction (c_m, ϕ_m), rather than the ultimate values:

$$\tau_{\text{mobilized}} \equiv \tau_m = \frac{\tau_{\text{ultimate}}}{F} \equiv \frac{\tau}{F} = \frac{c_u}{F} + \frac{\sigma_n \tan\phi_u}{F} \equiv c_m + \sigma_n \tan\phi_m \qquad (9.21)$$

The critical circle for steep slopes passes through the slope with the lowest point on the failure arc at the toe of the slope. When the friction angle ϕ_u is taken as zero, ϕ_m is also zero and, theoretically, the most critical circle (for slope angles less than 53° approximately) extends to infinite depth for uniform soils. This latter scenario is unrealistic and occurs because a $\phi_u = 0$ analysis is equivalent to assuming that the soil shear strength is constant for infinite depth. When the mobilized frictional angle exceeds zero the shear strength of the soil increases with depth and there is a limit to the depth to which the failure surface can penetrate. The depth factor D is used to represent the presence of a strong

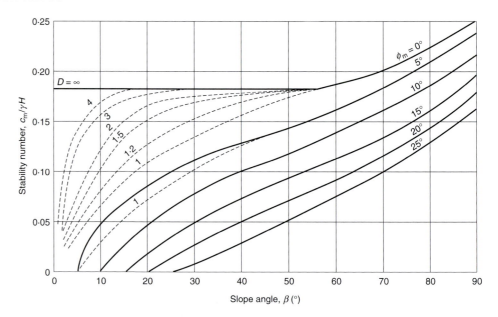

Figure 9.25. Taylor's stability chart

stratum (actual or imaginary) below which the failure circle cannot penetrate. D is equal to the depth of this stratum (below the crest of the slope) divided by the height of the slope. The shear strength of a uniform soil will automatically increase with depth because the effective stress will increase with depth (see Chapter 6, Section 6.6.4). Hence for real slopes it is unrealistic to take the case of D equal to infinity. In the absence of any information that indicates the maximum depth of penetration of a failure surface, an upper value of D of 2.0 may be assumed (see Section 9.4.2).

The Taylor chart is usually used in one of two ways:

- to select a suitable slope for a given Factor of Safety
- to determine the Factor of Safety of a given slope.

The first case is straightforward, because if the Factor of Safety is known then two of the parameters of the chart ($c_m/\gamma H$ and ϕ_m), can be calculated immediately and hence the third parameter β is read off the chart.

In the second case the unknown Factor of Safety F appears in two parameters. However, there is a unique solution, and so the problem is solved by induction (i.e. by systematically 'guessing' values of F). The process is continued until a Factor of Safety is found for which the values of the three parameters correspond to one point on the chart. The process is made easier to undertake by defining separate Factors of Safety for cohesion and friction (F_c and F_ϕ, respectively), such that $c_m = c_u/F_c$ and $\tan\phi_m = \tan\phi_u/F_\phi$.

The analysis is started by assuming that F_ϕ is unity. Thus ϕ_m is known and a point on the chart is defined from which the corresponding value of $c_m/\gamma H$ ($= c_u/F_c\gamma H$) can be read off. Hence the value of F_c required for equilibrium can be calculated. If this value of F_c is not equal to unity then the correct solution has not been obtained and the process must be repeated with a revised starting assumption for F_ϕ. This process is continued until the calculated value of F_c is equal to the assumed value of F_ϕ. At this stage the unique solution ($F = F_c = F_\phi$ has been obtained. To make the analysis rapid, values should be read off the chart by eye (rather than by scaling) and hence a better method is simply to select a range of values of F_ϕ (to cover the likely range of values of F) and determine the corresponding values of F_c. Then a graph of F_ϕ against F_c is plotted (a smooth curve should be drawn) and the point where F_ϕ is equal to F_c is located.

9.4.2. Effective stress analysis

Inspection of the Bishop simplified equation for the Factor of Safety indicates the functions and parameters that determine the value of the Factor of Safety in an effective stress analysis:

- slope geometry
- weight of the failing mass (as for undrained analysis)
- shearing resistance along the failure surface (depends on the effective strength parameters and the effective stress level).

For uniform conditions the pore pressure regime may be represented sufficiently accurately by the pore pressure ratio r_u. This parameter is the ratio of the pore pressure at a point to the total vertical stress at that point: $r_u = u / \sum \gamma h$. The advantage of representing the porewater regime by r_u is that, while the actual pore pressures on the failure surface may cover a wide range, the pore pressure ratio for most of the slip surface will be more or less constant. This is particularly true where there is a high water-table. The approximate range of r_u is from zero (very low water-table or low degree of saturation) to 0.55 (water-table at, or very close to, the ground surface). The Factor of Safety can thus be expressed as a function of four non-dimensional parameters: $c'/\gamma H$, $\tan\phi'$, $\tan\beta$ and r_u.

In a similar way to Taylor (1937), Bishop & Morgenstern (1960) analysed (using a computer) a variety of uniform slopes for practical ranges of the foregoing non-dimensional parameters. They found that the data could be represented by the following simple relationship between the Factor of Safety and the pore pressure ratio r_u:

$$F = m - nr_u \tag{9.22}$$

The stability coefficients, m and n, are analogous to Taylor's stability number and are dependent on the values of the non-dimensional parameters mentioned previously. Hence a series of charts was produced that give values of m and n for a practical range of $c'/\gamma H$, slope angle and ϕ' values (Figure 9.26). Charts are presented for discrete values of $c'/\gamma H$, with the interval between these charts being such that interpolation can be applied without major loss of accuracy. This method has been used for many years as a means of assessing the stability of slopes. It is sometimes claimed that the use of Bishop & Morgenstern charts is impaired by the small range of effective cohesion values that they cover and additional charts have been derived (O'Connor & Mitchell, 1977) (Figure 9.27). However, the reliability of contemporaneously mobilizing high values of c' over a large area, for long-term stability with relatively low Factors of Safety, is highly questionable (see Section 9.2.2).

In the Bishop & Morgenstern charts the depth factor D denotes the region within which the lowest point of the sliding surface occurs (e.g. for $D = 1.0$ the depth range is from zero to $1.0H$, for $D = 1.25$ the depth range is from $1.0H$ to $1.25H$). This approach for presenting the data for m and n is adopted because for given values of $(c'/\gamma H)$, $\tan\beta$ and ϕ' the depth to which the failure circle penetrates is dependent on the groundwater regime, or r_u (compare with the $\phi_u = 0$ case in Taylor's charts and the use of the depth factor D; see Section 9.4.1). The higher the water-table (i.e. the greater the value of r_u), the deeper is the failure surface. The Bishop & Morgenstern charts indicate an important point in this respect — even with the highest possible water-table ($r_u \approx 0.6$) a depth factor of 1.50 is sufficient to cover all cases (for sustainable c' values). Therefore, for undrained analysis of uniform slopes it is not necessary to consider a D value greater than 2.0, even with a $\phi_u = 0$ analysis.

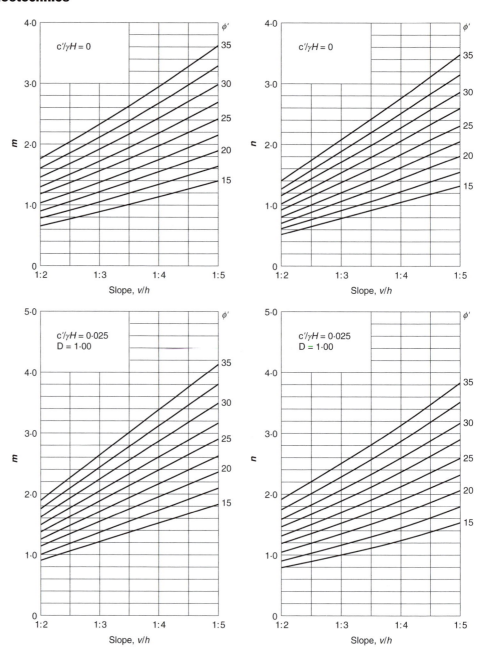

Figure 9.26. Bishop & Morgenstern stability coefficients

The first step in determining the Factor of Safety of a slope is to calculate $c'/\gamma H$ and identify the two charts that bracket this value.

For each value of $c'/\gamma H$:

- read off m and n for $D = 1.0$
- calculate the Factor of Safety from $F = m - nr_u$
- repeat the preceding steps for charts with other D values (if available)
- select the lowest predicted value of the Factor of Safety (this is the minimum for this particular $c'/\gamma H$ value).

Finally, interpolate, linearly, between the two preceding selected values to obtain the Factor of Safety for the actual value of $c'/\gamma H$ for the slope.

The form of the charts makes them eminently suitable for determining the Factor of Safety of a given slope. However, in order to design a slope for a given Factor of Safety then an iterative induction process has to be undertaken wherein

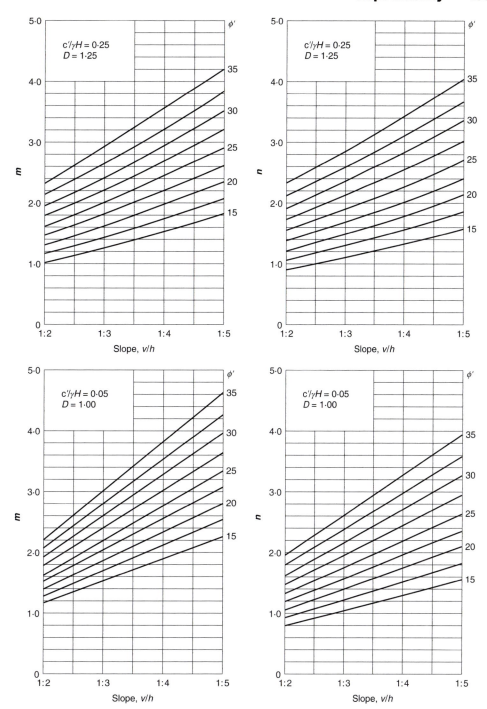

Figure 9.26. continued

the Factors of Safety of a range of likely slope angles are determined and the data interpolated for the specified value. Alternative slope stability charts, which lend themselves to speedy determination of an acceptable slope for a given Factor of Safety, have been produced by Mitchell (1983) (after the style of Hoek & Bray (1974)). An example is given in Figure 9.28.

Selection of an average value of r_u for a slope can be on the basis of past experience or it can involve detailed calculation (e.g. estimation of r_u values for the various parts of the slip surface and determination of an area-weighted mean). For preliminary stability analysis the average pore pressure ratio for the whole of the slope may be taken as $(1 - d/H)N$, where d is the depth of the water-table in

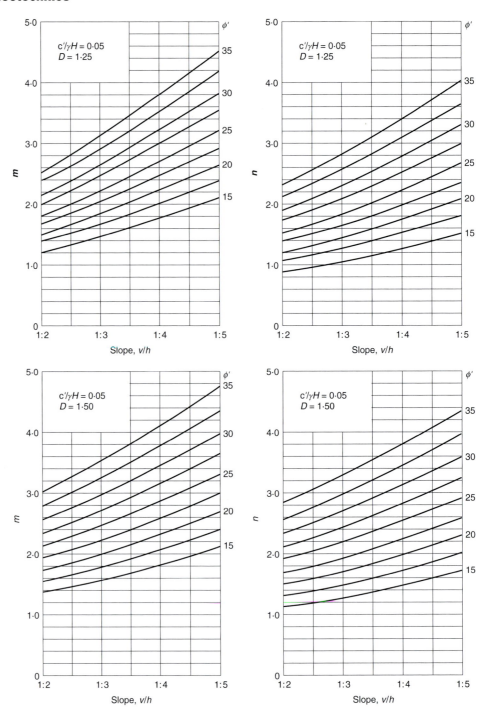

Figure 9.26. continued

the plateau behind the crest and H is the height of the slope. N varies with the slope angle (it is 0.39, 0.44, 0.47 and 0.51 for gradients of 1 vertical and 2, 3, 4 and 5 horizontal, respectively). An alternative way to assess r_u is to use the charts derived by Bromhead (1992) from a systematic, parametric study of seepage in slopes. These charts relate to slopes formed with two stratigraphies: a slope overlying an impermeable bedrock; and a slope underlain by a permeable soil that provides underdrainage of the slope. A different approach has been provided by Barnes (1995) who published slope stability charts for uniform slopes in which the pore pressure conditions were represented directly by a simple water-table. The form of the water-table is horizontal in front of the toe and behind the crest,

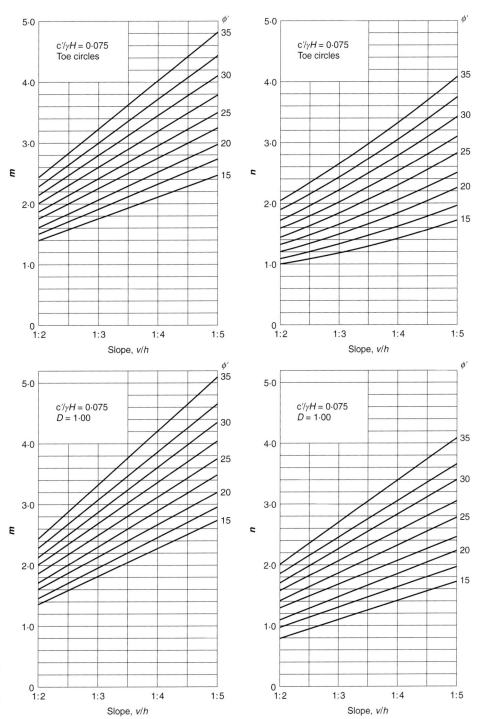

Figure 9.27. Extended Bishop
& Morgenstern stability
coefficients (after O'Connor &
Mitchell, 1977)

and inclined within the slope according to the depth of the water-table below the crest.

9.5. Wedge failure

In wedge failure the slip surface is composed of one or more straight lines that are not parallel to the ground surface. Wedge failures occur due to the presence of weak strata or the interface between materials of significantly different strengths lying in a direction that encourages sliding to occur. A classical example of slope failure by a multiple wedge mechanism was the failure of an earth embankment dam at Chingford (Skempton, 1990).

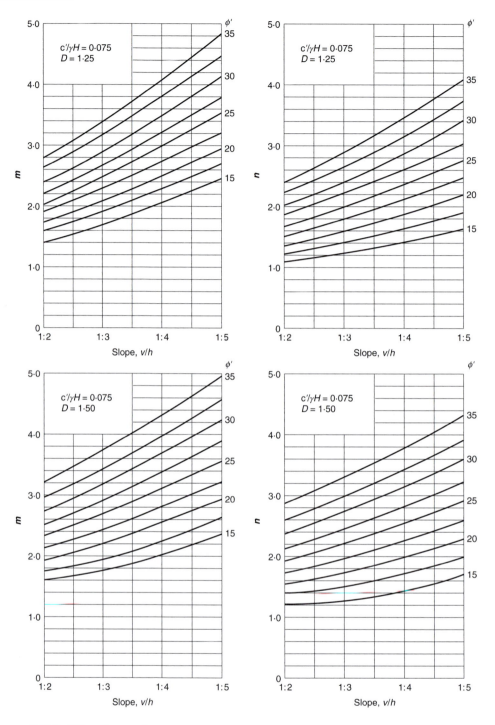

Figure 9.27. Continued

9.5.1. Simple wedge

For a simple wedge, wherein all the soil mass slides in a unique direction and the slip surface only passes through one material, the determination of the Factor of Safety is straightforward because it is simply the ratio of the ultimate shearing resistance along the failure surface to the force acting down the sliding plane, as in the case of the translational slide (see Section 9.2.1). In the absence of externally applied forces the only disturbing force is the component of the weight of the wedge in the direction of the sliding (Figure 9.29). Thus

$$F = \frac{T_{ult}}{W \sin\alpha} = \frac{\tau L}{W \sin\alpha} \tag{9.23}$$

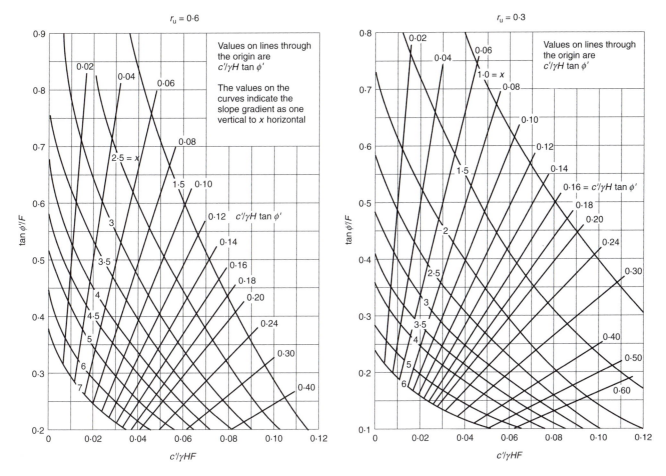

Figure 9.28. An alternative slope stability chart (from Mitchell, 1983)

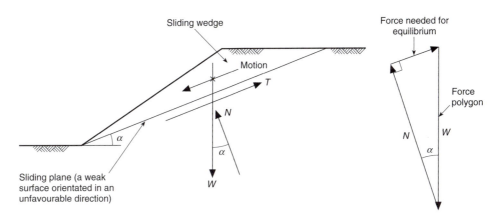

Figure 9.29. Simple wedge failure

The ultimate shearing resistance T_{ult} may be expressed in either total stress or effective stress terms according to the situation being considered:

$$T_{ult} = c_u L + L\sigma_n \tan\phi_u = c_u L + N \tan\phi_u \qquad (9.24)$$

or

$$T_{ult} = c'L + L\sigma'_n \tan\phi' = c'L + N' \tan\phi' \qquad (9.25)$$

Furthermore,

$$N = W \cos\alpha \qquad (9.26)$$

and

$$N' = N - (\text{mean pore pressure on the base of the wedge})L = u_m L \qquad (9.27)$$

In Figure 9.29 the indicated shearing resistance T must be the mobilized value:

$$T \equiv \frac{T_{ult}}{F} \equiv T_m \qquad (9.28)$$

For a total stress analysis

$$T_m = \frac{c_u L}{F} + \frac{N \tan\phi_u}{F} \equiv c_m L + N \tan\phi_m \qquad (9.29)$$

Likewise, for an effective stress analysis

$$T_m = c'_m L + N' \tan\phi'_m \qquad (9.30)$$

9.5.2. Multiple-wedge analysis

If the failure plane consists of a series of straight portions, or if it passes through different materials, then the sliding mass is divided into a series of wedges, or thick slices (as in the method of slices). The vertical boundaries of the slides occur where the failure surface changes direction or where the strength properties of the soil on the slip surface change (even if there is no change of direction at this point). In the method of slices, when the sliding mass is subdivided, interslice shear and normal forces must be introduced at the slice boundaries (see Section 9.3.2). In the wedge method of analysis these are conventionally represented by a total interslice force acting at some angle θ to the horizontal (Figure 9.30). At failure, the Factor of Safety equals unity and the full shear strength is mobilized at the interslice boundaries (otherwise sliding failure could not occur), and the relationship between the shear and normal forces is known. However, for other Factors of Safety the relationship between the interslice shear and normal forces can only be conjectured and a definitive value of θ is not known. Hence assumptions are made about this angle, typical assumptions being:

- $\tan\theta = \tan\phi'F$ (this has some logic for a soil with no effective cohesion)
- θ is the angle to the horizontal of the straight line joining the ends of the sliding surface of adjacent wedges (i.e. equal to ω in Figure 9.30)
- θ is the mean inclination of the bases of the wedges (i.e. $0.5(\alpha + \beta)$).

All three assumptions give similar values for the Factor of Safety. The second assumption is perhaps the most popular.

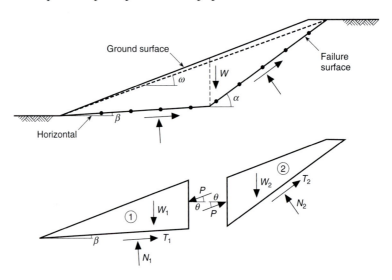

Figure 9.30. Multiple wedge failure

The stability of each wedge can be analysed either by using a polygon of forces or by resolution of forces with the soil shear strength being represented in either total or effective stress terms. With the polygon of forces method, a Factor of Safety is assumed and for each wedge, separately, the inclined force P required for equilibrium is determined (Figure 9.31). If the calculated interslice forces are not equal and opposite across each wedge interface, the correct Factor of Safety has not been assumed. So a new value is assumed and the process repeated until convergence is achieved. Alternatively, the difference between the magnitudes of the interslice forces, at each interface can be calculated and plotted against a range of Factor of Safety values. The point corresponding to a zero out-of-balance force is then obtained by interpolation.

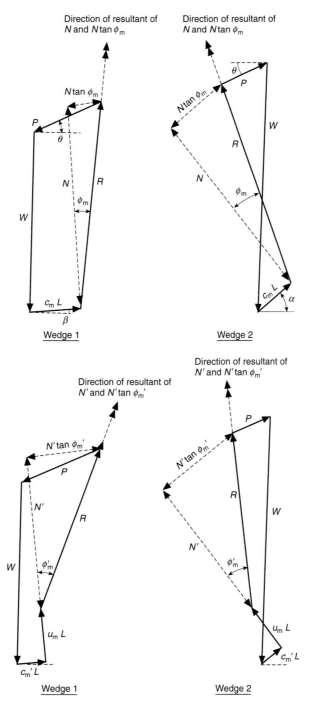

Figure 9.31. The polygon of forces for wedge failure: (a) total stress analysis; (b) effective stress analysis

Solution by resolution of forces involves the derivation of complicated equations, because both the interslice force P and the Factor of Safety F are unknown initially. For example, if wedge 1 (see Figure 9.31) is analysed using total stresses, the resolution of the forces in horizontal and vertical directions gives, respectively:

$$P \cos\theta = T \cos\beta - N \sin\beta \tag{9.31}$$

$$P \sin\theta = T \sin\beta + N \cos\beta - W \tag{9.32}$$

If T is replaced by $c_m L + N \tan\phi_m$ (from equation 9.29) and equations (9.31) and (9.32) are rearranged to eliminate N, then the following relationship is obtained:

$$P[\sin\theta + \cos\theta \tan(\phi_m + \beta)] = c_m L[\sin\beta + \cos\beta \tan(\phi_m + \beta)] - W \tag{9.33}$$

For wedge 2 the equations of equilibrium are:

$$P \cos\theta = -T \cos\alpha + N \sin\alpha \tag{9.34}$$

$$P \sin\theta = -T \sin\alpha - N \cos\alpha + W \tag{9.35}$$

Elimination of the unknown normal force N leads to the relationship:

$$P[\cos\theta + \sin\theta \tan(\alpha - \phi_m)] = W \tan(\alpha - \phi_m)$$
$$- c_m L[\cos\alpha + \sin\alpha \tan(\alpha - \phi_m)] \tag{9.36}$$

Equations (9.33) and (9.36) may be solved by trying different values of the Factor of Safety and comparing the resultant values of P for the two wedges. The actual Factor of Safety is that which gives the same value of P from both equations.

For an effective stress analysis the relationships for resolution of forces are the same as those for total stress analysis (equations 9.31, 9.32, 9.34 and 9.35). However, the mobilized shearing resistance is now defined by equation (9.30):

$$T = c'_m L + N' \tan\phi'_m \equiv c'_m L + (N - u_m L)\tan\phi'_m \tag{9.37}$$

The resultant equilibrium equations, for wedges 1 and 2, respectively, are:

$$P[\sin\theta + \cos\theta \tan(\phi'_m + \beta)] =$$
$$- WL(c'_m - u_m \tan\phi'_m)[\sin\beta + \cos\beta \tan(\phi'_m + \beta)] \tag{9.38}$$

$$P[\cos\theta + \sin\theta \tan(\alpha - \phi'_m)] =$$
$$W \tan(\alpha - \phi'_m)L(c'_m - u_m \tan\phi'_m)[\cos\alpha + \sin\alpha \tan(\alpha - \phi'_m)] \tag{9.39}$$

9.6. Practical aspects

The method used for repairing slope failures typically involves excavating the failed material to below the failure surface, usually with the incorporation of benches (steps in the final, excavated surface) and backfilling with a granular free-draining material such as gravel, brick rubble or crushed rock. Geosynthetics may be included in the backfill to provide reinforcement, as was done on the M4 motorway for reinstatement of a deep-seated failure (Perry, 1989). Sometimes failure of a previously repaired slope has subsequently occurred because excavation did not proceed beyond the original failure surface.

9.6.1. Improvement of stability

There are several alternatives for modifying a slope in order to improve its stability:

- grade to a uniform flatter angle
- place fill at the toe of the slope, creating a step (berm) in the section

- reduce the overall slope height while keeping the profile unchanged
- erect a retaining wall at the toe of the slope and fill behind the wall
- increase the strength of the material forming the slope.

Usually, the simplest and most effective way of improving the shear strength of soil is to reduce the pore pressures in it by some form of drainage.

Shallow, rubble- or gravel-filled trench drains are used extremely widely in slopes to prevent shallow translational slides. The term 'counterfort drain' is sometimes used to describe trench drains that penetrate into solid ground below the soil which is being drained, and which therefore provide some mechanical buttressing effect as well as their effect on the porewater pressure and hence shear strength in the drained soil. Deep drains act to modify the seepage pattern within the soil. While they are often much more costly than shallow drains, they are usually more effective because they remove water, and decrease the porewater pressures, directly at the seat of the problem. Drainage can be a very effective method of slope stabilization, but long-term maintenance will be required if drains are to continue to function.

Rising land prices, the scarcity of good-quality fill and the need to widen existing motorway cuttings and embankments provide incentives to create steep, stable slopes and to utilize marginal fills. Soil reinforcement and soil nailing are techniques for producing a strong, composite soil material by including relatively small quantities of tensile elements at locations within the soil to provide additional, lateral restraining forces (Jewell, 1996). When soil reinforcement is applied to slopes the strengthening elements are usually sheets of man-made materials (geosynthetics), and the shearing interaction between the geosynthetic sheets and the soil provides an anchorage for the potential failure mass. Soil nails are stiff rods 'fired' into the ground to intersect potential sliding surfaces, and stability is enhanced by the tensile and shear strength of the rods.

9.6.2. Rapid drawdown

When a water reservoir is emptied (drawndown) rapidly the porewater regime in previously submerged slopes can affect their stability. For slopes composed of slow-draining material a drawdown period of several weeks can be considered 'rapid'. Hence, during rapid drawdown the soil behaves in an undrained manner and minimal porewater pressure changes result (except for those that occur in direct response to total stress change). The reduction in total stress produces a more or less identical reduction in the pore pressure (see Chapter 6, Section 6.1.5) so that the effective stress does not alter and hence the shear strength of the soil does not change significantly. Hence the term T in the expression (equation 9.9) for the Factor of Safety against rotational failure ($F = \sum T / \sum W \sin\alpha$) does not change. Unfortunately, the 'buttressing' effect of the water in the reservoir (it counteracts rotational movement) is lost (Figure 9.32). The total disturbing moment (the $\sum W \sin\alpha$ term) increases and the Factor of Safety decreases. Rapid emptying of canals may cause failure of banks and walls for the same reason.

9.6.3. Earthquake effects

There are two possible ways in which the Factor of Safety may be reduced to unity or less during an earthquake, even though the static Factor of Safety is greater than unity:

- Earth dams and embankments, when subjected to strong ground motion, will be set into vibration, which will produce inertial forces on the structure. These, together with the pre-earthquake static forces, may bring the Factor of Safety to a value below unity. These forces may be incorporated into analyses that use the method of slices, by adding the appropriate vertical or horizontal force to each slice. The force is usually defined as the product of

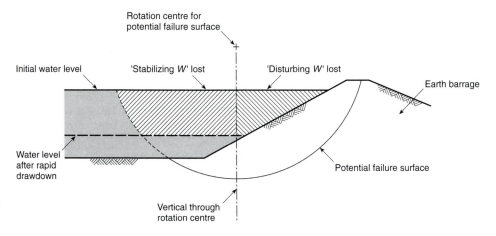

Figure 9.32. Rapid drawdown of a reservoir

the slice weight and the imposed acceleration (expressed as a proportion of the acceleration due to gravity). If the slope is composed of slow-draining soil then these additional forces will increase the total disturbing moment, but the resisting moment will be unchanged because the shear strength does not change instantaneously (no change in effective stress).

- Pore pressures within the soil mass of an earth dam or embankment are likely to increase during earthquake loading. Cyclic loading–unloading tests on slow-draining soils in the laboratory show that pore pressures increase with every cycle. If the applied shear stress is sufficiently high (around more than 60% of the soil shear strength) this increase continues until the sample fails by liquefaction (i.e. the effective stress reduces to zero) (see Chapter 15, Section 15.4.4).

Analytical methods that do not allow for the inclusion of horizontal force components can still be used if the acceleration field is simulated by simply 'tilting the slope'. This allows any analysis technique to be used with an existing slope stability computer program that otherwise has no seismic facilities, but of course it only works for uniform acceleration fields. Each part of the slope is rotated by an angle θ:

$$\theta = \tan^{-1}\left(\frac{g_\mathrm{h}}{1 + g_\mathrm{v}}\right) \tag{9.40}$$

where g_h and g_v are the horizontal and vertical components of the seismic acceleration, respectively. It should be noted that errors in the porewater pressure will arise if a piezometric line is rotated, or if r_u values based on the original, untransformed, section are used.

It has been stated that the usual concept of the Factor of Safety (on shear strength) does not provide a proper assessment of slope stability during strong earthquakes and that in such circumstances slope performance should be measured in terms of relative displacements. Accordingly, it has been postulated that it may be permissible to allow the 'conventional' Factor of Safety to fall marginally below unity during an earthquake, as this state will only exist for a short period.

Sarma (1975) extended a sliding block stability model (originally proposed by Newmark) by incorporating inertial forces and pore pressures to derive an expression for the Factor of Safety of a clay slope under earthquake conditions. The seismic motion was assumed to produce accelerations (and hence inertial forces) in both vertical and horizontal directions. The expression for the Factor of Safety was solved to determine a critical acceleration (i.e. that which would reduce the Factor to unity). The displacements that were induced depended on the amount by which the Factor of Safety fell below unity, but the major factor influencing the magnitude of displacements was the duration of the critical seismic pulse. Displacements were proportional to the square of this duration. Sarma's analysis indicated that both the Factor of Safety and the displacement magnitude were insensitive to the overall direction of the acceleration produced by an earthquake. It was concluded that a stability analysis that assumes that inertial forces are horizontal is subject to only a small error. Hence for this latter case Sarma produced charts that can be used to determine the dynamic Factor of Safety for a given seismic acceleration factor or to determine the critical acceleration for a specific combination of ϕ', pore pressure parameters (Skempton's A and B) and slope angle. Furthermore, it was concluded that, for a dynamic Factor of Safety less than unity but greater than 0.9, the resultant slope instability would lead to only very small displacements during a strong earthquake.

10 Retaining structures

10.1. Introduction

Permanent retaining structures are normally constructed in order to support a vertical or near-vertical face of soil. The structure provides a force which, when combined with the internal shear strength of the soil, is sufficient to maintain stability. Temporary ground support is also provided by retaining walls, which are often required to prevent or limit ingress of water into excavations below the water-table (see Chapter 7, Section 7.5.1). The force acting between an earth-retaining structure and the earth mass it retains is the summation of the lateral earth pressures acting and may be provided by various types of retaining wall (e.g. sheet-piling, sheeting of pits and trenches, bulkheads or abutments, basement or pit walls). The walls may be self-supporting, such as gravity or cantilever concrete walls, or be laterally supported by means of bracing or anchored ties.

The foundations for the principal structures at Sizewell B Power Station were constructed under dry ground conditions, even though their formation level is up to 14 m below the level of the adjacent North Sea. This was achieved by enclosing them within the perimeter of one of the largest diaphragm walls built in the UK, and lowering the groundwater within the enclosure by some 17 m. Diaphragm walls represent a method of constructing walls in the ground in a narrow trench filled with bentonite slurry. This diaphragm wall was 1259 m long, penetrated 56 m deep into the ground and was 0.8 m thick (Howden & Crawley, 1995).

Although, during the past two centuries, the magnitude of earth pressures and analysis of earth-retaining structures has been the subject of much research and analytical development, some of the theories formulated in the 18th and 19th centuries still remain as the fundamental analytical approaches, particularly for sandy soils. While current assumptions about pressure distributions on retaining walls, and the failure surface of backfills, may not be quite those depicted by these early investigators, substantial evidence exists that analysis and design based on their theories give acceptable results for most cases of cohesionless-type backfills. Comparison of measured and theoretical pressures leads to the general conclusion that the Coulomb and Rankine theories for the fully active case yield calculated pressures that are in reasonable agreement with the measured pressures on the upper portion of a retaining wall (Coyle & Bartoskewitz, 1976). However, the data further indicate that the measured pressures on the lower portion of the wall may approach the theoretical pressure that would exist for 'at-rest' conditions. Nevertheless, complete collapse of the older, marginally underdesigned walls is unlikely to occur if the wall can move outwards as a unit and mobilize shearing resistance within the retained fill (Figure 10.1).

The method of installation of a wall is of fundamental importance to the earth pressures generated, for instance; if a wall retaining a fill can undergo some outwards horizontal movement, then internal resistance of the soil is developed so that the pressure on the wall is reduced. On the other hand if backfill is placed behind a wall it will be compacted and this may induce high horizontal pressures behind the structure.

Figure 10.1. Cracks due to outward movement of a masonry retaining wall

A retaining structure can fail to perform satisfactorily because of failure of the structure itself, failure of the soil or unacceptable deformations. The general aspects of stability to be considered are:

- The structure should not overturn. The disturbing moments on the structure should not exceed the restoring moments and the bearing capacity of the ground must not be exceeded.
- The structure should not slide. The horizontal disturbing force must be less than the resistance to sliding on the base.
- The general stability of the soil around the structure should not be impaired. Excessive deformation of the wall or ground should not occur such that adjacent structures or services reach their ultimate limit state.
- Earth pressure should not overstress any part of the structure, to prevent failure of structural members, including the wall itself, in bending, shear or tension/compression.

From a geotechnical point of view analyses are undertaken only to determine the overall proportions and the geometry of the structure necessary to achieve equilibrium under the relevant earth pressures and forces. The disturbing forces to be taken into account are: the earth pressures on the retained side of the wall, loads due to the compaction (if any) of the fill behind the wall, surcharge loads, external loads and water pressures. Resistance to these disturbing forces may be provided by a variety of means such as:

- the pressure from the soil in front of an embedded wall
- base resistance to sliding

- struts and walings in trench excavations
- ground anchorages
- the stability of the building itself in basement construction.

10.1.1. Earth pressure states

Imagine a vertical retaining wall (line AB in Figure 10.2) which is inserted into the ground (without creating any disturbance) and held rigidly in position while the soil to the left-hand side of the wall is removed. At this stage the soil mass behind the retaining structure is exerting pressure on it but there is no lateral movement of the wall — this is the *at-rest condition*.

If the wall is allowed to move outwards, under controlled conditions, the earth pressure acting on the rear of the wall will drop because internal shearing resistance of the soil behind the wall is mobilised with displacement (see Chapter 6, Section 6.1.1). After a relatively small outward movement of the wall the lateral earth pressure falls to a minimum value, which remains constant even if the wall continues to move outwards. This state of affairs arises because the maximum internal strength of the soil has been developed, i.e. it has been brought to failure. The soil behind the wall is now in an *Active state* (the soil is actively causing failure of the wall) and the earth pressure on the wall is known as *Active pressure*. When soil is in an Active state of failure its shearing resistance is in opposition to the disturbing effect of the soil self-weight.

On the other hand, if the wall was pushed inwards the soil behind the wall undergoes a compression. Once again displacement of the soil will cause mobilization of its internal shearing resistance, but this time it will resist inwards movement of the wall, together with the self-weight of the soil. Thus the earth pressure acting on the back of the wall will rise. Eventually the soil will be brought to failure and the earth pressure becomes constant and further movement occurs without change in soil pressures. The soil behind the wall is now in a *Passive state* (the soil has had a passive role in resisting deformation) and the earth pressure on the back of the wall is termed *Passive pressure*.

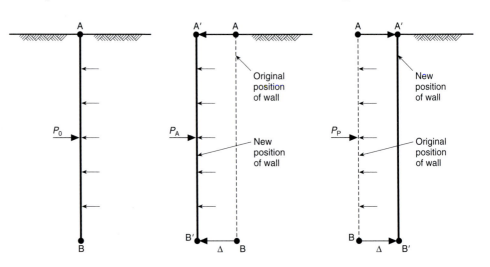

Figure 10.2. The effect of wall movement on earth pressures

Self-weight of soil adjacent to wall generates a lateral force P_0

AT-REST CONDITION
No lateral movement

As the wall moves outwards (the soil actively pushes the wall outwards) the internal shearing resists this movement so that the pressure on the wall is reduced below the original at-rest condition, i.e. $P_A < P_0$

OUTWARD MOVEMENT BY Δ
Soil is *Active* in causing outward displacement

As the wall moves inwards (the soil sits passively as the wall moves) the internal shearing resists this movement, together with the self-weight of the soil, so that the pressure on the wall is increased above the original at-rest condition, i.e. $P_P > P_0$

INWARD MOVEMENT BY Δ
Soil is *Passive* in resisting inward displacement

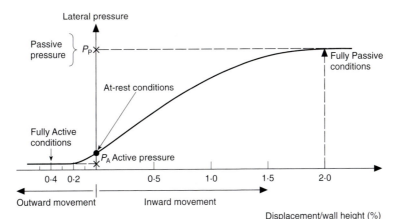

Figure 10.3. The development of earth pressure with wall movement

In general, much greater movement of a wall is required to mobilize the full value of the Passive pressure compared to the small movements required to achieve the full value of the Active pressure. A typical relationship for sands is shown in Figure 10.3. The strain required to mobilize the Active pressure behind a wall will mobilize only a portion of the Passive pressure in front of the wall, so the full amount of Passive pressure should never be relied on.

Most retaining structures are not rigid but bend under the applied loads. The deformation of an earth-retaining structure is important because it has a direct effect on the forces on the structure, the forces from the retained soil and the forces that result when the structure moves against the soil. Gravity walls, cantilever walls, sheet-pile walls and timbered walls are likely to yield sufficiently so that the full Active state is achieved. With some structures where yielding is restricted, such as bridge abutments, propped or anchored basement walls and rectangular culverts, the horizontal pressure acting could be greater than the Active pressure and nearer the at-rest condition.

In general, the lateral earth pressure depends on;

- the bulk unit weight and shear strength characteristics of the fill
- the interaction between the soil and the retaining structure at their interface
- the stiffness of the retaining structure (relative to the fill)
- the presence of additional imposed loading (such as surcharge loads).

10.1.2. Soil behaviour

Solutions to retaining wall problems can be obtained by applying slope stability methods of analysis, but specific solutions have been derived for walls because in the main they have more or less vertical interfaces with the soil. This leads to there being a unique failure surface (regardless of groundwater conditions) for a uniform soil mass, unlike the situation with slopes. Earth pressure theory and the behaviour of earth-retaining structures is dominated by considerations of shear strength. Hence it is vital to understand shear strength concepts and theory.

As with slopes, the changes in loading associated with the construction of a retaining wall may result in changes in the strength of the ground (see Chapter 6, Section 6.6.4). Where the mass permeability of the ground is low, these changes in strength may take some time to occur and consideration should be given to both short- and long-term conditions. Which situation will be critical depends on whether the changes in load applied to the soil mass cause an increase or decrease in soil strength. The long-term condition is likely to be critical where the soil mass undergoes a net reduction in load as a result of excavation, such as adjacent to a cantilever wall. Conversely, where the soil mass is subject to a net increase in loading, such as beneath the foundation of a gravity or reinforced concrete stem wall at ground level, the short-term condition is likely to be critical for stability.

When considering long-term earth pressures and equilibrium, allowance should be made for changes in the groundwater conditions and porewater pressure regime that may result from the construction of the works or from other agencies.

As with slopes, there can be a problem in deciding what analysis to use with clay (slow-draining) strata. The valid period for an undrained, total stress analysis can vary from a few days to several months. Clay strata may be laminated with bands of sand or silt, which can dramatically shorten the drainage time that would apply if the whole soil mass were pure clay. In these situations earth pressure calculations, for temporary retaining systems for excavations, which are made using undrained parameters, will be inadequate and may be unsafe. In such circumstances a drained, effective stress analysis may well be more appropriate than an undrained approach for that particular clay layer. There is a great similarity between walls and slopes in terms of the behaviour of the soil due to loading or unloading in the short-term and long-term states. Hence the information given in Tables 6.5 and 9.2 can be used to select appropriate shear strength parameters for the design and checking of a wall.

Shear-strength-based calculations for stability assume that all elements of the mass are simultaneously at failure — this is not often true. Consider the case of Passive resistance (where the failure surface is long) of a dense sand (or overconsolidated clay). Because of the peaked nature of the stress–strain curve, parts of the failure surface near to the wall will pass through peak resistance before this state is attained by points more distant from the wall. The overall effect is that the mass at failure behaves as if it had a strength considerably less than maximum measured in a compression test on an element — progressive failure occurs. If, on the other hand, the soil is a loose sand or normally consolidated soil, the stress–strain curve has a different form so that all elements on the slip surface should eventually be at failure at the same time. But in this latter case the deformation to achieve failure is so large that a failure analysis is inappropriate. Hence in some cases deformation or serviceability criteria might determine the earth pressures to be used, although more often the question of limitations of deformations is addressed by applying a Factor of Safety to the maximum forces or moments resulting from the soil.

10.2. Earth pressures

To evaluate the magnitude of the forces acting on retaining systems two main lines of approach are adopted: the Coulomb wedge (a force approach) and Rankine theory (defines pressure distributions).

10.2.1. Coulomb wedge analysis

Coulomb considered a rigid mass of soil sliding upon a shear surface which was a straight line set at an angle θ above the horizontal (Figure 10.4). He was well aware that the critical shear surface might not be planar, but noted that a straight failure surface was a good approximation to the real behaviour. If the soil behind the wall is in an Active state then the forces acting on the soil wedge can be arranged in a polygon of forces (of W, T, N and P) as shown in Figure 10.5. This polygon can be closed, so that the soil wedge is in equilibrium, by the addition of forces acting in a variety of ways, each of which corresponds to a different practical situation (e.g. smooth wall (P_A), rough wall (P_B), anchors (P_C)). It is readily apparent that a 'smooth' wall (no shear force acting on its back) must provide a greater horizontal force to maintain equilibrium than must a 'fully rough' wall (maximum possible shear force acting). Thus, the design of a retaining wall on the basis of it being smooth is a conservative approach, but it is not excessively conservative for Active conditions and it is often the most realistic model. This is because it is not the physical roughness of the wall in itself which is the crucial factor in determining whether or not there is a shear force acting on the

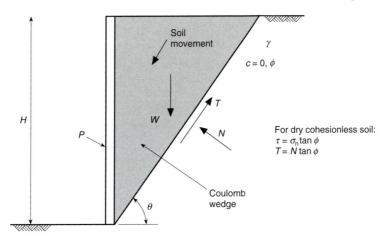

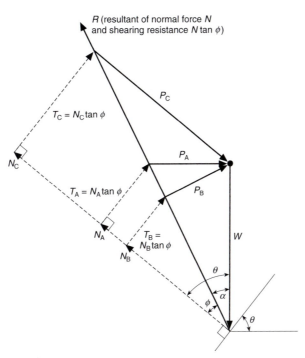

Figure 10.4. The Coulomb wedge

Figure 10.5. The polygon of forces for wedge failure

back of a wall. For a shear force to be generated (rather than it simply being a potential force) there has to be significant sliding motion between the soil and the back of the wall. In many practical situations this movement will not occur.

In general, the total Active thrust P_A on a retaining system is the sum of the thrust due to groundwater P_W and the thrust from the soil skeleton (the force resulting from the effective stresses P'_A):

$$P_A = P_W + P'_A \tag{10.1}$$

Water exerts the same pressure in all directions and thus the thrust that it generates is easily calculated from the groundwater regime (the pore pressure is usually hydrostatic on the back of a wall). The same is not true for the soil skeleton because of its internal shear strength.

An initial assessment of the behaviour of the soil skeleton can be obtained by undertaking an effective stress analysis of a unit thickness of dry, cohesionless soil with bulk unit weight γ. Because the pore pressure is zero $P_A = P'_A$ and $N = N'$ so that, at the point of limiting equilibrium, a force polygon composed of W, N', P'_A and T (the shearing resistance on the base of the wedge) must close (see Figure 10.5). Since the soil is cohesionless,

$$T = \tau \times \text{sliding area} = (\sigma'_n \tan\phi') \times \text{area} \equiv N' \tan\phi' \tag{10.2}$$

The effective normal force N' and the shearing resistance T can be combined to give a resultant force R as indicated in Figure 10.5. From this diagram, at failure (Factor of Safety equal to unity)

$$P'_A = W \tan\alpha \equiv W \tan(\theta - \phi') \tag{10.3}$$

If the wall is of height H,

$$W = \frac{H^2\gamma}{2\tan\theta} \quad \text{(per unit length of wall)} \tag{10.4}$$

Thus

$$P'_A = \frac{H^2\gamma \tan(\theta - \phi')}{2\tan\theta} \tag{10.5}$$

By choosing different values of θ the critical value of P'_A (i.e. the force needed to just prevent sliding of the soil wedge) can be estimated. Alternatively, the critical sliding plane can be identified by differentiating P'_A with respect to θ and setting the derivative to zero. Such an analysis will show that there is one particular critical value of θ:

$$\theta_{\text{crit}} = 45° + \frac{\phi'}{2} \equiv \frac{\pi}{4} + \frac{\phi'}{2} \quad \text{(radians)} \tag{10.6}$$

By substituting the foregoing value of θ_{crit} into equation (10.5) an expression for the minimum force that the wall has to provide (equal and opposite to the Active effective thrust from the soil) is obtained:

$$P'_A = \frac{1}{2}\gamma H^2 \tan^2\left(45° - \frac{\phi'}{2}\right) \tag{10.7}$$

A similar approach (force polygon) may be applied to cases where porewater is present and also when the soil exhibits effective cohesion. Again the critical plane is inclined at an angle of $(45° + \phi'/2)$ to the horizontal. The effective Active thrust is then given by

$$P'_A = \frac{1}{2}\gamma H^2 \tan^2\left(45° - \frac{\phi'}{2}\right) - 2c'H \tan\left(45° - \frac{\phi'}{2}\right) \tag{10.8}$$

If there are pore pressures in the soil behind the wall, the effective Active force and the water thrust must be determined separately and then added together to give the total thrust on the wall as indicated in equation (10.1).

A similar wedge analysis can be undertaken for the case where the soil behind the wall is in a Passive state (i.e. the soil self-weight and shear strength combine to resist the lateral thrust). In this case the base of the critical wedge is inclined at an angle of $(45° - \phi'/2)$ to the horizontal and the effective Passive thrust is given by

$$P'_P = \frac{1}{2}\gamma H^2 \tan^2\left(45° + \frac{\phi'}{2}\right) + 2c' \tan(45° + \phi') \tag{10.9}$$

The main drawback with the Coulomb wedge analysis is that the point of application of the thrust on the wall is not known, and if moments are to be calculated this point of action is needed. The point of application of the water force is known because the water pressure behind the wall increases linearly with depth. From the results of subsequent developments in earth pressure theory it can be assumed that the effective thrust (derived from the force polygon) acts two-thirds of the way down from the top of the wall for cohesionless soil and $0.6H$ down for soil with effective cohesion. To a certain extent the effects of wall friction, inclined wall back and sloping ground surface can be included in Coulomb wedge analysis by analysing a series of wedges with different basal inclinations and then taking the most critical value. However, such analyses are generally rendered unnecessary by the availability of means of calculating earth pressure distributions directly.

10.2.2. Effective earth pressures

Rankine extended earth pressure theory by deriving a solution for a complete soil mass in a state of failure, as compared to Coulomb's solution which considered a soil mass bounded by a single failure surface. The subsequent analysis by Bell (1993) incorporated the effect of cohesion on earth pressures.

Expressions for the earth pressures exerted on a smooth vertical wall by a cohesive–frictional fill (with a horizontal ground surface) can be derived from the Mohr–Coulomb diagram (see Chapter 6, Section 6.1.3 and Figure 6.7) by considering the behaviour of an element of soil immediately adjacent to a smooth wall (Figure 10.6) that has been installed without disturbing the ground (see Section 10.1.2). Prior to installation of the wall, the effective stresses on the element will be σ'_h and σ'_v in the horizontal and vertical directions, respectively. The ratio between the horizontal and vertical effective stress at this stage is defined as the coefficient of earth pressure at rest (K_0), whereby

$$\frac{\sigma'_h}{\sigma'_v} = K_0 \qquad (10.10)$$

A more fundamental definition of this coefficient would be $K_0 = \sigma'_3/\sigma'_1$, but practice is to ratio horizontal and vertical pressures, because for overconsolidated soils with unknown stress history K_0 can only be determined experimentally. For an overconsolidated clay, as the vertical stress reduces some horizontal stress remains 'locked-in', and so K_0 approaches or exceeds unity (values up to 2 have been recorded). For an undisturbed soil at rest the ratio of the horizontal to vertical stress depends on:

- the type of soil
- its geological history
- the temporary loads that may have acted on the surface of the soil
- the topography
- changes in ground strain or groundwater regime.

Therefore the ratio, σ'_h/σ'_v, should not be thought of as a fundamental property of the soil. Two frequently used expressions for K_0 are given in Table 10.1.

In the at-rest state the soil is not failing and so the stress conditions on any plane within the soil must lie below the failure envelope. At this stage the Mohr's circle for the material is something akin to that shown in Figure 10.7. If the lateral

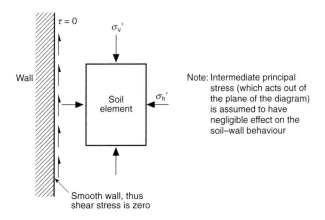

Figure 10.6. An element of soil adjacent to a 'smooth' wall

Table 10.1. Expressions for the coefficient of earth pressure at-rest

Author	Soil conditions	Relationship
Jaky (19??)	Normally consolidated granular material	$K_0 = \dfrac{(1 - \sin\phi')\left(1 + \dfrac{2}{3}\sin\phi'\right)}{(1 + \sin\phi')} \cong (1 - \sin\phi')$
Massarach (19??)	Fine-grained, normally consolidated soil	$K_0 = 0.44 + 0.42\left[\dfrac{PI\ (\%)}{100}\right]$

stress is now reduced (hence it must be the minor principal stress σ'_3) and the vertical stress is kept constant, then the Mohr's circle gradually increases in size, because the shear stress in the soil is increasing. Eventually the circle will just touch the Coulomb failure envelope (see Figure 10.7). At this stage there exists a plane within the soil on which the shear and normal stresses satisfy the failure criterion (i.e. the soil is just on the point of Active failure). The lateral stress cannot be reduced further because the Mohr's circle cannot get any larger. So the minor principal stress at this stage is the minimum lateral stress that the wall must exert to provide a Factor of Safety of unity against failure of the soil. For this situation, from equation (6.14)

$$\sigma'_1 = \sigma'_3\ K' + 2c'\ \sqrt{K'}$$

and

$$K' = \frac{1 + \sin\phi'}{1 - \sin\phi'} \tag{10.11}$$

For a smooth wall in contact with soil that is just on the point of failure under Active conditions

$$\sigma'_1 \equiv \sigma'_v$$
$$\sigma'_3 \equiv \sigma'_h \equiv p'_A \tag{10.12}$$

Thus the effective Active pressure on the wall is obtained from;

$$p'_A = \sigma'_3 = \frac{\sigma'_1}{K'} - \frac{2c'}{\sqrt{K'}} \equiv K'_A \sigma'_v - 2c'\sqrt{K'_A} \tag{10.13}$$

and

$$K'_A = \frac{1}{K'} = \frac{1 - \sin\phi'}{1 + \sin\phi'} \equiv \tan^2\left(45° - \frac{\phi'}{2}\right) \tag{10.14}$$

The resultant Active earth pressure diagram (the variation in lateral effective stress, with depth, on the back of the wall) is illustrated in Figure 10.8. The

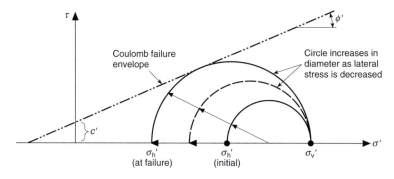

Figure 10.7. The Mohr–Coulomb diagram for Active failure

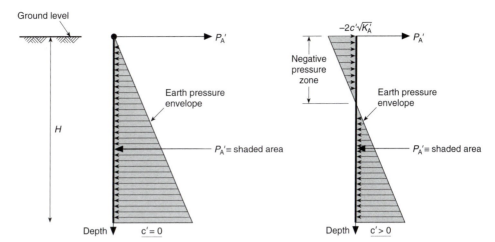

Figure 10.8. Earth pressure distributions

effective active thrust P'_A is the area of the foregoing diagram and, if there are no pore pressures, it is given by

$$P'_A = \frac{1}{2}\gamma H^2 K'_A - 2c'H \sqrt{K'_A} \tag{10.15}$$

This value is exactly the same as that predicted by the Coulomb wedge analysis (equation 10.8). Furthermore, equation (10.13) and Figure 10.8 enable the point of application of the effective active thrust to be determined. They show that the active pressure is a combination of a triangular pressure distribution (the $K'_A\sigma'_v$ term) and a rectangular distribution (the $-2c'\sqrt{K'_A}$ term), hence the comment towards the end of Section 10.2.1.

The preceding approach can also be applied to the determination of a relationship for the Passive case, only in this derivation the wall is pushed into the soil (i.e. the horizontal effective stress σ'_h is increased to bring the soil to failure). Since the vertical stress remains constant, the Mohr's circle is likely to decrease in size initially until σ'_h exceeds σ'_v, after which the Mohr's circle will grow in size until it touches the failure envelope (Figure 10.9). At this stage Passive failure of the soil occurs (both the self-weight and shrear strength have resisted deformation) and the following stress conditions apply, for a smooth wall with a vertical back and horizontal ground surface:

$$\sigma'_1 \equiv \sigma'_h \equiv p'_P$$
$$\sigma'_3 \equiv \sigma'_v \tag{10.16}$$

Thus the effective Passive pressure on a smooth wall is obtained from

$$p'_P = \sigma'_1 = \sigma'_3 K' + 2c'\sqrt{K'} \equiv K'_P\sigma'_v + 2c'\sqrt{K'_P} \tag{10.17}$$

and

$$K'_P = K' = \frac{1+\sin\phi'}{1-\sin\phi'} \equiv \tan^2\left(45° + \frac{\phi'}{2}\right) \tag{10.18}$$

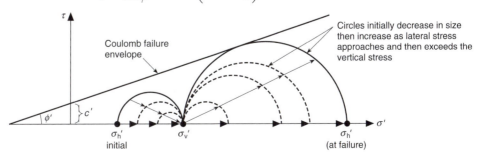

Figure 10.9. The Mohr–Coulomb diagram for Passive failure

The resultant Passive earth pressure diagram (the variation in lateral effective stress, with depth, on the back of the wall) is constructed using equation (10.17). The effective Passive thrust P'_P is the area of this diagram, and is thus given by

$$P'_P = \frac{1}{2}\gamma H^2 \, K'_P + 2c'H \, \sqrt{K'_P} \tag{10.19}$$

10.2.3. Earth pressure analysis

To obtain the total earth pressures (p_A or p_P), any water pressures in the soil behind the wall are added to the appropriate effective stresses. The overall force is then obtained by adding all the small increments of force that are applied over the height of the wall. If a diagram of total earth pressure versus depth is drawn, the overall force is the total area of the diagram. Layered soils are commonly encountered, and in this case the earth pressures immediately above and below the interfaces between strata are calculated using the earth pressure coefficients appropriate to each layer. This procedure is likely to generate abrupt changes in the magnitude of the lateral stress at the interface between strata. In fact, such sudden jumps in lateral pressures, while used in analysis, are unlikely to exist in practice. The overall earth pressure distribution and force acting on a wall is obtained by the process outlined in Figure 10.10, and a typical output is shown in Figure 10.11.

To allow for the fact that the rear face of a wall may not be completely smooth, or vertical, or that the retained fill does not have a horizontal upper surface, the

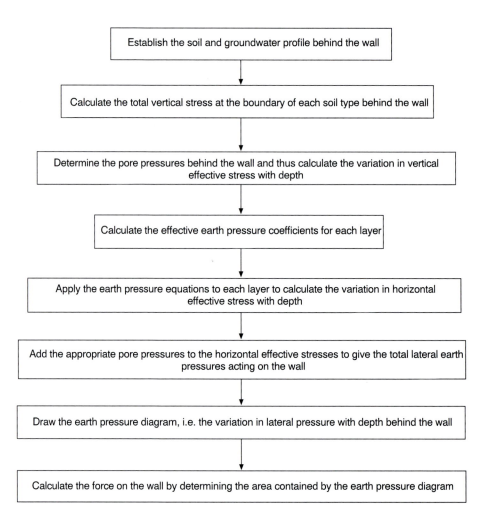

Figure 10.10. A flow chart for the calculation of earth pressures (effective stress analysis)

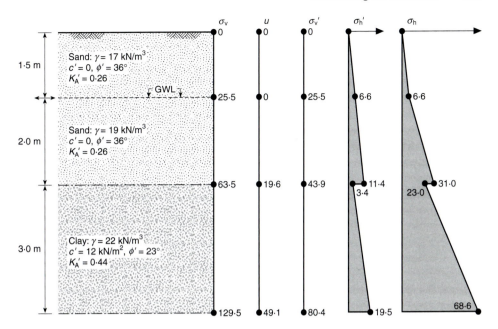

Figure 10.11. A typical earth pressure output

equations for active and passive pressure can be written in general terms:

$$p'_A = K'_A \sigma'_v - K'_{AC} c' \tag{10.20}$$

$$p'_P = K'_P \sigma'_v + K'_{PC} c' \tag{10.21}$$

The coefficients K'_A and K'_P depend on ϕ' and δ (the friction angle for the soil–wall interface), while K'_{AC} and K'_{PC} depend on ϕ', δ and c'_w (the adhesion between the wall and the soil). It must be remembered that the analysis of Active and Passive situations involves considerable simplification of some very complex processes and, because it is possible to postulate a variety of failure surfaces within the soil mass, there are a variety of published earth pressure coefficients. Typical values of the Active and Passive earth pressure coefficients (K'_A and K'_P) are presented in the charts in Figures 10.14 and 10.15. Values of K'_{AC} and K'_{PC} may be calculated from the following approximate expressions:

$$K'_{AC} = \sqrt[2]{K'_A \left(1 + \frac{c'_w}{c'}\right)} \tag{10.22}$$

$$K'_{PC} = \sqrt[2]{K'_P \left(1 + \frac{c'_w}{c'}\right)} \tag{10.23}$$

For the Active failure case there is a depth of soil z_0 over which the Active pressure is theoretically negative:

$$z_0 \approx \frac{c' K'_{AC}}{\gamma(1 - r_u) K'_A} \tag{10.24}$$

If wall friction and adhesion are ignored, equation (10.24) becomes

$$z_0 \approx \frac{2c'}{\gamma(1 - r_u)\sqrt{K'_a}} \tag{10.25}$$

Use of high values of c' for the retained soil results in a significant depth of theoretical negative Active effective pressure (around 1.7 m for an effective cohesion of 10 kN/m^2). However, for field situations where some ground movement is likely to occur, it is unwise to rely, in the long-term, on shearing resistance due to effective cohesion. Consequently, unless one is very confident about a c' value it is recommended that it should be taken as zero. This predicted negative effective pressure (equivalent to a tensile stress pulling the wall backwards) cannot be sustained in an actual field situation, as a tension crack would develop (see Section 10.2.8).

On the Passive side, the amount of movement required to fully mobilize Passive thrust will be large and progressive failure of the soil in the shear plane will occur, so taking c' as zero is the safest, albeit conservative, approach. For normally consolidated clays and for compacted clays the cohesion intercept should be expected to be zero.

10.2.4. Undrained analysis

When a slow-draining soil is subjected to rapid shearing then it may be assumed to behave in an undrained manner, so that a total stress analysis, using the undrained shear strength parameters (c_u, ϕ_u, δ and c_w), is applicable. The form of the resultant earth pressure equations (which predict directly the total pressures on the wall) is the same as for effective stress analysis:

$$p_A = K_A\sigma_v - K_{AC}c_u \quad \text{(for active conditions)} \tag{10.26}$$

$$p_P = K_P\sigma_v + K_{PC}c_u \quad \text{(for passive conditions)} \tag{10.27}$$

and

$$K_{AC} = \sqrt[2]{K_A\left(1 + \frac{c_w}{c_u}\right)} \tag{10.28a}$$

$$K_{PC} = \sqrt[2]{K_P\left(1 + \frac{c_w}{c_u}\right)} \tag{10.28b}$$

A total stress analysis is appropriate for an existing wall that is in the long-term condition because pore pressure equilibrium will have been achieved. A total stress analysis can be used for a wall forming an embankment by containment of fill because, in the long-term, consolidation of the fill will lead to an increase in strength of the soil. Caution should be exercised in applying undrained analysis to temporary excavations because of the danger of unforeseen drainage (see Section 10.1.3). Expansion of the soil in the Active state behind a wall is likely to open up any fabric (fissures, joints) present in the soil, thereby accelerating the softening process. Expansion may also lead to the development of vertical tension cracks, which subsequently fill with water.

For undrained analysis of Active conditions equation (10.26) will predict that there is a deep (5–10 m) zone behind the wall within which the total lateral pressure is negative. The analysis thus provides no information for the design or assessment of this part of the wall. Furthermore, this situation is unrealistic because, under these conditions, a tension crack would form and the soil would soften and expand to fill the crack so that it then exerted some force on the wall. An approach used to overcome this difficulty with undrained analysis is to assume that there is always some minimum Active pressure acting on the wall and that it increases linearly with depth. A common assumption for this minimum total active pressure is $5z\,\text{kN/m}^2$, where z is the depth in metres — this is the *half-hydrostatic* or *minimum equivalent fluid pressure* approach. This is a rule-of-thumb approach and cannot be justified theoretically (it concerns an extremely complex soil–structure problem), but the total stresses predicted are of the same order as would be obtained if it is assumed that the soil in the problem zone expands and porewater suctions dissipate. The process of calculating lateral earth pressures using an undrained analysis is outlined in the flow chart given in Figure 10.14.

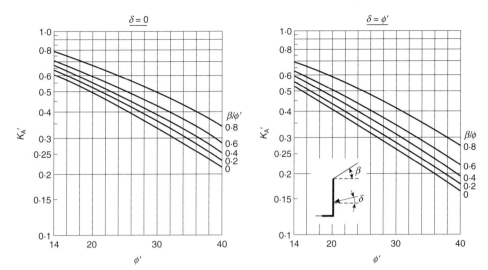

Figure 10.12. Active earth pressure coefficients (after Caquot & Kerisel, 1948)

If negative Active pressure is predicted the value of c_w should be taken as zero, because it cannot be relevant if there is a tension crack. Considerable care should therefore be exercised in use of c_w in a total stress analysis. The concept of a minimum total Active pressure is also applicable to an effective stress analysis (for heavily overconsolidated soils with high c' values the theoretical total Active pressure in the top part of a wall can be negative) and the half-hydrostatic limit can be applied in the same way as for an undrained analysis. Because an undrained analysis is a significant simplification of the actual situation, the values of K_A and K_P are often calculated ignoring wall roughness (i.e. taking K_A equal to $\tan^2(45° - \phi_u/2)$ and K_P equal to $\tan^2(45° + \phi_u/2)$). In many practical situations the soil is virtually saturated, so that ϕ_u approximates to zero and $K_A = K_P = 1$.

10.2.5. Groundwater effects

A rise in groundwater increases the total thrust on a retaining wall by a large amount. Thus an accurate assessment of groundwater conditions is vital for realistically assessing the forces on walls. If the equilibrium level of the water-table is well defined and measures are taken to prevent it changing during heavy rain or flooding, the design water pressures can be calculated from the position of the equilibrium water-table, making due allowance for possible seasonal variations. Otherwise the most adverse water pressure conditions that can be anticipated should be used in design. Where the groundwater regime is modified by drains and this modification is assumed in the design to be permanent, the drains should be designed, installed and maintained so as to function in the intended manner throughout the life of the structure.

A wide range of materials may be used as fill behind retaining walls. Cohesionless soils such as gravels and sands are easy to place and require little in the way of drainage provision. Cohesive materials may result in significant economies, by avoiding the need to import granular materials, but their compaction may be difficult (particularly in confined spaces) and effective drainage is required. Adequate drainage immediately behind retaining walls is important to reduce the water pressure on the wall. Without drainage, the water pressure can exceed the effective earth pressure. For cohesionless backfills of medium to low permeability (2×10^{-5} m/s or less) and for cohesive soils, it is usual to place a drainage layer behind the wall to prevent the build up of hydrostatic pressure (Figure 10.15). The drainage layer is usually vertical, as it is generally impracticable to form a coherent drainage layer at an angle even when fill is placed behind the wall. Various systems may be used for a drainage layer:

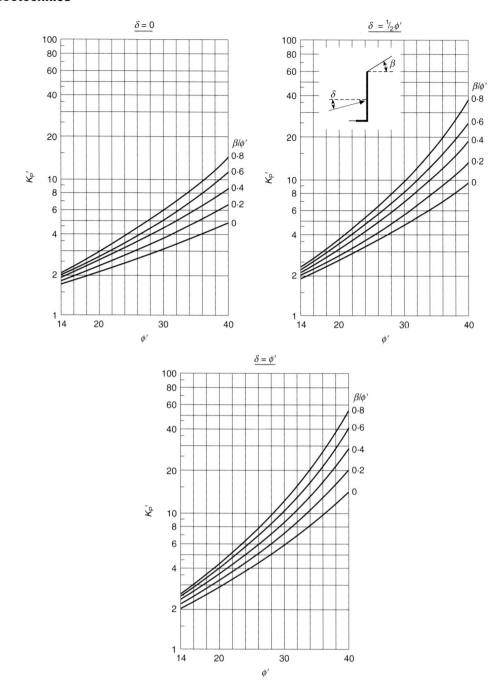

Figure 10.13. Passive earth pressure coefficients (after Caquot & Kerisel, 1948)

- hand-placed permeable blocks
- a graded filter drain, where the backfilling consists of fine-grained material
- a geotextile filter in combination with a permeable granular material
- a geotextile composite (a geotextile filter fixed to one or both faces of a permeable core)
- a blanket of coarse aggregate or clean gravel.

Account should be taken of seepage flow occurring around the structure where a difference in water pressures is likely to exist on opposite sides (e.g. cofferdams, excavations below groundwater level). The distribution of porewater pressures will not be hydrostatic and may be determined from a flow net (see Chapter 7, Section 7.4.3) representing the hydraulic and permeability conditions in the vicinity of the structure. Alternatively, the porewater pressure distribution

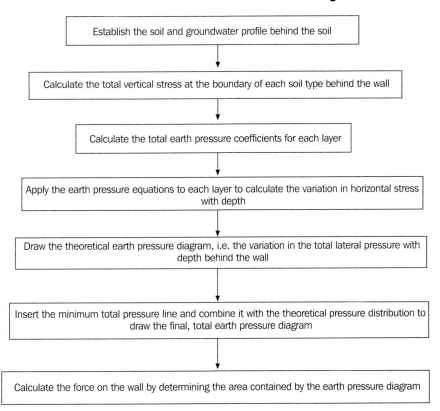

Establish the soil and groundwater profile behind the soil

↓

Calculate the total vertical stress at the boundary of each soil type behind the wall

↓

Calculate the total earth pressure coefficients for each layer

↓

Apply the earth pressure equations to each layer to calculate the variation in horizontal stress with depth

↓

Draw the theoretical earth pressure diagram, i.e. the variation in the total lateral pressure with depth behind the wall

↓

Insert the minimum total pressure line and combine it with the theoretical pressure distribution to draw the final, total earth pressure diagram

↓

Calculate the force on the wall by determining the area contained by the earth pressure diagram

Figure 10.14. A flow chart for the direct calculation of total earth pressures

can be calculated based on the simplifying assumption that the hydraulic head varies linearly down the back, and up the front, of the wall (Burland *et al.*, 1981). A linear dissipation of seepage pressure will give reasonably reliable results for retaining walls where the seepage flow upwards through the Passive zone in front of the wall is free to dissipate laterally as well as vertically.

10.2.6. Wall friction

The introduction of wall friction into earth pressure analysis modifies the stress field at the soil–wall boundary. For rough walls there will be a normal stress and a shear stress acting on the rear of the wall. Normally, the soil in an Active state moves downwards with respect to the wall and so applies a downward shear stress, while in the Passive condition the soil moves upwards. As a result, the principal stress directions are rotated and the plane of failure will be curved (Figure 10.16). However, for Active conditions this rotation only extends a short way back into the soil and most of the ground is unaffected by wall friction. The failure surface curves from the bottom of the wall, becomes straight and reaches

Table 10.2. Typical values of parameters for wall roughness*

Parameter	Value
δ'	ϕ' for fully rough walls
	$20°$ for wall surfaces with fine texture
	$\frac{2}{3}\phi'$ for Active conditions
	$\frac{1}{2}\phi'$ for Passive conditions
c_w	$c_w' = 0$ to $0.5c'$
	$c_w' = 0.5c_u$, but $c_w \leq 50\,\mathrm{kN/m^2}$ (Active case) and $\leq 25\,\mathrm{kN/m^2}$ (Passive case)

* Sources: Terzaghi (1966), Clayton & Mililtsky (1986), BSI (1994).

Figure 10.15. A drainage layer adjacent to a wall

the ground surface at the same angle ($45° + \phi'/2$ to the horizontal), as the failure surface for a smooth wall. For Active conditions stability solutions derived for planar failure surfaces are sufficiently accurate for engineering purposes.

The introduction of wall friction into classical solutions for a Passive planar shear surface causes unrealistically large increases in the predicted forces on the wall. It has been reported that when the angle of wall friction exceeds about 10° the classical solutions (planar sliding surface) give higher passive earth pressure coefficients than those based on a curved surface (Clayton & Milititsky, 1986). It is, of course, the minimum Passive resistance which is required for design.

The wall friction angle and adhesion depend on the shearing characteristics of the soil and the roughness of the wall. Typical values are given in Table 10.2. Wall friction in the retained or Active side should be excluded when the wall is capable of penetrating deeper (e.g. due to the vertical thrust imparted by inclined anchors, or where a clay soil may heave due to swelling as a result of outward movement of the wall). Wall friction on the Passive side should be excluded when the wall is prevented from sinking but the adjacent soil may fail to heave, or when the wall is free to move upwards with the Passive soil zone (as may happen with buried anchor blocks; see Section 10.4.7). Charts giving earth pressure coefficients for rough walls are shown in Figures 10.12 and 10.13. It is vital to remember that these coefficients give the horizontal component of pressure (i.e. the stresses normal to the rear face of a vertical wall).

10.2.7. Extraneous loading

The soil supported by many types of retaining structure may be subject to external loads (e.g. a quay wall may be loaded by dock traffic and freight placed on the fill behind the wall). The various surcharges imposed on retaining structures are classified in Table 10.3. External loads normally act to increase the horizontal stresses on a retaining wall. For wall design BS 8002 (BSI, 1994) requires a minimum surcharge of 10 kN/m^2 to be applied to the surface of the retained soil. Additional surcharge loading should be used in the design to take account of incidental loading arising from construction plant, stacking of

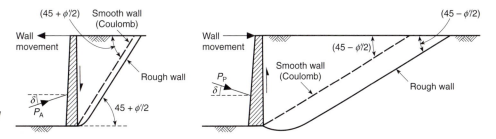

Figure 10.16. The effect of wall shear on the failure surface

materials and movement of traffic both during construction and subsequently, unless the nature or layout of the site precludes this from happening.

A number of methods exist to predict the influence of extraneous loads, but there are few reliable data to confirm the accuracy of these methods. The simplest case is that of uniformly distributed load q placed over the entire ground surface behind a retaining wall. This has the effect of increasing the vertical stress, in the soil behind the wall, by a constant amount q at all levels behind the wall. Thus, for any soil layer, if its lateral earth pressure coefficient is K, the lateral pressure increase at any elevation behind the wall is Kq.

For loads of limited extent the problem is much more difficult. A common method of estimating the horizontal stress on a structure is to use elastic solutions, either to obtain the horizontal stress directly or to derive the vertical stress increase down the back of the wall (e.g. by using the Fadum chart; see Chapter 8, Section 8.4.2) and multiplying by the appropriate earth pressure coefficient. Table 10.4 contains elastic solutions for the horizontal stress increase due to discrete surface loading (take Poisson's ratio ν as 0.5 for incompressible soils (fully saturated clay) and as 0.3 for other cases).

For line loads, Terzaghi & Peck (1967) suggested the simple empirical approach illustrated in Figure 10.17. The line AB is drawn from the line load W_L, at an angle of 40° to the horizontal, towards the wall. An equivalent line load of $K_A W_L$ is applied horizontally to the back of the wall at point B, where the inclined line hits the back of the wall. For loaded areas of limited extent the 40° line is constructed from the centre of the loaded area. If the length of the load (parallel to the wall) is L, and the distance between the back of the wall and the near edge of the loaded area is X, the resultant load on the back of the wall is assumed to be of length $(L + X)$. The horizontal thrust generated per unit length of the wall is then $K_A W_L/(L + X)$.

Many earth-retaining structures are built in a free-standing manner and then fill is placed and compacted behind them (Figure 10.18). However, the compaction plant not only compacts the fill but it also imparts a lateral pressure on the back rear of the retaining wall. In the upper levels of a wall these compaction stresses may greatly exceed the design earth pressures, particularly if heavy plant is used and excessive rolling is applied. For a shallow element of fill the stress path

Table 10.3. Surcharge loading of retaining structures

Loading category	Examples
Uniformly distributed	Roadways, materials stored on ground behind the wall, parking areas, rising ground surface behind the wall
Concentrated	Column footings, anchorage points near quay walls
Line	Strip footings, railway lines parallel to the wall
Dynamic	Safety barriers attached to wall or founded in the ground behind a wall, fenders for boats on quay walls

Table 10.4. Horizontal stress increase due to surface loading

Case	Horizontal stress increase p_h at point N
Point load	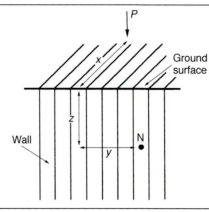 $$p_h = \frac{P}{2\pi A^2}\left[\frac{3B^2}{A^3} - \frac{(1-2\nu)}{A+z}A\right]$$ $$A = \sqrt{x^2 + y^2 + z^2}$$ $$B = \sqrt{x^2 + y^2}$$
Line load parallel to the wall	$$p_h = \frac{2W_L x^2 z}{\pi C^4}$$ W_L = vertical load per unit length $$C = \sqrt{x^2 + z^2}$$
Strip load parallel to the wall	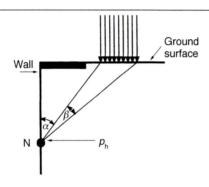 $$p_h = \frac{Q}{\pi}[\beta - \sin\beta \, \cos(\beta + 2\alpha)]$$ Q = vertical load per unit area

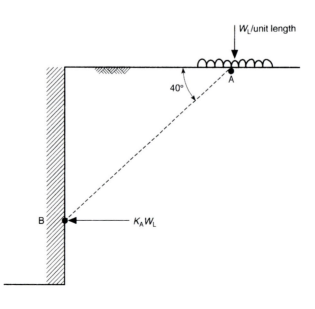

Figure 10.17. The surcharge due to a line load

Figure 10.18. Backfill behind a retaining wall

experienced during compaction can be simplified to that shown in Figure 10.19. Before passage of the roller, $\sigma'_{hi} = K_0\sigma'_{vi}$ (the fill is essentially normally consolidated; point A). When the roller is immediately above the element the vertical stress is increased (up to point B) and the horizontal stress is estimated on the assumption of no lateral yield so that $\sigma'_{hp} = K_0\sigma'_{vp}$. Once the roller has moved away the vertical stress decreases, but initially there is little change in the horizontal pressure. It is assumed that σ'_{hp} remains constant until the vertical stress is reduced below a critical value at which a Passive failure state will be approached (point C). Once this occurs, horizontal pressures are assumed to reduce linearly with σ'_v until the original vertical stress (due to self-weight of the fill) is reached (point D). For the path from C to D, $\sigma'_h = K_r\sigma'_v$ where K_r is the coefficient of earth pressure at rest for unloading (for shallow depths the residual horizontal stress is higher than the original horizontal stress). For deeper soil elements the vertical stresses are higher, and so during unloading Passive failure is not attained and the full maximum induced horizontal pressure is retained (e.g. path EFG). Hence if fill is placed and compacted in layers behind a retaining wall the resultant earth pressure envelope will be as shown in Figure 10.19.

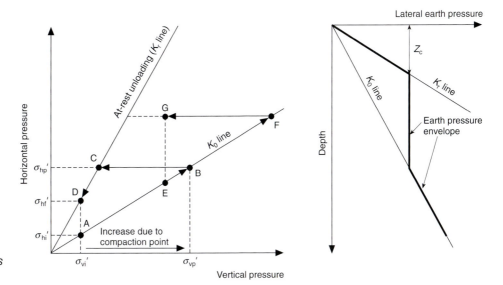

Figure 10.19. Earth pressures due to compaction

There will be a critical depth z_c below which removal of the compaction plant will not lead to a reduction in lateral pressure, and there will also be a depth below which the self-weight stresses of the fill will exceed those induced by compaction (see Figure 10.19). At the upper critical depth,

$$\sigma'_{vi} = \sigma'_{vc} = \frac{\sigma'_{hp}}{K_r} = \frac{K_0 \sigma'_{vp}}{K_r} \tag{10.29}$$

In practice, most backfill is free-draining and water levels are kept as low as possible, so $\sigma'_{vc} \approx \gamma Z_c$. Thus

$$z_c \approx \frac{K_0 \sigma'_{vp}}{\gamma K_r} \tag{10.30}$$

Ingold (1979) has produced a simplified version of the preceding analysis with K_a' substituted for K_0 and K_p' replacing K_r. The revised expression for z_c is $(K_A')^2 \sigma'_{vp}/\gamma$, and the maximum compaction stress is $K_A' \sigma'_{vp}$.

10.2.8. Practical considerations

There are three general major types of retaining structure instability to consider: outward/translational sliding, outward rotation and large-scale backwards rotation (Figure 10.20). The consequences of failure of a retaining wall are likely to be much more serious than for a slope, and higher Factors of Safety tend to be used. Factors of safety are also sometimes used as a means of limiting movements of retaining structures, and hence the range of values used is typically 1.5–3.0. There is much in common between the design considerations applying to all conventional types of retaining wall and also between temporary and permanent works. However, the design of temporary works differs from permanent works in the following ways:

* Negative excess porewater pressures, which maintain the original shear strength of the soil, are assumed to be present in slow-draining soil.

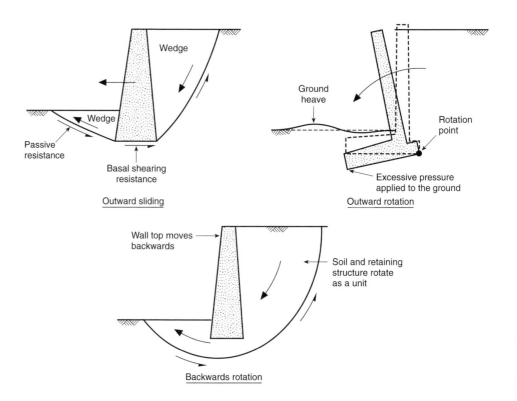

Figure 10.20. Retaining wall failure

- There is less control over loading conditions (e.g. unplanned surcharges are likely during construction but not in the long-term).
- Ground movements are likely to be a significant matter of concern for temporary ground support.
- The works may be subject to delays so that the actual support time is much longer than that originally envisaged.
- All walls should be designed for a minimum depth of additional unplanned excavation in front of the wall to provide for unforeseen and accidental events. Typical allowances are 0.5 m and 10% of the total retained height. Foreseeable excavations, such as service or drainage trenches in front of a retaining wall, which may be required at some stage in the life of the structure, should be treated as a planned excavation.

The selection of a particular form of earth-retaining structure depends on a variety of factors, as outlined in Table 10.5.

If the water-table is well below ground level, tension cracks may develop as a result of negative porewater pressures in the clay above the water-table (because of capillary rise), or as a result of desiccation of the upper layers. The depth of such cracks depends on factors that are difficult to quantify (e.g. the degree of access of surface water to the soil, the seasonal effects of droughts and vegetation). Hence tension cracks under long-term conditions should not be considered for design purposes. At the base of the tension crack (depth z_0 below ground level) the value of σ'_3 is zero and, from equations (10.20) and (10.25),

$$z_0 \approx \frac{2c'}{\gamma(1 - r_\mathrm{u})\sqrt{K'_\mathrm{a}}} = \frac{2c'/\sqrt{K'_\mathrm{A}} + u}{\gamma} \tag{10.31}$$

In reality little is known about the actual depth to which cracking occurs and, according to BS 8002 (BSI, 1994), in a situation where a tension crack could form adjacent to a wall, a design should be checked as follows:

- All clays: the end-of-construction condition, with soil parameter c_u, and the tension crack fully or partially filled with water.

Table 10.5. Factors influencing the selection of an earth-retaining system

Factor	Aspects
Wall location	Its position relative to other structures, noise and ground vibration limitations, the amount of space available, the necessity to confine the support system within the site boundaries, the need to provide temporary/permanent support for adjacent structures
Wall geometry	Proposed height of the wall, plan shape of the wall, topography of the ground (both before and after construction)
Ground conditions	The ground stratigraphy, groundwater and tidal conditions, chemical aggressiveness of the ground, need to maintain current groundwater regime, presence of contaminants
Movements	Extent of ground movement acceptable during construction and in-service, effect of movement of the earth-retaining structure on existing or supported structures and services, movement under live loading
Loading	Known dead and live loadings, surcharges (anticipated and unplanned), possible seismic effects, potential for changes in groundwater regime, dependence on long-term functioning of drainage systems
Regional location	The availability of materials, required external appearance of the retaining system, environmental constraints
Commercial	Design life, maintenance, costs, reliability/track record of the system

- Hard clays: the final equilibrium condition (c', ϕ') and the tension crack fully or partially filled with water, to a level higher than the equilibrium water level.

10.3. Gravity walls

The simplest retaining structure is the gravity wall, which can be made of mass concrete, stone or brick, reinforced concrete, gabions or soil. These walls are commonly economic up to 4 m high. They can be designed satisfactorily for greater heights, but as the height increases other types of wall become more economic. Gravity walls derive their stability from their self-weight (combined with the weight of any retained integral material). Usually the base is wide, so that the resultant of the active thrust and the self-weight falls within the middle third of the base (no tensile stress in the wall). Economy of wall material can be achieved by stepping or inclining the front or the back of the wall. A wall with a stepped back should be designed as having a 'virtual back', which corresponds to the vertical plane through the rear extremity of the base.

10.3.1. Types of wall

The reinforced concrete *cantilever wall* has a vertical cantilevered stem above a horizontal base which may extend in front of or behind the stem. The most common forms are an L or inverted T. Suitable preliminary design dimensions are shown in Figure 10.21. If sliding resistance is a problem, it is sometimes increased by a downward projection (key) below the base or by bolting the base to underlying strong ground (Figure 10.22).

Counterfort walls are similar to cantilever walls but they have ribs (counterforts) connecting the wall face and base, thus reducing bending moments and shear stresses. They are used for high walls (greater than 6 m approximately) or where there is a very high pressure behind the wall.

A *gabion* is a large rectangular cage (typically 2 m long by 1 m wide and 1 m high) made of metal or plastic mesh that is filled in situ with crushed rock or cobbles and used as a basic building unit (like a brick) for retaining walls, revetments and anti-erosion works (Figure 10.23). The major advantages of gabions are their flexibility and permeability, which make them suitable where the retained material is likely to be saturated and where the bearing capacity of the soil is low. The basic shape of gabion retaining walls is trapezoidal, but the outer and inner faces may be straight or stepped, the latter being more common. Gabions can blend with the environment, as they resemble open stonewalling.

Crib walls are built of individual units assembled to create a series of box-like structures containing compacted suitable granular, free-draining fill, to form a

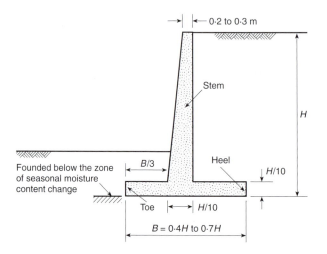

Figure 10.21. Preliminary cantilever wall dimensions

Figure 10.22. A reinforced concrete wall bolted to underlying rock

Figure 10.23. A wall formed from gabions

gravity retaining wall system up to 6 m high approximately (Figure 10.24). The units are so spaced that the fill material is contained within the crib and acts in conjunction with the cribwork to support the retained earth. There are timber cribs (usually for temporary works) and reinforced precast-concrete crib walls. The major advantage is that large movements can be tolerated without damage, since a crib wall is a flexible structure and the use of permeable fill assists drainage of the soil behind the wall. Less concrete is used than in a concrete gravity wall, and the crib wall is quickly constructed. The effect of friction on the rear of the wall will add to the stability and should be included in the design.

Figure 10.24. A crib wall

Reinforced soil is a means of producing a *mass gravity wall* whereby soil is used as the bulk material. Reinforcement (in various forms such as strips or grids) is included within soil fill (Figure 10.25), and shearing interaction (friction and cohesion) between the soil and the surface of the reinforcement provides resistance to lateral outwards movement of the soil fill. A variety of reinforcements have been used, the most common consisting of galvanized mild steel strips (plain or ribbed, 50–100 mm wide and up to 6 mm thick) and polymer geotextiles. The length of the reinforcement is typically about 80% of the wall height, although for low walls a minimum length of 5 m is usually required. The fill is usually a free-draining granular material. The 'wall' utilizes a front facing (often precast reinforced-concrete units) to stop local unravelling of the soil fill or local erosion. It must be remembered that the facing is not the wall; the wall is the reinforced mass soil block behind the facing. Major advantages of the system are:

Figure 10.25. Reinforced soil

- the flexibility of the wall
- the use of soil to create the mass of the wall
- the ability to accommodate large settlements and differential rotations without damage
- the strength of the wall increases (due to its self-weight) as it height increases
- structures often cost less than 50% of a conventional reinforced-concrete retaining wall.

With reinforced soil walls it is necessary to consider both internal and external stability, because the wall is not a solid structure (almost all of its mass is soil). There are two design methodologies: the 'tie-back' approach, and the coherent gravity structure method. The *tie-back* approach assumes that overall failure is via an Active Coulomb wedge and that local failure adjacent to the wall face is defined by Rankine earth pressure theory. The design approach in the *coherent gravity structure method* is based on experimental work (laboratory and field), which showed that at failure the boundary between the Active zone (immediately behind the face of the wall) and the stable zone (some way back from the face) was curved. The experimental work allowed the shape of this boundary to be defined (Figure 10.26). The reinforcement can be considered as holding the Active mass in place, and the total length of reinforcement required is equal to the sum of the length needed within the Active zone and the 'anchorage' length in the stable zone. Reinforcement lengths are typically 0.7–1.0 times the height of the wall (subject to a minimum value for low walls).

In the *anchored earth system* the flat or corrugated metal strips normally used to reinforce earth structures are replaced by anchors. The face of the wall consists of vertical panels attached to one end of the anchors. Support of these face panels, and retention of the ground behind them, is based not on the soil–element friction forces employed in earth reinforcement, but on the passive resistance developed at the curved ends of the anchors located within the stable zone.

10.3.2. Sliding stability

The outward Active thrust is balanced by a combination of shearing resistance on the base of the wall and Passive resistance in front of the wall (Figure 10.27). There is some doubt concerning the full mobilization of any Passive resistance from soil in front of the wall, as large movements are required. This resistance

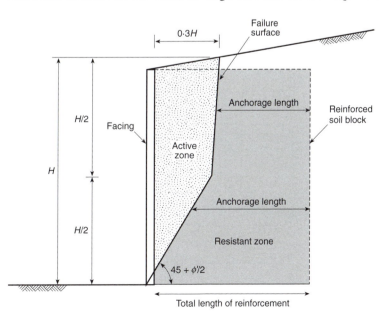

Figure 10.26. A coherent gravity structure failure surface

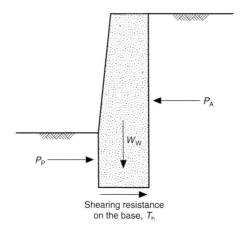

Figure 10.27. Failure by outward sliding

cannot be relied on if the soil shrinks or is excavated some time after construction. Typically, any Passive pressure resistance will be factored (typically reduced by 50%) to limit the movements needed to mobilize resistance.

Thus the Factor of Safety is given by:

$$\frac{\text{factored Passive resistance} + \text{basal shearing resistance}}{\text{horizontal Active force}}$$

or

$$F = \frac{P_P/f + T_b}{P_A} \tag{10.32}$$

Base resistance can be expressed in terms of either total or effective shear strength parameters (for the soil–wall interface), so that

$$T_b = c_w L + W \tan\delta \tag{10.33a}$$

or

$$T_b = c'_w + W' \tan\delta' \tag{10.33b}$$

Overall Factors of Safety of 1.5–2.0 are normally required with respect to sliding, and the factor f on the Passive resistance will usually be in the range 2.0–3.0. If the base resistance to sliding is inadequate it may be increased by either widening the base, inclining the foundation or providing a shear key, which should be located under the rear part of the base (Figure 10.28). The effect of the key is to mobilize more Passive resistance by providing a greater surface area in contact with the soil and also by utilizing the greater shear strength usually found at depth within a soil.

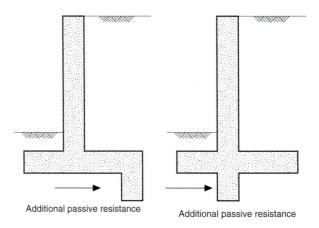

Figure 10.28. Shear key below retaining wall

Additional passive resistance   Additional passive resistance

10.3.3. Bearing failure

Rotation is assumed to occur about the heel of the wall (Figure 10.29). Consequently, the compression of the underlying soil increases linearly with distance from the heel and the vertical stress in the foundation soil is assumed to follow the same pattern. Hence the vertical pressure beneath the base of the wall follows a trapezoidal distribution (see Figure 10.29), with the maximum pressure beneath the toe and the minimum beneath the heel. If the overturning produces negative bearing pressures on the underside of the foot then the situation is reanalysed with a reduced base size (i.e. the length of the base sustaining negative pressures is ignored).

If the back of the retaining structure is assumed to be smooth (i.e. negligible shear stress is acting), then the wall can be analysed by taking the moment about the heel and resolving the vertical forces. For vertical equilibrium,

$$W_w = \frac{B(q+p)}{2} \tag{10.34}$$

The downward force W_w comprises the weight of the wall and any soil that acts integrally with it (if a significant shear force is acting on the rear of the wall then it must be added to W_w). Taking moments about the heel (point O) yields

$$\frac{qB}{2}\frac{2B}{3} + \frac{pB}{2}\frac{B}{3} = P_A a + W_w x \tag{10.35}$$

where a and x are the lever arms (about O) of the Active thrust and the wall weight, respectively. Any moment due to shear on the rear of the wall would act in the same sense as the bearing pressure from the underlying ground. Equations (10.34) and (10.35) may be combined and rearranged to provide an expression for the maximum vertical stress q developed within the soil beneath the base of the wall:

$$q = \frac{6P_A a + 6W_w x - 2W_w B}{B^2} \tag{10.36}$$

This maximum pressure cannot exceed the allowable bearing pressure for the soil beneath the wall (which is defined by the shear strength of the soil). Equation (10.36) is very useful for checking an existing wall or it can be used in design, where the dimensions of the wall are arrived at by a process of trial and error. Alternatively, the equation can be rearranged to define the base width that suits a given base pressure distribution. This is usually the requirement that the minimum pressure p must not be negative, and for this case

$$B \geq \frac{1.5(P_A a + W_w x)}{W_w} \tag{10.37}$$

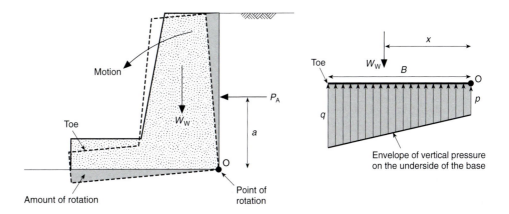

Figure 10.29. Failure by outward rotation

The preceding analysis may be adapted to suit the case where there is a significant and predictable shear force on the back of the wall, by adding its contribution into the force and moment equilibrium relationships (equations 10.34 and 10.35, respectively) as appropriate.

10.4. Embedded walls

The characteristics of embedded walls are:

- they are thin (essentially a thick membrane embedded in the soil)
- they are installed through undisturbed ground
- they do not have enlarged foundations
- they are supported primarily by the Passive resistance of the soil below excavation level
- they may be propped or anchored (in this case direct support is provided at high level)
- their self-weight is ignored
- they do not require a check for sliding or bearing capacity failure.

10.4.1. Types of wall

Sheet-pile walls consist of interlocking individual piles that form a continuous structure capable of retaining soil and, to some extent, water (Figure 10.30). Steel piling is used most frequently for anchored retaining walls (normally ranging in height from 4.5 to 12 m) and to make cofferdams. The speed at which sheet-piles can be pitched and driven makes them particularly appropriate for use at sites where construction time is limited. There is comparatively little displacement of soil during pile driving and suitable steel sections can be driven into almost any soil. Driving may be by impact (steam, air or diesel drivers), vibratory or hydraulic means. Impact methods generate the highest stresses during installation, but have the advantage of being suitable for virtually all soil types. Vibratory and hydraulic pressing systems do not impose high peak stresses in the piles during installation, but are not fully effective in certain types of soil. One of the problems with conventional vibratory hammers is that, as they run up to and down from operating speed, they may pass through the resonant frequency of the soil. This could potentially lead to liquefaction and large settlements and ground vibration problems with adjacent residents and structures.

Figure 10.30. A sheet-pile wall

Close bored or contiguous pile walls consist of piles installed at centres that are equal to or slightly greater than the external diameter of casing or lining. Borehole casing is required to retain cohesionless, fast-draining soils during boring. This type of wall is unsuitable for retaining water-bearing granular soils, which are liable to bleed through the gaps between the piles, unless special measures are taken to provide a seal between adjacent piles.

Secant piles are bored at centres less than the diameter of the casing (Figure 10.31). Alternate piles are thus of full circular section, while intervening piles are cut away in part during the construction process. A fully continuous, relatively watertight, retaining wall may be constructed, provided the tolerances of positioning and verticality are sufficient to eliminate gaps between piles.

Contiguous bored piles are frequently associated with both shallow and deep excavations for basements to buildings and with cofferdam work. They are used when a soil replacement rather than a soil displacement method of piling is required to form a retaining wall, and when there is a need to minimize noise and vibration. It is easier to overcome ground obstructions with this method than with sheet-piling or diaphragm wall construction.

Diaphragm walls represent a method of constructing deep, narrow, permanent walls in the ground. They compare favourably, in terms of watertightness, stiffness and mechanical strength, with cut-offs formed using steel sheet-piling, precast piles or in situ piles. They may be rigid (concrete) (Figure 10.32) or plastic (concrete and bentonite mixture) when load-carrying capacity is not required (e.g. for pollution containment) (see Chapter 13, Section 13.5.3). Diaphragm retaining walls are usually more economical when used as load-bearing walls within the permanent structure. Construction commences with the excavation of a shallow, narrow trench that is filled with a bentonite suspension. Excavation is then carried out through the slurry using narrow grabs (Figure 10.33), with the resultant slit trench being kept filled with slurry to maintain its stability. When the requisite depth of trench has been dug the bentonite suspension is displaced by concrete, placed by tremie pipe from the bottom of the trench upwards, which then hardens to form the diaphragm wall. Steel reinforcement is inserted into the concrete before it sets. Excavation and concreting is done in stages, and construction joints are formed by inserting a round stop-end pipe at the end of the excavation. This is withdrawn when the concrete has set, so forming a semi-circular joint against which the concrete of the next panel is placed.

Figure 10.31. A secant pile wall

Figure 10.32. A basement diaphragm wall

Figure 10.33. Diaphragm wall construction

The *soldier/king pile system* of temporary ground support consists of steel piles installed around the perimeter of the excavation with sheeting (which supports the ground) being placed between the piles as excavation proceeds. The sheeting spans either horizontally between the soldier/king piles (Figure 10.34) or vertically between horizontal walings. The system may be used to support deep, narrow, shallow or wide excavations in various materials, including clays and sands. This method is unsuitable for the exclusion of water, and if soil is washed out from behind the sheeting unacceptable settlement may be caused to adjacent structures or services.

10.4.2. Cantilevered walls

Cantilever sheet-piling depends on embedment into the soil below excavation level to provide a moment to resist the overturning moment from the Active thrust. These walls are economical up to 3–4 m height in soils with high shear strength. The deflections at the head of a cantilever wall are significant, and if services or foundations are located within the Active zone movement of the retained soil may cause them to be damaged. Figure 10.35 shows a case where wall movement caused settlement of the ground beneath the road and the sewer contained therein.

Figure 10.34. King piles with timber sheeting

Figure 10.35. Settlement of a ground surface due to pile head movement

Cantilever sheet-pile walls are considered to be rigid and to rotate about some point above the tip at failure, so that Passive pressures are developed at the tip (but behind the wall). This means that, at the point of rotation, there should be an instantaneous change from full Passive pressure on the front of the wall to full Passive pressure behind the wall. Bearing this in mind, a simplification of the earth pressure situation is widely used in practice.

A wall is analysed for equilibrium of moments about its tip (Figure 10.36) so that the only Passive thrust is in front of the wall and the soil behind the wall is completely in an Active state. A Factor of Safety is incorporated by reducing the horizontal Passive pressures by applying a factor of 1.5–2.0 to the passive earth pressure coefficients. Thus,

$$\frac{P_P b}{F} = P_A a \equiv \frac{P_1 b}{F} = P_2 a \tag{10.38}$$

The passive and active moments are functions of the depth of penetration d, and so equation (10.38) can be solved to give the embedment depth. The value of d is then increased by a small amount (typically 20%) to ensure that the actual point of rotation corresponds more or less to that assumed in the analysis. Thus the depth of pile below excavation level is around $1.2d$.

For simple, uniform ground conditions an expression for d may be derived from equation (10.38). For instance, if the soil is cohesionless and there is no

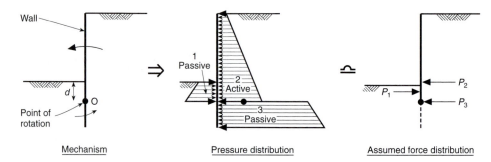

Figure 10.36. Analysis of a simple cantilever wall

Mechanism   Pressure distribution   Assumed force distribution

groundwater and the depth of excavation is H, then

$$P_1 = \frac{1}{2}\gamma d^2\, K_{\mathrm{P}}', \quad b = \frac{1}{3}d, \quad P_2 = \frac{1}{2}\gamma(d+H)^2 K_{\mathrm{A}}', \quad a = \frac{1}{3}(H+d) \quad (10.39)$$

Thus

$$\gamma d^3\, K_{\mathrm{P}}' = F\gamma(H+d)^3 K_{\mathrm{A}}' \tag{10.40}$$

Because of the uncertainty in the relative movement between the soil and a wall, it is common to treat the piles as smooth, since this is a conservative approach. Hence $K_{\mathrm{A}}' = 1/K_{\mathrm{P}}'$ and

$$\left(\frac{H+d}{d}\right)^3 = \frac{K_{\mathrm{P}}'^{\,2}}{F} \tag{10.41}$$

The preceding type of analysis may be used to produce diagrams or charts that provide preliminary estimates of the required pile length for simple conditions (Figure 10.37). Seepage round the wall may be approximated to a linear water head distribution on either side of the sheet piles (Burland *et al.*, 1981).

10.4.3. Anchored walls

Anchored or propped walls derive their support from both embedment in the ground and the use of anchors near the top of the wall, and can be used for great heights (Figure 10.38). The anchor or prop acts to reduce the lateral deflection, bending moment and depth of penetration, as compared with a simple cantilever wall. The use of more than one line of anchors or props may be necessary with

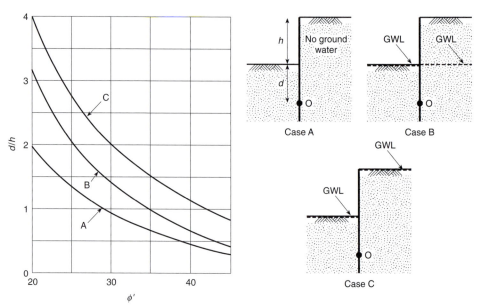

Figure 10.37. A preliminary estimate of pile penetration (after Clayton & Milititsky, 1986). GWL, groundwater level

Figure 10.38. Anchored sheet piling

high walls. Usually, the sheets are installed to full depth and the anchors installed progressively as soil is excavated in front of the wall. Ground movements produced around deep basements by the removal of vertical and horizontal stresses must be minimized, particularly if there are existing structures nearby. The top–down method of construction has been developed to ensure minimal ground movements.

Because of the crucial part played by the anchorage, the calculated tie force is normally increased (by 10–30%), regardless of method of analysis, to allow for:

- applications of unforeseen surcharges
- unequal yield of anchors leading to horizontal arching between anchors
- the catastrophic consequences of the failure of an anchor.

When designing an anchored wall there are two possible approaches: free-earth and fixed-earth analysis. The difference between the two approaches lies in the different fixity conditions assumed for the embedded lower part of the wall.

In the *free-earth support method* there is insufficient embedment to prevent rotation of the toe of the wall (Figure 10.39). Nevertheless, the wall is capable of being in equilibrium with the idealized pressure distribution. The piles are supported by ties at the top of the wall and by the soil at the base of the wall, in a manner similar to a vertical beam with simple supports. The wall is assumed to be rigid, rotating about point B where support is provided by an unyielding anchor

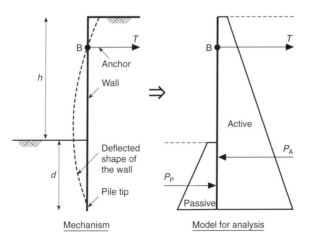

Figure 10.39. Free earth support analysis

(which provides a horizontal force T). The depth of pile embedment d is calculated on the basis of achieving moment equilibrium about the anchor level (point B). It is common to increase the depth of penetration by 20% to allow for unintentional excavation, erosion, etc. The anchor force is then calculated on the basis of horizontal force equilibrium. There are a number of different ways of incorporating a Factor of Safety in the free-earth support analysis. The methods in current use for the design of small- to medium-sized retaining walls are summarized in Table 10.6, which is based on a parametric study by Potts & Burland (1983). The free-earth support method is still widely used, and often gives an economical design, despite the age of the method.

Table 10.6. *Ways of incorporating a Factor of Safety into the free-earth support method*

Factor of safety applied to	Method	Comments
Passive parameters	The soil parameters are reduced by a Factor of Safety before the earth pressure coefficients are calculated (for $F = 1.5$ the method is roughly equivalent to applying a Factor of Safety of 2.0 to the Passive earth pressure coefficient)	The method is analogous to the approach used in the calculation of slope stability. The effect is to increase K_A, reduce K_P, and modify the earth pressure distribution. The method should be used only to determine the depth of penetration. The method has the merit of factoring the parameters that frequently represent the greatest uncertainty in design, although the results are quite sensitive to the value of the Factor of Safety chosen.
Moments	The geometry is adjusted such that restoring moments exceed overturning moments by a prescribed margin, using fully mobilized strength parameters. The components of the expression for the Factor of Safety (restoring moment and overturning moment), may be expressed in different ways.	Wall pressures calculated in this way relate neither to a physically meaningful collapse mechanism nor to a working load distribution. The three principal methods give values of the Factor of Safety that vary significantly, as geometry and strength parameters are varied (unless the Factor of Safety is 1.0, when all the methods agree). There is no rigorous justification for judging between methods.
	Gross pressure. This method consists of only factoring the gross Passive pressure diagram. Water pressures are not factored. Net water pressures are calculated and included in the bottom line of the equation for Factor of Safety.	There is often some justification in factoring the restoring moments in that the total Passive force is only partially mobilized at working load. The method may be illogical for soft clay when an undrained analysis with $\phi_u = 0$ is considered, because F_p (the factor on Passive resistance) exceeds K_P/K_A (which is equal to unity). When the method is used with effective stress analysis of clay, it is conservative for permanent works if the conventional value of $F_P = 2.0$ is adopted. Recommended values of F_P are: 2.0 for $\phi' > 30°$, 1.5–2.0 for $\phi' = 20°$ to 30°, 1.5 for $\phi' < 20°$ (Padfield & Mair, 1984). The method has been used for many years and, despite its inconsistencies, it continues to be used with success and is popular because of its simplicity.
	Net total pressure. All balancing loads are eliminated from the equilibrium equation, leaving only the unbalanced loads. The Factor of Safety F_{NP} is defined as: $(moment)_{net\ passive}/(moment)_{net\ active}$.	To obtain a conventionally satisfactory design consistent with the other design methods, very large values of F_{NP} have to be used. This approach diverges so much from accepted practice that dangerous errors are likely to result. Consequently, this method is not recommended.

Net available passive resistance. The Factor of Safety is applied to the moment of the net available Passive resistance (i.e. the difference between the gross Passive pressure and those components of the Active pressure that result from the weight of the soil below the dredge line). A vertical line is drawn on the Active pressure diagram from the level of the excavation, and the increase in Active pressure below this level is subtracted from the Passive resistance to give the net Passive resistance. The resulting Factor of Safety F_r is defined as (moment)$_{\text{net available passive}}$/(total moment)$_{\text{retained material}}$.

In effect, the dead weight of the soil below the dredge line, on both sides of the wall, is factored. This method requires different Factors of Safety F_r to be used over the range of values for ϕ'. The method has partially overcome the anomaly in the gross pressure method.

| Embedment | By adjusting the depth of embedment of the wall, a geometry can be found to represent moment equilibrium with fully mobilized active and passive soil pressures (i.e. the wall is theoretically at the point of collapse). It is then necessary to increase the depth of embedment by an empirically determined factor F_d. | The only pressure distributions that may be predicated with reasonable certainty are those that occur at collapse of the wall. Full active and full passive pressure distributions are then acting, regardless of the flexibility of the wall, the stiffness of any props or anchors, etc. Because of the empirical nature of F_d, it is incorrect to treat it as a Factor of Safety against overall rotation of the wall. It is recommended that the method should always be checked against one of the other methods. |

In the *fixed-earth support method* of analysis it is assumed that penetration of the piles is sufficient to ensure not only that the Passive pressures in the front of the wall resist forward movement but also that the rotation of the toe is restrained. Because of the presence of an anchor, the wall deflects as indicated in Figure 10.40, and so it contains a point of contraflexure (i.e. a point where the bending moment is zero, point O). In the method devised by Blum (1930), the wall is divided into two 'equivalent beams' at the point of contraflexure (see Figure 10.40). By taking moments about point O for the upper section the magnitude of the anchor force is determined, and horizontal resolution of forces gives the shear force R at the hinge. Moments can now be taken for the lower 'beam', about the tip of the wall (point A), to provide an expression from which the wall penetration below the hinge can be calculated.

For uniform ground conditions the hinge point lies approximately level with the point of zero net pressure, and this approximate coincidence is sometimes

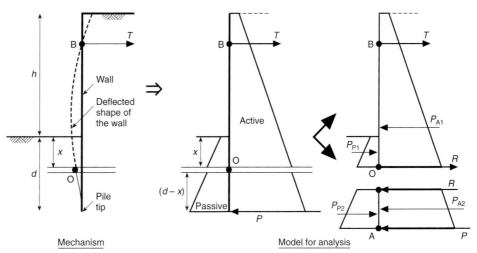

Figure 10.40. Fixed earth support method

Mechanism $\Rightarrow$ Model for analysis

assumed for the analysis of non-uniform soil conditions. Terzaghi (1966) conducted elastic analyses of anchored walls and found that the depth of the hinge below excavation level (as a proportion of the wall upstand) depended on ϕ'. On this basis it is often assumed that the point of contraflexure is at a depth of $0.1H$ (which corresponds to an effective friction angle of approximately 30°).

A fixed-earth analysis results in a much greater depth of pile embedment below excavation level (so that much longer piles are needed) than the free-earth method. However, it reduces significantly the maximum bending moment in the wall, so that piles with a smaller cross-section can be used. However, the increased depth of driving may mean that a heavier pile has to be used to accommodate driving stresses. The major benefit resulting from a fixed-earth design is likely to be in the reduction of outward deflection of the wall and smaller movements in the ground behind the piles (which may be critical for adjacent sensitive structures and buried services).

10.4.4. Wall flexibility

Unless prestressed struts are installed during the excavation process, movements of most flexible walls are sufficient to develop Active conditions behind the wall. Structures that should be designed for an enhanced value of earth pressure coefficient (up to K_0) are those that are effectively rigid (and require little Passive resistance for stability) or are unable to move to mobilize the shearing resistance of the soil behind the wall (e.g. basement walls propped by horizontal floor slabs, particularly when constructed from the top downwards).

The simplifying assumptions made in design, concerning the progressive increase with depth of Active and Passive pressures, take no account of the interaction between the soil and the structure. In a flexible propped wall (e.g. a sheet-piled structure), the mid-span is able to flex outward and this lowers the pressure at mid-height at the expense of higher pressures behind the prop and the base of the excavation. This phenomenon is often referred to as *arching*. Numerical studies (Potts & Fourie, 1986) and model tests (Rowe, 1952) have shown that the wall–soil interaction has a significant influence on the distribution of earth pressures in service and on the resulting bending moments and prop forces.

The distribution of earth pressures is affected by the deflected shape of the wall, which is a function of the flexibility of the wall relative to the soil. Redistribution results in an increase in the disturbing force at the waling and at the toe of the wall and a reduction in the disturbing pressure in the centre of the span. For a relatively flexible structure, such as an anchored sheet-pile wall in dense sand, the effect of wall deformation will enhance the pressure acting above the anchor position, with reduced pressure behind the wall at lower levels, where the greatest deflections occur. The net result is a reduction in the maximum bending moment compared with design based on a linear increase in limiting pressure with depth. This is, however, accompanied by an increase in the anchor loads.

In Rowe's design method for sheet-pile walls the design bending moment is obtained by reducing the maximum bending moment obtained by conventional analysis by a factor that depends on the relative flexibility of the sheet-piling with respect to the soil. For further details see Barden (1974).

10.4.5. Multilevel anchored walls

For high walls anchors may be installed at several different levels to limit the maximum bending moment or wall deformation. Multilevel anchored walls present a problem for analysis because of the need to investigate the various stages of construction and also because of the degree of redundancy of support. The general methodology is to analyse the wall progressively (i.e. as excavation progresses downwards to each level of anchorages), and assume that wall deflection is sufficient to generate Active conditions behind the wall. The force that is estimated to exist in a line of anchors after excavation to the next level of supports is then assumed to remain constant, regardless of further excavation. The analysis is conducted as follows:

(1) From the ground surface down to the level of the highest support to be installed, treat the wall as a cantilever and check that the pile section can withstand the maximum bending moment.

(2) Analyse the wall as an anchored structure for the stage where excavation has reached the level at which the second anchorage support is to be installed. Calculate the anchor force needed using the free-earth method.

(3) For analysis of the situation at the next level of anchors, assume that the force in the upper level of support remains unchanged during excavation and supporting. Choose an arbitrary position for the point of action of the resultant force of the two levels of anchors. Take moments about this point, assume free-earth support to calculate the required depth of embedment, and then determine the force carried by the lower level of support from horizontal force equilibrium. If the assumed point of action of the resultant anchor force is significantly in error, repeat the process until there is agreement.

(4) For the next line of anchors repeat the process undertaken in step (3). Now there will be two, constant, high level anchor forces.

(5) Proceed in the foregoing way until the bottom of the excavation is reached, and then modify the anchorage pattern to optimize the design.

10.4.6. Braced excavations

Props or struts supporting a wall hamper construction operations (Figure 10.41). Although anchorages allow unrestricted access in front of a supported wall, they penetrate and affect adjacent ground. Thus potential disturbance of buildings, services and other works in the vicinity must be considered.

Figure 10.41. Propping of temporary walls

Braced excavations usually consist of deep trenches with vertical walls, which are kept stable by internal propping. Analysis should be undertaken to prevent:

- collapse due to excessive strut loads
- instability due to failure of the base of the excavation
- excessive movements due to excavation and dewatering (particularly for sensitive structures).

The construction procedure for braced or strutted walls is essentially the same as that for anchored walls, but the form of support provided is different. Braces (and struts) are rigid, structural members inserted between the face of the wall and some unmoving point. The rigidity of the bracing system means that a wall is unable to move outwards by any significant amount. Hence the soil behind the wall only mobilizes a small proportion of its internal shearing resistance and the earth pressure coefficient cannot fall to the Active value. Furthermore, a strut carries a compressive load and, if its load capacity is exceeded, it will fail by buckling. When this happens, all the support from the strut is lost, causing progressive failure of the system as load is thrown onto other struts. On the other hand, if a ground anchorage is overloaded then it will move outwards, thereby mobilizing the full shearing resistance within the soil, and it will still be able to carry a significant load.

The magnitude of the lateral earth pressure at various depths of a strutted cut is determined by the specific method of construction and the resultant deformation condition of the sheeting. Because each level of bracing is essentially unyielding, the deformation of a propped wall differs from that of a conventional wall in that, in the former case, rotation is about the top of the wall. Hence Rankine's theory will not give the actual earth pressure distribution.

There are a number of empirical methods for designing strutted walls. Most are based on the semi-empirical procedure proposed by Terzaghi & Peck (1967), which was derived from back-analysis of strut load from various sites. Lateral pressure envelopes exist for braced cuts in loose sand, soft to medium clay and stiff clay (Figure 10.42). Strut loads may be determined by assuming that the vertical members are hinged at each strut level, except the top-most and bottom-most ones. The major concern is to assess the strut loads at each level of props. The positions of struts are usually selected to prevent excessive deflections during construction of the cofferdam and to suit the construction sequence for the works within it.

With free-draining soils the foregoing pressure envelope gives the effective lateral pressure, and the pressure from any groundwater must be accounted for separately. In clay soils the calculations are based on the undrained shear strength of the clay soil. The earth pressure coefficient K_A depends on a stability number

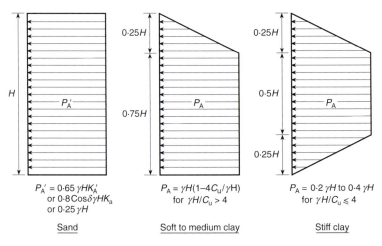

Figure 10.42. Earth pressure envelopes for braced cuts

$P_A' = 0.65\,\gamma H K_A'$
or $0.8\cos\delta\,\gamma H K_a$
or $0.25\,\gamma H$

Sand

$P_A = \gamma H(1 - 4C_u/\gamma H)$
for $\gamma H/C_u > 4$

Soft to medium clay

$P_A = 0.2\,\gamma H$ to $0.4\,\gamma H$
for $\gamma H/C_u \leqslant 4$

Stiff clay

$N (= \gamma H / c_u)$, which is related to the stability of the clay beneath and around the excavation. When N is less than 4 the soil around the excavation is still mostly in a state of elastic equilibrium. When N exceeds 6, movements of the sheet-piling and ground become significant because plastic zones are beginning to form near the base of the excavation. When N exceeds about 7 for a long excavation, or about 8 for a circular or square excavation, then complete shear failure, base heaving and extensive collapse of the excavation is imminent. For deep excavations in clays, Terzaghi & Peck (1967) showed that considerable variations in strut loads can be obtained (up to $\pm 60\%$ from the average load).

10.4.7. Anchorages

An anchorage for an earth-retaining structure is a system installed in the retained ground mass to provide a tensile form of support to the structure. Anchorage systems permit a clear excavation and are often used in preference to struts. They may be used solely during construction, may form part of the permanent structure or may be designed to perform a dual function (Figures 10.31 and 10.38). The anchorage should not yield by moving forward any significant distance and the Factor of Safety for the anchorage system and the individual members should be not less than 2. In addition, a check should be made for large-scale instability (such as rotational failure) with a failure surface enclosing both the wall and any anchors. There are two primary types of support: ground anchorages and horizontal tie rods.

Ground anchorages include rock and soil anchors. A ground anchor consists of a stressing tendon of a 'fixed length' (akin to a rough cylinder), which is anchored within the ground, usually by grouting into relatively strong soil or rock at depth (Figure 10.43). The fixed length is prevented from being pulled out of the ground by shearing resistance along the surface of the 'cylinder' and passive resistance acting on its front face. For sands and gravels a cased hole is drilled and, after installing the anchor cable, grout is pumped in under pressure and the casing is withdrawn. For clay soils the hole may be drilled without casing and the pull-out resistance of the anchor is enhanced by under-reaming to increase the effective diameter of the fixed length. It is usual for ground anchors to be drilled inclined below the horizontal to reach more competent ground. Account should be taken of the effects on soil loadings (both behind and in front of the wall) due to the component forces arising from the inclination of the anchorage.

Horizontal tie rods (or tendons) are attached to some anchorage point. Most commonly, the anchorage point is provided by a deadman consisting of a

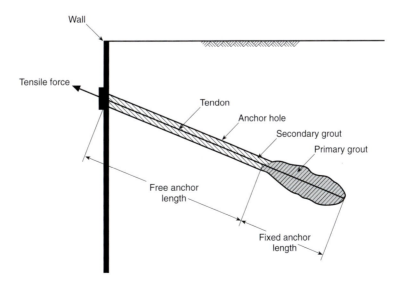

Figure 10.43. A schematic ground anchor

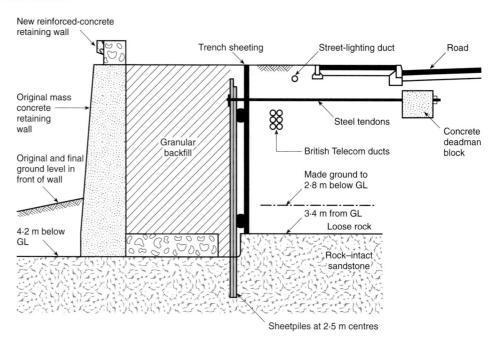

Figure 10.44. Temporary support using a deadman anchor

concrete block or a continuous concrete beam. This is the most suitable anchorage for relatively good ground. Figure 10.44 illustrates the use of a deadman to anchor temporary ground support during replacement of a permanent retaining wall. Piles, sheet-piles or vertical reinforced slabs can be used to provide an anchorage point instead of a deadman. The deadman should not undergo excessive settlement or rotation in relation to the tendons. This requirement is seldom significant in undisturbed granular soils, but for uncompacted fill or weak soils it may be necessary to provide a foundation or use an alternative system. Where the Passive soil zone in front of the deadman anchor wall does not interfere with the Active soil zone behind the main retaining wall (each zone is taken as that defined by the smooth wall condition) (Figure 10.45), the design of each component in the system can be undertaken independently. The resistance to forward movement of a deadman anchor wall is the difference between the Passive resistance of the soil in front of the anchor wall (ignoring any surcharge or live load on this ground) and the Active force on the back of the anchor wall (including any surcharge or live load in this ground). For discrete anchor blocks, shearing resistance on the sides of the block is also taken into account.

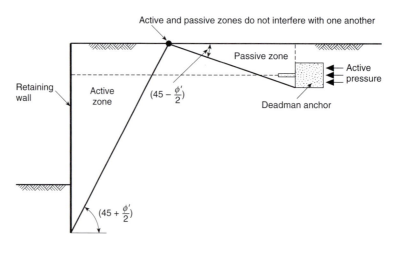

Figure 10.45. Earth pressure zones for anchored walls

11 Instrumentation and monitoring

11.1. Introduction

The geotechnical engineer works with a wide variety of naturally occurring heterogeneous materials, which may be altered to make them more suitable for construction, but which do not possess precise, unique numerical values of their engineering properties. Ground investigation can only test a small part of a real site. While laboratory or field tests are performed on selected samples of geotechnical materials to obtain values for engineering properties, these tests only define a band of possible values, and 'unexpected ground conditions' are always a possibility. Field observations (including quantitative measurements) aid the geotechnical engineer in designing a safe and efficient project, despite the foregoing difficulties, and provide a means for ensuring safe and economical execution of the works.

Ground investigations for the Littlehampton bypass showed the presence of soft alluvial soils of varying thickness (3.0–10.5 m). The alluvial soils adjacent to the River Arun were composed of soft to very soft grey silty clay and clayey silt, the undrained shear strength of the alluvial soil being of the order of 15 kN/m^2. At some depths samples were too soft for testing in the triaxial apparatus (Anandan & Piercy, 1991). The undrained shear strength of the alluvium was far too low to permit construction of the bypass embankment to its full height of 6 m without allowing time for consolidation to occur before the construction period. The alternative solutions for dealing with this soft alluvial material would have been to remove it by deep excavation (adjacent to the existing road on an embankment) or to use large quantities of imported lightweight fill. Comparative estimates indicated that the least costly solution was to install vertical wick drains (see Chapter 8, Section 8.5.2) at close centres (1 m) in order to expedite the process of consolidation of the soil. This solution was chosen.

To monitor lateral movement of the embankment associated with possible rotational failure, eight vertical inclinometer tubes were installed. Standpipes were placed within the embankment so that the permeablility of the soil could be measured under the correct stress level. As vertical drain installation neared completion, areas became free to allow installation of the soil monitoring equipment, namely piezometers, extensometers (settlement gauges) and inclinometers. For the extensometers, the positions of the settlement markers, which were situated just below existing ground level, were recorded using a magnet-sensitive electronic probe, and water levels in the piezometers and standpipes were recorded using a water-sensitive electronic probe. Settlements of up to 1 m were recorded. Inclinometers did not require a settling-in period as they did not come into operation until after embankment filling had commenced. Changes in ground profile were measured by inserting an inclinometer probe into the vertical tubes and recording how far out of vertical the probe was at various depths within the tube. Throughout construction the pore pressures remained low and at an acceptable level so that the Factor of Safety against rotational failure never fell below 2.5. Furthermore, the inclinometer readings indicated negligible lateral movements of the embankment.

Satisfactory completion of any geotechnical construction relies on the integration of the behaviour of a large number of material elements (which the engineer cannot control precisely). Consequently, field instrumentation is a vital geotechnical tool and engineers must have more than just a passing awareness of it.

11.1.1. Purpose of monitoring

Not all field instruments are installed for the sole purpose of checking the safety of a structure or construction operation or to confirm design assumptions. Some instruments are used to determine the initial conditions of the ground or the environment, such as observations of groundwater (ground investigation) or ambient sound level (site investigation), prior to construction. In addition, advances in knowledge require large-scale or full-scale observations. This is particularly true for geotechnical engineering, which utilizes a number of well-established empirical relationships derived from field monitoring.

Before deciding on or selecting a programme of field instrumentation, it is first necessary to define in detail the overall and specific purposes of the monitoring. Even though instrumentation has a vital role to play in geotechnical construction, it is not an end in itself and it cannot guarantee good design or trouble-free construction. If instrumentation is inappropriate or incorrect or it is in the wrong place the information provided will, at best, be confusing and at worst it may distract attention from significant signs of real trouble. Consequently, every instrument installed on a project should be justifiable and it should be selected and placed to assist in answering a defined, specific question.

The primary uses of field measurements are summarized in Box 11.1. In general, the quantities to be monitored are:

- Pore pressure/groundwater: this relates to effective stress, shear strength, consolidation, settlement.

Box 11.1. Primary uses of field measurements

In situ determination of parameters
To define site conditions during the design phase of a project, e.g. parameters (shear strength, permeability, consolidation, etc.), initial ground conditions (water pressures and their fluctuations, in situ stresses).

Performance monitoring
There is a scarcity of data on the actual performance of geotechnical works. Uncertainties in design or specifications for complex, major or innovative construction may require large-scale testing (e.g. trial embankments or test piles) to be conducted to verify the adequacy of a design. Many advances in geotechnical engineering have resulted from field measurements.

Construction control
Most of the applications fall into this category. Uncertainties in engineering properties or behaviour often affect construction procedures or schedules. Simple records (such as groundwater level changes, movements, etc.) provide data in case of unforeseen happenings, which enable appropriate modifications to be derived and which provide a basis for settling contractual claims. For major, novel works it may be necessary to monitor actual behaviour during construction so that procedures or schedules can be modified in accordance with actual behaviour.

Safety of the works
To confirm that construction complies with specifications, and hence it should function correctly. Provided that threshold limits (acceptable, action, danger, etc.) have been set for relevant parameters, instrumentation programmes provide a forewarning of any untoward behaviour or adverse effects of construction.

- Total stress: this is usually the total pressure measured at a boundary, although stresses within a soil mass can be important.
- Movements: settlement, differential settlement, lateral deflection.

11.1.2. Instrumentation requirements

Most measurements are essentially point values. These are subject to any variability in geotechnical or other characteristics, and therefore depend on the local characteristics of that zone. A large number of measurement points may be required to have confidence that the data represent overall conditions. On the other hand, individual devices can be located so as to respond to movements within a large and representative zone.

Ideally, the presence of a measuring instrument should not alter the value of the parameter being measured. However, instruments are discontinuities that do not have the same properties as the surrounding ground and usually their presence alters the very quantities they are intended to measure. Furthermore, in order to measure quantities such as stress and strain some actual movement (however small) needs to occur at the location of the measuring instrument. This localized deformation will alter the ground conditions near the instrument. Furthermore, the act of drilling a borehole or compacting fill around an instrument can affect the conditions within the geotechnical material significantly. Hence the very presence of instrumentation within the ground changes the situation (the change may be significant or negligible) and so the equipment will not read the original, true value.

Accuracy is the closeness of approach of a measurement to the true value of the quantity being measured. Accuracy is synonymous with degree of correctness. When selecting instruments to have appropriate accuracy for a given situation, the entire system must be considered, including the accuracy of each component and each source of error. *Precision* is the closeness of approach of each of a number of similar measurements to the arithmetic mean. Precision is synonymous with reproducibility and repeatability. Thus an instrument reading may be precise but not necessarily accurate. However, if the error is consistent and quantifiable then the reading can be corrected so that the data can be considered to be both accurate and precise.

The question of which parameters are the most significant should always be addressed and an estimate has to be made of the likely values of monitored parameters so that appropriate working ranges are selected for the instruments. These numerical values will often be in terms of rate of measured change, rather than absolute magnitude. Inherent in the use of instrumentation for construction purposes is the absolute necessity for deciding, in advance, the range of values that is considered likely and realistic so that prompt warning is given of anomalies and potential problems. Furthermore, it should be borne in mind that geotechnical instrumentation has to perform in hostile environments (damp, dirty, under stress, etc.) and apparatus must be robust, as well as sensitive, so that early signs of distress are detected. The two prime requirements for field instruments are:

- reliability — to ensure that dependable data can be obtained throughout the period when the observations are needed
- sensitivity — this should be sufficient to indicate significant changes promptly.

Usually, the most dependable devices are the simplest, although simple instruments are sometimes inappropriate and more complex ones may have to be used. If there is a conflict between high accuracy and high reliability then the latter should always be given preference. Reliability is usually associated with good quality, established instruments (i.e. those that have a good record of past performance, particularly in the installed environment).

Measurements are by either direct or indirect means, relative to some local, temporary datum or reference station. A stable reference datum is therefore a prime requirement. Many geotechnical instruments monitor by indirect means using a device (a transducer) that converts a physical change into a corresponding output signal, a data recording system, and a communication system between the two (Box 11.2). The good functioning of the instrumentation depends not only on the sensor but also on the means of transmitting the data to the read-out unit, and the latter unit itself. With certain instruments, if a reading can be obtained, that reading is necessarily correct, while with other instruments their calibration can be verified after installation. Either feature is very desirable. Normal wear and tear, misuse, creep, moisture ingress and corrosion can all cause changes in the output of a monitoring system, and calibrations or function checks of all parts of the system (sensor, transmitting agent, read-out) are required during service life. In addition, systems should be such that readings can be taken quickly, and the output should be easy to process or interpret.

Attention must be given to two easily neglected factors and their associated costs:

- Protection of the instruments. Damage can arise from construction operations and vandalism. Access points that project above ground level are more prone to accidental damage than those set flush with ground level. Padlocked protective caps, or gauge houses, may be an open invitation to vandals.
- Reading the instruments. This is particularly important where instruments with a slow response/settling time are used, or if a long-term record is required. The costs of reading the instruments (including travel and staff time) can easily exceed the costs of installation.

11.1.3. Instrumentation planning

The task of designing a monitoring programme begins with defining the objectives and ends with defining how the measurement data will be implemented (Figure 11.1). This requires some notion of the construction procedures likely to be followed, a thorough knowledge of the capabilities and shortcomings of the instruments themselves, and an appreciation of the practical problems of installation. It is also very useful to know how the results of the observations will be used.

Box 11.2. Components of a monitoring system*

Transducer
In general, transducers can be placed in the following order of decreasing simplicity and reliability: optical, mechanical, hydraulic, pneumatic, electrical. The transducer must have proven longevity to suit the application.

Data reading
These systems include direct, individual readings taken manually, simple portable read-out units, and complex automatic on-line set-ups. The point of extraction of the data may simply be a lockable port in the ground, or it may be a permanent, secure monitoring building (gauge house) with appropriate services (electricity, water, telephone, etc.). A non-moving reference datum/reference station may be needed.

Communication system
Cables, tubes and pipes are used to connect the transducer to its read-out unit. They must be able to survive imposed pressure changes, deformation, water, sunlight effects, corrosion, etc. Special attention should be paid to routeing, wherever possible avoid traversing zones of large differential strain and provide slack at any such transitions. Horizontal tubes and all cables are normally installed in trenches surrounded by a medium sand.

* Adapted from Dunnicliff and Green (1994).

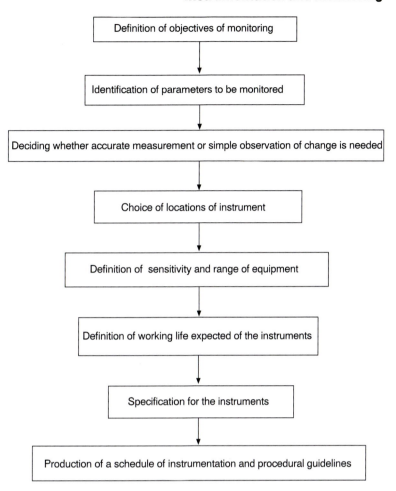

Figure 11.1. A flowchart for the production of an instrumentation scheme

Prior to preparing a programme of instrumentation, a working hypothesis should be developed for mechanisms that are likely to control the behaviour of the works. Simple, quantitative analysis will suffice. Estimation of the maximum possible value of a parameter (or the maximum value of interest) leads to a selection of instrument range — this estimate requires substantial engineering judgement. Estimation of the minimum value that has significance leads to selection of instrument sensitivity.

Instrument locations should reflect the anticipated behaviour of the works and should be compatible with the method of analysis to be used to interpret the data. Cross-sections that represent the behaviour of the whole works are instrumented to provide comprehensive performance data. Additional instrumentation is installed in zones of particular concern (e.g. where high porewater pressures are expected).

Survival of the instruments in critical zones is very important, and it is vital that complete coverage is provided over the monitoring period, especially during the initial period of the works. This will require duplication of instrumentation (particularly at key locations) and the incorporation of redundant instruments. There have to be enough instruments to allow not only for the inevitable losses (from malfunction and damage by construction activities), but also to provide a meaningful picture despite the inevitable scatter in the results (due to inherent variations in geotechnical materials and construction procedures). Wherever possible, locations should be arranged to provide cross-checks between instrument types.

Detailed procedures for installation of instruments should be produced and should aim to produce minimum disturbance to the surrounding ground

Box 11.3. A checklist for planning instrumentation*

1.	Define the project conditions.
2.	Predict mechanisms that control behaviour.
3.	Define the geotechnical questions that need to be answered.
4.	Define the purpose of the instrumentation.
5.	Select the parameters to be monitored.
6.	Predict magnitudes of change.
7.	Devise remedial action.
8.	Assign tasks for design, construction and operation phases.
9.	Select instruments — plan for high reliability.
10.	Select instrument locations.
11.	Plan recording of factors that may influence measured data.
12.	Establish procedures for ensuring reading correctness.
13.	List the specific purpose of each instrument.
14.	Prepare budget.
15.	Write instrument procurement specifications.
16.	Plan installation and prepare step-by-step installation procedure well in advance.
17.	Plan regular calibration and maintenance.
18.	Plan data collection, processing, presentation, interpretation, reporting and implementation.

* After Dunnicliff and Green (1994).

conditions, minimum interference to construction, and minimum access difficulties during installation and reading. Selection of appropriate backfill materials for boreholes and avoidance of non-standard construction in the vicinity of instruments will help to prevent the generation of anomalous data.

There are also questions of who is responsible for installing, monitoring and maintaining the equipment and what is the appropriate level of supervision and competence of the associated personnel.

A checklist for instrumentation planning is given in Box 11.3.

11.1.4. Data handling

Written procedures for data collection, processing, presentation and interpretation should be prepared beforehand. When assigning tasks for monitoring, the party with the greatest vested interest in the data should be given direct responsibility for producing it accurately. Ideally, data should be recorded under the direction of an experienced geotechnical engineer.

During instrument installation and reading, all factors that may influence measured data need to be recorded, including:

- construction details
- progress
- ground conditions
- weather conditions
- environmental factors
- any problems
- extraneous actions/effects in the vicinity of the instrumentation.

Duplicate sets of the original data should be kept along with detailed notes or records of installation and monitoring sessions.

Many instruments take a few days to stabilize after installation. Similarly, the ground needs time to achieve equilibrium. Instruments should therefore be installed as early as possible and monitoring should be commenced immediately after installation. Several initial values should be taken and the repeatability between these readings should satisfy the expected tolerance (more readings may need to be taken to define an appropriate monitoring regime). The subsequent frequency of data collection will be related to the construction activity (and hence the rate at which the readings are changing) and to the need for assessment of the output. If too many readings are taken then the data may not be scrutinized properly, whereas if too few readings are taken important events may be missed.

Data need to be analysed, plotted and considered quickly. There should be a procedure for comparing the latest readings with the previous values so that any significant changes are identified immediately. During the planning of the instrumentation limits are set as to what values are anticipated and what values are the limits of acceptability. The latter limits define hazard-warning levels. For each hazard warning level there must be a plan of action. After the data have been processed the output should be summarized and presented to show trends and to compare observed and predicted behaviour for the benefit of future design or construction operations.

Every measurement involves error and uncertainty. Gross errors (misreading, misrecording, computational errors, incorrect installation, etc.) are caused by carelessness, fatigue or inexperience. Gross errors are avoidable and can be minimized by taking duplicate readings, checking present readings against previous readings, etc. Systematic errors are usually caused by improper calibration or by alteration of calibration with time such that readings are consistently high or low. These errors can be minimized by periodic recalibration and by checking readings against standards and dummy gauges kept in the laboratory or in the field. Even when errors are recognized and remedied, readings will still show variation due to random error. Data can be treated mathematically using statistical analyses so that measurements are presented as average values with a standard deviation and confidence limits. However, average values should be used with caution, because extreme values may in fact be the most significant data and may indicate the development of a critical situation.

11.2. Vertical ground movements

Vertical movements can be determined by construction site surveys carried out using engineering levelling procedures. Measuring points on structures may be simple bolts cemented into shallow holes. Benchmarks on permanent structures are normally stable. However, verification should be made that the structure is not affected by conditions such as groundwater lowering and seasonal thermal effects. For ground surface movement a measuring point set below the zone of frost heave and seasonal moisture changes may suffice. This type of settlement monitoring gives localized values at the ground surface and may not be a feasible method of monitoring during construction. Furthermore, construction activities may cause a near-surface benchmark to settle by subsoil densification from blasting or pile driving, by consolidation from nearby loading or drawdown, or as a result of extension strains directed toward an excavation. The emphasis for geotechnical monitoring is instrumentation within the ground.

11.2.1. Hydraulic overflow settlement cell

This instrument records settlement (the vertical component of displacement) at a particular location. The actual overflow cell is a sealed container cast in a concrete block, which is buried or installed at the location at which settlement is to be monitored (Figure 11.2). This cell is connected via a flexible, water-filled tube to a graduated standpipe in a gauge house remote from the monitoring location and which is accessible for the duration of the monitoring period. The water level in the standpipe in the gauge house is raised so that water is flushed through the system and overflows at the open end in the cell. The system will come to equilibrium with the water surfaces in the standpipe and the cell at the same elevation — this level is recorded by reading the standpipe. As the cell settles (due to the movement of the surrounding soil) the elevation of the open

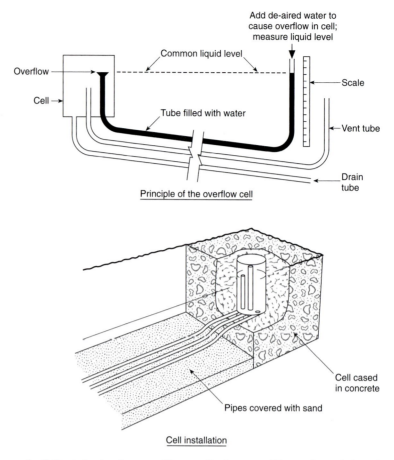

Figure 11.2. Hydraulic overflow settlement gauge

end of the tube in the overflow cell changes. Hence by raising the level of water in the gauge house and allowing equilibrium to be re-established the new elevation of the cell can be determined and hence the settlement is known. The cell is also connected to tubes that allow drainage and equalization of air pressure (to atmosphere). To remove water from the cell, compressed air is applied via one lead and water is allowed to exit via the other.

Normally, the read-out system (i.e. the standpipe in the gauge house) would be at a similar elevation to the overflow cell. However, significant elevation differences between the cell and the measuring standpipe (more than 2.5 m approximately) can be accommodated by pressurizing the air in the cell, using the flushing line, so that there is a known pressure differential between the free water surfaces at the two ends of the system. The air pressure is converted into a water head to give the elevation of the cell. Alternatively, for a cell at a higher elevation than the read-out unit, a known 'back-pressure' may be applied to the water surface in the monitoring standpipe.

The apparatus is straightforward and reliable for long-term use, particularly because there are no moving, mechanical elements and the instrument is unaffected by temperature. The instrument provides accurate settlement values (about ± 0.5 mm) over a wide settlement range — up to around 2 m. With this system there are no vertical rods, tubes, etc., to interfere with construction, and measurements can be made beneath structures and at inaccessible locations. In general, liquid level gauges are sensitive to liquid density changes caused by temperature variation, to surface tension effects and to any discontinuity of liquid in the liquid-filled tube. The greatest potential source of error is discontinuity of liquid caused by the presence of gas, and great care must always be taken to ensure absence of gas. Overflow cells are commonly used for measurements in embankment dams and under embankments on soft ground.

11.2.2. Settlement profile gauge

This apparatus provides continuous settlement readings along a line that is more or less horizontal. It is thus eminently suited to monitoring settlements beneath embankments on soft ground. The principle of the system is that the pressure in a liquid is proportional to the head (or elevation) at a point. If the liquid is contained within a sealed system (e.g. a tube), when one end of the system is raised or lowered the pressure at the other end will vary according to the amount of vertical movement. The actual profile gauge comprises a gas-filled probe (the end of the tube that senses the settlement) containing a flexible bladder, which is connected via a water-filled tube to a pressure transducer at the reading position (Figure 11.3). A second tube connects the gas in the probe to a pressurization system at the ground surface, so that negative pressure in the water-filled tube is avoided by operating the whole system under an external, elevated pressure. The magnitude of the gas pressure is not used in calculations, because it is applied to both ends of the system (i.e. to the probe and the pressure transducer). The change in the pressure transducer reading (due to vertical movement of the probe) divided by the specific gravity of water gives the vertical deformation directly.

The system uses a horizontal access tube into which the probe (and attached pressure lines) is fed (Figure 11.4). The probe is pushed, in stages, from one end of the tube to the other and a series of readings are taken. The distance of the instrument from one end of the pipe is established from graduations on the pressure line. The system thus gives a continuous profile of settlement along a

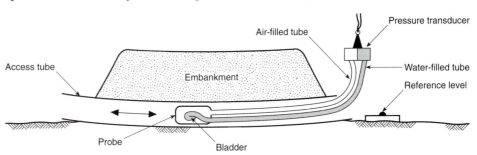

Figure 11.3. A settlement profile gauge

Figure 11.4. The insertion of a settlement profile gauge probe into an access tube

horizontal line (normally discrete values are taken at 0.5 m spacings). To obtain absolute elevation the relative level of the transducer must be determined or the probe must be 'zeroed' at a reference station. These gauges are particularly appropriate where vertical deformation is likely to be non-uniform, such that many single-point gauges would otherwise be required.

The apparatus is relatively simple, apart from the pressure transducer, and long-term reliability is obtained if the equipment is serviced regularly. The probe calibration can be checked at any time by varying the gas pressure surrounding the bladder by a known amount and reading the output from the transducer. Because the transducer becomes unstable if it is set to very high sensitivity and because of variation in the orientation of the probe in the access tube, the accuracy is limited to ±5 mm vertical displacement.

11.2.3. USBR settlement gauge

In essence the system consists of a tape measure that records the depth of reference points (typically at a vertical spacing of 0.5 m) from the top of a more or less vertical access tube (Figure 11.5). The access tube consists of a series of telescoping pipe sections with alternate, narrower bore sections anchored to the surrounding ground by horizontal steel channel cross-arms (in fill) or spring clips (in boreholes) at 1–3 m intervals. The cross-arms ensure that the pipes move together by an amount equal to compression of the intervening fill.

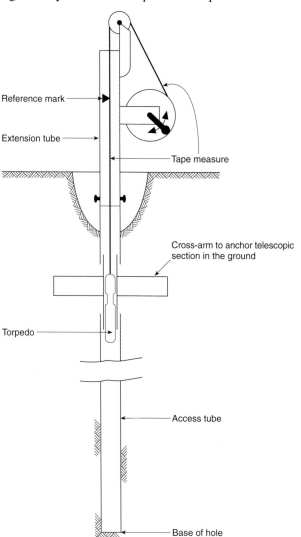

Figure 11.5. A USBR settlement gauge

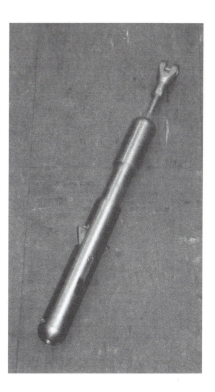

Figure 11.6. A USBR settlement gauge torpedo

To take measurements, a heavy torpedo is lowered down the access tube on a graduated metal tape. The torpedo (Figure 11.6) contains tapered, spring-loaded pawls that close as they pass through the narrow tubes on their downward travel. When an attempt is made to withdraw the torpedo, the pawls will 'snag' on the end of a small-diameter tube. Working from the top downwards the torpedo is progressively passed through each small-diameter tube and then pulled back against the pawls so that the taut tape measure can be read against a reference datum at the top of the access tube. This process is repeated down the whole tube. After the torpedo has passed the lowest reference point it is allowed to fall to the bottom of the access tube. This activates a mechanism to retract the pawls and the torpedo can then be drawn out of the access tube.

The difference between successive tape readings gives the separation of adjacent cross-beams, and the change in cross-beam spacing indicates the vertical distribution of settlement. Movements can be referenced to either the base of the access tube (if it is installed in solid ground) or to a top datum (attached to the top of the tube). In either case the absolute elevation of the reference datum should be established, usually by levelling.

The system is simple, robust, reliable and the only moving mechanical component is the arrangement of the spring-loaded pawls. The settlements are usually determined at 1 m intervals down the tube. The access tubes are relatively inexpensive and one torpedo services many tubes. The tubes do not need to be precisely vertical, and even at deviations from the vertical of $\pm 25°$ the system has been found to work satisfactorily (Dunnicliff & Green, 1994). The system can accommodate large vertical movements (but lateral movements must be small if the system is not to be damaged), and it is thus applicable to the monitoring of earth and rockfill dams, foundation settlements, behaviour of retained fills, etc.

However, installation must ensure that the top of the access tube can always be reached for levelling or attachment of the datum tube. In fill situations, installation commences by forming a starter hole in stable foundation material and inserting the base tube. The hole is then grouted or backfilled. The first small-bore tube is usually positioned with the cross-arm at original ground level. Further sections of large and small base are added as filling proceeds. As each successive cross-arm is placed, the elevation of a reference point on the uppermost pipe section is determined to the nearest 3 mm by optical levelling. The top of the access tube is either always kept above the fill, or it is covered and filled over and subsequently exposed by digging. When the system is located in boreholes the whole access tube is lowered into the ground with compressed, spring-loaded clips at the locations of the measuring points. With the tube in the correct position the clips are released to press against the sides of the borehole and hold the tube in place.

11.3. Horizontal movements

Methods that can be used to measure overall, gross movements include:

- Triangulation to determine a change in the lateral position (accuracy around ± 2 mm).
- Lasers are used for alignment measurements and levelling (accuracy around ± 3 mm).
- Electronic distance measurement (EDM), either to determine distance change directly or to determine lateral position change by trilateration (accuracy ± 5 mm).
- Global positioning system (GPS), whereby two or more GPS receivers simultaneously receive radio signals from the same set of satellites (accuracy around ± 20 mm).

Horizontal displacements can also be monitored directly by conducting direct measurements of the horizontal separation of pairs of points. The tape extensometer is used to measure the distance between rigidly mounted bolts. The measuring tape is always tensioned to the same degree, to eliminate errors due to tape extension. This method requires uninterrupted connection between points and access to both the monitored location and some fixed point. Hence the technique is used mainly with underground excavations such as tunnels in rock, or to monitor displacement of solid structures.

For measurements of movements within the mass of soil there are two principal methods: the magnetic probe extensometer and the inclinometer.

11.3.1. Magnetic probe extensometer

This system is a method of measuring directly between two points using a tape measure or a rigid rod attached to a vernier measuring scale. While the system is best suited to measuring horizontal movements, it can also be used in the vertical direction. The points between which distance measurements are made are ring magnets threaded onto an access tube. These magnets are intimately connected to the ground either by a leaf spring or by a wide mounting plate, and they can slide freely along the access tube to follow ground movements (Figure 11.7). The magnets are located and detected by inserting a steel probe (which incorporates reed switches at one or both of its ends) into the access tube. When a reed switch enters a magnetic field an electrical circuit is completed so that a buzzer or light is activated.

For vertical measurements the probe may be lowered on the end of a tape, which is simply read against a datum mark (Figure 11.8). For horizontal measurements the steel probe is inserted into the access tube and, by progressively coupling on 1 m long, rigid, steel rods, the probe is pushed along the access tube (see Figure 11.8). At the end of the tube where the probe is inserted the whole rod passes through a clamping device, which is attached to a fixed, stable reference point. The clamping device incorporates a means of accurately measuring horizontal displacement (a vernier gauge), and when the jaws of the clamping device are tightened around the steel tube the whole rod

Figure 11.7. A magnetic probe extensometer: access tube and magnets

may be advanced into the access tube at a slow and precise rate. To measure the distance between adjacent magnets the probe is inserted into the access tube so that one of the reed switches is activated. The jaws of the clamping device are now tightened onto the steel rod and the rod assembly is advanced until the second reed switch is activated. The horizontal movement applied is read from the vernier scale on the clamping device and hence the separation between adjacent pairs of magnets is known accurately (the reed switches are set 1 m apart). By comparing this value of separation with the value obtained previously the relative movement of the magnets is obtained. The total relative displacement of any points is derived by summing the relative displacements of the relevant magnets. For determination of absolute deformation data the location of the reference point must be tied into some stable datum.

Typical applications of probe extensometers are monitoring lateral deformations of fill within retaining structures and embankments, and measuring compression within embankments or embankment foundations and heave at the base of excavations.

The apparatus is reliable and relatively simple (particularly the access tube and magnets). One probe can be used with many tubes. High accuracy is achievable (around 0.1 mm consistently) over long distances (tens of metres). However, distortion of horizontal access tubes, due to settlement, etc., may mean that the stiff rod assembly cannot be pushed along the tube. For low-accuracy applications (particularly settlements, when the extensometer is used vertically), the distance of the probe along the access tube is best measured directly (to 0.5 mm) by using a tape measure.

11.3.2. Inclinometer

Boreholes that penetrate the shear surface of an active slide will be sheared off at the level of the slip surface, and the borehole may be subsequently plumbed to find the depth at which shearing has taken place. Unfortunately, slow movements may take an unacceptably long time to do this and progressive failure may occur.

The inclinometer system is used to monitor horizontal displacements along a vertical line, and often its function is to define the location of any deforming zone

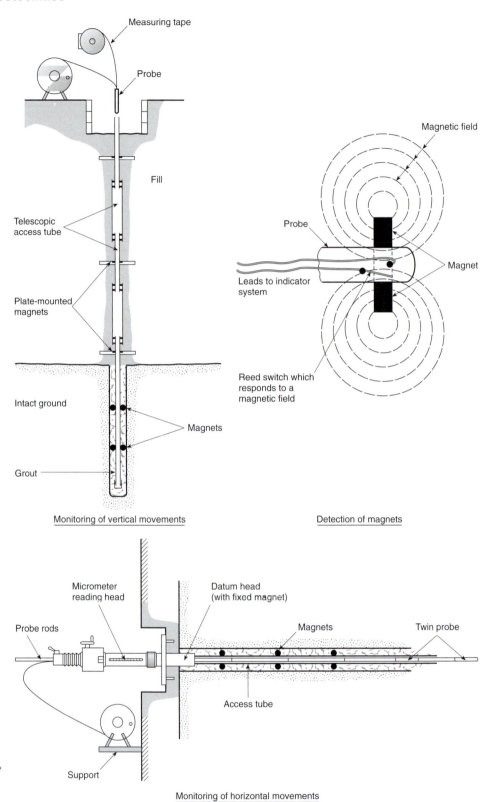

Measuring tape

Probe

Fill

Telescopic access tube

Plate-mounted magnets

Intact ground

Magnets

Grout

Monitoring of vertical movements

Magnetic field

Probe

Leads to indicator system

Magnet

Reed switch which responds to a magnetic field

Detection of magnets

Micrometer reading head

Datum head (with fixed magnet)

Magnets

Twin probe

Probe rods

Access tube

Support

Monitoring of horizontal movements

Figure 11.8. A magnetic probe extensometer (from Soil Instruments Ltd)

and to allow an evaluation of the movement of that zone as time progresses. The system comprises; a small-diameter casing (aluminium or plastic), which is installed more or less vertically in the ground, and a probe (the inclinometer), which is lowered down inside the casing (Figure 11.9). The probe (Figure 11.10(a)) contains a gravity-sensing transducer (an out-of-balance force sensor)

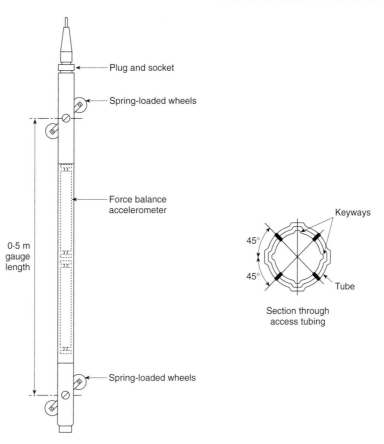

Figure 11.9. The inclinometer (from Soil Instruments Ltd)

Plug and socket

Spring-loaded wheels

Force balance accelerometer

0·5 m gauge length

Spring-loaded wheels

Keyways

45°

45°

Tube

Section through access tubing

designed to give a voltage output that is directly proportional to the inclination from the vertical. The access casing can be installed in a borehole or in fill and it contains two sets of internal, orthogonal grooves running along its length. These grooves fix the orientation of the inclinometer probe as it is moved up and down the casing. Figure 11.10(b) shows access casing (fastened to a retaining wall to measure outward deformation of the face).

After installation of the access casing, the probe is lowered in stages (usually in steps of 0.5 m) to the bottom, with an inclination reading being made at each position. Thereby data are provided for determining initial casing alignment. The deflected shape of the access tube can be estimated by summing the offsets along the tube. A measurement set is obtained by holding the probe stationary at each depth interval throughout the casing and recording the depth and inclination. This is done for all four orientations of the inclinometer in the grooves in the casing. Systematic errors are then eliminated by combining the two readings obtained with the wheels of the inclinometer in the same grooves (i.e. the inclinometer is rotated by 180° after taking the first set of readings). The differences between the initial readings and a subsequent set define any change in alignment. At the time of initial reading, survey measurements should be made of the top of the casing to establish its lateral position to within the accuracy required of inclinometer measurements.

Provided that one end of the casing is fixed against translation or that translation is measured by separate means, the readings allow calculation of absolute horizontal deformation at any point along the casing (accuracy is in the order of ±1 mm). Typical applications include monitoring movements of slopes, cuts and retaining walls. However, the access tube can become kinked at even small shear deformations, and so serious consideration should be given before the system is used in active landslides.

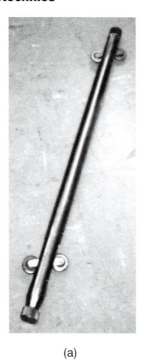

(a)

(b)

Figure 11.10. (a) A uniaxial inclinometer. (b) An inclinometer tube for measuring the movement of a retaining wall

11.4. Porewater monitoring

There are various reasons for monitoring the water within the ground:

- to provide information on the location of the water-table (which may influence temporary works)
- to measure pore pressures (during construction and subsequently) in order to estimate effective stresses and thus shear strength
- to assess settlement rates and the degree of consolidation
- to measure permeability (usually during the ground investigation stage)
- to check for water quality.

The equilibrium groundwater regime and transient pore pressures are usually measured using some form of piezometer. Water quality is checked by extracting a sample of water and subjecting it to a suite of appropriate chemical tests (the latter form of monitoring is not covered in this chapter). A piezometer consists of a buried porous element that is in intimate contact with the groundwater (so that uninterrupted flow between the two is possible). This sensory element incorporates some means of sending a signal about the groundwater to a receiver above the ground surface. The porous element should be sealed within the ground so that it responds only to groundwater pressure around itself and not to groundwater at other elevations. Piezometers can be installed in fill, sealed in boreholes, or pushed or driven into place.

During the determination of the water-table or the measurement of pore pressure changes, there has to be some flow of water between the piezometer and the surrounding soil. Until this flow ceases the pressure or head recorded will not be correct. The time taken for equalization depends primarily on soil permeability, the volume change needed for the measuring system to respond, and the compliance of the system. The presence of air, and suctions within a soil, makes accurate measurement of pore pressures very difficult. Errors can also arise due to clogging of the piezometer tip or sedimentation or erosion adjacent to the tip. Ideally, a piezometer should:

- be able to record both positive and negative pressures
- be stable with respect to time and the surroundings
- cause minimal disturbance when installed
- respond quickly.

11.4.1. Standpipe piezometer

This system is widely used in highly permeable soils, and basically it consists of an open pipe connecting a ground zone with the ground surface. The lower end is open or perforated (Figure 11.11) and acts as the sensing element, which communicates directly with the ground surface via the bore of the pipe (Figure 11.12). The groundwater level is determined directly by measuring the level of the water in a standpipe by lowering a water-detecting probe on a tape measure down the tube. Usually the probe contains electrical contacts and when the probe touches the water a circuit is completed and a signal is given out (a buzzer or a light). The overall groundwater regime can be determined by using several standpipes, each with its tip sealed into the ground at a different elevation.

Because of the large bore of the tube and the size of the sensing element, a large volume of water has to move in or out of the standpipe to record a change in groundwater level. Hence the system has a large time lag and is not suitable for measuring rapidly changing transient pore pressures. The standpipe piezometer obviously can only register positive pore pressures, and with the simple reading probe ('dipper') the top of the tube must be permanently accessible. Thus the instrument is not useable in fills or when the top might become submerged. When embankment fill is placed around the standpipe, nearby compaction tends to be inferior and the standpipe is prone to damage by construction equipment. Nevertheless, this is a very useful piezometer for site investigation (it can be used to conduct rising and falling head tests for permeability; see Chapter 7, Section 7.3.2) and the size of the pipe bore means that it is possible to sample the groundwater to check for pollution.

The major advantages of the standpipe-type piezometer are its simplicity and robustness. The device is self-de-airing, and any air bubbles that do collect in the sand filter or 'piezometer cavity' escape of their own accord up the tube.

The 'purge bubble' principle can be used to take remote readings with this type of piezometer (Figure 11.13). In this system a length of plastic tubing is inserted in the standpipe so that its lower end is below the water surface; the elevation of this end is measured. If a charge of compressed air is fed into the tube at ground level it will emerge at the lower end and form bubbles. If the upper end of the tube is now closed, air will continue to escape until the air pressure is equal to the water pressure. If the residual air pressure is measured, it can be converted into the head of water above the lower end of the tube.

11.4.2. Casagrande piezometer

This type of piezometer consists of an isolated sensing element connected to the ground surface via a small-diameter pipe, which reduces the response time of the

Figure 11.11. A porous plastic tip of a standpipe piezometer

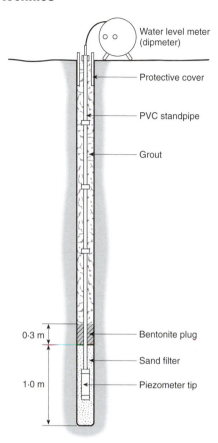

Figure 11.12. A standpipe piezometer (from Soil Instruments Ltd)

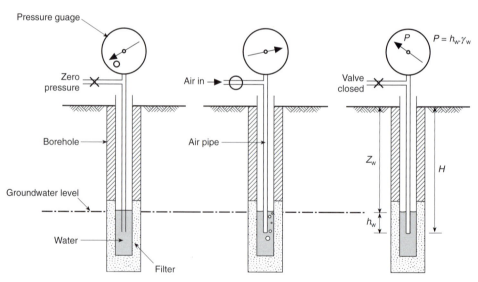

Figure 11.13. Automatic monitoring of a standpipe piezometer

system. Originally the sensing element (the porous tip) was made of a ceramic, but nowadays tips made of porous plastic are usual. If the soil is soft the porous tip can be driven into the ground. For push-in installation it is essential that good contact is obtained between the tip and the surrounding material — a small thickness of loose soil adjacent to the tip can prevent the piezometer from functioning correctly. Push-in piezometers are affected by the potential for an inadequate seal, for smearing or clogging, and false readings caused by gas generation. Standpipe piezometers are placed in a borehole and, after pulling back the casing, sand is poured in to form a filter surrounding the porous tip. Compressed, dry bentonite balls are rammed into place on top of the sand filter to seal in the sensing element and the remainder of the hole above the filter is filled and sealed (typically with a cement and bentonite grout). At the ground surface the pipe from the sensing element may be open to the atmosphere (it forms a manometer) or it can be attached to a pressure gauge (if artesian conditions exist).

11.4.3. Hydraulic piezometer

The porous sensory element is connected to a remote read-out station by twin flexible plastic tubes filled with de-aired water (Figure 11.14). The water pressure in the tubes at the read-out position is measured by a pressure gauge or manometer. The difference in elevation of the piezometer tip and the read-out station is established at the time of piezometer installation so that the water pressure at the tip can be calculated. The small volume of the sensing element, the stiffness of the plastic tubes and the incompressibility of the de-aired water in the system mean that there is rapid response to pore pressure changes in the ground. High air-entry tips are often used to prevent air passing from the ground to the water-filled tubes although the tip and tubes can be flushed at any time so that the system remains filled with water. This direct connection between the groundwater and the water in the tubes mean that the system can be used to measure permeability in situ.

The twin-tube hydraulic piezometer was developed for installation in the foundations and fill during construction of embankment dams. The apparatus is reliable, with a long working-life, and it is very suitable for remote reading (distances of 500 m between tip and surface station have been used). Generally there is no limit for the depth of the tip. However, the read-out or any of the connecting tubing must not be more than 5 m above the piezometric level, otherwise the resultant negative pressure in the water in the tubes will cause it to 'boil' so that air comes out of solution and continuity of the water thread is lost. Pipes should be laid below the freezing zone. The main disadvantage is that a terminal enclosure is needed to contain the read-out and flushing arrangements. This enclosure must be protected from freezing and from vandalism.

11.4.4. Pneumatic piezometer

The problem of elevation difference and cavitation in water in the system (which exists with the hydraulic piezometer) may be eliminated by using air as the means of transmitting the pressure sensed at the tip to the ground surface. This is the

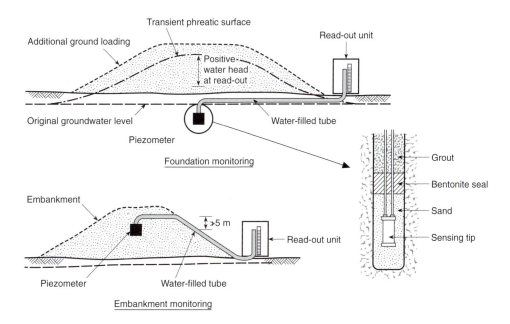

Figure 11.14. The hydraulic piezometer

approach in the pneumatic piezometer, wherein the tip (Figure 11.15) comprises a porous element integral with a diaphragm so that groundwater pressure acts on the diaphragm. Twin plastic tubes go from a void behind the diaphragm to the monitoring station at ground level (Figure 11.16). The diaphragm closes off the end of one of the tubes (which is connected to a compressed gas supply at ground level), and when this air pressure exceeds the groundwater pressure the diaphragm moves and allows air to be vented back to the ground surface via the other tube. If the air pressure is now allowed to drop (due to air being vented through the return tube) then the diaphragm will move back to seal off the air inlet in the piezometer. The porewater pressure is indicated by the cut-off air pressure value.

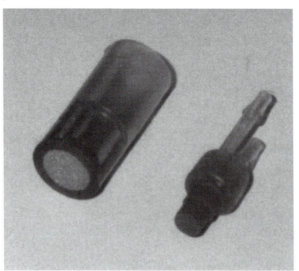

Figure 11.15. A pneumatic piezometer tip

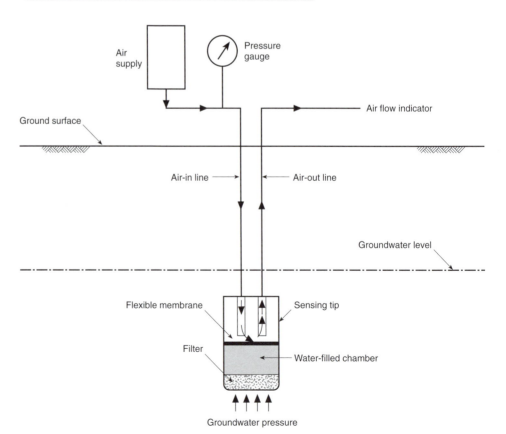

Figure 11.16. The pneumatic piezometer

The apparatus has a fast response because of the small volume of liquid that undergoes pressure changes (i.e. the water in the porous tip only). The tubing elevation is not critical and there is no correction needed for elevation because of the pneumatic conductor and very long leads may be used. Unfortunately, it is not possible to de-air the tip and thus there are problems with suctions in unsaturated soils (compacted embankments may never achieve full saturation). Hence this type of piezometer is best suited for foundation measurements below the water-table.

11.4.5. Vibrating wire piezometer

With this system (Figure 11.17) the signal is transmitted from the sensory tip to the ground surface via pulses of electricity so there is no elevational effect. The tip comprises a porous element integral with a diaphragm to which is attached a length of steel wire which is free to vibrate (Figure 11.18). The frequency of vibration varies with the wire tension and is thus affected by small relative movements of the ends of the wire. An electrical coil is mounted near the wire and application of voltage pulses to the coil creates a magnetic attraction that causes the wire to vibrate. The coil then becomes a listening device because wire vibrations cause an alternating voltage to be induced in the plucking coil. The frequency of this voltage is identical to that of the vibrating wire. The voltage signal is transmitted along the signal cable to a frequency counter, which is used to measure the time for a predetermined number of vibration cycles. This frequency change is directly related to the vibrating wire tension and hence the

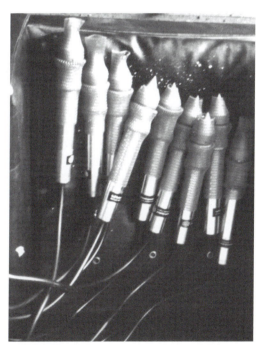

Figure 11.17. Vibrating wire piezometers

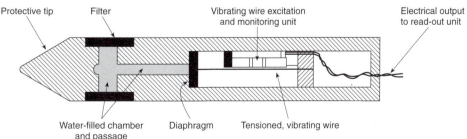

Figure 11.18. A section through a vibrating wire piezometer

pore pressure at the tip. The relationship between vibration frequency and tip pressure is obtained by calibration in the laboratory.

Armoured cable connects the transducer to the read-out unit. Since the output signal contains the required information in the form of a frequency rather than as a magnitude of a resistance or voltage, undesirable effects involving cable resistance, contact resistance, leakage to ground or length of cable are negligible. Very long cable lengths are acceptable. However, the cost of the armoured cable becomes very high. If leads are severed they can simply be reconnected. This type of instrument usually exhibits very good long-term stability, although it is desirable for the instruments to undergo a period of 'ageing' after manufacture. Vibrating wire piezometers are accurate, robust and of rapid response, because of the very small volume change required. The use of this type of piezometer for long-term measurements in compacted fill is questionable because, despite the use of saturated high air-entry filters, the longevity of tip saturation is uncertain as gas may enter by diffusion.

11.5. Ground pressures

Total stress measurements in soil fall into two basic categories: measurements within a soil mass (embedment cells) and measurements at the face of a structural element (contact pressure cells). While the stress vector at a point will generally contain both normal and shear average components, 'routine' measurements are only made of the normal stress. The primary reasons for use of earth pressure cells are to confirm design assumptions and to provide information for the improvement of future design. Such cells are not really used for construction control or other reasons. The particular items of interest are at the stresses at the boundary of the soil mass and the initial stress state.

To measure pressure some measurable change must occur at the soil–instrument interface. The force developed as a result of the movement of a flexible diaphragm which is in contact with the soil is often the monitorable change, but sometimes it is the deformation of the diaphragm itself. Insertion of a relatively stiff instrument into the soil also causes modification of the natural stress field due to arching and redistribution (Figure 11.19). Generally a cell will overpredict the free-field stress value. If this overregistration is essentially constant then the cell can be calibrated and a reliable method of measurement exists. Installation effects are an additional source of error. For fill situations it is likely that cells will be surrounded by a more compressible zone of soil than will the remainder of the fill, because of differences in the compaction energy applied to the different sections (fill placed over instruments will initially be compacted manually or by light machinery). If a hole is excavated to insert a cell to measure internal stresses then the very act of forming the hole will release internal stress and create non-uniformities. Some success has been achieved in measuring horizontal stress in soft soils by pushing specially designed earth pressure cells into natural ground.

Measurements of total stress against a structure are not subject to so many of the errors associated with measurements within a soil mass. It is possible to measure total stress at the face of a structural element with high accuracy.

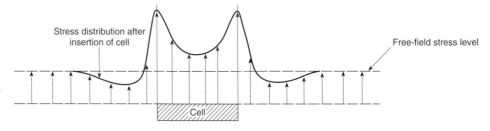

Figure 11.19. The effect of the inclusion of an earth pressure cell on the vertical stress distribution in the ground

Table 11.1. Requirements for the measurement of total stress

Requirement	Consequences
An embedment cell should not appreciably alter the state of stress within the soil mass because of its presence	The cell should be as thin as possible. The error resulting from insertion of the cell can be minimized by designing for high stiffness and an aspect ratio (cell thickness/ diameter) of less than 1 : 10. A stiff outer (or guard) ring to reduce mechanical cross-sensitivity
A large enough sensing area to record an average value (not just a point value as might occur if a large stone was in direct contact with the active face)	It is recommended that the diaphragm is at least 50 times the size of the largest soil particle. Always encase the cell (or cover the active face) with sand
The inaccuracy in the pressure reading is a constant, definable value	The central deflection of the diaphragm must be small — it should not exceed around 0.05% of the diameter. The pressure sensing element (the diaphragm) should be less than 45% of the total face area of a cell. Calibration of cells is needed
A method of installation that will not seriously change the state of stress	Development of push-in embedment cells. In fill situations protect the active face of cells with a layer of sand around 250 mm deep, then place other fill and compact as normal. Ensure that the sensitive face of boundary cells is absolutely flush with the interface of boundary surfaces. Increase the number of duplicate cells to account for damage losses

General requirements for earth pressure cells, and the consequences of these conditions, are outlined in Table 11.1.

11.5.1. Acoustic pressure cell

This embedment cell consists of a circular or rectangular flat jack formed from two sheets of stainless steel welded around their periphery (Figure 11.20). The gap between the sheets is filled with fluid of comparable compressibility to that of the ground (i.e. oil for soils and mercury for rocks). The layer of liquid is thin (0.5–5 mm) to give a cell stiffness that is similar to that of the surrounding soil, and the installed cell experiences minimal effects due to thermal expansion and contraction of the liquid. The fluid-filled chamber is connected to a vibrating wire transducer (i.e. a diaphragm connected to a vibrating wire system, as described in Section 11.4.5) by a short length of steel tubing, thereby forming a closed 'hydraulic' system.

Cells should be calibrated to ensure that they are functioning correctly and not leaking. Unless installations are to be made in soft clay, fluid pressure calibrations are insufficient. If measurement accuracy must be maximized, each cell should be calibrated in a large calibration chamber where it is in contact with the soil in which it will be embedded.

The cells are accurate and have long-term stability and a low aspect ratio (ratio of thickness to diameter or length) which minimizes stress redistribution. Unfortunately, the cells can be easily damaged during construction. The typical

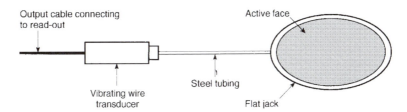

Figure 11.20. An embedment earth pressure cell

Output cable connecting to read-out

Active face

Vibrating wire transducer

Steel tubing

Flat jack

diameter of cells available for field use is 230–300 mm. Cells can be installed in clusters within the soil mass to try to identify principal stresses.

11.5.2. Pneumatic and hydraulic earth pressure cells

These instruments are very similar to acoustic earth pressure cells, the only difference being in the method of quantifying the pressure and transmitting a signal to the read-out unit at ground surface. The flat jack is connected to a pneumatic or hydraulic transducer to form a closed system, such that the pressure acts on one side of a flexible diaphragm (like the pneumatic piezometer arrangement; see Section 11.4.4). The chamber on the other side of the diaphragm has two tubes entering it. To obtain a value of the pressure in the flat jack, air, nitrogen or a liquid is supplied from the read-out unit and acts on the flexible diaphragm. When this latter pressure slightly exceeds that from the flat jack, the diaphragm moves to allow fluid flow along the return tube to a detector in the read-out unit. The supply pressure is then allowed to reduce until the diaphragm closes off the vent tube. The residual pressure is the flat jack pressure.

These cells are simple and robust but they are sensitive to temperature changes. They can be installed inside fill or at the boundaries, or by being pushed into soft natural ground. Remote measuring is feasible and cells can be connected by long lengths of tube (500 m or more) provided that they are well protected.

11.5.3. Electrical resistance boundary cell

A stiff circular membrane, fully supported by an integral stiff edge ring is deflected by the external soil pressure. Electrical resistance strain gauges (their resistance changes directly in proportion to applied strain) are attached to the membrane. The strain gauges are arranged in a Wheatsone bridge circuit and energized by a constant-voltage source. Diaphragm deflection causes a change in the out-of-balance voltage across the bridge, and the relationship between this voltage and pressure applied to the membrane is obtained by calibration of the cell. This type of cell is cross-sensitive (i.e. stress acting in the plane of the membrane will produce an error in the voltage reading).

Moisture penetration (both into the cell and into the wires) is a major problem, as is the change in bond between the strain gauges and the membrane with time.

11.5.4. Vibrating wire boundary cell

The cell comprises a rigid housing and a circular flexible membrane to which is connected a vibrating wire monitoring system (the same as for vibrating wire piezometers; see Section 11.4.5) (Figures 11.21 and 11.22). When an external pressure is applied the frequency of vibration of the wire changes and this is recorded and converted to a pressure using a calibration chart.

The cell and output are not affected by water (although filling the cell with water will damp the oscillations of the vibrating wire, the frequency of vibration is unaltered). Temperature-change effects are controlled by matching the coefficients of expansion of the diaphragm and the vibrating wire element. Usually a dummy calibration wire is incorporated to assess age effects in the properties of the wire and the cell.

This type of instrument has been in use for decades and has proved very successful.

Figure 11.21. Vibrating wire earth pressure cells (to be covered with sand)

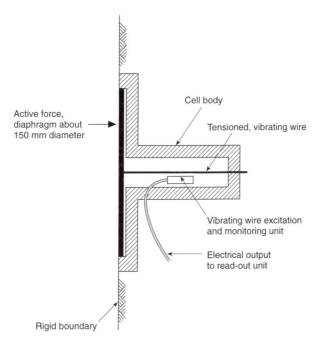

Figure 11.22. A boundary earth pressure cell

Cell body

Tensioned, vibrating wire

Active force, diaphragm about 150 mm diameter

Vibrating wire excitation and monitoring unit

Electrical output to read-out unit

Rigid boundary

12 Waste disposal by landfill

12.1. Introduction

The material placed in a waste disposal site (landfill) is very mixed in composition and suffers from organic decomposition and physico-chemical breakdown after deposition. Liquids are usually present in wastes and rainwater enters during the landfilling process. This water both promotes biological activity and acts as a transport mechanism for contaminants that dissolve in the water. If this polluted water (leachate) is allowed to enter the underlying groundwater, contamination results. Carbon dioxide, methane and hydrogen sulphide are also produced during the decomposition process.

The early morning of 24 March 1986 was bright and dry in south-east Derbyshire, although the region was experiencing an exceptionally rapid fall in atmospheric pressure. At about 6.30 a.m. the bungalow at 51 Clarke Avenue, Loscoe, was destroyed by an explosion. Happily there were no fatalities, although the bungalow was completely wrecked and all three occupants were trapped in the rubble (Figure 12.1). Gas samples subsequently taken from the garden and the underfloor space of the bungalow showed characteristics typical of landfill gas.

Some 60 m from the bungalow is an old quarry which was worked for brickmaking clay, although coal seams were also worked as they occurred. From 1973 until 1982, the disused quarry was operated commercially as a waste disposal landfill site into which was dumped domestic and commercial wastes (including more than 6000 t of putrescible material). When tipping finally ceased, 100 t of refuse was being dumped each day. After the site had been filled, it was covered with a light, permeable cover and seeded.

Derbyshire County Council mounted a public enquiry into the gas explosion and arrived at the following conclusions:

Figure 12.1. The bungalow at Loscoe wrecked by the explosion of landfill gas (courtesy of Derby Evening Telegraph)

- The incident was caused by an explosion of a mixture of landfill gas and air in the underfloor at the bungalow.
- The landfill gas had migrated from the old quarry, which had been used as a landfill site. The natural fissuring of the sandstone and other local strata had been amplified by the effects of many years of rock blasting and by coal-mining activity, and this allowed the lateral migration of landfill gas to take place.
- The unusual and sudden drop in atmospheric pressure during the early morning had enhanced the release of landfill gas from the old quarry. The gas accumulated in the underfloor void of the bungalow so that an explosive mixture of methane and oxygen was produced.
- The explosive gas mixture was ignited by the burner of the pilot light of the central heating boiler in the bungalow.

In the UK, waste disposal by landfilling remains an integral part of the approach to waste management. Each year approximately 250 Mt of controlled wastes are placed in landfills. Current government strategy is to promote disposal practices that will achieve stabilization of landfill sites within one generation.

12.1.1. Waste disposal
The hierarchy of solid waste disposal which is shown in Table 12.1 indicates that the volume of wastes can be reduced significantly. Nevertheless, there will always remain a residue that has to be dumped and this will have to be disposed of as landfill or landraise.

The magnitude of the problem associated with the management of waste materials only really becomes apparent on examination of the actual quantities of these materials that are produced each year. It is currently estimated that approximately 435 Mt of waste is generated in the UK each year, of which approximately 245 Mt is controlled waste (i.e. refuse subject to the provisions of the Environmental Protection Act). Sources of waste are outlined in Table 12.2. Approximately 70% of the controlled waste (excluding sewage sludge and dredged spoils) is disposed of directly to landfill. Based on these statistics, and assuming a compacted waste density of $1.0 \, t/m^3$, the estimated volume of space required annually for landfill of controlled waste is approximately $170 \, Mm^3$. The waste-generation situation is similar in most other countries, with each inhabitant producing between approximately 0.5 and 2.0 kg of disposable refuse each day. The overall way of disposing of this waste varies from country to country (in some instances almost all waste is incinerated, in other locations major recycling and energy-recovery programmes are in operation). Nevertheless, even with these processes a significant final residue will exist and this is then landfilled.

Landfill sites are subject principally to control under three acts:

- The Town and Country Planning Act (1971) — provides the planning background to which all development of land is subject
- The Control of Pollution Act (1974) — Part 1 concerns waste collection and disposal
- The Environment Protection Act (1990) — Chapter 43 (Part II) deals entirely with Waste on Land.

However, it was the introduction of the CEC Directive on 'The Protection of Groundwater against Pollution by Certain Dangerous Substances' which has driven the development of properly engineered landfill sites.

12.1.2. Waste composition
Waste comprises a heterogeneous mass of material that varies widely (according to source and time particularly). The typical composition, by weight, of

Table 12.1. The hierarchy of waste management

Position in hierarchy	Action	Comments
1	Waste reduction	The best action is to avoid the creation of waste materials. Waste reduction has two components: reducing the amount of waste produced, and reducing the hazard of the waste produced. Re-use involves putting objects directly back into use (after necessary 'rehabilitation') so that they do not enter the waste stream.
2	Recycling	Recycling is the recovery and re-use of products that would otherwise be thrown away. In some instances it simply involves separating particular items (such as scrap metals or plastics) and then re-using them or returning them to their suppliers for processing and re-use, in other cases much greater processing is needed. However, recycling can only be undertaken a finite number of times and there is always some residual waste.
3	Revalorization, utilization, stabilization	Revalorization is the process of treating or using waste materials in a way that recovers some value from them (or at least the process reduces their negative 'value'). At the same time they may be stabilized (e.g. energy recovery by incineration, composting). Incineration liberates the energy within waste in the form of heat, which can be used directly or can be converted into electricity. However, mixed wastes do not have high calorific value and there are limited uses for heat, which is in essence a by-product of another process and is therefore fixed in location. Furthermore, the process of converting heat into electricity has low efficiency and currently there is a surplus of low-cost energy. The real benefit of incineration is that it reduces the volume of the refuse significantly (by up to 90%), thus minimizing the volume of material that is finally placed in landfill. Composting is the controlled aerobic degradation of organic wastes and produces a material which may be suitable for landscaping, landfill cover, or soil conditioning. To achieve this degree of usefulness the refuse needs to have been carefully sorted (to remove deleterious and non-compostable material) or individual types of waste must be segregated at source. From the viewpoint of waste disposal the real value of composting is that it reduces the waste volume by 40–75% (Polprasert, 1989).
4	Disposal (dumping)	Landfill/landraise is the least desirable option for solid waste disposal and yet approximately 80% of all solid waste is currently landfilled (Ray, 1995).

Table 12.2. Annual waste generation in the UK*

Waste type	Category	Approximate annual quantity (Mt)	Total annual quantity (Mt)
Controlled	Domestic	20	245
	Industrial	70	
	Construction and demolition	70	
	Commercial	15	
	Sewage sludge	35	
	Dredged materials	35	
Non-controlled	Agriculture	80	190
	Mining and quarrying	110	

* Adapted from DOE (1995b).

household refuse in the UK during the past 50 years is illustrated in Figure 12.2. The 1956 Clean Air Act resulted in a major reduction in the use of solid fuel, and hence in the ash content of refuse (it fell from over 50% in the 1940s to around 8% in 1990). The loss of this inert, denser constituent and the increase in volume

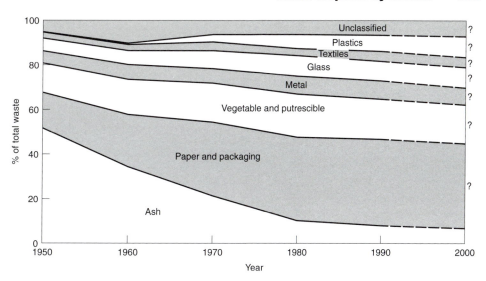

Figure 12.2. Variation in waste composition with time

of paper, vegetable matter, rag and plastics has had a significant effect on the biological and geotechnical properties of refuse fill. Furthermore, the elimination of domestic burning of refuse led to a drastic increase in the volume of household refuse, or municipal solid waste (MSW), which needs to be disposed of in a controlled manner. Recent domestic refuse contains a high proportion (around 55%) of organic material. Commercial and mixed industrial wastes can contain even higher proportions of organic material (this categorization includes both vegetable and putrescible matter and paper-based waste). As wastes are progressively minimized, re-used, recycled, processed and recovered, the characteristics of waste arriving at a landfill site will change further. It is not possible to predict the effects that the foregoing changes will have on refuse arriving at a particular site, and accordingly the containment works and operational methods should be designed to cope with this uncertainty.

The composition of landfill refuse is made more complex by the presence of 'difficult wastes' and the practice of co-disposal. Difficult wastes have some characteristics that require specialist handling at the site (although they are acceptable for disposal in landfill). Such wastes might include asbestos (both fibrous and bonded), fine particulate material, very large objects, animal carcasses, sludges and very light materials. Co-disposal is the joint disposal of putrescible municipal or commercial wastes with difficult wastes to achieve treatment of industrial, commercial, liquid and solid wastes by interaction with biodegradable wastes. The principle is that organic components of many co-disposed wastes are degraded by the same micro-organisms that are responsible for the biodegradation of municipal wastes. This latter material has a considerable capacity to neutralize acids and alkalis. Wastes that have been successfully co-disposed include: animal and food industry wastes, detergents, fats and greases in water, industrial effluent treatment sludges, acids and alkalis (DOE, 1993a).

The density of waste fill varies from about 0.12 to 0.30 t/m^3 when tipped. After compaction (Figure 12.3) bulk densities range from a low of 0.4 t/m^3 to a high of around 1.2 t/m^3; a value of 0.65–1.0 t/m^3 may be assumed for most analytical purposes. The bulk density of waste in a landfill varies widely due to:

- the large variation in waste constituents
- the degree of compaction
- the state of decomposition
- the amount of daily cover
- the total depth of waste
- the depth from which a sample is taken.

Figure 12.3. Waste spreading and compaction

Waste density may change with age as significant mass is lost by the formation of landfill gas and leachate, to an extent governed by the corresponding settlement. Refuse composition and compacted densities may also change as greater use is made of pretreatments such as shredding or pulverization and baling. Not only do shredding and pulverization reduce the bulk volume of wastes (by reduction of voids space), but they are also used to separate out the organic and inorganic fractions. The organic part of the waste can then be made into compost or refuse-derived fuel.

The moisture content of refuse, as placed, ranges from 10% to 50% approximately, and the average specific gravity of the solids is 1.7–2.5.

12.1.3. Development of landfilling

Until the late 1970s there was little engineering input into landfilling practices and little concern over the effects of waste disposal on the environment. Landfilling operations commonly involved the uncontrolled infilling of natural depressions and man-made excavations. The underlying, implicit principle at these sites was one of 'dilute and disperse'. This assumed that within the groundwater the concentrations of any contaminants derived from landfill would reduce to acceptable levels as they dispersed and were diluted under natural processes. However, dilution is not a mechanism by which leachate constituents are chemically altered or attenuated by the soil, it simply reduces the concentration of leachate constituents. Dispersion is a mechanical phenomenon that has no effect on the toxicity of pollutants. Many dilute-and-disperse sites were constructed in former sandstone quarries where the high permeability of the rock mass ensured the necessary groundwater interaction. Subsequently it has become accepted that most dilute-and-disperse sites represent potential environmental hazards.

In the 1990s landfill philosophy moved to the objective of total containment and isolation of wastes and there was a major upsurge in the development of engineered waste disposal by landfill. The current preferred method of landfilling is in individual cells (formed by low-permeability earth bunds 2–3 m high) because it encourages progressive filling and restoration. The waste is thus deposited within preconstructed containment areas (Figure 12.4). Design measures that are commonly adopted in current landfill sites are outlined in Table 12.3 and illustrated in Figure 12.5.

It is now recognized that total containment is unattainable and that it may be more responsible to design for controlled release, rather than to attempt indefinite isolation, so that natural attenuation can be utilized. This is a process by which the concentration of pollutants is reduced to an acceptable level by natural processes such as (Bagchi, 1994):

Figure 12.4. A cell within a landfill site

Table 12.3. Common elements of current engineered landfills

Design measure	Purpose
Low-permeability lining system	To minimize leachate egress and prevent ground pollution. Typical forms are clay mineral liners, geomembrane liners and bentonite–enriched soil liners
Underdrainage/leachate detection system	It is placed beneath the lining system to detect any leachate that has breached the liner and to allow for its subsequent control
Leachate drainage and control system	This overlies the lining system to ensure the maintenance of a low head of leachate above the liner and to allow efficient leachate recirculation
Low-permeability capping layer	Covers the waste to prevent water ingress into the waste and therefore to limit future leachate generation
Gas ventilation system	To control the movement and concentrations of landfill gas within the landfill and to mitigate against potential explosive and/or asphyxiation hazards. Systems may be passive or active
Leachate and landfill gas monitoring system	Both within the landfill and outside of the site, prior to, during and on completion of construction. This monitoring is vital for the early detection of environmental pollution

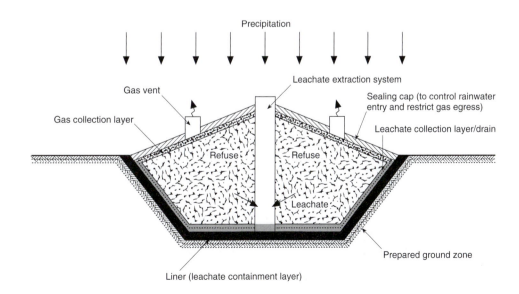

Figure 12.5. A schematic representation of an engineered landfill

- adsorption
- biological uptake
- cation- and anion-exchange reactions
- dilution
- filtration
- precipitation.

The principle issue is therefore how to predict, and subsequently to measure, the time at which the generation rates of leachate and gas, and the resultant strengths of these products, are so low that the emissions may be assimilated by the local environment.

As the number of 'holes in the ground' in which to tip waste dwindles then disposal practice moves from landfilling to landraising or landforming. This change provides certain benefits:

- drainage of leachate to the outside of the site is by gravity
- leachate drainage systems can be made more accessible for inspection and maintenance
- it is probably easier to form a proper basal liner
- sites are chosen on their technical merits (rather than by the location of former mineral workings, etc.)
- ponding of water on a completed site due to settlement is less likely
- some technical aspects become more straightforward (e.g. avoidance of problems of lining vertical faces).

However, the formation of hills of refuse (Figure 12.6), which will undergo significant decomposition and settlement, has a number of disadvantages and it requires even greater geotechnical input into engineered construction of such waste disposal facilities:

- The shear strength of the refuse, and the way it changes with time, will become very important because of slope stability aspects.
- Slope instability may occur not only in the side slopes of the refuse mound but also at the interface between the refuse and the basal liner, in the form of a wedge failure (see Chapter 6, Section 6.1 and Chapter 9, Section 9.5), due to the low shear strength of this interface.
- Settlement or distortion of the covering over the refuse will become very important, because of the potential for collapse of this cap, and the development of perched water-tables, leachate springs in the side slopes, etc. (due to infiltration of rainwater through a cracked cap).
- Long-term functioning of drainage will be vital to prevent the development of large lateral forces from fluid within the refuse.

Figure 12.6. Landraise disposal of waste

- The prominence of waste heaps leads to pressures to cover them with vegetation to improve them aesthetically and to prevent surface erosion. Unless this vegetation and the resultant habitat are carefully managed the sealing cap over the refuse requires substantial engineering to prevent damage from tree roots, burrowing animals, etc.

The overall considerations for design and operation of engineered landfills are determined by the expected time to achieve stabilization and thus whether the works have to function for decades or for centuries. If materials are needed to last for very long periods of time aspects to be considered include: the location of the material within the landfill, its susceptibility to changes in physical loads and/or chemical attack, and the effects of variations in operational or maintenance practice. The complete design, construction, operation and restoration of a refuse disposal site is a major civil engineering project with a high geotechnical content.

12.1.4. Landfill containment

The primary performance objectives for a landfill containment system can be summarized as follows:

- to control or prevent seepage of leachate from the landfill so that it does not cause contamination of groundwater
- to control or prevent the migration of landfill gas and collect it so that it can be used or rendered harmless
- to remain stable and operate efficiently for the required design life, and to be compatible with the expected leachate and gas composition and temperature
- to ensure that the completed landfill site does not fill up with liquid so that there is no leachate overflow.

The design concept for the base and sides of a containment-type landfill consists of restricting leachate and gas seepage into the surrounding ground. To achieve this aim, waste disposal sites are generally lined with mineral layers or synthetic membranes (or a combination of both) and a leachate collection system is installed. Some leakage through the base of a containment landfill is unavoidable. However, it can be reduced to practically zero. Synthetic materials may, in theory, allow less leakage but are difficult to protect from damage, whereas clay liners are not easily damaged but are inherently more permeable. Furthermore, mineral barriers are orders of magnitude less effective at restricting gas flow than leachate seepage, whereas geomembranes have very low gas permeability.

Flow through a liner occurs by two mechanisms: advective flow (physical flow under a hydraulic gradient as determined by permeability; see Chapter 7, Section 7.1.2) and diffusion (a physico-chemical phenomenon, whereby chemicals travel as a result of a concentration gradient). At high flow rates advective flow is the predominant transport mechanism, but when the flows become very small (corresponding to a coefficient of permeability of around 10^{-10} m/s) diffusion tends to become at least equally important. If the liquid head inside the refuse is higher than the water-table in the surrounding ground then advective flow will be out of the landfill; if the head difference is the opposite way round advective flow will consist of groundwater moving into the refuse. Diffusion will always act to move the contaminants out of the landfill site because of the higher concentrations in the waste than in the groundwater. If the direction of diffusive transport is the same as the direction of advective flow (i.e. out of the landfill), then it will increase the amount of contaminant transport and decrease the time it takes for contaminants to move to a given point away from the source. Because diffusion will invariably be out of the site it is possible for contaminant to escape from a landfill, even though the groundwater flow is directed into the landfill. However, it is also important to recognize that for an engineered landfill the concentration gradient across the liner will decrease with time.

A landfill liner system may comprise a combination of barriers and fluid collection layers, plus mineral or synthetic components (fulfilling a separation or protection function). The principal types of liner system that can be fabricated from these components are illustrated in Figure 12.7:

- A *simple liner* comprises a single primary barrier. This is typically overlain by a leachate collection system with an appropriate separation or protection

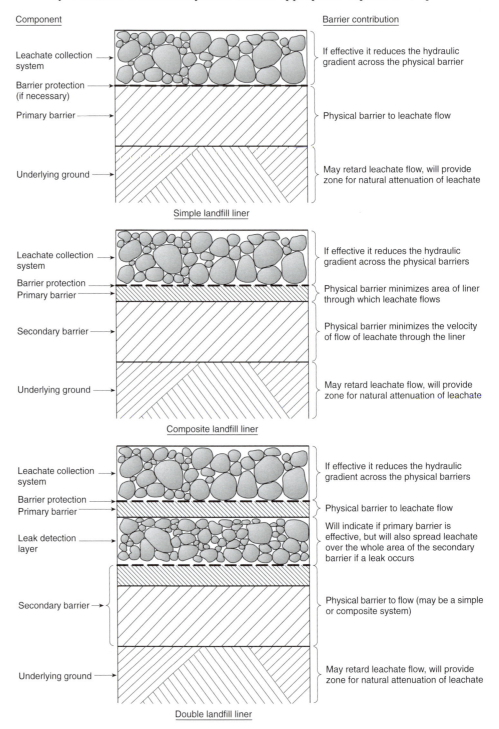

Figure 12.7. Types of liner system

layer. This type of liner system is only used in situations with low sensitivity to pollution.

- A *composite liner* includes two separate barriers (primary and secondary), usually made of different materials, placed in intimate contact with each other to provide a synergistic effect. Typically, a primary geomembrane is placed above a secondary low-permeability mineral barrier. Again, a leachate drain is typically placed above the primary barrier. Such a liner system combines the advantages of the two materials with their different physical and hydraulic properties. This type of liner system is in common use in medium-sensitivity situations. The performance and robustness of this liner system should be much better than that of a single liner.

- A *double liner* includes primary and secondary barriers with an intermediate high-permeability drainage layer to monitor and remove liquids or gases from between the barriers. The primary barrier is again overlain by a basal collection drain. This type of liner system allows the monitoring and removal of any fluids that may seep through the primary barrier. However, the intermediate high-permeability layer promotes the spread of leachate if there is leakage through the primary (upper) barrier. Hence the system is not recommended.

Lining systems are required to provide short-, medium- and long-term protection of the environment, and therefore must be robust, durable, resistant to chemical attack and puncture, and be constructed to the highest engineering standards. There are procedures available for testing individual liner elements for some of these properties and some systems for the detection of holes in installed geomembranes (described in later sections).

The leachate concentrations and the contaminating lifespan of a landfill are directly related to the movement of water through the cover. The design and construction of the final cover on a landfill are thus as important as the liner at the base of the landfill — the barrier layer in the final cover should be the same as, or better than, the primary liner at the base of a landfill. This provides a major engineering challenge, as covers over refuse must be capable of withstanding stresses caused by differential movements and settlements. The stability of the final cap is also of prime concern, especially on steep restoration profiles or those incorporating low-friction layers such as flexible membranes. Erosion control may be required on steep slopes where large volumes of run-off can be expected. Such controls may include the early establishment of grass cover or the provision of drainage channels. Cracking of the capping layers may arise as a result of differential settlement at phase boundaries, across the site or on the site perimeter. In extreme cases of settlement, ponding may occur or surface water may flow into the underlying waste through cracks in the cap.

As outlined in Table 12.3 and Figure 12.5, a landfill containment system generally comprises:

- a liner (very low-permeability layer) on the base and sides of a landfill
- an effectively impermeable cap over the top of the waste
- a leachate abstraction system below the waste
- a gas collection layer over the refuse.

Such systems may incorporate, either individually or as part of a composite construction, the following materials:

- natural minerals (e.g. compacted clay or shale)
- manufactured construction materials (e.g. bentonite, asphaltic concrete)
- geomembranes (also known as flexible man-made liners or (FMLs)) (e.g. polypropylene, polyethylene).

12.2. Waste decomposition

At the present time it is not possible to predict with confidence, nor support an estimation with experience, how long it will take for landfill decomposition to be effectively completed. Typical untreated waste is a highly heterogeneous material with variable permeability and water-retention properties, which limit access for biodegradation by the micro-organisms. The micro-organisms rely on leachate movement for colonization and to bring nutrients to them. The ability to achieve rapid flushing of leachate uniformly throughout the waste is thus a key factor in the promotion of waste degradation. Recirculation of leachate within a landfill enhances fermentation and degradation and removal of chemical constituents. The hydraulics of landfilled wastes are poorly understood, but often appear to be affected by 'short-circuiting' (most flow passing through a few major channels) and by barriers to flow. Some form of pretreatment is required prior to landfilling to produce a much more uniform waste mass. Placement of the waste at a degree of compaction that will optimize permeability and flushing rate is more problematical because it will produce a looser, more compressible body of fill.

In theory, encapsulation can keep waste dry and thus prevent decomposition and leachate formation. However, some leakage through the cover over a landfill is certain (for both synthetic membranes and clay liners) because of the nature of the works, and leachate formation is unavoidable in the long run. Encapsulation merely delays the process. Thus, encapsulated, partially decomposed waste represents a long-term threat to groundwater and the environment.

12.2.1. Biodegradation

It has become conventional practice since the 1970s to level and compact landfilled refuse as soon as it is discharged at the working area. Rapid compaction enables the maximum amount of waste to be emplaced in the minimum time and reduces the impact from litter, flies, vermin, birds and fires. Greater understanding of landfill biological processes has shown, however, that there are also disbenefits which should be taken into account where the rate of degradation is an issue. High waste densities may inhibit biodegradation by restricting leachate and landfill gas movement and cause perched water-tables within sites. A bulk density of about $0.8\,t/m^3$ seems to be the optimum for the biodegradation processes in mixed household waste. If this is so, excessive use of steel-wheeled compactors, which can produce in situ bulk densities in excess of $1.2\,t/m^3$, should be avoided.

Commonly, a thin layer of inert material or soil (daily cover) is placed on top of the refuse at the end of each day in order to:

- prevent escape of windblown refuse
- reduce odour emission
- deter scavengers, birds and vermin
- improve the visual appearance of the site.

Unfortunately, after compaction, daily cover can have a relatively low permeability, which results in the partial containment of each layer of waste unless it is broken up before the next layer of fill is placed. If not broken up, leachate becomes perched and difficult to extract and landfill gas moves preferentially sideways giving greater potential for migration off-site.

Biodegradation processes are enhanced if the waste mass is made uniform in particle size and density. Waste pretreatments that would be beneficial include:

- screening excavation and demolition waste to separate fines and aggregate for use on site
- composting vegetable waste to create restoration material
- pulverizing biodegradable waste to create a more homogeneous feedstock for accelerated stabilization of landfill.

Furthermore, there is evidence that the addition of sludge to mixed waste accelerates chemical and physical stabilization of the landfill, while occupying almost negligible void space.

The process of refuse decomposition and the microbiology of landfills is relatively well understood (Senior, 1990). Nevertheless, it is impossible to extrapolate the definition of individual degradation events within the refuse to predict the overall behaviour because of the heterogeneity of the waste. However, for engineering purposes it is the outputs from decomposition that are important and so it is sufficient to describe the process in an overall way (Figure 12.8). Following its deposition in a landfill site the organic fraction of the waste will begin to undergo degradation through chemical and microbiological action. The organic fraction of wastes characteristically comprises carbohydrates such as cellulose, lipids, proteins and fats. In domestic wastes, these account for approximately 55% (by dry weight), while for commercial and mixed industrial wastes the proportions are around 65% and 60%, respectively.

Degradation of the organic fraction of waste materials within a landfill may be described in a simplified way as a five-stage process (Table 12.4). The first and fifth stages occur under aerobic conditions (i.e. in the presence of oxygen), while the remaining stages take place under predominately anaerobic conditions (i.e. oxygen is absent). Each stage of the process has an impact on the characteristics of the intermediate and final breakdown products and the quality and rate of generation of leachate and landfill gas. Within a landfill site as a whole, all the stages of degradation may be occurring simultaneously at different rates because of the extended duration of waste emplacement, the different biodegradabilities of the various refuse components, and the spatial variability in the physical and chemical environment of the waste materials.

12.2.2. Landfill leachate

The infiltration of rainwater and surface water into a landfill, coupled with biochemical and physical breakdown of wastes, produces a polluted liquor (leachate). Leachate consists of water which carries, in solution or suspension, chemicals, metal contaminants and organic matter. The main problem is generally caused by the continuing decomposition of the suspended organic matter, which absorbs oxygen in this process. The total oxygen demand for respiration and breakdown of organic material is the Biological Oxygen Demand (BOD) of a volume of water. If sufficient oxygen is available, aquatic animals can breathe and organic particles in the water can be oxidized to form carbon dioxide. BOD is one quantity that is used to indicate the 'strength' (or 'quality') of leachate. If untreated leachate is discharged directly into a watercourse it will absorb oxygen from the water to complete its decomposition, and unless the dilution is such that the watercourse can cope with this demand the level of oxygen in the water can fall below that necessary to support fish and animal life

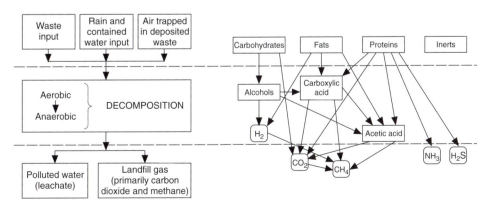

Figure 12.8. The general process of refuse decomposition and outputs

Table 12.4. Phases of refuse decomposition

Phase	Stage	Comments	Outputs/outcomes
I	Aerobic	Generally of limited duration due to the high oxygen demand of waste relative to the limited quantity of oxygen present. Proteins are degraded into amino acids, then into carbon dioxide, water, nitrates and sulphates. Carbohydrates are converted to carbon dioxide and water and fats are hydrolysed to fatty acids and glycerol. Cellulose (constitutes the majority of the organic fraction) is degraded into glucose. The landfill may reach elevated temperatures (80–90 °C). As the anaerobic conditions develop, the anaerobic micro-organisms, which include the methane generating organisms (methanogens), gradually become established	Major decomposition products of this process are water and carbon dioxide, and the characteristic odour associated with this stage of the process is mainly due to the presence of organic esters. This phase is short and no substantial leachate generation will take place as the waste matrix is not fully saturated
II	Anaerobic	First stage involves acid fermentation (acetogenesis). The complex organic compounds convert first to simpler organics, then to organic acids, and finally to methane and carbon dioxide. Acetic acid serves as a primary pathway to the end-products. Decrease in pH is caused by the high production of volatile fatty acids and the high partial pressure of carbon dioxide. Nitrogen is displaced by carbon dioxide and hydrogen to form leachate with a high ammoniacal nitrogen content. The temperature drops to 30–50 °C. Anaerobic bacteria aid the breakdown of materials and reduce the redox potential so that the methanogenic bacteria can grow	Decrease in leachate pH, high concentrations of volatile acids and considerable concentrations of inorganic ions (e.g. Cl^-, SO_4^{2-}, Ca^{2+}, Mg^{2+}, Na^+). Typically 6–12 months after deposition of the waste substantial quantities of methane will be produced
III	Anaerobic	Intermediate anaerobic stage starts with the slow growth of methanogenic bacteria. Conversion of fatty acids	Methane concentration in the landfill gas increases, while hydrogen and carbon dioxide decrease. Concentration of sulphate decreases. Increase in pH values and alkalinity, and a consequent decrease in the solubility of calcium, iron, manganese and heavy metals
IV	Anaerobic	Final anaerobic degradation is characterized by methanogenic fermentation, bacteria metabolize acetate and formate produced during the other degradation stages to form methane and carbon dioxide. The pH range tolerated by methanogenic bacteria is extremely limited and ranges from 6 to 8. Solubilization of the majority of organic components has decreased at this stage of landfill operation, although the process of waste stabilization will continue for several years. At large, deep landfills the production of methane (at maximum concentration) can be expected to continue for periods in excess of 10 years	Leachate is characterized by almost neutral pH values, low concentrations of volatile acids and total dissolved solids. Landfill gas has a methane content of 50–60%, most of the remainder is carbon dioxide. Leachates are characterized by relatively low BOD values and low BOD/COD ratios. Ammonia continues to be released
V	Aerobic	The methane production rate is so low that air will start diffusing from the atmosphere, giving rise to aerobic zones (in the upper region of a landfill) and zones with redox potentials too high for methane formation. As a consequence the generation of decomposition products reduces and the site becomes stabilized. Only the more refractory organic carbon remains in the landfilled wastes	Production of landfill gas and contaminated leachate decreases

BOD, biological oxygen demand; COD, chemical oxygen demand.

and these will suffocate. A further problem that can occur is the production of ammonia, which can also adversely affect fish and plant life. Once groundwater has become polluted it may be unsuitable as a source of potable water for many years.

An important factor affecting leachate generation is the absorptive or moisture-retention capacity of the waste. The absorptive capacity of waste varies according to its type, pretreatment and degree of compaction. Shredding and pulverization may increase absorptive capacity by providing a more uniform fragment size, a higher porosity and a higher surface area for attack. In contrast, baling or increasing the emplacement density reduces the rate of percolation but increases the risk of water flow being concentrated within narrow zones of high permeability.

Expressions have been formulated to estimate the volumes of leachate that arise before site closure L_v and after completion of the site L_p:

$$L_v = P + W - E - A \tag{12.1}$$
$$L_p = P - E - R - A - S \tag{12.2}$$

where P is the precipitation, W is the liquid squeezed out of the waste, E is the loss due to evaporation, A is the absorption within the waste, R is the surface run-off and S is the absorption within the cover.

Equation (12.1) is difficult to use because of the lack of accurate data for S, E and W. Hence use is made of typical rates determined from various field measurements. Published leachate production rates range from 15% to 55% of annual rainfall (Table 12.5). For British climatic conditions this range corresponds to a daily leachate production rate of around 5–15 m^3/ha. General trends that are discernible from the data are that well-compacted landfill produces a little less than half the amount of leachate of loose waste, and an uncovered deposit produces approximately 50% more leachate than one covered with soil and vegetation.

Equation (12.2) is only applicable to landfills where a relatively highly permeable layer of soil is used as final cover. Nowadays covers are composed of low-permeability layers and the amount of water infiltration will be much less. Hence such a water balance equation is no longer appropriate and more sophisticated predictive models are needed (e.g. that described by McDougall *et al.*, 1996) if the general values obtained from past experience are too imprecise.

Table 12.5. Leachate generation rates

Source	Proportion of annual precipitation (uncapped cell)
Campbell (from Gettinby, 1999)	50–55%
DOE (1995a)	30%
Ehrig (1988)	15–25% if steel-wheeled compactors are used 25–50% if crawler tractors are used on the waste
Hoeks & Harmsen (1980)	25% (for Dutch climatic conditions)
Newton (from Gettinby, 1999)	39% if the cell has a vegetated surface 55% if there is no vegetative cover
Robinson & Maris (1979)	36% for a vegetated surface 52% for a bare soil surface
Stegmann & Spendlin (1989)	15–25% for landfilled waste with a bulk density greater than 0.7 t/m^3 25–50% for landfilled waste with a bulk density less than 0.7 t/m^3

In general, current practice is to construct a final cover over a landfill as soon as the waste reaches the designed final grade, in order to minimize leachate production. However, there is a growing body of opinion in favour of delaying construction of the final cover (in which case equation (12.2) would be relevant). The advantage of prompt construction of a final, properly designed cover is a significant reduction in leachate quantity within a short period of time after the construction of a final cover. Vegetation growth in the topsoil of a final cover reduces the moisture available for infiltration significantly by evapotranspiration, and a low-permeability layer in the cover reduces percolation. The disadvantage of prompt construction of a final cover is that leachate which will need treatment is likely to be produced for a number of years after closure.

The composition of leachate depends, in a general way, on the stage of degradation and the type of waste within the landfill. Table 12.6 shows the range of values obtained for 35 leachate parameters from a compilation of 48 published sources of data on landfills. The reported values for each parameter cover such

Table 12.6. Leachate composition*

Parameter	Total range[†]
Aluminium	0.27–2.7
Ammoniacal nitrogen	0–3000
Alkalinity	0–20 850
Arsenic	0.021–0.13
BOD	2.0–57 700
Boron	4.2–7.4
Cobalt	0.01–95
Cadmium	0.001–17.0
Calcium	5–7200
COD	0–89 250
Chloride	4.7–4816
Chromium	0.002–18
Conductivity	400–50 000
Copper	0–9.9
Fluoride	0.27
Iron	0.09–2500
Lead	0–12.3
Magnesium	13.3–15 600
Manganese	0.06–1400
Mercury	0.002–19.5
Nickel	0.002–79
Nitrate	0–1300
Nitrite	0–25
pH	3.7–9.1
Phosphorus	0–154
Potassium	2.8–3770
Silica	12–34
Sodium	0–7700
Strontium	0.94–72
Sulphate	1–2000
Sulphide	0–30
Suspended solids	0–700
TOC	0–28 500
Total dissolved solids	0–44 900
Zinc	0–1000

BOD, biological oxygen demand; COD, chemical oxygen demand; TOC, total oxygen consumption.

* From Gettinby (1999).
[†] All values in mg/l, except for pH and conductivity (μS/cm).

wide ranges that this information is of very limited use. The main factor causing this wide spread of values is the age of the individual landfills (i.e. they are at various stages of refuse decomposition from the waste being 'fresh' to it being more or less stabilized). By just dividing data from landfill sites according to their age or decompositional stage, narrower bands are identifiable for various leachate parameters (Table 12.7) and general behavioural trends can be defined:

- Leachates generated from recently emplaced wastes, which are in the acetogenic phase of degradation, contain high concentrations of organic compounds. Such leachates are also characterized by: depressed pH values; the production of organic compounds, the volatility of which may also produce unpleasant smells; and dissolved carbon dioxide. Acetogenic leachates also contain ammonium ions, which result primarily from the breakdown of proteins and other nitrogenous compounds in the wastes.
- Leachates generated during the early stages of anaerobic degradation are characterized by: high concentrations of volatile fatty acids; acidic pH (5 to 6); a high BOD and BOD/chemical oxygen demand (COD) ratio; and high levels of ammoniacal nitrogen and organic nitrogen. Ammonia is largely generated as a result of the degradation of proteinaceous materials. The low redox potential of this leachate facilitates the production of metals such as chromium, iron and manganese. As the pH rises, these metals are precipitated as sulphides, hydroxides and carbonates.
- The 'strength' of leachate reaches a peak value after a few years (see Table 12.7) and then gradually declines. The various contaminants attain their peak concentration at different times after infilling and the time versus concentration variation plots of all contaminants from the same landfill may not be similar in shape.

The magnitude of particular parameters for leachate do vary from site to site because of climatic effects, waste composition, different site practices, etc. However, by taking an overview of the large amount of archival data available, it

Table 12.7. The effect of landfill age on leachate composition (domestic and commercial waste)*

Pollutant	Upper bound value (mg/l) at different ages of a landfill				
	5 years	10 years	20 years	30 years	40 years
Chloride	2400	2600	2700	2800	2800
Iron	540	380	100	40	20
Ammoniacal nitrogen	800	700	590	580	570
BOD	2000	1200	350	100	70
Potassium	580	570	440	380	350
Cadmium	1.1	0.7	0.12	0.10	NS
Manganese	1.5	2.3	3.2	2.8	1.6
pH[†]	5.5–8.7	5.5–8.7	5.8–8.5	6.2–8.1	6.5–8.0
Nickel	3.5	1.3	0.33	0.20	0.10
Zinc	3.5	1.8	0.40	0.30	0.20
Calcium	NS	500	300	200	NS
Magnesium	100	200	150	100	NS
Sodium	2000	2000	1200	500	NS
Chromium	0.50	0.35	0.15	0.12	0.10
Lead	0.55	0.50	0.20	0.10	NS
Copper	0.80	0.40	0.15	0.10	0.06
COD	8000	4000	2000	500	NS

BOD, biological oxygen demand; COD, chemical oxygen demand; NS, insufficient data available for a recommendation to be made.

* After Gettinby (1999).

† The pH values represent the extremities of acidity and alkalinity and are not given in units of mg/l.

Table 12.8. An example of possible landfill completion criteria

Pollutant	Upper bound value (mg/l)
Chloride	2000
Iron	2
Ammonia	5
Potassium	120
Cadmium	0.05
Manganese	0.5
pH[†]	6.5–8.5
Nickel	0.5
Zinc	1
Calcium	1000
Magnesium	500
Sodium	1500
Chromium	0.5
Lead	0.5
Copper	1
Phenols	0.005
PAHs	0.002
Mineral oils	0.1

PAHs, polyaromatic hydrocarbons.

* From DOE (1994c).
† The pH values represent the extremities of acidity and alkalinity and are not given in units of mg/l.

is possible to define realistic upper bound values for leachate constituents at various times from the start of landfilling (see Table 12.7). Using such data, and possible completion criteria (e.g. Table 12.8) an estimate may be made of the time needed before a landfill becomes benign in terms of various substances within the leachate (Table 12.9).

12.2.3. Landfill gas

As with leachate, the composition and quantity of gas emitted from a landfill varies with degradation stage and so it varies with time (Figure 12.9). Gas

Table 12.9. Time periods for leachate components to decline to the completion value given in Table 12.8

	Time (years) after landfill closure						
	0–5	5–10	10–15	15–20	20–30	30–40	>40
Chloride	•	•	•	•	•	•	•
Iron	•	•	•	•	•	•	•
Ammoniacal nitrogen	•	•	•	•	•	•	•
BOD	•	•	•	•	•	•	•
Potassium	•	•	•	•	•	•	•
Manganese	•	•	•	•	•	•	•
Cadmium	•	•	•	•	•	•	
pH	•	•	•	•	•	•	
Nickel	•	•	•	•			
Zinc	•	•	•				
Sodium	•	•	•				
Lead	•	•					
Chromium	•						
Magnesium*							
Calcium*							
Copper*							

BOD, biological oxygen demand.
* Never exceeds the criterion.

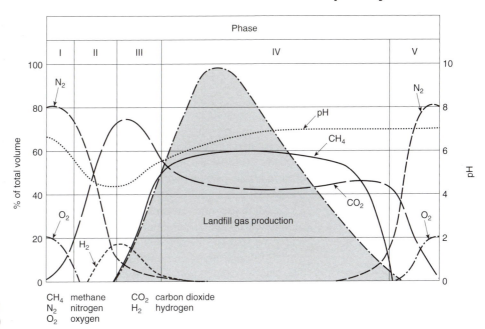

Figure 12.9. Landfill gas production with time (adapted from Farquhar & Rovers, 1973)

CH₄	methane	CO₂	carbon dioxide
N₂	nitrogen	H₂	hydrogen
O₂	oxygen		

production may be speeded up by adding sewage sludge or agricultural waste, removing bulky metallic objects from the waste stream, or using less daily cover soil. Factors that determine the gas production rate are:

- the physical dimensions of the site
- the type of refuse and its input rate
- the moisture content of the waste
- the landfill pH
- the fill temperature and density
- the site operational practices.

During the first aerobic degradation stage (phase I in Table 12.4) oxygen levels become depleted, with a concomitant rise in levels of carbon dioxide within the gas expelled from the decaying refuse. As hydrolysis and fermentation occur (phase II) levels of carbon dioxide and hydrogen increase and nitrogen levels fall — in some instances hydrogen may account for up to 20% (by volume) of total gas yield. The concentration of methane present in the gas from the infilled refuse gradually rises as the acetogenic and methanogenic stages develop (phase III). It may take as little as 2 months, or more than a year, for significant quantities of methane to be produced, but this generation rate can subsequently continue for more than 20 years. When steady-state methanogenic conditions are attained (phase IV) the resulting gas composition is about 60–65% methane and 40–35% carbon dioxide. Typically the methane production rate ranges from around 1 to 6 m³/year per tonne of wet waste. In theory total yields of around 400 m³ of methane per tonne of household/commercial refuse are possible (Crawford & Smith, 1985). Actual yields are much lower due to the conditions for anaerobic digestion being less than optimal in landfill, and outputs ranging from 40 to 200 m³ per tonne of wet refuse over 10- to 20-year periods have been reported (DeWalle *et al.*, 1978; ETSU, 1993). As the degradable organic fraction within the waste becomes exhausted, methane and carbon dioxide generation decline (phase V).

While, over the life of a landfill site, a large number of gases are produced in varying proportions, the term 'landfill gas' is usually used to denote that gas issuing during phases III and IV (i.e. essentially a mixture of methane and carbon dioxide). The gas can make its way off site (migration distances of up to 400 m have been recorded) and into buildings and services. The problems that arise as a result of this uncontrolled migration include:

- explosion or fire due to build-up in confined spaces
- asphyxiation of persons entering culverts and manholes in the vicinity of the site
- harmful effects on vegetation
- odour.

Methane is flammable at concentrations of between about 5% and 15% (by volume) in air; these are the lower and upper explosive limits, respectively. Carbon dioxide is non-flammable, but stimulates the respiratory and central nervous systems at high concentrations. At a level of 3% the breathing rate is doubled, and higher levels lead to headaches and exhaustion. An individual may collapse if subjected to concentrations above 10% and death usually results above 25%. Methane and carbon dioxide are colourless and odourless, but they are mixed with very small quantities of other gases (primarily hydrogen sulphide) that give rise to odours. The precise density of landfill gas is dependent on the actual proportions of its major components, but it is usually about the same as air. The gas is usually saturated with moisture.

The re-use of derelict land for redevelopment has led to considerable development on or near to previously closed landfill sites. Gas monitoring (Figure 12.10) should be undertaken at the edge of a landfill site, in the vicinity of the nearest property at risk and at an intermediate point. In the absence of nearby property monitoring up to a distance of 250 m from the fill materials is recommended (DOE, 1991a). Although there is no universal stipulation for sampling holes it is generally accepted that monitoring points should be sunk to the approximate depth of the fill.

12.2.4. Waste settlement

Accurate prediction of settlement is difficult because there are few detailed long-term records of landfill settlement. Laboratory-scale tests are not feasible or particularly useful because of the difficulties of obtaining reliable and representative test specimens and of defining appropriate testing regimes. Field measurements are essential for developing methods for estimating the settlement of landfills. Available data (Table 12.10) indicate that most settlement takes place over 30 years, with the major proportion occurring in the first 5 years. Likely values of total settlement (as a percentage of depth of fill), which are often quoted as rules-of-thumb, are:

- household refuse, 20%
- commercial waste, 10%
- industrial waste, 5%.

Figure 12.10. A gas monitoring borehole

Table 12.10. Observed settlement behaviour of landfilled refuse

Source	Waste type	Settlement (% of depth of waste)	Time period (years)
Edgers *et al.* (1992)	Municipal solid waste	25–50%	20
Edil *et al.* (1990)	Municipal solid waste	5–30%	Most occurs in 2 years
Frantzis (1981)	Household refuse	Up to 20% (compacted to about $0.6 \, t/m^3$) Up to 10% (compacted to maximum in situ density of $\approx 1.2 \, t/m^3$)	65% of total occurred in 3 years, most complete in 15 years
Hurtric (1981)	Household refuse	Overall 15–20%	Around 20
Jessberger (1994)	Mixed landfill	About 20%	15–20
Nobel *et al.* (1988)	Household refuse	20%	20
Sarsby (1987a)	Household and commercial	6–9%	5
DOE (1994c)	Household refuse	Overall 10–20% Initial 9–18%	30 5

Furthermore, most of the settlement (70–80% of the total) occurs within the first 5 years after placement.

Settlement of the surface of the refuse will inevitably occur as a result of degradation and deformation of the 'skeleton' of the waste within a landfill. Five mechanisms have been suggested as being responsible for settlement of waste fills (Sowers, 1973):

- mechanical — due to distortion, crushing, material reorientation
- ravelling — due to the washing of finer particles into voids between the larger particles
- physico-chemical change — due to corrosion, oxidation
- biochemical decay — due to both aerobic and anaerobic degradation
- interaction — between products of decomposition that 'trigger' displacements.

Initial settlement occurs predominantly due to the physical rearrangement of the waste material after it is placed in the landfill. Later settlement mainly results from biochemical degradation of the waste, which in turn leads to further physical settlement. Inevitably there is an overlap of timing in these two processes which makes them impossible to separate. Biological activity will affect the compression characteristics of the infilled material. As a result of utilization of organic material within the landfill by micro-organisms, the structure of the matrix will be weakened. This will increase the compressibility, and also change the permeability of the landfill matrix as a result of the gas generation. With refuse, by far the largest compression is due to the slow and continuous process of secondary settlement, where the particle structures begin to break down. The foregoing factors mean that refuse settlements cannot be predicted to a high degree of accuracy. On the other hand, the very nature of the situation being considered means that an accuracy of the order of ±50 mm is probably acceptable.

Typical settlement–log(time) curves for covered landfill are shown in Figure 12.11. These curves are similar to those for peat or organic materials. However, the settlement theories developed from fine-grained materials are not strictly applicable to landfill. The mechanics of refuse settlement are many and complex, even more so than for a soil due to the extreme heterogeneity of, and large voids present in, the refuse fill. Some immediate settlement will occur very quickly and be concurrent with the construction operations that are creating the load.

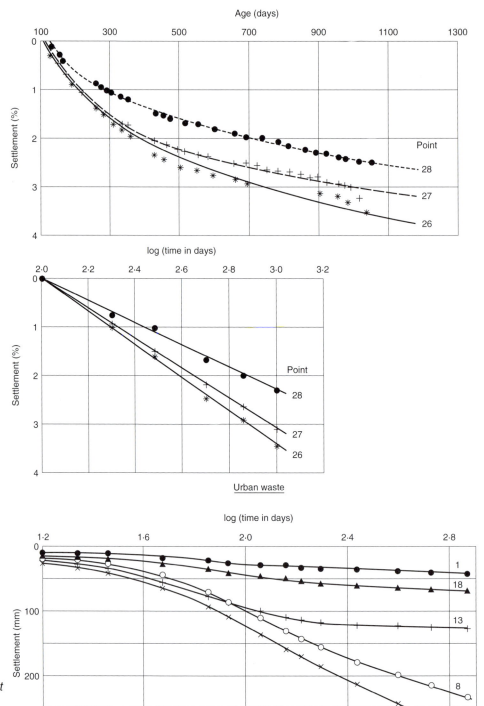

Figure 12.11. Typical settlement behaviour of a landfill over time: urban waste (after Sanchez-Alciturri et al., 1995) and mixed waste (after Sarsby, 1987a)

Subsequently, the settlement develops rapidly as in primary consolidation, but then it tapers off as in secondary compression. Since the refuse is not always saturated, the initial settlement is called *primary compression* and not consolidation. However, the time for completion of primary compression is in-keeping with C_v values in the range 4–160 m²/year. Settlement data show a large initial compression followed by long periods of delayed compression. It has been proposed that the primary settlement S_{ps} occurring in a waste layer due to imposed stress can be represented by an equation similar to that for normally consolidated soils (Yen & Scanlon, 1975; Morris & Woods, 1990):

$$S_{ps} = H \frac{C_c}{1 + e_0} \log \frac{P_0' + dP'}{P_0'} \equiv HC_c' \log \frac{P_0' + dP'}{P_0'} \tag{12.3}$$

where H is the initial thickness of the layer, C_c' is the primary compression index, P_0' is the existing effective overburden pressure acting at the mid-level of the layer, and dP' is the increment of overburden pressure at the mid-level of the layer. There is a scarcity of published data on the value of the primary compression index, but available information indicates C_c' to be in the range 0.2–0.5 for commercial or household waste.

The settlement of a landfill continues after the primary compression and, from the point of view of performance of a cap over a landfill, this is the most important component of settlement. The long-term settlement S_{sc} appears to be linear on a log(time), scale and can be represented by the following equation:

$$S_{sc} = C_\alpha H \log t \tag{12.4}$$

where C_α is the secondary compression index and t is the time at which the settlement value is required. Quite a few data have been published for C_α (Table 12.11), and for the design of landfilled refuse a value in the range 0.03–0.15 would seem reasonable.

For a landfill with a non-uniform base, differential settlement may lead to stresses and breaks in the final cap. The large settlement that normally takes place means that landfill slopes can change significantly with time, negating careful contouring and drainage provisions (Figure 12.12) so that water ponds in depressions, thereby increasing the hydraulic head and consequent flow through the cover. With the use of geomembrane sheets an additional consideration is that the straining of the geomembrane over the settlement may initiate holes or ruptures. The formation of tension cracks in the underlying compacted or natural clay barrier is also a possibility under some extreme circumstances. Increasing use of landraising can be expected to promote the occurrence of problems related to differential settlements.

Table 12.11. Values of the coefficient of secondary compression index C_α for waste fills

Source	C_α	Situation
Bromwell (1991)	0.13–0.32 at first reducing to 0.01–0.02 after 10 years	Field records
Druschel & Wardwell (1991)	0.14–0.95 (same for old and young landfill)	Field measurements
Edgers et al. (1992)	0.01 (early times) to 0.5 (long-term)	Field records
Landva & Clark (1990)	0.002–0.03 0.06–0.09	Laboratory tests (large samples) Field records
Sanchez-Alciturri et al. (1995)	0.02–0.06	Field measurements
Sarsby (1987a)	0.01–0.07	Field measurements
Sowers (1973)	0.03–0.09	Field records
Watts & Charles (1990)	0.005–0.01 uncompacted open-cast mining backfill 0.1–0.23 fresh refuse	Field measurements Field measurements
Yen & Scanlon (1975)	0.06–0.07	Field records

Figure 12.12. A playing field over refuse (after settlement)

12.2.5. Strength characteristics of waste

Quantification of geotechnical properties of waste materials can obviously be very difficult, particularly in the case of heterogeneous materials such as urban refuse and contaminated fill. Hence the shear strength characteristics of refuse material, or even the means of determining them, are still ill-defined. Caution should be exercised in applying soil mechanics principles and Mohr–Coulomb failure theory to refuse because of the major difference in the strains associated with shear failure in soils and those that are needed to produce shear failure in refuse. Nevertheless, there have been a number of determinations of refuse shear strength parameters, mainly using the direct shear box (but triaxial tests and in situ tests have also been used) (Table 12.12). The use of standard geotechnical strength tests to define the behaviour of refuse is questionable for the following reasons:

- the waste is likely to release gas during the test
- putrescible waste degradation will lead to a change in the shear properties
- most waste deposits are extremely heterogeneous
- small samples are unlikely to be representative of the waste structure.

However, in the absence of any better representation of the shear strength characteristics of refuse an interim method using a quasi-geotechnical approach represents a way forward and is considered acceptable. Considering the perceived variability of landfilled refuse the reported shear strength parameters are relatively consistent. For design purposes, reasonable values would seem to be:

- a cohesion value of 0–30 kN/m^2 and a friction angle in the range 20–35°
- for freshly placed waste, take a low cohesion with high friction (the refuse is akin to a granular material)
- for aged fill, take high cohesion with a low friction angle (to account for increased 'fineness' of waste due to degradation).

Table 12.12. Shear strength parameters for landfilled refuse

Source	Cohesion (kN/m^2)	Friction angle (°)
Duplancic (1990)	0	34
EMCON (1987)	18–35	14–20
Fang (1977)	~60	~18
Jessberger (1994)	0–30	17–42
Landva & Clark (1990)	0–23	24–41
Saarela (1987)	0–70	19–33
Sanchez-Alciturri *et al.* (1995)	0	28–35
Siegel *et al.* (1990)	0–70	39–50
Stoll (1971)	0	~44

The values reported in Table 12.12 were obtained from drained tests or trials on unsaturated materials. Hence it would seem appropriate to regard them as being equivalent to effective stress parameters.

The stability of a number of slope situations needs to be assessed, particularly for landraise. There are four likely modes of failure:

(1) shallow rotational/translational failure within the face of the lateral containment
(2) translational failure of the side slope of the cover
(3) rotational failure through the landfill and the foundation material
(4) rotational failure of the waste face.

Failure may be due to excessive settlement or due to lack of shear strength. A check for both settlement and slip failures should be made. Both short-term and long-term analyses need to be undertaken, with appropriate assumptions being made about the pore fluid regime.

As most waste deposits are blatantly heterogeneous, it could be argued that the use of conventional slope stability approaches is inappropriate. However, in the foregoing cases 1 and 2, the slip surfaces will not pass through refuse and will consist of soil-on-soil and soil-on-other-material (e.g. geomembrane) zones, as illustrated in Chapter 6, Section 6.1. Furthermore, it must be remembered that the geotechnical materials within slopes are not uniform or homogeneous on an absolute scale. Hence a slip surface will pass through a variety of solid-to-solid contacts and void spaces. However, because of the sheer number of contact points and void spaces on a real slip surface it is acceptable to apply average strength parameters to bands of material that appear to be more or less uniform when viewed at the appropriate scale (e.g. when the engineering situation is drawn on a sheet of A4 paper). Deposited refuse is likely to be inherently more variable than geotechnical materials, but where cases 3 and 4 are possible the potential sliding surface will be significantly longer than most run-of-the-mill earth slopes to which slope stability analyses would be applied. Hence it is believed that conventional geotechnical slope stability analyses can be applied to waste deposits, but higher Factors of Safety are required to account for material variability and to limit the deformations needed to mobilize the shearing resistance.

While it is possible to form apparently stable steep slopes (1 : 1 or steeper) in refuse (Figure 12.13 shows a cut made through old refuse), this is the result of unquantifiable and unreliable effects (e.g. unknown suctions and viscosity in the pore fluid, reinforcement due to the tensile strength of long pieces of refuse, physical interlocking). For long-term stability, more suitable (and more justifiable from an engineering viewpoint) maximum slope inclinations are probably in the

Figure 12.13. A cut slope in landfilled refuse

region of 1 (vertical) to 3 or 4 (horizontal) (Singh & Murphy, 1990). However, it should be possible to engineer steeper slopes through the inclusion of soil reinforcement or the provision of other forms of support.

12.3. Clay liners

Traditionally, waste disposal sites have been created within naturally occurring depressions or man-made excavations. Many such sites are located in low-permeability strata as a result of excavation of clay for brickmaking, pottery manufacture, tilemaking, etc. However, these natural clay deposits are not acceptable as the only liner because of their potential to contain fissures and fractured zones. It is necessary to construct a clay liner by reworking the existing soil (if it is suitable) to produce a low-permeability, uniform material.

12.3.1. General requirements

Although the detailed requirements for compacted clayey barriers vary, the following criteria commonly apply:

- the coefficient of permeability (hydraulic conductivity) of the liner must be 10^{-9} m/s or less
- the clay layer must be free of natural or compaction-induced fractures
- the minimum thickness of a compacted clay liner for a domestic or commercial waste facility should be 1 m
- a minimum clay content (particles smaller than $2 \, \mu$m) of 10–20%
- a plasticity index in the range 10–65% and a liquid limit not greater than 90%.

There is no consensus on the upper limit of acceptability for coarse material. Requirements for a gravel content of less than 20–30% and a maximum particle size of 25–30 mm are typical. Large particles are acceptable (they themselves are impermeable) provided that they do not prejudice the integrity of the liner (i.e. they are well dispersed throughout the fines matrix).

To evaluate a source of lining clay an appropriate suite of assessment tests would be:

- natural moisture content
- liquid and plastic limits
- particle specific gravity
- particle size distribution
- compaction curves (Standard and modified Standard)
- permeability (at likely compaction moisture contents)
- organic content
- mineralogy
- undrained shear strength (wet of optimum)
- attenuation/retardation capacities (if possible).

For development of this assessment process it would be useful if a series of 'standard' laboratory leachates could be defined to assess the compatibility of the soil with potential leachates.

Shale, mudstone and marl are composed of clay minerals and, therefore, are potentially suitable for use in the formation of a low-permeability liner. However, material from one source, or even from a single stockpile, can show a high degree of variability in terms of induration, cementing and weathering. Material processing should be considered where the 'as-dug' material is too variable or is not quite suitable by virtue of any of the following:

- stone content too high
- clay content too low

- clod size too large
- mudrock pieces too large.

Material processing could comprise screening, bentonite addition, milling/comminution, and even chemical treatment to encourage breakdown of the mineral complexes in the clay. The latter technique has been used in Germany, where clay-like materials are milled to break down the clods and the moisture content then altered to the required value by the addition of water in on-site, undercover batching plants. The result is pellets of clay with a plasticine-like consistency, thus providing a very high degree of material quality control which enables as-laid permeabilities as low as 10^{-10} m/s to be consistently achievable.

A vital part of achieving an effective compacted clay liner is specification of the earthworks procedures to be followed and the associated construction quality control. Most civil engineering earthworks contracts are aimed at achieving a specified strength or density. Although these factors remain important the primary objective in landfill applications is that a specified maximum permeability should not be exceeded at any location within the placed material. To attain this it is essential that:

- the material is capable of achieving the required low permeability
- the material is placed to the specified thickness in a series of thin layers
- the clay has an appropriate moisture content
- the placed material is subjected to sufficient compactive effort.

Generally speaking, liner compaction specifications are based on Standard or modified (heavy) Standard compaction (see Chapter 5, Section 5.2.1).

12.3.2. Permeability behaviour of compacted clay

The compaction characteristics of clay soils are well understood. The accepted behaviour is illustrated in Figure 12.14. The dry unit weight of a soil varies with the water content at which it is compacted and the maximum dry unit weight is obtained at an optimum water content (see Chapter 5, Section 5.2.1). Figure 12.14 also shows the corresponding relationship between the coefficient of permeability and water content for compacted soil. A very important trend evident in this latter relationship is that permeability decreases (by more than an order of magnitude) from high values when the soil is compacted at water contents less than optimum, to a minimum value at water contents 2–4% above the optimum. The permeability of soil compacted wet of optimum is not greatly different from the minimum value. Another consideration for formation of a clay liner is the shear strength of the soil (for trafficability by plant), which decreases markedly when the water content increases significantly beyond optimum (see Chapter 5, Section 5.2.2).

The permeability of soil is controlled mostly by large pores, the distribution and size of which depend on the fabric of the soil. In compacted soil the fabric observable under the electron microscope shows that the clay particles are aggregated together to form 'assemblages' that are packed tightly to form a larger 'clod'. The pore sizes within a clod are much smaller than those between the clods. Compacted soil is formed into an integral mass by kneading the individual lumps (clods) of soil to force them into intimate contact and reduce the size of interclod voids (Olsen, 1962). At water contents drier than optimum the soil clods are hard and difficult to force together resulting in large interclod macropores. On the wet side of optimum the soil clods are soft and plastic. Thus the reduced permeability wet of optimum is attributable to reduction and dispersal of interclod voids because of the remoulding of the soil mass, thereby providing a more tortuous flow path for water to follow. The effect on permeability of 'welding' clods together is very important for clay liners — reduction of clod size

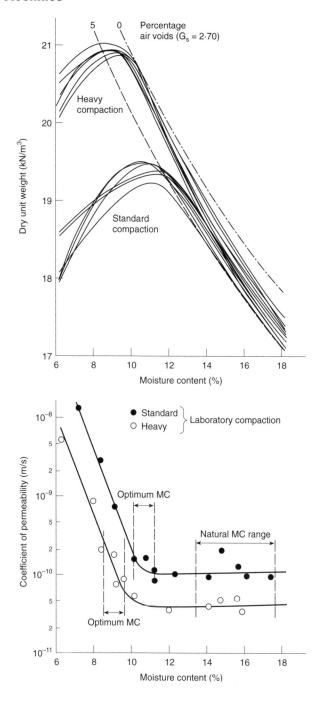

Figure 12.14. The compaction–permeability characteristics of clay soil. MC, moisture content

by a factor of four before Proctor compaction has been shown to reduce laboratory-measured permeability values by several orders of magnitude (Elsbury *et al.*, 1990). Such extreme variations in the coefficient of permeability indicate the presence of open voids between hard clods that must be destroyed by field compaction procedures. Since the lowest values of permeability are usually obtained with kneading compaction, it is common to make a great effort in the field to knead the soils repetitively, with many passes by pad-foot or wedge-foot rollers that destroy soil clods and interclod macropores. The author believes that the superior type of compactor for forming clay lining layers is the grid roller.

The traditional laboratory method of measuring the permeability of compacted clay has been to apply falling head permeability tests (see Chapter 7, Section 7.2.2) to the soil while it remains in its compaction mould. No attempt is made to restrict swelling of the soil on the assumption that the soil will be free to swell in the field. Back-pressure is normally not used. This type of permeameter is the simplest and most economical for testing compacted clays. However, it is claimed that the apparatus has the following limitations:

- the soil will not be saturated if back-pressure is not used
- the stresses that act on the soil are not controlled
- side-wall leakage can occur between the clay and the rigid walls of the mould
- the hydraulic gradient that can be applied to the soil is limited.

There have been a number of investigations into the influence of various factors on the value of the coefficient of permeability as measured in the laboratory:

- *Fixed or flexible confining boundary around the soil.* Flexible-wall devices (triaxial cell permeameters; see Chapter 7, Section 7.2.3) tend to minimize side-wall leakage and are convenient for testing with back-pressure, for measuring volume change within the soil specimen, and for controlling both the horizontal and the vertical effective stress.
- *Use of the oedometer.* In consolidation cell permeameters (see Chapter 7, Section 7.2.4) back-pressure can be applied. A vertical pressure simulates field conditions and squeezes the soil against the rigid permeameter wall (although the potential for side-wall leakage will still apply).
- *Double-ring permeameter* (Anderson *et al.*, 1985). This is a modified compaction permeameter wherein a thin vertical wall is built into the base plate to separate the outflow that occurs through the central portion of the soil from the outflow that occurs near or along the side-wall. If there is significant side-wall leakage, the rate of flow into the outer collection ring will be much greater that the rate of flow into the inner ring.
- *Hydraulic gradient.* Under field conditions the maximum hydraulic gradient applied to a compacted clay liner is likely to be of the order of 5–10.
- *Degree of saturation.* Compacted lining clay (in both the field and the laboratory) will be unsaturated initially. The coefficient of permeability can be affected greatly by the presence of air within the voids of a soil.

The results of the various investigations are summarised in Table 12.13. More is known about how the type of permeameter affects the permeability of compacted clay than any other type of soil:

- Comparative data on compacted clays indicate that, with careful attention to sample preparation and compaction, the various types of permeameters yield virtually identical values of the water permeability (for the soil moisture content range applicable to liners).
- The coefficient of permeability is essentially independent of hydraulic gradient until it becomes so large (greater than 100, approximately) that erosion of particles occurs within the soil matrix.
- While low degrees of saturation decrease the permeability of clay soils by several orders of magnitude, a clay liner should be compacted wet of optimum moisture content (so that degrees of saturation after compaction are typically around 90–95%), and thus the compacted liner is very close to the zero air voids line (full saturation).

12.3.3. Stability of compacted clay

The attenuation and retardation properties of mineral liners (and the geological strata below the lining system) can play an important role in mitigating the potential pollution impact of migrating leachate. Natural clays have a structure

Table 12.13. Influence of test conditions on the coefficient of permeability of compacted clay

Source	Test conditions	Findings
Edil & Erickson (1985)	Soil compacted into rigid-walled and flexible-walled permeameters	Both apparatuses gave very similar values
Daniel *et al.* (1985)	Conventional permeameter and triaxial cell	The compaction permeameter is acceptable for low-overburden situations; flexible-walled cell recommended for high-overburden situations
Peirce *et al.* (1986)	Oedometer	The consolidation apparatus was recommended because it could simulate field conditions
Brunelle *et al.* (1987)	Compacted clay in conventional permeameters, oedometers, double-ring permeameters	Permeability values all essentially the same
Bagchi (1994)	Conventional permeameter and triaxial cell	No significant differences indicated, but triaxial cell with back-pressure preferred
Sarsby *et al.* (1995)	Falling head tests in normal permeameter and triaxial cell	Tests gave very similar permeabilities for samples compacted wet of optimum
Campbell (1997)	Conventional permeameter and triaxial cell	Both apparatuses gave essentially the same permeability values
Beeuwsaert & Sarsby (2000)	Conventional permeameter and triaxial cell	Both apparatuses gave essentially the same permeability values

that absorbs cations and anions (particularly heavy metals). However, these cations and anions may not be permanently fixed, and if environmental conditions change they may be released. At this time a more important aspect of the interaction between a clay soil and chemicals (in particular, landfill leachate) is whether there is an adverse effect on the containment properties of clay liners. Interaction between clay soils and industrial leachate or wastes, particularly those containing certain organic chemicals, could cause dehydration (leading to shrinkage and cracking), dissolution of certain components (to give increased permeability), and increased brittleness (due to ionic exchange). The interaction between pore fluid and clay minerals is due to absorption, cation exchange and the intrusion of water and other molecules into clay mineral sheets. One would expect lime-rich clays to be rendered more permeable by acid attack.

Much research has been conducted to investigate the effects of various permeants on the permeability of compacted clay (as used to form landfill liners). This research is summarized in Table 12.14. Variables that have been considered include:

- acidic and caustic permeants (pH 1–13)
- organic chemicals (from full strength to weak solutions)
- leachate or chemical mixtures.

From the research work to date it can be concluded that inorganic permeants do not affect the permeability of compacted clay unless the chemicals are present in very high concentrations. When water-soluble organics invade a clay–water system, water and other cations in the diffused double layer around a clay particle may be replaced by the organic compounds. The organic molecules can enter the interlayer spaces, causing the clay layers to swell. This will have the effect of reducing both shear strength and permeability. However, it is only when water-

Table 12.14. Effect of non-aqueous permeation on the permeability of clays

Source	Soil and permeant	Apparatus	Comments
Griffith *et al.* (1976)	Clay soil and leachate	Conventional permeameter	Some decrease in permeability compared to water permeation
Brown & Anderson (1983)	Water followed by concentrated organic chemicals on compacted clay	Conventional permeameter	Concentrated chemicals increased permeability by factor of 100–1000*
Daniel & Liljestrand (1984)	Compacted clay and 'real' leachate	Conventional permeameter	No significant effects
Acar *et al.* (1985)	Compacted kaolinite and organic fluids (weak and concentrated solutions)	Flexible-wall permeameter	All tests at low concentrations exhibited slight decreases in permeability[†]
Lentz *et al.* (1985)	Kaolinite, kaolinite–bentonite mixture. Hydrochloric acid (pH 1–5), tap water, sodium hydroxide (pH 9–13)	Triaxial cell, conventional permeameter	No significant effect[‡]
Bowders & Daniel (1987)	Compacted kaolinite, illite–chlorite specimen Organic chemicals in pure form, at full strength and various dilutions	Rigid-wall and flexible-wall permeameters	Dilute organic chemicals (less than 80% by volume in aqueous solution) had little effect[§]
Mitchell & Madsen (1987)	Clay soil and chemicals	Literature review	High concentrations of water-soluble hydrocarbons may cause large increases in permeability
Brunelle *et al.* (1987)	Natural clay liner material and actual leachate	Flexible-wall permeameter, consolidation cell, double-ring compaction mould	Leachate appeared to have very little effect; permeabilities for leachate and water were essentially equal
Fang & Evans (1988)	Clay soil with landfill leachate and tap water	Conventional permeameter	Practically no difference between permeabilities for the two fluids
Peterson & Gee (1985)	Clay liner materials and acidic (pH 4) uranium tailings solutions	Conventional permeameters	Permeability decreased with duration of permeation
Gettinby (1999)	Compacted clay soil and laboratory 'leachate'	Conventional permeameter	Sustained leachate permeation (12 months) had negligible effect on clay permeability

* The increase in the coefficient of permeability was due primarily to soil shrinkage and consequent side-wall leakage.
[†] It was found that side leakage due to shrinking and consequential cracking of specimens permeated with pure organic fluids might give apparent increases in permeability values obtained in rigid wall permeameters.
[‡] Only pH 13 sodium hydroxide caused a significant change (reduction) in permeability due to precipitation of magnesium hydroxide in the pores. Reactions that occurred, such as chemical precipitation, particle dispersion and subsequent pore plugging, tended to decrease the permeability.
[§] The increase in the permeability for the concentrated chemicals was attributed to shrinkage of the double layer surrounding the clay particles, causing cracks to develop.

soluble organics are present at high concentrations that permeability is affected adversely. Low concentrations (below 70% of full strength) are likely to induce decreases in permeability because of the increased viscosity. On the other hand, organics with low water solubility, such as hydrocarbons and related compounds, are not able to displace the porewater and will have little effect on the soil properties. Inactive soils, the clay minerals of which consist of illites and chlorites, have been shown to be insensitive to leachate from municipal wastes. Consequently, at this time, it can be concluded that it is unlikely that leachate from household or commercial wastes will have any significant effect on the permeability of compacted clay that has been placed as an engineered liner.

Liquid and plastic limits are used as an index of the consistency of a soil (see Chapter 3, Section 3.3.2). The mechanical properties of clay depend on several interacting factors:

- mineral composition
- percentage of amorphous material
- distribution and shape of particles
- pore fluid chemistry
- soil fabric
- degree of saturation.

Quantitative prediction of soil behaviour based on the preceding factors and double-layer theory is impossible because of the inadequacy of the available physico-chemical theories and the difficulties of taking into account in situ environmental factors. However, experimental studies have been undertaken; the effects of permeation by organic compounds on the liquid and plastic limits are illustrated in Table 12.15. Concentrated organic liquids seem to have much more effect on the Atterberg limits than their aqueous solutions (even when at high concentrations) and a clay soil is likely to lose some of its plasticity. However, weak solutions do not seem to have any detrimental effects and it has been recorded (Gettinby, 1999) that sustained permeation with artificial, organic 'leachates' either had no effect on, or increased slightly, the plasticity index and plasticity classification of clay suitable for barrier formation.

Experience to date would suggest that with good engineering practice and quality control, high-quality, low-permeability liners can be constructed using compacted inactive clays. Furthermore, it would appear that, while some laboratory tests utilizing very concentrated organic wastes have produced increases in the permeability of clay, significant changes are not expected from exposure of liner clay to actual landfill leachate during the working life of a containment site.

Table 12.15. The effect of chemical permeation on the physical properties of clay soil

Source	Comments
Torrance (1975)	Generally, chemicals caused some increase in the W_L (liquid limit), in proportion to their concentration, but dispersing agents caused a decrease in the W_L
Dascal & Hurtubise (from Gettinby, 1999)	Addition of lime increased the P_L (plastic limit) and W_L
Eklund (1985)	Laboratory 'leachates' had no significant effect on the plasticity of clay liners
Yong (1986)	Sensitive clay was subjected to leaching tests with acidic, alkaline and neutral fluids. In cases where partial removal of soluble salts resulted there were significant increases in both W_P and W_L
Gettinby (1999)	Sustained permeation of clay soil by laboratory 'leachate' had a negligible effect on the W_P but produced slight increases in the W_L

12.3.4. Field compaction and control

Stringent Construction Quality Control (CQC) and Construction Quality Assurance (CQA) is required during barrier compaction. In addition, a suitable post-placement protocol is required to prevent drying or swelling, since this can be critical in the final performance at the compacted clay liner.

The ability to compact a certain clay to form a landfill liner, and the adequacy of the liner itself, depends on the maximum permissible coefficient of permeability, the minimum shear strength required for trafficability, and the acceptable lower limit of field compaction. Typical values of these parameters might be 10^{-9} m/s, 40–50 kN/m^2 and 95–100% of Standard compaction, respectively. The foregoing requirements can be used with laboratory compaction–permeability–strength data to define an admissible range of states of the soil in a compacted clay liner (e.g. the dark shaded area in Figure 12.15). Subsequently this diagram can be used as a quality control tool by taking field measurements of density and moisture content within the liner (these items can be determined rapidly). The resultant data should lie within the target area. It is then inferred that the permeability will be similar to that predicted by the diagram. However, it is unwise to rely solely on the density–moisture content–permeability relationship established in the laboratory. The primary criterion for the acceptability of a clay liner is that its overall permeability is not greater than the limit set, and this can only be assessed confidently by undertaking a certain amount of direct verification of the as-laid permeability.

To undertake the necessary checks on the quality of a liner (in terms of its permeability) it is vital to consider five items:

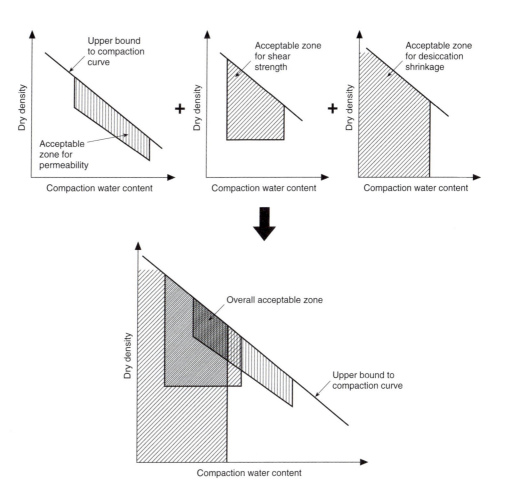

Figure 12.15. The admissible state for a compacted clay liner

- the use of both direct and indirect quantification
- the specific parameters which have to be measured
- the number of samples to be taken from the liner and the frequency of testing for particular parameters
- the selection of an appropriate method for laboratory determination of the coefficient of permeability for samples taken from the liner (this has already been addressed in Section 12.3.2)
- in situ permeability testing of the liner and correlation of the data with laboratory results.

Recommended frequencies for quality control tests vary widely (Table 12.16) and there is no consensus of opinion on methodology. Usually a large number of density and moisture content tests are specified during construction to ensure uniformity, with fewer permeability tests (laboratory or field) being ordered. However, while it is recognized that it is easier (and cheaper) to conduct classification, density and moisture content tests than permeability tests, it must be remembered that liner permeability is the crucial characteristic. Hence the practice of conducting a much reduced number of permeability tests (compared to classification) is not appropriate. The testing regimes recommended by Beeuwsaert (1998) were derived from statistical analysis of a very large number of tests of five lifts of an actual landfill liner and must therefore be given considerable weight. If it is assumed that a compacted clay liner is formed in lifts not thicker than 0.25 m (to ensure good compaction) then taking samples on a grid pattern, with centre-to-centre spacings of 20 and 30 m, corresponds to one test per 100 and 225 m^3 of material, respectively. Accordingly, it is recommended that an appropriate sampling and testing frequency for permeability is at 20 m centres, on a square grid, with density and moisture content (for indirect estimation of permeability) being determined on a 10 m square grid.

Table 12.16. Recommended testing regimes for compacted clay liners*

Source	Field testing regime
Attewell (1993)	One test per 20 m^3
Bagchi (1994)	PSD, classification, compaction curve: determine for each 3800 m^3 of soil placed to check the variability of the source
Beeuwsaert (1998)	Recommended (based on field study of an actual liner) one test per 40 m^3 of soil for a 95% confidence level in permeability or one test per 100 m^3 for a 68% confidence level
Benson (1991)	Assessment of an area of 0.15 m^2 will incorporate the variability in permeability
Daniel (1990)	Moisture content and dry density: one test per 120–600 m^3 Classification: one test per 750 m^3
North West Regulation Officers Technical Subgroup (1996)	Moisture content, dry density and classification: one test per 250 m^3 Permeability: one test per 500 m^3
Rogowski et al. (1985)	Dry density and moisture content: a 30 m grid per lift (with a maximum of 12 tests/Ha/lift) PSD, classification, permeability: test 25% of the samples collected from the 30 m grid (subject to a minimum of 5 tests/Ha/lift)
UK Consulting Engineers (from Beeuwsaert, 1998)	Moisture content and dry density measured on a 20–30 m grid across the site or one test per 150–250 m^3 of compacted soil. Classification assessed on a 25–40 m grid or one test per 400–500 m^3

PSD, particle size distribution.
* After Beeuwsaert (1998)

There are two main in situ methods for determining the permeability of compacted clay liners: the large fixed-ring infiltrometer and the large-scale field lysimeter.

In *large fixed-ring infiltrometers* a ring is embedded in the compacted soil and then filled with water. The permeability of the soil is calculated by measuring the rate of change of the water head within the ring. Unfortunately a test lasts from several weeks to several months. In situ permeability values (of compacted clay liners) obtained by this method are usually in good agreement with values obtained by the other main in situ method (i.e. the lysimeter) (Table 12.17). For good-quality liners the apparatus seems to give permeability values up to one order of magnitude higher than laboratory values (Elsbury *et al.*, 1990). These devices provide accurate average values of inferior liners that are poorly compacted, fissured by desiccation, etc. in which case values may be up to 10 000 times the laboratory value for an intact clay sample. A double-ring infiltrometer (similar to the laboratory version) is also available.

The second method is the use of a *large-scale field lysimeter*. A lysimeter is essentially a large void, lined with material of very low permeability (such as a synthetic membrane) and backfilled with highly permeable material, which is installed below or within a clay liner. Water is collected in the void, and liner permeability is calculated from the measured flow rate. The time to complete a test is likely to be several months. Lysimeter values of permeability have been reported to agree closely with data from other in situ tests (Day & Daniel, 1985). Furthermore, good agreement has been obtained with laboratory permeability values (for good-quality liners).

12.3.5. Performance of compacted clay liners

The permeability of clay liners is significantly influenced by clod size when the soil is compacted dry of optimum and if the compaction plant is too light (Elsbury *et al.*, 1990). The soil clods are destroyed and large inter clod pores are eliminated by wetting to above optimum and by using a suitably large compactive effort. In the UK the natural moisture contents of many clays are well above the optimum value (Standard compaction). Hence it is recommended that an overall requirement for the formation of clay liners is that the soil must be compacted at,

Table 12.17. In situ measurement of liner permeability

Source	Test arrangement	Findings
Day & Daniel (1985)	Ring infiltrometers applied to actual liners	Ratio of field and laboratory permeabilities was 5 good liners–10 000 (very poor-quality liners)
Rogowski *et al.* (1985)	Field tests of infiltrometers and lysimeters	Good correspondence of data between the two tests
Trautwein & Williams (1990)	Infiltrometer applied to liner	Measured permeability up to 10 times greater than laboratory values
Elsbury *et al.* (1990)	Field trials of infiltrometer and lysimeter	Excellent agreement between the two apparatuses, values very similar to those from the laboratory
Bracci *et al.* (1991)	Studied field liners using infiltrometer	Ratio of field to laboratory permeability was 2 good liners–1000 (poor-quality liners)
Lahti *et al.* (1987)	Field tests using a lysimeter	Field permeability values correlated well with laboratory values
King *et al.* (1991)	Field permeability by lysimeter	Field permeability initially greater than laboratory value by a factor of 10, but decreased with time

or above, the Proctor optimum moisture content. Grid or sheepsfoot rollers are recommended for compacting clay liners. If smooth rollers are used the surface of the previous lift should be scarified prior to emplacement of subsequent fill to ensure adequate interlayer bonding.

It has been proposed that field compaction trials should be conducted prior to the formation of an actual liner. The purpose is to show that the soil condition obtained during the source evaluation stage (of a lining clay) can be achieved in the field and to provide the information needed to prepare a detailed liner construction method statement. It is believed that a better way to ensure good-quality liner production is to develop site understanding and an appreciation of the behaviour of geotechnical materials, and to employ high-quality, intensive site supervision.

Desiccation cracking (Figure 12.16) occurs when compacted soil is exposed to the atmosphere so that porewater evaporates and negative pore pressures develop in the soil. These changes increase the effective stress with consequential reduction in volume, shrinkage and cracking. If the soil is wetted with water, it will swell and tend to close the cracks, resealing the liner. However, if foreign material is washed into cracks then permanent zones of higher permeability may result. Freezing of a clay soil can result in the creation of fissures due to the formation of ice lenses. Cracking due to desiccation and freezing can be prevented by avoiding significant moisture movements (by covering the liner). If a liner is left exposed to the natural elements for a year the mean permeability of the top 200–300 mm of soil can be expected to increase by an order of magnitude at least.

Figure 12.16. Desiccation damage to liners

An item that is often forgotten about in the construction of a good-quality containment liner is the construction of the liner up the sides of a site. If the gradient of these sides is too steep problems will arise from the compaction plant not being fully effective or from plant being unable to operate on the slope. In general the following methodology will achieve a satisfactory compaction of the sides of the containment:

- If the gradient is less than 1 : 3 (vertical to horizontal), compaction plant can be run up and down the sides.
- If the gradient is in the region of 1 : 3 to 1 : 1 the side lining can be formed by compacting the clay in lifts, with compaction plant running along the slope. In this case bonding between consecutive lifts is very important because otherwise interlayer boundaries (more or less horizontal) will provide high-permeability paths through the liner.
- If the gradient is steeper than 1 : 1 it would be best to use a geomembrane to provide lateral containment.

Where mineral liners are constructed above the level of groundwater, they essentially comprise part of the unsaturated zone. However, seepage rates through mineral liners are typically approximated to flow according to Darcy's law in saturated strata and, if the groundwater level is assumed to be at the underside of the liner, then

$$q = k\frac{(h + d)}{d} \tag{12.5}$$

where q is the seepage rate, h is the leachate head above the top of the liner, and d is the thickness of the liner.

Increasing the permeability of the mineral liner or the head of leachate, or decreasing the thickness of the liner, will cause an increase in the seepage rate through the liner. An assessment of flow rates through mineral liners should also involve serious consideration of whether (and how) to take account of the potential for localized flaws in the liner.

When the advective flow is very small (k is less than 10^{-10} m/s approximately), diffusion flows through a liner can become significant. The mass flux f transported by diffusion alone can be written as

$$f = -nD\frac{\partial C}{\partial x} \tag{12.6}$$

where n is the porosity of the soil, D is a diffusion coefficient (typically in the range 10^{-9} to 10^{-10} m^2/s) and x is the distance.

The presence of soil particles, particularly adsorptive clay minerals and organic matter, complicates the diffusion process. For monomineralic systems, steady-state diffusion is reached fairly rapidly compared to a natural heterogeneous soil where extensive cation exchange, precipitation–dissolution and biodegradation may occur. Use is then made of Fick's second law, which describes the rate of change of concentration C with distance x and time t as follows:

$$\frac{\partial C}{\partial t} = D\frac{\partial^2 C}{\partial x^2} \tag{12.7}$$

A solution of the preceding equation, for an infinite layer with a constant surface concentration C_0 and zero advection, has been given by Ogata (1970):

$$\frac{C}{C_0} = \text{erfc}\left(\frac{x}{2\sqrt{Dt}}\right) \tag{12.8}$$

If

$$X = \frac{x}{2\sqrt{Dt}} \tag{12.9}$$

then

$$\frac{C}{C_0} \cong 1 - \frac{2}{\sqrt{\pi}}\left(X - \frac{X^3}{3} + \frac{X^5}{10}\right) \quad \text{for } X \leq 0.5 \text{ approximately} \tag{12.10}$$

or

$$\frac{C}{C_0} \cong \frac{e^{-x^2}}{X\sqrt{\pi}} \quad \text{for } X > 1.0 \text{ approximately} \tag{12.11}$$

The variation in the relative concentration with depth and time (for a typical diffusing chemical, e.g. chloride) is shown in Figure 12.17 along with a plot showing the change in velocity of the 50% relative concentration front $(C/C_0 = 0.5)$ with time. This figure shows that initially the rate of diffusion is quite rapid, but it decreases quickly with time due to the reduction in $\partial C/\partial x$ as x increases.

It would appear that for well-designed and properly constructed clay liners, there is generally good correlation between laboratory and field permeability values (see Table 12.17). Close supervision of the site work is essential. Laboratory tests are unlikely to account for effects such as large uncompacted clods or desiccation cracking and these have to be controlled by good field supervision. A number of field studies have demonstrated that low-permeability (10^{-9} m/s) clay liners can be constructed successfully (Rowe *et al.*, 1995). However, it should be recognized that some variability in hydraulic conductivity permeability is to be expected and should be considered in both the design and the specifications.

12.4. Geo-membranes and composite liners

Advances in plastic and synthetic technology have meant that man-made 'plastic' materials (geomembranes) have been developed that can provide lining and capping systems which are at least as good as engineered mineral lining systems. Geomembranes are a category of 'geosynthetics' (the family of man-made materials that are used in intimate contact with geotechnical materials). In general terms, there are three other major geosynthetic categories: geotextiles, geogrids and geocomposites. The majority of geosynthetic materials are man-made polymers produced from a combination of hydrocarbons.

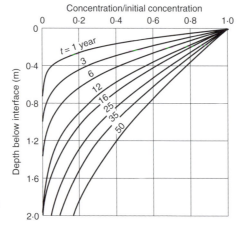
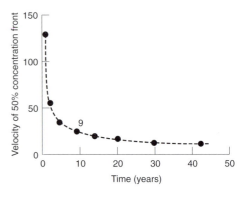

Figure 12.17. Progression of a chemical front due to diffusion (after Ogata, 1970)

12.4.1. Geomembranes

Although geomembranes are thought to be impermeable, they are not strictly so. Water-vapour transmission tests show that the coefficient of permeability of geomembranes is in the range 10^{-12} to 10^{-15} m/s and hence they are 'essentially impermeable'. Contaminant transport can, however, still occur through intact geomembranes by diffusion mechanisms. If geomembranes are intact, they can be assumed to be virtually impermeable, so their thickness is not a major factor with respect to advective transport mechanisms. Degradation due to ultraviolet rays and puncture resistance are determining factors in choosing liner thickness, and in general synthetic membranes 1.5–3 mm thick are used for lining landfills. A geomembrane is usually fabricated with a fabric reinforcement (a 'scrim') between the individual plys of the material. This results in a three-ply laminated geomembrane and provides dimensional stability to the material together with a major increase in mechanical properties (i.e. tensile strength, modulus of elasticity, tear resistance). Important factors relating to the selection of a geomembrane for landfill lining are summarized in Box 12.1, and geomembrane materials considered for use in waste containment are outlined in Box 12.2.

In most instances geomembranes do not replace whole lining systems, but rather they are installed as an addition to or replacement for part of a lining system (i.e. they are components in a composite lining system). Composite liners use two or more separate lining techniques together in one lining system. These can be a combination of a variety of materials (e.g. clay, geomembrane, bentonite-enriched soil). Geomembranes or flexible man-made liners (FMLs) are now widely used in engineered landfills, and a substantial body of experience has been built up with regard to their suitability, construction, CQA, performance and limitations.

12.4.2. Geomembrane liner formation

Geomembranes are manufactured as long sheets (generally up to 10 m wide) and adjacent sheets must be joined (seamed) in the field in such a way that the effectiveness of the barrier is not compromised. The geomembranes cannot be 'tailored' exactly to the shape of the containment cell, and seaming is frequently carried out under difficult circumstances (on steep slopes, during wet and windy conditions, etc.) as illustrated in Figure 12.18. Selection of an appropriate jointing method and good workmanship on-site are crucial elements in the formation of an effective geomembrane barrier system. The three main seaming methods (Box 12.3) are:

- extrusion welding — used extensively on geomembranes made from polyethylene
- thermal fusion/melt bonding — there are three methods, which can be used on all thermoplastic geomembranes
- adhesive (solvent) jointing.

There are a variety of factors to be considered when designing a geomembrane lining system:

- materials selection (many different polymers are available)
- ease of seaming during installation

Box 12.1. Important attributes for geomembranes for landfill lining

- **Physical:** density, mass per unit area.
- **Containment capacity:** water vapour transmission capacity, resistance to impact, puncture resistance.
- **Strength:** tensile behaviour, thermal properties.
- **Durability:** stress cracking, chemical resistance, ultraviolet light resistance.
- **Constructability:** behaviour of seams.

Box 12.2. Geomembrane materials

Butyl rubber

This was first used to line water reservoirs more than 25 years ago. It is a thermoset plastic. While it has excellent resistance to swelling and permeation by water, it is susceptible to attack by hydrocarbon solvents and oils (even in trace concentrations) and therefore is not suitable for landfill containment.

Chlorosulphonated polyethylene (Hypalon)

Available in both thermoplastic and vulcanized compositions. Normally reinforced with a polyester or nylon scrim (to improve tear strength and puncture resistance). Good resistance to microbial growth, ozone and chemicals. Poor resistance to hydrocarbons. Seaming is relatively easy and generally good, undertaken by either solvent cements or heating.

Low-density polyethylene (LDPE)

Only available in thin sheets, which can be difficult to handle in the field. The sheets have poor resistance to puncture and mechanical damage. It can be seamed by any of the available methods and offers very good resistance to a wide range of industrial chemicals, including oils and solvents.

High-density polyethylene (HDPE)

The thickness (up to about 3.5 mm) gives increased strength and damage resistance, also it can withstand stresses due to refuse/subgrade settlements. The resultant disadvantage is limited flexibility. It can be seamed by any of the available methods and offers very good resistance to a wide range of industrial chemicals including oils and solvents. One of the most widely used lining materials in the UK. HDPE has very good chemical resistance, very low permeability, a long track record in environmental protection works, good ultraviolet resistance, medium creep resistance.

Polyvinyl chloride (PVC)

Widely used for lining waste treatment lagoons. Good tensile strength, resistant to mechanical damage, resistant to many inorganic chemicals, but weakened by hydrocarbons, solvents and oils. Its plasticizer may be slowly leached out or volatilized, so the material becomes brittle and prone to cracking. Rate of plasticizer loss is unlikely to affect the properties of PVC if the geomembrane is covered within a reasonable period of being installed. It is lightweight and dual-track hot wedge welding techniques can be used.

Polypropylene (PP)

Has properties that make it suitable for use as a landfill liner under a wide range of conditions. It offers very good chemical resistance, only highly oxidizing reagents are likely to attack it. It has the highest resistance of all thermoplastics to organic chemicals. It is very flexible, with a low flexural modulus and medium creep resistance.

Figure 12.18. Seaming a geomembrane on a steep slope

Box 12.3. Methods of joining geomembrane sheets

Extrusion welding

A strip of molten polymer is extruded over the edge of (fillet welding), or in between (flat welding), the two surfaces to be joined. The hot extrudate causes the surfaces of the sheet material to melt and the entire mass then cools and fuses together.

Thermal fusion

A seam is formed by melting the opposing surfaces of overlapping sheets and fusing the surfaces together with external pressure being applied without the addition of new polymer. The 'hot wedge' is the most frequently used technique, in which a heated element (in the shape of a wedge) travels between the two sheets to be sealed. As it melts the surface of the sheets, the machine moves forward and the sheets converge at the tip of the wedge. The two sheets are then pinched together by a roller which applies pressure to create the seam. The degree of melting has to be carefully controlled, since too much weakens the geomembrane, and too little results in poor seam strength. The dual hot-wedge weld forms a continuous channel between the welds which can then be used for testing the bond continuity (see Box 12.4). The ultrasonic and hot-air methods use vibrational energy and air, respectively, to cause melting.

Adhesive jointing

A solvent or adhesive is applied between the two geomembrane sheets to be joined. After waiting a few seconds for the geomembrane surface to soften, pressure is applied to make contact and bond the sheets together. Too much solvent will weaken the adjoining sheets and too little will result in a weak seam.

- compliance tests to ensure that the material supplied is of the quality specified
- tests to ensure that the membrane is not damaged during installation
- protection of the geomembrane
- appropriate CQA/CQC systems.

Most lining projects that utilize geomembranes are subject to CQA with an associated plan which is developed to ensure that installation works are undertaken in a controlled manner, and in accordance with specification. The CQA plan should contain details of all materials to be used in the lining system, including material and method specifications, material compliance and conformance testing requirements for each component of the system, allowable welding techniques and weld test requirements for geomembranes, etc. The CQA plan should also provide a course of action to be taken should any item not meet the required specification (determined by testing or inspection) and any other details considered essential during design. Leachate leakage through the intact membrane is negligible, but imperfections can arise from several sources, including poor installation, material instability and puncture.

12.4.3. Geomembrane performance

Improper installation is the primary reason for failure of a synthetic liner. Problems that may be encountered include:

- installation of damaged liner material
- inadequate preparation of site
- inadequate preparation of sheets for seaming
- use of an inappropriate seaming method
- perforation of the liner by plant/machinery.

Ideally, seaming two geomembrane sheets would result in no net loss of tensile strength across the two sheets, and the joined sheets would perform as one single sheet. However, due to stress concentrations resulting from the seam geometry, current seaming techniques may result in minor tensile strength loss compared to the parent geomembrane. A thicker membrane is less likely to be weakened by the seaming process, but such a liner is heavier and may require special equipment to handle it. Testing techniques are utilized to evaluate the integrity and strength of each seam. Where any fault is located, it should be repaired by welding a patch over it (Figure 12.19) or by removing it from the liner and replacing it.

There are a number of in situ, non-destructive tests for checking the adequacy of seams, ranging from simple visual inspection to sophisticated electronic monitoring (Box 12.4). There are essentially only two types of destructive seam test: the shear test and the peel test. For shear testing, a sample is subject to a shear force applied across the seam. It is normally a requirement that the geomembrane sheet fails before the seam. However, the sheet must fail in a ductile manner to prove that the sheet has not been altered to a great extent by the high-temperature seaming process. If the seam failed in a brittle manner, it may be possible that the liner would fail while in service, at low stress, by a process called stress cracking. For peel testing, a sample is subject to a force applied in such a way as to pull the geomembrane sheets apart. The test is a measure of the adhesive force of the seam (i.e. between the two welded sheets or between the extrude polymer and the sheets). Again, it is normal to require that the geomembrane sheet fails before the seam fails by separation (Smith, 1996).

Geomembranes are susceptible to puncture, localized stress and point stress concentrations due to indentation, which can lead to stress cracking. A range of materials, including granular materials, thick geotextiles (Figure 12.20), shredded tyres, etc., can provide appropriate protection to man-made membranes. Medium-to heavyweight needle-punched non-woven geotextiles with a weight per unit area of more than $200 \, \text{g/m}^2$ are recommended to provide sufficient protection against puncturing of a thin membrane. Where granular materials are used they are generally of sand or similar sized particles, in a layer 300 mm or more thick, overlain by the leachate collection system (which is usually composed of rather coarse granular material). Placement of the protection layer requires particular care to ensure that the placement machinery itself does not damage or overstress the liner.

The integrity of a membrane after placement and covering may be checked by using geophysical testing methods — to date electrical resistivity systems have proved the most successful (Anon., 1999). This system is based on the principle that a polymeric geomembrane acts as an electric insulator, so that if an electric potential is applied across such a liner there should be no electric flow (current). However, if the geomembrane contains a hole then an electric circuit will be completed and there will be an associated electric current. In order for the system

Figure 12.19. A patch welded onto a geomembrane

Box 12.4. Non-destructive methods for testing geomembrane seams

Visual observation
Some unbonded or heat-affected areas can be detected by visual inspection only. All seams should be visually inspected for signs of damage.

Mechanical point stress
Unbonded areas can be checked using a blunt instrument, which is passed along the edge of a seam. The tool can easily lift unbonded areas of thin geomembranes, but cannot be used to find unbonded areas within seams.

Air lance
A high pressure (~350 kN/m^2) air jet is used to try and separate the edges of seams. If an unbonded area is present the air jet will inflate the seam. Again, this method cannot locate an unbonded area within a seam, and is not generally used for stiff geomembranes (e.g. high-density polyethylene).

Pressurized dual seam
This method can only be used when a dual track weld has been formed. A pressure (~200 to 300 kN/m^2) is applied to the air gap at one end of a seam, and the pressure is monitored using a pressure gauge connected to the other end of the seam. The pressure is applied and monitored for a certain time, typically about 10 min. If the pressure drops by more than a prescribed amount, then a leak is deemed to be present.

Vacuum chamber technique
A vacuum is applied to a soaped area of seam. If bubbles appear, there must be a leak. The vacuum is applied through a chamber equipped with a vacuum gauge, a clear glass view panel and a soft rubber gasket around the edges. In some locations the vacuum box cannot be positioned suitably to test the seam, but in general it is a fairly quick and reliable test method.

Ultrasonic pulse echo
A high-frequency sound wave is passed through a seam overlap to detect discontinuities. Fairly complex electronic equipment is required for this method, with viewing on a monitor on-site essential. Due to its sophistication, it has not been used to any great extent in the field, and is therefore currently of more use in research than in production seaming.

High voltage electric
This can only be undertaken when a continuous metallic tape or wire has been inserted between the two sheets during the seaming process. A high voltage (15–30 kV) is applied at the end of the metallic conductor and a set point moved along the seam. A seam defect results in a spark passing to the set point. The test method is limited to locating reasonably large unbonded areas (<0.7 mm not usually detected).

Figure 12.20. Geotextile protection of a geomembrane

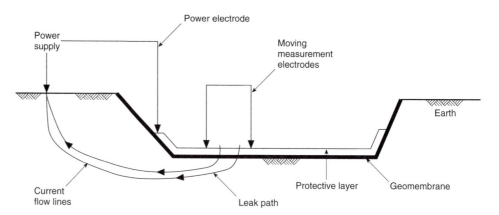

Figure 12.21. Mobile leak location surveying (from Environment Agency, 1999)

to operate, the soil adjacent to the liner needs to be moist (the natural moisture content of most soils is sufficient for this purpose).

In the mobile survey method, a fixed electrode is placed in the ground outside the lined area and an electric potential is applied between this electrode and a set of electrodes which are moved about (on a grid pattern) within the lined area (Figure 12.21). As a moveable electrode approaches a defect (along a particular grid line) the measured potential gradually increases. As the defect is passed the polarity of the signal reverses and the potential decreases.

With fixed detection systems a permanent array of conductors is installed beneath the geomembrane. Examples of such conducting systems are: a network of electrodes buried in the soil beneath the liner and connected together by wires; wires woven longitudinally into a geotextile (two layers of which are buried, at right-angles to one another, beneath a liner). Leakage through the geomembrane will induce a flow of electricity between conductors, and this can be detected.

Seepage through geomembrane liners is controlled by the number, size and location of any defects. The number of flaws likely to be present in a geomembrane can be related to the effectiveness of the CQA programme (Giroud & Peggs, 1990). Typical values are given in Table 12.18. An alternative way to define flaws is in terms of the hole size likely to be present (e.g. pinholes, 0.1–5 mm^2; small holes, 5–100 mm^2; large holes, 100–10000 mm^2). CQA programmes will preferentially detect the large holes. The number of such holes will generally be small; suggested distributions of holes of different sizes are given in Table 12.18.

Deterioration of a geomembrane liner is an irreversible process in which useful polymer properties degenerate when exposed to the environment. Sources contributing to deterioration include:

- heat
- ultraviolet light
- high-energy radiation
- environmental stress
- biological organisms

Table 12.18. Potential flaws in geomembranes

CQA quality	No. of small holes per hectare	Suggested hole distribution per hectare*
Very good	2–3	6 P, 2.5 S, 0.2 L
Average	10–20	38 P, 15 S, 1.1 L
Poor	30–50	100 P, 40 S, 6 L

* P, pinhole (0.1–5 mm^2), S, small (5–100 mm^2); L, large (100–10000 mm^2).

- chemicals
- oxygen.

Because high-density polyethylene (HDPE) liners do not contain plasticizers and other additives that act as foodstuffs for biological organisms, they are not susceptible to decay through biological activity. Environmental stress cracking results from contact with surface-active agents in conjunction with low-level physical stress. Extensive testing has provided much information on factors affecting the ageing of synthetic membranes, and if properly installed and protected there is no reason why the membranes should not last for very long periods (Koerner, 1993). The average guaranteed service life of a synthetic liner is about 25–30 years.

12.4.4. Composite liners

It is generally agreed that all liners leak to some degree (Parkinson, 1992). Single membranes are therefore insufficient on their own and require some form of back-up. The advantages and disadvantages of simple liners made solely from either mineral materials or membranes are compared in Table 12.19.

A composite liner combines the good qualities of two materials, typically geomembranes and clays, to minimize contaminant transportation. The clay layer also provides potential long-term attenuation of any contamination which may still remain after the landfill has been decommissioned and has gone beyond the post-closure maintenance period. This may, in fact, be the most important role of the mineral barrier. The amount of attenuation will depend upon the proportion and type of clay minerals in the mineral layer. The presence of a high proportion of expanding lattice clays of the montmorillonite type will increase the efficiency of attenuation. With a composite liner the total thickness of the barrier can usually be reduced, although the thickness of a clay layer should not be less that 600 mm (which represents a reasonable minimum thickness to provide robustness and durability).

Table 12.19. Comparison of simple mineral and membrane liners

Parameter	Mineral liner	Membrane liner
Attenuation capacity	Low to high depending on the contaminant	None
Storage capacity	Large volume for retention of leachate	None
'Intact' permeability	Low to very low	Essentially impermeable
As-laid permeability	Points of high permeability are likely to exist. Extent depends on the quality of the CQA	Small holes will be present. The number and size depends on the quality of the CQA
Manufacture	On-site, traditional earthworks practice applies (but emphasis needs to be placed on importance of permeability)	In factory, but assembled on-site, close supervision of seaming and jointing is needed
Chemical stability	Not attacked significantly by leachate	Usually very resistant to chemicals
Climatic effects	Presence of lying water or rainwater can make liner formation difficult, desiccation is likely to cause cracking	Substantial exposure to sunlight can cause damage, high winds can lift geomembranes
Robustness	Unlikely to be physically damaged	Easily punctured and torn
Availability	Site dependent, transport of suitable materials over large distances is not feasible	Readily available virtually everywhere
Tolerance of deformation	Bending can cause cracks; significant resistance to shear	Generally good

A geomembrane–clay liner works well because the geomembrane is a barrier to pressure-driven mass transfer, while the underlying clay liner forms a barrier to concentration-driven mass transfer. The clay layer directly underneath a geomembrane means that the chemical concentration gradient is reduced across the geomembrane, since diffused chemical species accumulate in the clay pores at the clay–geomembrane interface. With chemical concentrations roughly equivalent on either side of the geomembrane, the driving force for diffusion is eliminated.

A hole in a geomembrane has infinite permeability. However, the cross-sectional area through which flow occurs is only the area of the hole, and this is much less than the total leachate–liner contact area. If the geomembrane is in good contact with an underlying low-permeability layer then this latter layer will determine the rate of flow through the hole in the geomembrane. This combination of two limiting systems makes a composite liner potentially the most effective way of minimizing leachate flow out of a landfill site.

However, it must be remembered that the effectiveness of the composite system is highly dependent on intimate contact between the geomembrane and the underlying layer, so that negligible lateral flow of leachate occurs beneath the geomembrane. In practice, the reality of intimate contact is questionable (Figure 12.22). 'Wrinkles' in the geomembrane, which are present after installation, cannot be eliminated entirely by pressure from the overlying waste without the development of creases (which will induce cracking), and the geomembrane will tend to span local depressions in the clay layer or it may undergo localized overstretching. Furthermore, a leakage detection layer cannot be interposed between the geomembrane and mineral liner, otherwise any defects in the

Figure 12.22. Wrinkles in a geomembrane placed over a clay layer

geomembrane will cause the leachate head within the waste to be transmitted over the full surface area of the mineral liner (i.e. the system will revert to a simple mineral liner).

An estimation of the leachate flow rate through a flawed composite liner may be made using empirically derived equations such as the one given by Giroud & Bonaparte (1989):

$$Q = C \, A^{0.1} \, H^{0.9} \, k^{0.74} \tag{12.12}$$

where Q is the flow rate (m^3/s), C is a constant depending on the contact between the membrane and the subsoil (0.21 for good contact, 1.15 for poor contact), H is the head of leachate (m), A is the area of holes (m^2) and k is the coefficient of permeability of the subgrade (m/s).

12.4.5. Practical aspects of composite liners

For composite liner systems particular care is required to prevent desiccation of the underlying clay liner (especially on the side slopes) due to potentially high temperatures at the underside of the geomembrane while it is exposed. A crucial factor is the nature of the material underlying the clay layer. The risk of desiccation is high if the clay is underlain by a permeable stratum — in a field trial an 11% reduction in volumetric water content was observed within a 2-year period (Gottheil & Brauns, 1997). The risk of this desiccation can be reduced by covering the geomembrane with some insulating material (e.g. the protection layer and the leachate collection system), as quickly as possible after installation.

A properly constructed composite double-liner system of the type used at Kettleman Hills (see Chapter 6, Section 6.1) provides protection against transport of leachate out of the containment system. At the same time, however, the system contains a number of low-strength interfaces that may act as potential surfaces of sliding whenever elevation differentials exist. The low shear resistance of the individual elements will be compounded when they are in intimate contact because the mineral liner will be designed to have a high moisture content. This will be exacerbated by any water that is allowed to accumulate during liner construction. The sliding resistance between soil and man-made materials with relatively smooth surfaces is typically around 50–70% of the shearing resistance of the soil itself. However, the shearing resistance between geosynthetics (e.g. between a geomembrane and its geotextile protection layer) may be only about 30–50% of the shearing resistance of the soil. Actual values of interface shear strength parameters will vary with the soil type and the actual geosynthetics, but the following values are typical: soil-on-membrane, $\phi' = 15$–25°; soil-on-geotextile, $\phi' = 20$–30°; geotextile-on-membrane, $\phi' = 5$–10°. In all cases the effective cohesion is essentially zero. It is now common practice to either incorporate geosynthetics with specially manufactured 'rough' surfaces within liner systems, or to attach or bond 'rough' surfaces to geomembranes to provide a greater interface shearing resistance.

Drainage layer material and waste that is piled up against the side walls of a geomembrane-lined cell will impose a down-the-slope force on the liner. To withstand such forces, and the down-slope self-weight of the geomembrane, these liners are anchored within a trench dug around the containment cell (Figure 12.23). The dimensions of the trench need to be calculated so that pull-out does not occur. If the pull-out force F_R is resisted by only frictional resistance between the soil and the membrane, then

$$F_R = \gamma h b \tan\delta + w L_A \tan\delta \tag{12.13}$$

where γ is the unit weight of the soil, b is the trench width, h is the trench depth, δ is the effective friction angle between the membrane and the soil, w is the weight per unit area of the membrane and L_A is the anchorage length.

Figure 12.23. A geomembrane anchor trench

12.5. Alternative lining materials

Alternative lining materials may be used for several reasons:

* lack of suitable lining clay in the vicinity of the site
* saving on 'air-space' (i.e. the liner occupies less volume, therefore there is more tipping space available on-site)
* direct cost benefits
* environmental benefits such as the use of low-grade, or waste, materials.

12.5.1. Geosynthetic clay liners

Geosynthetic clay liners (GCLs) consist of a layer of clay with a high swelling potential (usually bentonite in powder or granule form, unless the clay has been prehydrated in which case it will have the consistency of plasticine) held between two geotextiles or attached to a geomembrane (Figure 12.24). In landfill engineering applications the product is most often used as a low-permeability barrier on the side walls or base of a landfill or as part of its capping system (Darbyshire, 1996).

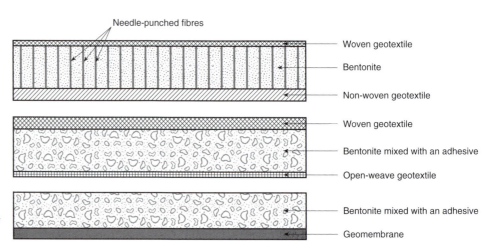

Figure 12.24. Examples of commercial geosynthetic clay liners

GCLs are delivered to site in rolls and simply unrolled onto a prepared surface. The strips are 'joined' by overlapping their edges and applying some clay (usually bentonite) in between to provide a suitable seal. When a synthetic clay liner comes into contact with water it swells, forming a continuous layer of bentonite 12–25 mm thick. High-swell clay has the capacity to seal around small punctures and holes, providing a degree of confidence that the effect of any small penetration during installation will be minimal. The permeability of synthetic clay liners varies between 1×10^{-9} and 1×10^{-11} m/s and may decrease if the effective stress increases. When properly installed a synthetic clay liner can replace about 600 mm of the clay thickness in a conventional compacted clay liner.

Since synthetic clay liners contain bentonite, it is preferable not to expose them to leachate directly. Bentonite is susceptible to attack by strong ionic concentrations, which cause collapse of the clay structure and a corresponding increase in permeability. However, the effect of pore fluid on the permeability of bentonite can be reduced by treating it with special polymers.

Differences between GCL liners and compacted clay layers are indicated in Table 12.20.

12.5.2. Bentonite-enriched soil

The particles of a granular soil are solid and totally impermeable, but the mass has high permeability because of the relatively large, interconnected voids between particles. If these voids are filled with finer particles (as is the case with a Boulder clay) the overall permeability is much reduced. This is the principle of Bentonite-Enriched Soil (BES). When mixed into soil and wetted, bentonite can swell up to 15 times its initial volume, thereby filling voids and creating a low-permeability barrier. Even when mixed with sandy soils, bentonite can provide a barrier with a permeability of less than 10^{-9} m/s.

Table 12.20. Comparison of geosynthetic clay liners (GCLs) and clay liners*

Characteristic	GCL	Compacted clay layer
Thickness	Typically 12–25 mm, thus uses up very little tipping space within the site	Typically 300–1000 mm
Coefficient of permeability	$\leq 5 \times 10^{-11}$ m/s	$\leq 10^{-9}$ m/s
Construction	Actual installation is a relatively simple process, but significant advance ground preparation is required	Construction is relatively slow because of the quantity of material to be placed
Potential for damage from puncture	Thin, very vulnerable; protection is required	Thick; cannot be punctured accidentally
Susceptibility to climatic effects	GCLs placed dry cannot suffer desiccation during construction; completeness of subsequent hydration is questionable	Difficult to place in wet weather; can desiccate and crack after construction if not protected; freeze–thaw action is unlikely to be a problem if layers are covered properly
Availability	Materials can be relatively easily transported to site	Suitable materials are not always readily available, and they cannot be transported over large distances because of the large quantity required
Long-term stability	May become brittle due to chemical interactions	Essentially inert
Experience of usage	Limited	Has been used for many years and appropriate site techniques have been developed

* After Daniel (1995).

In landfill applications bentonite powder or pellets are mixed or rotovated into a soil (usually sand), with the moisture content within the treatment depth (normally around 200 mm) being carefully controlled. The soil is then compacted to a known density and allowed to swell to produce a low-permeability seal. Alternatively, the BES can be produced in a mixer (to give high control over bentonite dosing and material moisture content) and then laid and compacted like asphalt.

The advantages of bentonite–soil admixtures include:

- the layer is effectively self-healing if punctured
- a flexible mat is produced
- the permeability of the treated soil is controlled by the bentonite content
- a suitable lining material can be created where no local sources of natural clay exist
- the layer may also provide some degree of leachate attenuation.

It is difficult to handle bentonite powder on site in windy weather, and if dosing is not even zones of high permeability will exist in the liner. The use of a thin layer (300–450 mm) of bentonite-amended soil as the sole containment liner is not generally recommended, mainly because of the relatively high probability of deterioration due to desiccation cracks and chemical incompatibility with the leachate. Construction of a clay layer above the bentonite liner can provide protection against both these effects. However, it should be constructed immediately after laying the bentonite-amended soil layer in order to minimize drying cracks.

12.5.3. Other materials

Colliery shale (see Chapter 16, Section 16.2) has been used both as a material to protect liners and as a liner in its own right. With good compaction a permeability as low as 10^{-9} m/s can be achieved, and there are very large, readily available, quantities of this material (Hird *et al.*, 1998; Lee, 2000). Its main disadvantage is that it degrades by weathering if left exposed after placement. The weathering may cause an increase in permeability and initiate production of acidic leachate high in sulphides or sulphates. Such effects can be minimized by using weathered material from spoil heaps, so leaching has already taken place.

Pulverized fuel ash (PFA) is a silt-sized material that is a by-product of the generation of electricity in coal-fired power stations (see Chapter 16, Section 16.4). Thus large quantities of it are readily available. When simply dumped in lagoons, PFA usually achieves a permeability of around 10^{-5} m/s, but compaction can bring this value down to 10^{-7} m/s. Permeabilities lower than 10^{-9} m/s have been obtained for compacted PFA–cement and PFA–lime mixtures prepared for landfill lining (Versperman *et al.*, 1985). Unfortunately, the setting action which lowers the permeability also reduces flexibility, so layers made from such mixes are brittle. Furthermore, unless PFA is conditioned by lagooning, soluble boron compounds are likely to appear in run-off water. Consequently, the use of PFA to form a primary liner is currently not recommended.

The use of asphalt or asphaltic concrete is a technology more familiar in dam construction than in landfill lining. However, it has been used successfully in Europe, notably in Switzerland (where space for refuse disposal is at a premium and where hard rock is often close to the ground surface) and Germany. The asphaltic material is used to form a lining layer in a similar way to flexible pavement construction: a road paviour lays a strip of the material, which is then rolled to produce a dense, low-permeability layer about 60–100 mm thick. A complete liner would typically comprise two such layers, in conjunction with a granular stabilizing underlayer and a mastic seal coat over the top. Asphaltic liners have been used because they occupy little air space (they are around

350 mm total thickness) and they do not require clay for their formation. However, this type of liner is expensive and needs a very firm foundation since it is not as flexible as soil and may not be able to accommodate stresses induced by significant differential settlement and bending. Furthermore, there may be cause for concern in the long-term in that solvents from decomposing waste may affect the stability and absorption capacity of asphaltic liners (Jessberger, 1995).

12.6. Fluid control

12.6.1. Leachate collection

The movement of contaminants through a barrier depends on both the advective velocity and the diffusion rate. The advective velocity will, in turn, depend on the hydraulic gradient. Modern landfills have a leachate underdrain system that reduces the leachate head (and thus the hydraulic gradient) acting on the base of the landfill. Furthermore, by removing leachate from the landfill the mass of contaminant available for escape through the liner is reduced.

Recirculation of collected, and possibly treated, leachate may be a key feature where accelerated stabilization is to be achieved. There will be a need to consider the means of distributing the recirculation liquids through the waste to ensure that there is overall wetting of the waste and flushing of the leachable contaminants, that preferential pathways are not established, and that downward flowing leachate does not impair the landfill gas-collection system.

Waste is usually underlain by a drainage blanket constructed from gravel or demolition rubble to give a permeability of 10^{-2} to 10^{-4} m/s. A system of large-bore, slotted collection pipes is contained within the drainage blanket, and this leads the leachate to the abstraction point (collection sump). Leachate is removed from this sump by various means, such as pumping from vertical wells or chimneys, side-slope risers located on the site perimeter (Figure 12.25) and gravity drains.

There is increasing awareness of the possible fall-off in efficiency of drainage blankets and pipes due to microbial growths and solid deposits. Clogging may be a result of a combination of blockage by particles, chemical precipitation and biofilm growth. Once severe clogging of a drainage blanket has occurred the majority of lateral flow towards the drains will occur in the waste. However, a clogged drainage blanket will still be substantially more permeable than the underlying liner. The likelihood of clogging occurring can be reduced by maximizing the flow velocity in the drain and the void size in the blanket, and by minimizing the surface area available for biofilm growth. One disadvantage of rapidly constructing an effective cover over a landfill site is that it may be many years before full leachate generation occurs. Hence only small quantities of

Figure 12.25. A leachate drain and an extraction pipe

leachate may flow through the collection system, which is then prone to siltation and biological clogging.

12.6.2. Landfill gas collection

All non-inert landfill sites will produce gaseous emissions over a long period of time. General problems that arise from landfill gas are summarized in Box 12.5. Landfill sites are often located in areas where coal and other minerals have been mined and where ancient methane gas of geological origin may be present. Furthermore, these areas could contain underground mine-workings where disused tunnels and shafts can act as conduits for gas. Other major man-made pathways for gas migration from a site are underground service ducts used for electricity, telephone, TV cables, street-lighting cables, water and gas pipes, sewers, drains and land drains. At some sites where no control measures have been installed, landfill gas has migrated up to 300–400 m outside the site.

A landfill gas management system is intended to prevent the preceding problems, and it will generally contain the following features:

- a containment system that will retain gas within the site and prevent off-site migration
- a system for landfill gas venting with adequate back-up facilities
- a separate system to control gas migration on the site perimeter
- gas monitoring boreholes outside the waste boundary.

Three types of system may be used to control lateral migration of landfill gas (either individually or in combination): passive venting, physical barriers and pumping systems.

Passive venting systems rely on the inherent ability of gases to move from a location with high pressure and/or concentration, to one where pressure and/or concentration is lower. Means of achieving this are venting columns, drilled vents and stone-filled trenches. The reduction in efficiency with depth means that simple venting trenches will not effectively prevent lateral migration at depths of more than about 3 m. Vertical vents consist of a perforated pipe surrounded by a gravel packing (Figure 12.26). Passive venting systems should only be used in situations where the rate of gas generation is low (e.g. biologically 'old' sites, and inert waste sites).

Physical barriers are low-permeability insertions into the ground such as flexible geomembranes, bentonite–cement slurry walls (see Chapter 7, Section 7.5.1) and piled cut-offs (see Chapter 13, Section 13.5.1). Bentonite–cement walls and other clay cut-offs are not fully effective against gas migration unless they incorporate a geomembrane.

Pumping systems depend on suction to remove landfill gas from the waste. They are the most complex of the gas-management systems and have five main components: gas wells/drains in the waste; connecting pipework; condensate traps; pumps to remove gas; and gas flares or utilization plant. The control system should be sized, and appropriate venting, flaring or utilization equipment should

Box 12.5. Potential problems from landfill gas

- **Explosions or fires.** Due to gas collecting in confined spaces, such as buildings, culverts, manholes or ducts on or near landfill sites.
- **Asphyxiation.** People entering culverts, trenches or manholes on or near landfill sites.
- **Waste ignition.** When released through fissures at the surface, landfill gas may be ignited and thereby initiate a fire within the waste.
- **Phytotoxicity.** The growth of crops and vegetation on or adjacent to landfill sites may be adversely affected.
- **Human effects.** Risks to health. Nuisance problems, especially odour.

Figure 12.26. A gas venting column

be specified. A typical gas well is illustrated in Figure 12.27. As a general principle, the performance of a well improves with increasing diameter and boreholes are now being drilled up to 1m diameter with the well pipe being around 150mm internal diameter.

On the basis of measured landfill gas production rates the following empirical equation has been developed for the extraction rate from a gas well located within domestic refuse (Blanchet, 1977):

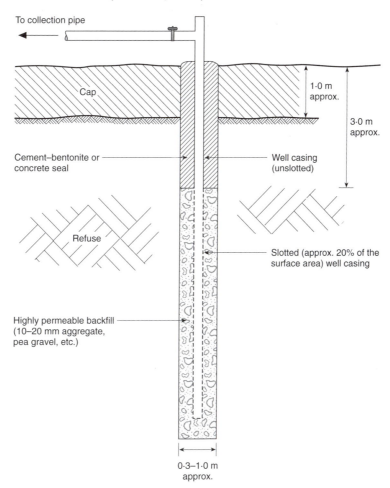

Figure 12.27. A gas well

$$Q = 14.1 \frac{Ah}{R^2} \tag{12.14}$$

where Q is the mean production rate (m^3/day), A is the landfill area (ha), h is the fill depth (m) and R is the radius of influence of the gas well (m).

Gas-collection systems are subjected to very considerable forces as the waste compacts, degrades and settles, and this should be taken into account when the system is designed and constructed. Substantial downward drag forces are imposed on the relatively incompressible vertical wells and stone columns which, in extreme cases, could result in them being punched down through the landfill base liner. Stone columns should have a substantial foundation pad designed to withstand the forces likely to be imposed. Lateral forces will result in the well structure being deflected out of vertical alignment and may result in horizontal pipes being crushed. All gas-extraction systems are subject to a progressive loss of performance due to the build up of silt, chemical precipitates and biomass. Small-diameter wells (less than 150 mm bore approximately) are particularly affected and can rapidly (within a few years) become unserviceable.

12.6.3. Landfill gas monitoring

Regular monitoring must be undertaken at all sites that contain controlled wastes to provide evidence of whether landfill gas is giving rise to a hazard or nuisance. Prior to undertaking any measurements or analysis a desk study should be carried out on the site and its environs. The study should aim at obtaining all available details of the geology including any man-made features, the location of houses or any other development and underground services within at least 250 m of the site boundary. For new sites comprehensive monitoring is needed to establish background levels. Monitoring should continue until the gas concentration has fallen to an acceptable level (e.g. the maximum concentration of flammable gas is consistently below 1% by volume and carbon dioxide is below 0.5% by volume, over a 24-month period (DOE, 1991a)).

A standard method of monitoring needs to be adopted so that results are comparable, reliable and reproducible, e.g. the guide developed by the Buiding Research Establishment (BRE, 1987). There are several techniques for carrying out landfill gas monitoring, including subsurface probes, backfilled pits or trenches and gas monitoring boreholes (Table 12.21). While for exploration at

Table 12.21. Field monitoring for landfill gas

Method	Methodology	Comments
Spiking	A metal spike is pushed into the ground and removed, creating a hole from which a gas sample can be taken	The method is very quick, cheap and easy to use. However, it has poor accuracy, is limited in depth (about 1 m), and confirms gas presence but not absence
Driven probe	A hollow casing tube with solid nose cone is driven into the ground, a monitoring pipe is installed in the casing which is then extracted leaving the nose cone behind	The method is straightforward, able to penetrate deeper than spiking and allows minimal ground disturbance. But it is not possible to examine strata, and obstructions cannot be penetrated
Borehole	This can be produced by hand auger (for shallow holes), or by percussive or rotary boring	Great depths are attainable and the strata can be sampled and inspected. The process is expensive and relatively slow
Backfilled trial pit	After excavation of a trial pit (a widely used technique for site investigation) a gas probe is inserted and the pit is backfilled	Generally considered as a short-term, shallow technique to accompany or precede a borehole investigation, acting both as gas monitoring of installations and, equally importantly, as a method of inspecting the physical state of the waste

shallow depths simple methods are used to create a void in the ground the best method for measurement in landfill or its surrounding strata is by the installation of boreholes or wells. These should be established in the wastes to provide information on gas quality, temperature and pressure and be installed around the landfilled zone to detect migration. Boreholes outside the landfilled area should extend below the base of the site. Hole spacing typically ranges from 50 m or more (for uniform strata with no adjacent development) to 5–20 m (near developments or where fissured strata adjoin the site).

There are various instruments available for detecting the major components of landfill gas (Table 12.22). Sampling errors associated with gas monitoring and measurement can be large. The usual procedure is to use portable field instruments for frequent checking and periodically collect a gas sample for analysis in a laboratory, either by gas chromatography (GC) or infrared (IR) analyses. An important factor in assessing the gas regime is the apparent emission

Table 12.22. Detectors for measuring gas concentrations

Detector	Principle	Range	Comments
Catalytic oxidation	Uses catalytic sensing elements to detect low levels of flammable gases as a percentage of the lower flammability limit of the gas	Usually requires oxygen concentrations in excess of 12% by volume to ensure complete oxidation of the gas. At lower levels of oxygen the instrument may not respond to the flammable gas	Instrument can be accurate only when measuring the particular gas for which it has been calibrated. Sensing agent can deteriorate with age and the instrument should therefore be regularly calibrated
Thermal conductivity	Measures the total concentration of all flammable gases present by comparing the thermal conductivity of the sample against an internal electronic standard representing normal atmospheric air. This type of detector is not affected by low oxygen concentrations	Can measure the full 0–100% range of gas concentrations. The sensitivity at low levels, below the lower flammability limit of the gas, is poor	With landfill gas the mixture of methane and carbon dioxide can cause response problems as each gas affects the thermal conductivity cell differently. Suppliers will calibrate their instruments using mixtures of the two gases. For accurate measurements it is essential that the sample being measured contains only the two gases for which the instrument has been calibrated
Gas chromatography	The sample is placed on a separating column and is washed through with an inert gas. The column selectively retards, and thus separates, the substances	The most reliable method for analysing the major components of landfill gas	Should be employed to confirm measurements made by portable equipment. The instrument is not portable or rugged enough for use on site, but it can be set up in a field laboratory
Indicator tubes	The gas sample is drawn through a tube containing a reagent which reacts, producing a colour change, the amount of colour change corresponding to the concentration of the gas	Provide a simple, crude indication of the presence of several components present in landfill gas	The potentially large number of minor components present in landfill gas could produce interference effects on the indicating reaction. Values obtained with these tubes should only be used as an indicator, with little weight being attached to them

rate of the gas from the source. This must obviously be measured in situ, and as such must involve portable instruments. Actual emission measurement, however, is more complicated than for concentration, and to date there are no fully reliable or proven techniques. Direct measurement involves volume flow rate measurement by devices such as rotameters, vane anemometers or bubble-flow meters.

Resolving gas levels to the accuracy available with GC is not necessary for routine checking. However, laboratory analysis of gas samples is essential where it is necessary to identify the source of the gas. Analysis of gas composition may allow distinctions to be made between mains gas and landfill gas, or landfill gas and gas of geological origin. This is typically done by GC, but will often include the use of mass spectrometry to give an accurate identification of trace gases, which may be indicative of a specific gas source. Carbon-14 dating, radiometric dating, mass spectrometric dating and stable isotope measurements can all be used to help identify the age of the gas, and thus distinguish, for example, between landfill or marsh gas and geological gas from mines or gas mains (Waldron *et al.*, 1995).

12.7. Covers and capping

Three types of cover may be used in engineered waste disposal: daily cover, intermediate cover and final cap. The general functions of these covers are management of the refuse as it is dumped and control of the products of waste decomposition. Covers will also influence the waste decomposition process (see Chapter 12, Section 12.2). Even though one of the prime functions of the final cap should be the regulation of biodegradation, in general practice this influence is usually regarded as simply a consequence of covering the waste.

12.7.1. Daily and intermediate cover

Daily cover consists of a layer of inert material (typically 100–150 mm thick) placed over the refuse at the end of the day's tipping. Daily cover performs many functions:

- facilitates access
- improves site aesthetics
- stops light debris (paper, plastics, etc.) from being blown off-site
- reduces risk of disease transmission (by birds, rats, etc.)
- contains odours
- minimizes fire risk
- provides a medium for partial attenuation of leachate (after burial of the daily cover).

Sandy soils and ashes have been the most commonly used daily cover material. If clay is used then, prior to the placement of the next lift of waste, the cover layer should be scarified or removed to avoid perched leachate levels. If soil or ash is used to form the daily cover it uses up a significant volume of tipping space (about one-fifth to one sixth of the total available). Hence greater use is being made of thin sheets of man-made material which are stretched over the lift of refuse at the end of a working day. The material is removed the next day prior to the resumption of tipping. The same piece of synthetic material (often a geosynthetic) can be reused several times over.

An intermediate cover is placed over portions of a landfill that are not full but where filling has been halted and the site is expected to remain open for a long period of time (2 years or more). Such a cover performs the functions of a daily cover, but also helps to reduce the volume of leachate produced. However, the cover will not be as substantial as the final cap (it will be around 250–500 mm thick) and it will not be engineered to the same degree.

12.7.2. Final capping

The primary functions of the final cap are:

- to minimize leachate production by reducing the ingress of rain and surface water into the underlying waste
- to prevent uncontrolled escape of landfill gas from decomposing refuse
- to provide a foundation for site rehabilitation and re-use after the waste has stabilized.

A major problem with the construction of a final cap is that it has to be placed over a medium that is heterogeneous, compressible (and resilient) and will degrade. Thus, if an attempt is made to form a compacted clay seal over the waste, much of the clay placed at first will be pushed into the waste. Even when a coherent clay layer has been formed it will be very difficult to compact it to achieve low permeability because it is not seated on a firm base. In addition, a geomembrane cannot be placed directly over the waste to form a sealing layer because it will be punctured by sharp objects or stones and will be overstretched (leading to tearing failure) where it spans over local voids and depressions.

Even if it is possible to construct a suitable sealing layer that initially permits very little infiltration into the landfill, the effectiveness of the barrier could disappear with time due to factors such as:

- settlement of the waste causing cracking or overstretching
- cyclic freeze–thaw action
- drying and desiccation
- penetration by animals, plants or humans
- erosion
- degradation of the synthetic material with time
- end-use of the finished landfill.

Hence it is widely accepted that a final cap over refuse needs to be composed of several layers, each of which performs a specific function so that the complete system operates as required. A suitable form of construction is illustrated in Figure 12.28 with the functions of the individual layers being outlined in Table 12.23. Final caps are illustrated in Figures 12.6 and 12.29.

12.7.3. Cap performance

A cap placed over refuse will undergo distortion as a result of construction operations and the behaviour of the underlying waste. This distortion may cause cracking, and enhanced overall permeability, of the sealing layer. Over the lifetime of a sealing layer three main distortion phases may be identified:

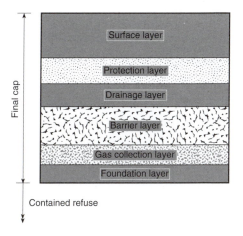

Figure 12.28. A composite final cap

Table 12.23. The functions of individual layers of a final cap

Layer	Functions	Materials
Surface	To support vegetation so that the completed site is aesthetically acceptable	Topsoil is most commonly used for the upper part of this layer. The soil should have sufficient water-retention capacity for vegetation to survive during dry weather. The layer thickness depends on the after-use of the site, but is typically 0.5–0.8 m.
Protection	To prevent damage to the underlying layers by burrowing animals, plant roots and human intrusion. To act as a frost-protection layer	A stony soil (e.g. gravel, quarry waste, mining spoil) that can be compacted to form a dense, hard layer. A typical thickness is around 0.2 m
Drainage	To remove excess percolating water, thereby reducing the water head on the underlying barrier layer and minimizing pore pressures in the upper layers. To prevent desiccation of the underlying barrier (if compacted clay is used for this barrier)	Sand or gravel with some fines, thick geotextiles, a geonet with a geotextile filter/separator. A mineral layer is about 0.3 m thick, whereas a geosynthetic layer might be only 0.1 m thick
Barrier	To stop the escape of landfill gas and to minimize unregulated influx of water into the waste	This layer is often regarded as the most crucial component of an engineered final cover system. Compacted clay has usually been used to form this layer — there can be problems due to cracking as a result of desiccation and flexure of the layer. Geosynthetics and geocomposites are also being used. These may be damaged during on-site fabrication and installation (because of the angularity of underlying materials) and also stretching due to waste settlement may cause damage. The thickness of the layer is likely to be between 0.3 m (geomembrane composite) and 1.0 m (clay only) approximately
Gas collection	To collect the landfill gas and direct it towards a treatment point for subsequent discharge or usage	This layer must not become plugged by fine-grained materials. Suitable materials are; sand or gravel (0.3–0.5 m layer thickness) or a thick geosynthetic with high in-plane permeability
Foundation	To level the surface of the waste (so that holes are infilled) and to provide a firm base for construction of the other capping layers	The layer should be sufficiently permeable for gas to pass through it without creating a large pressure head. The layer may be made of; granular material with limited fines content, selected waste materials such as minestone, furnace bottom ash and demolition rubble. Overall thickness of the layer is likely to be in the region of 0.3 m.

Figure 12.29. A multilayer final cap

- Initial spreading and rolling of the sealing layer material and trafficking by construction plant. Because of the compressible nature of the underlying refuse the rolling operations and plant movements produce a 'wave-like' motion of the barrier (i.e. it is subjected to bending). A soil sealing layer will be bent while in an undrained state (because there will have been insufficient time for consolidation), and thus the relevant strength for any analysis will be the original, undrained value.

- Placement of drainage, protection and surface layers over the sealing layer. Once again the barrier will be subjected to a motion that produces zones of hogging and sagging. At this stage a soil sealing layer may have consolidated by a significant amount (because of the relatively short drainage path length). When drainage occurs the strength of the barrier is likely to increase because of a decrease in pore pressure. Furthermore, the drainage and cover layers apply a surcharge to the sealing layer, which induces compressive stresses that act to close existing cracks and prevent other cracks from opening. However, if cracks developed in the preceding phase then some cohesionless drainage material would be likely to enter them, so that they would remain as highly permeable conduits.

- With time the confined refuse will decompose and settle so that the top sealing layer is also subjected to settlement. It is highly unlikely that waste settlement will be completely uniform. However, if the refuse has been placed in a controlled manner it is reasonable to assume that there will not be drastic differences in the magnitudes of settlement at places adjacent to one another. Thus the sealing layer will be subjected to differential settlement and bending.

An appropriate analysis for the long-term performance of the completed cap would appear to be a drained or effective stress approach (for a clay sealing layer), taking account of any surcharge. However, if the clay barrier layer is already cracked (due to bending during installation) and if the cracks have been penetrated by relatively free-draining material, then such an analysis is inappropriate and the critical case is one of undrained bending. Jessberger & Stone (1991) carried out a series of tests on compacted clay liners and concluded that the liners cannot withstand settlement-to-length distortions greater than approximately 0.05–0.1 without cracking. For a 1 m length of liner these limits correspond to radii of curvature of 1.3 and 2.6 m. An analytical model for the cracking behaviour of cohesive sealing layers subjected to bending has been developed by Sarsby & Wu (as reported in Wu, 1999). Such a deformation is predicted to produce a limited number of more or less uniformly spaced cracks (because of the natural variation in tensile strength of the barrier material). According to the theoretical analysis these cracks have to penetrate more than 75% of the barrier thickness before the flow rate through the cracked layer becomes more than twice that for the intact barrier (Figure 12.30). For a 0.5 m thick sealing layer this is predicted to be equivalent to a radius of curvature of about 2 m. In contrast to the sensitivity of compacted clay layers to cracking due to differential settlements, it is claimed that geomembranes can withstand large tensile strains, even when stressed three-dimensionally (Daniel & Koerner, 1993).

An increase in the permeability of a fine-grained soil is likely to result from freeze–thaw effects, typically by one to one and a half orders of magnitude from the initial value (Zimmie et al., 1992).

The depth of the soil cover will be related to the site after-use and landscaping requirements. The soil depth should be sufficient to prevent desiccation of a clay layer or physical damage (e.g. by ploughing or from tree roots). Protection from damage is a requirement for both clay and membrane caps.

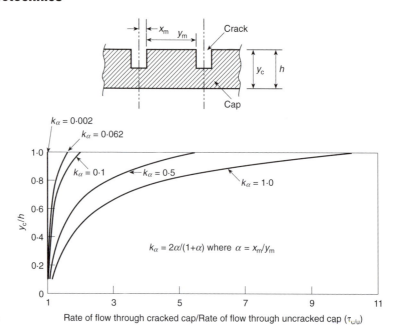

Figure 12.30. Leakage through a cracked sealing cap

The surface gradients of the final cap should be acceptable from a geotechnical stability point of view — a maximum side slope of 1 vertical to 3 or 4 horizontal will generally provide an acceptable Factor of Safety. However, such steep slopes can suffer erosion unless provided with either temporary surface protection or a good cover of vegetation (long-term). The final cap gradients (after settlement due to waste degradation) should be no flatter that 1:25 to ensure adequate surface water run-off and to avoid water-ponding problems created by local differential settlement. Particular attention should be paid to final cap gradients at the edge of a landraise, especially where this overlies an area of significant change in depth.

The use of geosynthetics in closure covers generally results in flatter slope angles than with covers using only soil materials. The use of a textured geomembrane (i.e. one formed with a 'rough' surface), on steep portions of closure covers is one technique that can be used to construct the steepest possible geosynthetic slope.

The Factor of Safety against instability of the final cover components can be assessed using the infinite slope method (see Chapter 9, Section 9.2) because of the relative thinness of the cap and the planar nature of the interfaces between different materials. Undrained parameters are often applicable to represent the construction period prior to covering of final caps and prior to filling over base liners. Drained parameters are then applicable to long-term modelling of the finished system.

13 Contaminated land

13.1. Introduction

Contamination of land has arisen from many different kinds of human activity and is essentially a legacy of our recent industrial history, although much older examples exist, such as from Roman lead mining nearly 2000 years ago. Sources of contamination include the deposition of waste products, industrial-operation spills and leakages, airborne contaminated dust and repeated raising and levelling of land as one industrial use supersedes another. Contaminants may be in solid, liquid or gaseous form and can adversely affect susceptible targets such as humans, rivers, aquifers and abstraction wells, buildings and the environment.

The most famous uncontrolled hazardous waste site where ill-effects have been demonstrated is probably Love Canal in New York State, USA. The Love Canal was an attempt by William Love to link the upper and lower sections of the Niagara River, above and below Niagara Falls, by a canal in 1896. The project was abandoned, leaving a 3 km trench, which was eventually purchased by a chemical company for use as a dump. From 1942 to 1953 this company disposed of about 22 000 t of chemical waste in the trench. At that time the area was a sparsely populated portion of the south east corner of the city of Niagara Falls. Shortly after closure of the landfill in 1953 the site was sold to the Niagara Falls Board of Education for a price of only $1, but with an important and unambiguous caveat that the site should not be disturbed by building works. The board built an elementary school on a centrally located portion of the land, and by 1972 the rest of the land had been covered by residential development. In the winter of 1975, heavy precipitation resulted in unusually high local groundwater and this caused portions of the landfill to subside, created ponds of surface water heavily contaminated with chemicals, and transported chemical wastes to the basements of nearby residences. Subsequently, residents began to experience medical problems such as asthma, renal disorders, miscarriages, hyperactivity and skin rashes.

In August 1978, the New York State Commissioner of Health declared a health emergency at Love Canal. After investigation, the government authorities eventually bought large numbers of houses, all nearby residents were relocated (Figure 13.1) and public access to the site was restricted. The US Environmental Protection Agency (USEPA) was ordered to complete an in-depth environmental study at the site. In total, 248 different chemicals were identified, including benzene, carbon tetrachloride, polychlorinated biphenyls and trichlorophenols. As part of its programme, the USEPA reviewed all existing hydrogeological studies and undertook a test drilling programme to define the geology of the study area and to identify the direction and rate of movement of groundwater in the study area. A geophysical investigation used a variety of remote-sensing techniques, such as ground-penetrating radar and electromagnetic conductivity.

Remedial action was taken to prevent further entry of rainwater into the buried canal and spread of the pollutants, and a health surveillance programme was conducted on the population living near the site. Various clinical effects were observed, but it was difficult to demonstrate cause–effect relationships using classical epidemiological techniques because many of the population had

Figure 13.1. A house abandoned because of contamination from Love Canal

workplace exposure to the chemicals concerned and the absolute numbers showing adverse effects were small compared with the total population considered at risk on the basis of proximity to the site. In addition, comprehensive monitoring showed that there was a general level of air and groundwater pollution associated with the local chemical industry. However, good 'forensic engineering' showed that there was a pattern of illness among those living close to the courses of buried streams, and hence that just a small part of the population was subject to a very high risk.

To date more than £50 million has been spent on site remediation, resident relocation and environmental and human health investigations at Love Canal (Deegan). However, the actual Love Canal site has still to be remediated (Figure 13.2) and the surrounding area still remains blighted and abandoned (Figure 13.3).

13.1.1. Background

Contaminated land can be defined as land which:

- because of its former use, now contains substances that present hazards likely to affect its redevelopment
- requires an assessment to determine whether redevelopment should proceed without some form of remediation of the site.

At the end of the 1980s a Department of the Environment (DOE) survey identified some 40 000 ha of derelict land in the UK, most of which was likely to be contaminated to some degree. For instance, current estimates suggest that the UK contains between 50 000 and 100 000 contaminated sites. Gasworks sites are

Figure 13.2. The Love Canal site

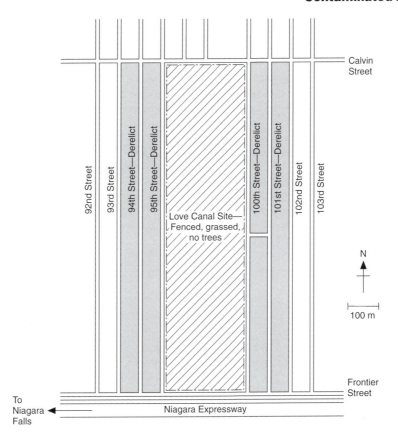

Figure 13.3. The current situation at Love Canal

particularly prevalent (about 3000 of them) and they are often close to the centre of towns. Old gasworks are very prone to leave a legacy of contamination because the coal carbonization process produced impure coal gas, ammoniacal liquor and tar, and a solid residue of coke.

Growing pressure on land resources and greater environmental awareness has led to an increasing need to reclaim and recycle contaminated land. Some 50% of all development takes place on previously used land ('second-hand land'), some of which will be contaminated. Contaminated sites need to be identified and characterized in terms of their nature and the extent of contamination and the consequential threat posed by such sites to human health and the environment. Unfortunately, once this has been done developers are reluctant to rehabilitate contaminated land because of the high cost of clean-up. Greater governmental intervention is needed to encourage re-use of second-hand land.

In North America and parts of Europe, contaminated land policies have evolved during the past 20 years. Fundamental to the UK policy framework is the implementation of the 'suitable for use' approach. This requires that any unacceptable risks to health or the environment from contamination should be dealt with and, where practicable, land that is already contaminated should be brought back to a suitable standard, taking into account the actual or intended use of the land development. This approach is consistent with the principle of sustainable development.

Contaminated land is found typically, but not exclusively, in urban or industrial areas. Generations of works owners have developed sites sequentially, often by simply laying hardcore covers over contaminated areas and founding their new structures on these platforms. In the times when bulk haulage was difficult and costly, the foregoing form of development usually entailed the use of available local materials, and often these were solid wastes. Thus slags, cinders, casting sands and such chemical wastes as gasworks spent oxides were often

- Gasworks, coal carbonization plants and ancillary by-product works.
- Metal mines, smelters, foundries, steelworks and metal-finishing installations.
- Scrapyards.
- Paint and dye works, textile industries.
- Industries making or using wood preservatives or pesticides.
- Chemical works, pharmaceutical plants.
- Oil refineries, petroleum storage and distribution sites.
- Asbestos works.
- Tanneries and plating works.
- Landfill and other waste disposal sites.
- Sewage-treatment works.
- Coal-burning electricity-generating power stations.
- Railway land (especially large sidings and depots).
- Munitions production and testing sites, military establishments.
- Ship-breaking.
- Paper and printing works.

imported to act as cover materials. Inevitably this added a secondary, and often very significant, extra source of soil contamination to that created by the industries themselves. Land developed in this way will contain substances likely to have detrimental effects on both living organisms and materials, and this distinguishes contaminated land from other categories of disused or derelict land.

The main industries and activities that give rise to contaminated land problems are indicated in Box 13.1.

13.1.2. Sources of contamination

The bulk of contaminants occur naturally, in varying concentrations. However, it is mankind who has led to their presence in the ground in potentially hazardous concentrations as a result of spillages, waste by-products and indiscriminate dumping. Figure 13.4 shows an area of land that will not support vegetation on a 'hill' formed from chemical wastes. Demolition of old works without adequate decommissioning and decontamination has significant potential for the release of

*Figure 13.4. Exposed
chemical waste in an old tip*

Figure 13.5. Pollution leaching into a stream

hazardous substances, as well as on-going hazards through the distribution and use of contaminated demolition products. During levelling operations, contaminated residues may be spread indiscriminately around a site. It is important to recognize that contamination can be moved many kilometres from the industrial process that created it, by natural effects such as wind and water movements. The dark area in the water near to the lower stream bank in Figure 13.5 is chemical pollution from an adjacent, old, industrial tip. The actual form of the hazardous substance will vary according to the former site processes, and may include gaseous, solid or liquid phases. In addition, many sites have physical hazards such as instability. Details of the industries that have the potential to cause significant ground contamination, and their associated pollutants, are given in Box 13.2.

Certain classes of contaminant substance are associated with specific industries or activities (regardless of their location), and are commonly found in land associated with that industry — Table 13.1. Because of this causal relationship between particular past activities and the types of soil contamination created, investigators can often identify discrete contaminant zones on larger sites, where space allowed a number of diverse production units to co-exist (Cairney, 1995). The full range of contaminants actually present can only be established by a full chemical site investigation, because:

- many sites (particularly those in city or urban locations) have a long industrial history
- process changes or the local practice or use of proprietary materials may have introduced unusual substances onto sites
- detailed local records of the operations carried out on a site may be lacking.

13.1.3. The nature of contaminants

The effect of exposure to a contaminant depends on a number of factors, including the size of the dose, the chemical properties of the substance, and the manner in which the dose is administered (e.g. inhalation, ingestion, skin absorption). Persons exposed to contaminants can suffer various levels of damage to their health. Symptoms can appear dramatically, with acute exposures leading to nausea, unconsciousness and even death, while chronic ill health can develop over a longer time-scale through repeated exposure to lower levels of toxic substances. The main concern with buildings is the durability of structural building materials (concrete, brickwork) and other materials used underground in construction (e.g. steel, lead, copper, zinc (as galvanized steel), rubber, bitumen, plastic) (Nixon *et al.*, 1979).

Box 13.2. Details of industries likely to produce contaminated land

Gasworks

These sites are often quoted as a major source of industrial pollution. Not only are there heavy metal rich ashes, but there are also substantial tarry residues that are rich in coal tars, acids and phenols and which are potentially carcinogenic. The high sulphate concentrations in the ground will usually require sulphate-resistant cement to be used in structures. Coal gas was purified using iron oxide (which contained free and complex cyanides, and elemental sulphur, sulphate and sulphide). This inevitably yielded a large number of by-products such as ammoniacal liquors, sulphate enriched lime ('foul lime') and cyanides ('Blue Billy'). The bulk of these were merely dumped on any vacant area of the site and remain today as a significant hazard to redevelopment. On some of the larger sites, these by-products were used in subsidiary industries such as tar and paint works, fertilizer and acid manufacture and even soap and toothpaste manufacturing. The problems associated with gasworks are also present on any site where coal carbonization has taken place, such as coke ovens at collieries and steelworks.

Non-ferrous metal-mining

The relics of mines and processing plants still exist in Devon, Cornwall, the Pennines, Mid- and North-Wales and parts of Ayrshire. The major contaminants are usually the metals that were actually mined (e.g. copper, lead, zinc) or the associated metals that occurred within the mineralized deposit (e.g. arsenic often occurs naturally with tin, and cadmium usually accompanies lead and zinc). In addition, the water that issues from these mines and tips is frequently heavily loaded with metals, causing pollution of watercourses for many miles downstream. Coal mines also can contain high levels of metals in the spoil heaps and, furthermore, the mine waters are frequently quite acidic and sulphate rich, which can cause problems of aggressiveness to concrete in the ground.

Scrapyards

These are among the most commonly encountered examples of contaminated land. The nature and degree of contamination depends on the activities carried out on the site, but can commonly include cyanides, sulphates, acids and alkalis and various organic compounds. Metals will also be found in high concentrations (lead, zinc, copper, cadmium, nickel). Waste oils are often present.

Dyeworks

These are generally accepted as being among the more difficult sites to redevelop safely. The petrochemical industries supply the primary materials for dyestuffs manufacture. These include many organic acids and solvents that are converted into dyestuff intermediates by reactions with dichromates, lead, nitric acid, sulphuric acid and some reactive gases. Dyestuffs are then produced by further chemical treatment of the intermediates. The chemicals used in dyeworks are principally organic compounds, and many are potentially carcinogenic. It is their presence that constitutes the greatest contaminant threat, rather than the presence of metals. If, however, the dyestuffs were used on site to dye materials, then common substances used to make fast the dyes may also be present on site. These include chlorates, acetates, and sulphides of barium, chromium, copper, lead and nickel.

Pharmaceutical works

Activities encompass the production of a wide range of products, including medicines, dental materials, toiletries, cosmetics, hormones, vitamins, hairsprays, deodorants, shampoos and perfumes. The vast majority of wastes from this manufacturing are liquid effluents, which are discharged to rivers and estuaries or pipes to coastal waters or sewers. The majority of solid wastes are disposed of in landfill. Practically every commercially available organic solvent is used in pharmaceutical manufacturing, and all could technically be present on land occupied by such an industry.

Chemical works and oil refineries

These are an obvious source of organic and inorganic contaminants. Most major industrial sites are likely to have had an oil spillage at some time, whether as a result of a burst pipe or leaking tank.

Box 13.2. Continued

Tanning and fellmongering

Leather is essentially animal skin protein combined with tars, oils, dyes and other finishing agents. The range of potential wastes is large and includes both organic and inorganic compounds (e.g. white spirit and kerosene for skin degreasing, chromium sulphate for tanning). The presence of organic chemical contaminants poses the greatest problems in site redevelopment, although contamination by metals such as chromium may pose additional problems.

Landfill sites

The main problems are gas and leachate production as a result of the biological breakdown of deposited wastes. Gases produced include carbon dioxide, hydrogen sulphide and methane, which inhibit growth of vegetation and can migrate into buildings. The leachate usually has high BOD, COD and ammonia content, and can cause serious pollution of groundwater. If all wastes are fully broken down and gas production has ceased, the presence of metal contaminants, which usually remain, becomes more important (Alker *et al.*, 1995).

Sewage works and sewage farms

Sewage sludges can contain a wide range of metal contaminants, which are related to the various types of industrial sewage input. Common contaminants include lead, zinc, copper, cadmium and nickel, while almost any metal, including arsenic, boron, mercury and manganese can occur less frequently and at lower concentrations.

Ironworks and steelworks

Much of the land is covered by iron, steel and blast-furnace slags containing elevated concentrations of manganese, boron, chromium, lead, zinc and cadmium. The actual metal content of the slag depends on the type of iron or steel produced. As the metals are usually fused into the slag, their impact depends on the likelihood of the slag being weathered and breaking down to release the metals into the environment. This does not often occur, and the impact of such wastes is generally low.

Railway land and sidings

These occur in all parts of the UK and most are not contaminated. In some areas, however, the trackways were constructed of waste ash and clinker from local industrial sources, and may thereby be contaminated. Common contaminants include lead, zinc and cadmium, with high concentrations of copper, nickel and manganese. Chromium and arsenic occur less commonly, as do contaminants similar to those found on gasworks sites.

BOD, biological oxygen demand; COD, chemical oxygen demand.

Table 13.1. The association of contaminants and hazards with particular industries

Contaminant/hazard	Likely locations
Toxic metals (e.g. cadmium, lead, arsenic, mercury) and other metals (e.g. copper, nickel, zinc)	Metal mines, ironworks and steelworks, foundries, smelters, electroplating, anodizing and galvanizing works, engineering works, scrapyards, shipbreaking sites
Combustible substances (e.g. coal, coke dust)	Gasworks, power stations, railway land
Flammable gases (e.g. methane)	Landfill sites, filled dock basins
Aggressive substances (e.g. sulphates, chlorides, acids)	Made ground, including slags from blast furnaces
Oily and tarry substances, phenols	Chemical works, refineries, by-products plants, tar distilleries
Asbestos	Industrial buildings, waste disposal sites

Toxic effects can generally be divided into those resulting from short-term (acute) exposure to a substance and those due to doses administered over a long period of time (chronic exposure). Many hazardous substances exhibit clear warning signs, such as distinctive odours, colours or physical forms. Others can be less obvious, although in some instances they can give rise to physiological symptoms that should alert one to the presence of a hazard (e.g. breathing difficulty or headaches suggesting asphyxiant conditions). Likely hazards are described in Box 13.3.

The major contaminants can be divided into five main groups:

- metallic
- non-metallic
- organic
- bacterial
- gaseous.

Information on the main contaminants is given in Box 13.4. For in-depth coverage of individual contaminants, see Barry (1991) and Steeds *et al.* (1996).

13.1.4. Mobility of ground contamination

Contaminants can exist in several forms (primarily solids, liquids (Figure 13.6), dusts and gases, and pathogens), and these affect the potential for harmful interactions:

Box 13.3. Hazards from contaminated ground

Carcinogenic substances
Some substances are known or are suspected to induce cancers (malignant tumours). Examples of proven and probable human carcinogens are arsenic oxide, asbestos, benzene, beryllium, cadmium oxide, coal tars and nickel salts.

Corrosive substances
Principal hazards are damage to human tissue and degradation of building services and materials. Effects can range from allergenic sensitization, to mild skin irritation, through to permanent physical damage. Tissue damage is generally associated with short-term exposure. Building materials that are potential targets include concrete, ferrous metal pipes and structures, and plastic pipework.

Combustible materials
Such materials are capable of burning or being set alight. The calorific ash value is the main criterion for determining the potential combustibility of a substance. Sources of combustion are a critical consideration, and the source will probably be an external agent rather than spontaneous combustion (which is considered to be extremely rare). Main hazards associated with combustible materials are release of toxic, asphyxiant or noxious gases and physical damage to humans.

Inflammability
This property relates to the ease with which a material will ignite. *Inflammability* will depend on the concentration of a gas in air and the concentration of oxygen present. *Explosiveness* generally relates to the rapid propagation of a flame in a confined space. The consequent effects are due to the high energy generated.

Asphyxiant gas
This causes unconsciousness and death by depriving an organism of oxygen. Some gases (e.g. carbon dioxide) can be both toxic and asphyxiant, which of these properties is critical being dependent on the concentration and the exposure period. Oxygen deficiency or gassing from hydrogen sulphide or sulphur dioxide, or other toxic gases, may occur during excavation work of trenches, manholes, etc., and during entry into basements, tanks, etc.

Box 13.4. Main contaminant groups

Metallic substances
Contaminants such as cadmium, chromium, lead, mercury, arsenic and selenium are all organic toxins that attack the central nervous system, causing gradual paralysis of the body. Many also cause damage to the liver and kidneys, and some can give rise to damage to the brain tissues. Other metals, primarily copper, nickel and zinc, affect, in the main, plant growth and are called *phytotoxins*, although the exact mechanism of their effect is not known. There are also other metals (e.g. iron) which, although causing a strong discoloration, do not have any significant toxic effects unless they are present in exceedingly high concentrations.

Non-metallic substances
Contaminants such as sulphides, sulphates and cyanides generally arise as by-products of manufacturing industry. Sulphates have a propensity to attack concrete structures and foundations within the ground. In certain conditions sulphides and cyanides can liberate highly toxic hydrogen sulphide and hydrogen cyanide gases, which the lungs absorb more easily than oxygen and which cause congestion of the lungs and death by asphyxiation. Asbestos is well known for its ability to cause cancer because of the physical irritant nature of the fibres.

Organic compounds
The effects depend on the components present. The contamination pathway is via inhalation and skin contact. Inhalation of volatile aromatic compounds presents a toxicity hazard (e.g. benzene and toluene have narcotic properties). Inhalation of polycyclic aromatic hydrocarbons (PAHs) may cause cancer.

Bacterial agencies
These are capable of causing disease to man, animals and plants (some may also attack construction materials such as concrete). These organisms include pathogens, mould and fungi, viruses, parasites, algae and parasitic worms. Exposure to these can lead to the contraction of disease and parasites, depending on the organisms involved. Relevant sources include sewage, hospital waste, laboratory waste, biodegradable domestic waste and disease/burial pits. The principal danger is to site investigations and redevelopers through skin contact, inhalation (of airborne spores, etc.) and ingestion (of contaminated dust/soil particles). After-users of the site could acquire pathogens through gardens.

Gases and vapours
The principal hazard from gases or vapours is usually in the build-up of concentrations within confined spaces such as cavities, basements, small rooms, trenches, drains and manholes. Landfill gases (methane and carbon dioxide) are particularly common, but gases or vapours can also arise from biodegradation processes, spontaneous combustion, combustion of waste materials and chemical reactions in the soil through increased acidity. Volatile organic compounds can also exist on sites with relevant spillages, storage or disposal areas.

- *Solids*. The principal direct hazard is associated with ingestion, either directly or indirectly. Material may enter the mouth, often from the fingers if smoking and eating occur on site. For the indirect route, part of the toxic fraction is either soluble or 'available', so that leaching (the mound in the centre of Figure 13.7 is leached chromium) into water supplies or uptake by plants is possible. For example, the aggressiveness of 'solid' sulphate compounds is fundamentally related to their solubility (and thereby their capacity to produce acidic conditions).
- *Liquids*. The principal hazard is their corrosive or aggressive nature (skin contact may produce burns, dermatitis, etc.) and their potential for water pollution. Toxicity is also an obvious potential, but consumption of such liquids is rare. Splashing of exposed skin by some substances (e.g. phenols)

Figure 13.6. A lagoon at a contaminated-land site

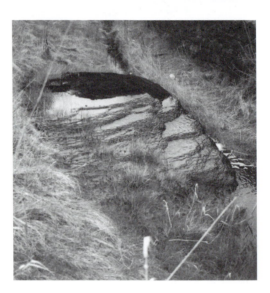

Figure 13.7. Metal pollution leaching from the ground

may lead to toxicity problems via percutaneous absorption (skin absorption). Skin absorption of a range of materials may have a general effect if the material is absorbed into the bloodstream, or may have a localized effect and cause skin irritation, dermatitis or skin cancer. Some liquids exhibit phytotoxicity (i.e. they are toxic to plant life).

- *Dusts and gases.* The principal hazard is from inhalation. This is generally likely to be a short-term exposure risk, and can be created by demolition and the removal of wastes from site (e.g. asbestos dusts), through the burning of contaminated material on site (e.g. nickel fumes) or through exposure to organic solvents. Also, some gases (e.g. hydrogen sulphide, sulphur dioxide) form acidic corrosive aerosols in the presence of water.

- *Pathogens.* The principal hazard is infection. The people most at risk are site investigation workers on former landfill sites and where sewage sludge or hospital wastes have been deposited. Disease-producing (pathogenic) organisms such as bacteria, viruses, or the eggs or cysts of parasites may be present. Sewage treatment significantly reduces the number of pathogens present and the incidence on old sites is likely to be small.

Table 13.2. Pathways for hazards

Hazard pathway	End-uses leading to hazard	Contaminants
Direct ingestion of contaminated soil by children	Domestic gardens, recreational and amenity areas	Arsenic, cadmium, lead, free cyanide, polycyclic aromatic hydrocarbons, phenols, sulphate
Uptake of contaminants by crop plants	Any uses where plants grow	Cadmium, lead, sulphate, copper, nickel, zinc, methane
Contact with building materials and services	Housing developments, commercial and industrial buildings	Sulphate, sulphide, chloride, tarry substances, phenols, mineral oils
Seepage through the ground, stockpiling or dumping of materials on the ground, leading to fires and explosions	Any uses involving the construction of buildings and services	Methane, sulphur, potentially combustible materials (e.g. coal dust, oil, tar, pitch, rubber)
Contact with contaminants during demolition clearance and construction	Hazard is mainly short-term (to site workers and investigation teams)	Polycyclic aromatic hydrocarbons, phenols, oily and tarry substances, asbestos, radioactive materials
Leaching, diffusion and seepage through the ground, leading to contamination of water	Any operation that may lead to run-off or leaching	Phenols, cyanide, sulphates, metals

The four principal exposure routes are:

- inhalation
- ingestion
- absorption
- surface contact.

Hence the principal groups affected directly, or otherwise, by the hazards of a contaminated site are:

- site investigation operatives
- demolition and construction workers
- after-users of the site, particularly children
- maintenance workers
- animal and plant life, including aquatic life
- building structures.

The pathways by which various contaminants reach targets are indicated in Table 13.2. Further specific advice can be found in a Health & Safety Executive booklet (HSE, 1991).

In summary, the development of contaminated land involves not only conventional ground engineering considerations, but also consideration of the hazards arising because of the potential for:

- the presence of toxic substances
- the existence of combustible materials
- chemical attack on building materials and other site services
- emissions of toxic or flammable gases
- groundwater contamination.

13.2. Site investigation

The investigation of a site on contaminated land is similar to an investigation of a site underlain by unpolluted natural materials, and as such it involves various facets:

- A desk study, to identify the previous processes or industries on the site and hence to infer the materials or chemicals to be encountered.
- The design of a site sampling and analysis programme to detect and/or confirm the location of previous structures and ancillary items, foundations and other obstructions, and the distribution of materials and contaminants.
- The identification of materials underlying the site, through on-site sampling, observation and testing.
- The measurement of the geotechnical and chemical properties of the ground materials.
- An analysis of materials taken from the site.
- The interpretation of the analytical data in order to assess the concentrations of contaminants both on and below the surface and to draw up a plan of contaminant 'hot spots'.

For contaminated land the consequences of inadequate site investigation are potentially much more serious than for construction on normal land. Late discovery of contamination may require a complete redesign of the project. Bearing this in mind, it has been proposed that it would be appropriate to undertake the site investigation in three stages: desk study, preliminary ground investigation, and second stage of ground investigation. Stages in the site sampling and analysis process are illustrated in Figure 13.8.

The desk study should identify the history of the site and former owners or users, and should be an initial reconnaissance of the area. In the first stage of ground investigation samples should be taken on a relatively broad grid and at any locations regarded as particularly vulnerable. If the ground investigation and the desk study show no contamination or potential for contamination, then development could proceed without further investigation. If at this stage there is any indication of the presence of contamination, a second-stage ground investigation should be conducted to assess the distribution of pollution. This latter investigation should be particularly aimed at providing information for the design of a site remediation scheme.

Figure 13.8. Stages in site investigation of contaminated land

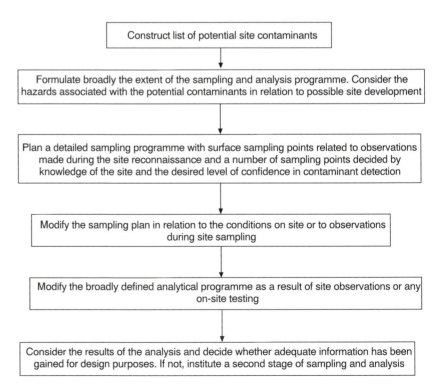

13.2.1. Preliminary work

One of the principal differences between the investigation of uncontaminated sites and those that are potentially contaminated is the much greater consideration which must be given to health and safety issues in the latter case. For this reason, and to maximize the quality of information obtained, no on-site investigation (fieldwork) should be done until preliminary research (on the history of the site and existing conditions, etc.) has been carried out. With potentially contaminated land it is very important to have prepared a detailed hypothesis of the likely site conditions before undertaking any physical investigation. The on-site investigation can then be designed to test the hypothesis and check for random 'hot spots', thereby increasing the chances of confirming the presence of contamination and permitting economy of effort.

As with conventional ground investigation, the preliminary work has two main components: the collection and examination of information on the site history, and a site reconnaissance or inspection. During the latter activity it is useful to contact local residents and collect any information that they have on the site. Local authorities, inspectorates and other organizations should also be contacted as indicated in Table 13.3.

Various sources of preliminary information and data relating to the geotechnical nature of sites are outlined in Chapter 4 (Section 4.1.2). However, when dealing with contaminated land it is also necessary to acquire historical data for the site (to obtain some appreciation of contaminants likely to be present and their location) and environmental data (to identify current levels of contamination and constraints with regard to investigation or remediation). There are a variety of sources of historical information, including:

- Ordnance Survey maps
- estate maps
- enclosure maps
- parish and town plans
- public health board plans
- parliamentary plans (railway building)
- books on local and regional history
- back-copies of local newspapers
- contaminated land surveys.

There are also many sources of environmental data, as indicated in Chapter 2 (Section 2.2.2).

As a result of the desk study a picture of the site should emerge in terms of:

- previous industrial history
- processes carried out at the location
- as-built layout of industrial works

	Category	Organization/section
Table 13.3. Contact points for information on land usage	Local councils	Environmental health, planning, waste regulation, engineering
	Regulatory authorities	Hazardous Substances Authorities, Environment Agency (Waste Regulation Authorities, HM Inspectorate of Pollution), Health & Safety Executive, Scottish Environment Protection Agency, Environmental & National Heritage Agency
	Other	British Coal, British Coal Opencast Executive, British Steel, water companies, PowerGen, National Power, gas companies, English Partnerships, Mines and Quarries Executive, Scottish Enterprise

- building or process modifications undertaken
- the nature and quantities of materials handled on the site
- the nature of surrounding land use
- physical features
- geology and hydrology
- surface and groundwater regime.

When all the foregoing information has been collated it should enable a fair prediction to be made of the likely ground conditions, without the need to carry out any excavation work at this stage. This hypothesis and the resultant anticipation of hazards, will serve to protect construction personnel, site investigators and the future occupiers of a contaminated site.

During the site walk-over inspection the whole area should be traversed, preferably on foot. Particular attention should be paid to the surface topography and site layout, as both can provide some guidance as to the types of contaminants likely to be present. The following are particularly useful indicators of contamination:

- Vegetation: the absence or poor growth of vegetation may indicate the presence of phytotoxic substances, the prevalence of particular species may suggest the presence or absence of particular substances or an abnormal pH value.
- Surface materials: unusual colours may be due to chemical wastes and residues.
- Fumes and odours: these are often readily detectable at very low concentrations, and many harmful gases have very distinctive aromas.
- Drums and similar containers: these may contain hazardous substances.
- Infilled areas: old plans, aerial photographs, etc., will often indicate such locations.

Differences between the current conditions and the information obtained from the site history (e.g. changes in the positions of boundaries, buildings, roads) should be looked for and recorded. All structures above and below ground, including foundations, tanks and pits, should be inspected and recorded. Any fill material, made-ground and materials such as ashes, slag, scrap and industrial or chemical waste should be inspected and recorded, as should any signs of settlement, subsidence or disturbed ground.

Samples of solids or liquids should be collected and tested in order to identify any hazards that may affect later stages of the investigation. Portable instruments may suffice to detect and measure some contaminants (e.g. emissions of flammable or toxic gases). However, care is needed in selecting and using portable instruments and in interpreting the results they give. Even under ideal laboratory conditions up to 50% of the content of volatile organic compounds can be lost during the sampling and subsequent storage and handling of the samples.

13.2.2. Ground investigation

Having identified the history of potential contamination on a site, it is necessary to plan a physical ground investigation accordingly. The primary objectives of the ground investigation are to:

- determine the nature and extent of on-site contamination of soil and groundwater
- identify likely pathways for contaminant movement
- assess potential targets
- set reference levels to judge the effectiveness of remedial treatments
- determine the nature and engineering implications of features of the site
- identify geotechnical effects that might be initiated by the remedial works.

The ground investigation is conducted primarily by invasive means (trial pits and boreholes) and the items listed in Box 13.5 need to be defined at the outset. For direct determination of the physical distribution of contamination, one of the easiest and best methods is the excavation of trial pits using manual or mechanical means (see Chapter 4, Section 4.2.1). This method allows a visual assessment of the in situ relationships and state of the various materials in the ground, as well as allowing a detailed description of the materials when brought to the surface. Extreme care should be exercised, as pit or trench collapse is possible even for shallow depths of excavation, and unstable ground will require shoring up.

For a deep investigation it is necessary to turn to boreholes. Hand auger holes are usually practical, especially when working in confined spaces, but there is a major problem with penetrating variable materials with hard obstructions. These latter materials can often be penetrated speedily and relatively cheaply with cable-percussive or rotary coring techniques (see Chapter 4, Section 4.2.2) and intact samples can be retrieved. Although the amount of sample required for modern analytical techniques (whether by wet chemistry, chromatography or

Box 13.5. Sampling considerations

Sample location
The choice of the horizontal distribution or pattern of exploratory holes will depend on the information and data found during the historical search and walk-over study, and whether the distribution of contaminants is expected to be uniform or biased in some way (orientation, location, etc.).

Number and type of samples
The number of samples will depend on the degree of confidence being sought with regard to the quantification of contamination, the comprehensiveness or level of detail of the site data obtained from the historical survey and walk-over inspection, and the funding available. The type of sample will depend on the anticipated nature of the contaminants (solubility, volatility, etc.) and the level of analytical precision required.

Sampling depth
This depends on the nature of the likely contaminants (density, miscibility, viscosity, etc.), the general nature of the ground (clay, sand, etc.) and the position and shape of the water-table.

Stage sampling
At some locations the contaminant concentrations may alter with time (due to natural attenuation, progressive movement of a contaminant plume, etc.), so samples may need to be taken at intervals over a period of time.

Sample retrieval and storage
Different means of sample retrieval (disturbed parings from an auger, intact cores from a borehole, etc.) can cause some loss of contaminant from specimens. Similarly, losses can occur as the specimen travels from the ground to the testing location. On the other hand, cross-contamination (through multiple use of 'unclean' tools) can increase apparent pollutant levels. Decontamination of equipment after each use is vital.

Sample size
This needs to be appropriate to the level of accuracy required for the estimation of contaminant concentration and the number of tests to be performed.

Sample testing
It is necessary to define the contaminants that are expected to be present (so that samples can be stored accordingly, appropriate analyses can be conducted, relevant safety procedures can be invoked, etc.) and the degree of precision required for the measurements.

atomic absorption spectroscopy) is small, it is preferable for samples to be as large as practicable (so that they are statistically more representative of the ground) and the analysed volume reduced by riffling. The description of all materials encountered should be detailed, and should include all the items observed (e.g. changes in colour, iridescence, odour).

The following factors may have an influence on the investigation programme:

- The concentrations of contaminants may vary both horizontally and vertically, according to the nature of the former activities of the site.
- Contaminants deposited on the ground surface may migrate downwards, and some contaminants may have been released at depth as a result of underground storage. An undisturbed clay layer will present a general barrier to contaminant movement. If this layer is relatively close to the ground surface then it would generally indicate the maximum appropriate sampling depth.
- Whether sampling should be restricted to the surface or near-surface layers, or whether it should be carried out to greater depth, depends on the types of contaminant present and the hazards they may pose.
- It may be sufficient to sample only the surface and near-surface layers of land that is destined for uses such as hard-surfaced vehicle parking areas, or if the site is contaminated by relatively immobile substances (e.g. metals).
- If garden or allotment development is proposed (or other areas where plants are to be grown), then samples should be taken throughout the rooting depth of likely plant species.
- Gaseous emissions and combustible materials should be determined to the full depths at which they are present.
- In general, substances should be investigated to the full depth of any excavations, pits, trenches or other structures that will be required for the purposes of the development.
- As a general rule, for many common types of development (e.g. housing, schools, recreational areas) a maximum sampling depth of 2.5 m, or to undisturbed clay, would be appropriate.
- For filled areas samples should be taken at ground surface and at several depths and/or in reaction to observations made during sampling.
- A minimum of three samples per sampling point is generally recommended: one specimen to represent the surface and near-surface layers (taken somewhere up to 200 mm deep), a second sample to represent the greatest depth of interest, and a third sample at a random intermediate depth.
- The individual samples should be collected and analysed separately. They should not be bulked or composited.

Ground investigation operatives have often been unaware of the toxic substances that they might encounter because, in many cases, they have been given little information about the previous usage of the sites on which they were working. However, before invasive ground investigation is undertaken it is important to carry out a thorough assessment of the potential human hazard posed to staff involved in the investigation and to provide appropriate protection. Two important guidelines for ground investigation (British Drilling Association, 1991; ICE Site Investigation Steering Group, 1993) detail a methodology for assessing the likely hazards and the appropriate precautions in terms of three categories of health and safety risk: green, yellow and red (Table 13.4). Unless there is firm evidence to the contrary, caution should be exercised in cases of doubt and the higher risk classification should be adopted.

Basic safety equipment that should always be available during the investigation includes gloves, safety boots, appropriate respiratory equipment and protective clothing. Disposable gloves are the minimum requirement for

Table 13.4. Ground investigation risk categorization for contaminated land*

Risk category	Likely situation and consequences
Green	Sites containing 'substantially inert materials (including hardcore)' which pose no health hazard
Yellow	Materials present are 'unlikely to cause serious impairment to health but constitute some risk'. Among the materials that could be present are 'vegetable matter, animal carcasses, sewage sludge and household waste'
Red	Materials present 'can bring risk of death, injury or impairment of health'. Contaminant substances within this category include 'a wide range of chemicals, toxic metal and organic compounds, pharmaceutical waste, micro-organisms, asbestos and cyanides as well as flammable, explosive and carcinogenic substances'

* Adapted from British Drilling Association (1991) and ICE Site Investigation Steering Group (1993).

those involved directly in fill or water sampling. In known corrosive conditions heavy-duty synthetic rubber gloves should be worn. Disposable body suits are useful for many site investigations, but chemically resistant suits should be used where exposure to acids or alkalis or substances absorbed through the skin is likely. On sites where gas or vapour release is possible trenches should not be entered until they have been adequately ventilated. Entry into a trench deeper than 1.2 m may only be made after it has been structurally supported.

13.2.3. Sampling strategy

A sampling programme must aim to obtain sufficient data to identify what are statistically the highest levels of contamination and to delineate the distribution of contamination both in plan area and vertically. If too few samples are taken the chances of finding small local 'pockets' of contamination are lessened and this may cause difficulty later when the site is being developed. The number of samples taken is usually a compromise between that which is desirable and that which is possible given the limits imposed by time and cost. The number of sampling points required depends on:

- the size and topography of the site
- the likely distribution of the contaminants present (this should be indicated by the preliminary investigation)
- the degree of confidence required
- the funding available for the investigation.

Sampling patterns (Figure 13.9) can be put into three categories: random, regular and biased.

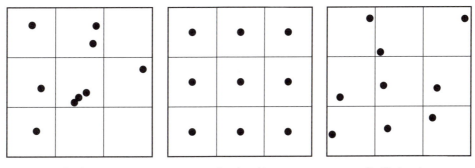

Figure 13.9. Sampling patterns

Random   Regular grid   Stratified random

Random sampling (non-systematic) techniques offer a vehicle for the estimation and prescription of sampling error or the probability of locating hot spots using sampling theory. In the simple technique the points are located randomly over the whole area of the site. Simple patterns tend to produce 'clumping' of sampling points, and significant areas of the site may therefore remain unsampled. Only after a very large number of points have been chosen may there be anything like a 'complete' average of the site. Other weaknesses of a simple system are that it ignores the information gained from initial surveys (concerning hot spot location) and that it allows no alteration of strategy based on experience or observation. The problem of 'clumping' can be overcome by using a modified technique. For example, in stratified random sampling the site is divided into cells and a number of samples are chosen at random within each cell. Stratified random sampling is close in concept to the most commonly used sample pattern that achieves overall coverage of a site — regular grid sampling. Unfortunately, it is complicated to set out on site.

In *regular sampling* the use of a grid enables the sampling points to be accurately located over the whole site. This helps to establish the distribution of contaminants more fully. However, in the systematic method it is possible, in theory, to miss a 'hot spot' in the mesh of the sampling grid. It is necessary, therefore, to design the sampling grid according to the assessed risk of minimizing such a spot. The closer the grid spacing, the greater the number of samples and therefore the greater the amount of data yielded, with a better spread of information and a lower risk of missing hot spots.

In *biased sampling* the distribution of sampling points takes account of prior information. At its extreme this approach could comprise identification of potential areas of significant contamination (by means of a desk study and a walk-over survey), with sampling only of these worst spots. The argument in favour of this approach is that the preliminary studies should have provided strong evidence for the likely location of contaminants, and the areas that categorize the degree of contamination of a site (i.e. hot spots) would be identified straight away. However, with such selective sampling it is a distinct possibility that other unrecorded, and not visibly apparent, hot spots may be missed. In practice, a modified approach is usually the most appropriate, comprising a systematic grid of sampling holes with reduced grid spacing in identified high-risk areas and with additional holes to cover specific hot spots.

There is no universally agreed method for determining the spatial distribution of ground sampling. There are wide variations in recommendations, with suggested horizontal spacings for sampling points typically being 15–30 m on a regular grid. Wider and closer spacings are usually qualified as being for areas assumed not to be contaminated, or for the investigation of suspected hot spots.

Bell *et al.* (1983) related the number of samples and the probability of locating hot spots, at different random sampling frequencies, on the basis of a binomial distribution of the hot spots. A significant attempt to apply statistical or probability techniques to sampling strategy was made by Ferguson (1992) — various sampling patterns were tested in a series of computer experiments to assess the efficiency of the patterns with various target shapes (circle, square, rectangle, ellipse, plume). Most experiments were performed with the target area set at 5% of the total site area. The target was located randomly for each trial. It was claimed that, overall, a herringbone pattern (Figure 13.10) was the best performer and was only weakly affected by target orientation. The simple random pattern was claimed to be the worst performer of all. The curves defining the number of sampling points required for a herringbone pattern could be represented by the equations:

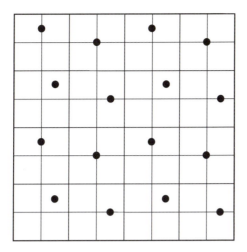

Figure 13.10. Herringbone sampling pattern (after Ferguson, 1992)

circular target $\qquad\qquad\qquad\qquad\qquad$ $N = 1.08/f$ $\qquad\qquad$ (13.1)

elliptical target $\qquad\qquad\qquad\qquad\qquad$ $N = 1.80/f$ $\qquad\qquad$ (13.2)

unknown target shape (conservative) $\quad$ $N = 1.40/f$ $\qquad\qquad$ (13.3)

where N is the number of samples and f is the target area as a proportion of the total site area.

When there are a number of hot spots, each is assumed to have the same area, so that the fractional size is known. By selecting the number of targets to be hit, a value of the parameter M is obtained (Figure 13.11), whereby the required number of samples is calculated as

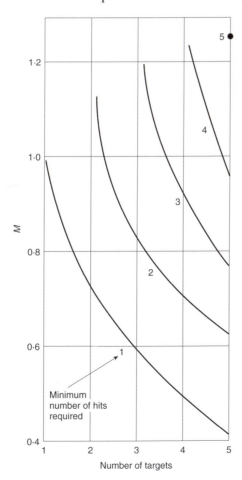

Figure 13.11. Sampling for multi-source sites (after Ferguson, 1992)

$$N = \frac{1.08\,M}{f} \tag{13.4}$$

It should be noted that Ferguson's comments on sampling patterns related to a target area of 5% of the whole site. As the target area became smaller the results from all the patterns converged — for a target area of 1% there was very little difference between the efficiencies of grid, random, stratified random and herringbone patterns.

Data such as those produced by Bell *et al.* (1983) and Ferguson (1992) should be treated with caution because they define the extent of contamination as a percentage of site area and deal with the probability of detecting a 'hot spot' within an otherwise uncontaminated area. It must be remembered that the sampling process has two functions:

- to detect any unknown contamination that might exist in zones for which there is no clear evidence of pollution (on the basis of the desk study)
- to establish a reasonably accurate picture of the concentration of pollutants within a known contaminated zone.

For the first situation there are several important points to bear in mind:

- After the desk study and walk-over investigation stages it should be possible to divide the site into clean, contaminated and uncertain zones.
- A comprehensive and thorough desk study and site walk-over should mean that the uncertain zone is a small portion of the whole site.
- A definition has to be made of the possible size of an unknown, significant area of contamination. For instance, it is unlikely the target would be larger than 10 m in diameter (otherwise some plan would have shown it or somebody would have remembered it), and it is average pollution concentrations (not point values) that matter, so an appropriate minimum area for detection may be around 3 m in diameter.
- Since the preliminary investigation has shown no reason for the uncertain area to be contaminated, it may be acceptable to accept a relatively low probability (say 50%) of hitting a 'hot spot'.

A sampling programme can then be chosen, rationally, using the aforementioned points and the data in Table 13.5.

For the second situation the following points are relevant:

- On the basis of the preliminary work, the contaminated zone can either be divided into sections (each with different, but uniform, characteristics) or the contamination has to be regarded as more or less uniform.
- The smallest plan area of ground which can be treated on its own is probably around 5 m × 5 m. Hence a concentration value is needed from each such parcel of land.
- Since point values of concentration are not applicable it may be assumed that the concentration values from each 5 m × 5 m parcel fit some sort of statistical distribution.

Table 13.5. *The percentage probability of locating contaminated areas by random sampling*

Contaminated area as % of total site area	No. of samples per hectare/centre spacing (m)			
	10/32	30/18	50/14	100/10
1	10	26	39	65
5	40	79	92	100
10	65	96	99	100
25	94	100	100	100

Hence it is proposed that for contaminated zones a sampling pattern is derived by dividing the area into squares of a suitable size, and then choosing sufficient evenly distributed squares to cover the zone or to provide enough concentration data points (say around 12) to enable the upper limit to contaminant concentration to be estimated statistically.

13.2.4. Sampling and measurement

Personnel engaged in sampling should be appropriately qualified and experienced, and the methods adopted for sampling soils, waters, gases, etc., from a contaminated site must ensure that:

- the material of interest is sampled
- the sample remains stable until analysed
- the sample does not undergo contamination from the sampling equipment
- there is no cross-contamination between samples
- the state of the sample is compatible with the method of analysis to be used
- samples of sufficient size are collected and transported speedily to the testing laboratory (since time delays affect the validity of some analytical results)
- there is careful handling and storage of samples at all times (including maintaining an appropriate storage temperature, generally 2–4 °C)
- there is careful and accurate field description of collected samples
- relevant regulations for the transport of hazardous materials are complied with.

Containers for soil samples should be sealable, water-tight, easy to carry and have large openings to facilitate filling and emptying. The material of the containers should not react with the sample and should be robust to avoid damage during handling and transport. Stainless steel or propylene trowels can be used to sample soil or fill (about 1 kg) from trenches, pits or the bucket of an excavator. Volatile solvents present perhaps the greatest difficulty in sampling because the soil may begin to lose some hydrocarbons by volatization as soon as the contaminated fill is exposed by excavation (Siegrist & Jenssen, 1990).

For most liquids, glass or plastic (polypropylene or polyethylene) bottles or jars are suitable (for specific details, see CIRIA, 1995–98, Volume VIII). Containers should be thoroughly cleaned and dried before use and rinsed with the fluid being sampled before they are filled. Care should be exercised when sampling water from trial pits to prevent contamination from the ground surface. Water in completed boreholes may be sampled by baler, or in situ by depth-controlled mechanical sampling equipment. Samples of water and other liquids deteriorate more rapidly after collection than do solid samples. Some forms of deterioration can be prevented by adding chemical preservatives, using pressure vessels or by chilling.

Metal containers are suitable for most gases and vapours. Plastic or rubber containers are not recommended for samples of this type because they are relatively permeable to many gases and they readily absorb vapours. Containers should be flushed through with the gas before final filling for analysis.

In most investigations the samples collected from the site are sent to a laboratory for detailed examination. There are, however, some purposes for which qualitative and quantitative, or semi-quantitative, testing may be needed on the site itself, such as:

- the detection and initial assessment of contaminants found during the reconnaissance visit and which may present hazards for further work on the site

Box 13.6. Portable test equipment

- **Detector tubes** (e.g. Draeger, Gastec). Available for a wide range of gases and vapours. They are of low accuracy.
- **Colorimetric field test kits.** A wide range is available. Soil test kits are not generally available, but the possibility exists of extracting with water to give a test medium.
- **Dedicated gas monitoring instruments.** The range available includes methane, sulphur dioxide, carbon monoxide and hydrogen sulphide.
- **Dispersive infrared analyser.** Enables the measurement of a wide range of gases and vapours and can analyse mixed component samples. It has a wide and sensitive detection range. Is cross-sensitive with hydrocarbons.
- **Flame ionization.** Enables the measurement of wide range of volatile materials, and can analyse mixed component samples. It is very sensitive. Erroneous readings arise in the presence of significant carbon dioxide.
- **Miscellaneous.** For example, water pH, qualitative test kits and conductivity of water.

- determination of properties and contaminants that can change rapidly with time after the sample has been collected (e.g. pH, dissolved oxygen, turbidity of liquid samples)
- rapid analysis of soil or fill materials excavated during clearance or development of the site, in order to decide whether they should be disposed of or retained
- to reduce the costs of analysis (i.e. by avoiding transport costs).

A range of portable analytical equipment exists such as that indicated in Box 13.6. Many gases and vapours can be detected with portable instruments with sufficient accuracy and sensitivity for safety control purposes and to locate and identify hazards.

The two major disadvantages of on-site analysis are likely to be the lack of specificity of some of the analysis techniques and the lack of accuracy of some of the techniques. However, in assessing the errors due to the preparation of test specimens and the subsequent analyses it is important to appreciate that the errors associated with analytical methods are usually insignificant in relation to those associated with sampling (BSI, 1998). For many purposes the accuracy of individual analytical results is of secondary consideration in the assessment of contaminated sites, since what is required is an overall picture of site conditions rather than detailed information on particular locations. An estimate of the order of magnitude of a particular contaminant may be sufficient.

Testing of samples should aim to determine the concentration of the likely contaminants in the ground. The design of the testing schedule for the samples should take into account the history of the site and the likely range of contaminants, since to test for every possible chemical, bacterium or spore would be prohibitively expensive. However, a standard suite of tests, such as the one shown in Box 13.7, provides a starting point that can be expanded on as the results become available.

Box 13.7. Initial suite of chemical tests for assessment of contamination

- Common toxic heavy metals (e.g. arsenic, cadmium, chromium, mercury, lead)
- Phytotoxic metals (e.g. copper, nickel, zinc)
- pH
- Sulphates
- Sulphides
- Cyanides
- Phenols
- Mobile organic compounds
- Mineral oils
- Total polycyclic aromatic hydrocarbons (PAHs)

It is recommended that a range of civil engineering soil tests is also carried out on samples, as these properties can affect the selection of methods of remediation. Appropriate tests are:

- Atterberg limits and particle size distribution, which are useful in assessing soil types
- compaction, which influences the methods of replacement of the soil if cleaned ex situ
- permeability, to indicate the mobility of treatment chemicals through a soil mass.

13.3. Site assessment and treatment selection

Having obtained a representative range of test data, it is necessary to assess the hazards posed by the site. This site categorization process has several components, which relate principally to assessment of the degree of contamination, the potential pathways for the contaminants, and the proximity of targets. The process leads to the selection of appropriate methods of dealing with the contaminated land. The overall objectives are:

- to identify and assess the hazards to various targets, which include personnel working on the site, subsequent users or occupiers and building materials
- to determine if any immediate action is needed or whether any action may be necessary in the future
- to provide the information needed to design protection against any hazards
- to identify those materials that may need to be removed to licensed waste disposal sites
- to provide a reference against which the performance of remedial actions may be measured
- to enable the relative advantages and disadvantages of different sites to be compared.

The targets at risk and their degree of exposure will depend on the use to which the land is to be put. For example, inorganic pollutants such as lead and cadmium will present no immediate hazard to humans if the soil is covered by concrete, (they will however present a problem if there is a change in land use in future years), but would be of major concern if present in garden soils. While the concept of burying immobile or relatively immobile contaminants beneath land development seems a generally acceptable and economic way to proceed, it is much more difficult to accept where the pollutants are mobile and free to move beneath the ground.

13.3.1. Trigger concentrations

Historically there have been two general approaches to site assessment (i.e. the identification of the remediation consequences for a particular level of contamination): fit for purpose and multifunctional after-use. With a fit-for-purpose approach, the basic tenet is one of reclamation to a state consistent with an intended use of the site. This assessment method has been exemplified by the UK (ICRCL, 1987) approach. Appropriate acceptable contaminant concentrations are specified for different types of site re-use. The perceived risk is greatest for domestic use and declines with decreasing human contact down to the least risk for open storage (e.g. a lorry park) (Box 13.8). In the multifunctional after-use approach, land that has been identified as unacceptably contaminated must be cleansed to a state whereby it is fit for any future use. This methodology has been typified by the Dutch approach to contaminated land. In general, the Dutch policy on contaminated land has been to clean it up. One of the main

Box 13.8. Sensitivity of land usage to pollution

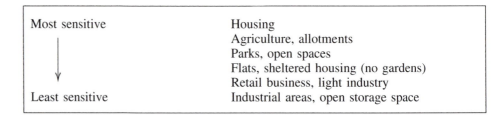

Most sensitive	Housing
↓	Agriculture, allotments
	Parks, open spaces
	Flats, sheltered housing (no gardens)
	Retail business, light industry
Least sensitive	Industrial areas, open storage space

reasons for this approach is that throughout most of the country the water-table is close to the ground surface. Therefore the contamination risk is primarily to the groundwater, which is not constrained in its mobility in the same way as is soil, and consequently the Dutch have considered it necessary to treat all sites equally, regardless of end-use.

To produce a definitive, precise set of guidelines identifying acceptable concentrations of contaminants in the ground and groundwater, extensive and detailed medical knowledge of the effects of varying levels of toxins (as well as a physical and chemical knowledge of the reaction mechanisms and the pathways of their release) is required. Human health risks associated with exposure to chemicals and other contaminants (in the context of their presence on a contaminated site) are not known with certainty. This situation reflects a lack of basic toxicological data relating damage effects to dose (dose–response relationships) and a lack of quantification of typical or anticipated routes of exposure. There may be substantial variations in response between individuals, threshold effects and, possibly, a long time lag between exposure and appearance of symptoms. Also, there may be strong synergistic effects (i.e. the total effect of a combination of constituents is greater than the sum of their individual effects) or antagonistic effects (the opposite of synergistic).

Because the risks posed by contamination are difficult to quantify, indirect methods based on 'threshold' and 'action' trigger concentrations have been devised (ICRCL, 1987) to assess the findings of site investigations. The concept is intended as an aid to professional judgement when the need is to:

- evaluate the significance of contaminants found on sites, and assess the suitability of land for various types of development
- decide whether remedial action is required
- formulate appropriate measures.

The concentration at which contamination poses a significant risk is defined as the threshold trigger concentration for that contaminant. As not all site uses are at equal risk from the hazards, it follows that the threshold trigger value varies with the actual or proposed use of the site. The action trigger value is the contamination at which the risks of the hazard are sufficiently high that the

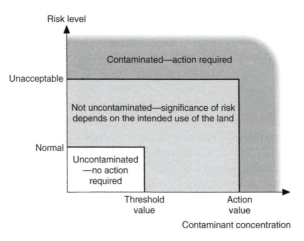

Figure 13.12. Interpretation of trigger concentrations

Contaminant concentration	Consequences
Less than the threshold trigger value	In this zone a contaminant is found only in relatively low concentrations, and it can usually be disregarded because there is no significant risk that hazards will occur. The site can be regarded as uncontaminated for that end-use and therefore no remedial action is needed, even though the concentrations present may be above the normal background values typical for the area
Between the threshold and action trigger values	Values in this zone do not automatically mean that the risk of the hazard is significant. They merely mean that there is a need to make an informed judgement as to whether the presence of the contaminant justifies taking remedial action for the proposed use of the site
Equal to, or greater than, the action trigger value	For areas falling into this zone, action of some kind, ranging from minor remedial treatment to changing the proposed use of the land entirely, is unavoidable

Table 13.6. Categorization of ground contamination by trigger values

presence of the contaminant has to be regarded as undesirable or even unacceptable. The trigger values thus define three general zones and associated actions, as indicated in Figure 13.12 and Table 13.6.

13.3.2. Assessment criteria

Table 13.7 contains an example of trigger concentrations formulated by the ICRCL (1987). Values are only available for a limited range of contaminants, although these are generally the most important. The trigger value concentrations are based on professional judgement after taking into account the available information. For the purposes of setting trigger concentrations, contaminants may be divided into three categories:

- Those which may present a hazard even in very low concentrations (e.g. methane, asbestos). For these substances any measurable concentration requires action to be considered or taken. Their threshold concentration is therefore effectively zero.
- Those for which a given concentration in the soil produces a measurable effect on a target (e.g. sulphates, attack on building materials; phenol and organic compounds, contamination of water supplies; phytotoxic metals (zinc, copper, nickel); cyanide, toxic through ingestion).
- Those for which no dose–effect relationship, between concentrations in the soil and their effects, has been determined experimentally. Most of the contaminants of importance to man's health, whether through uptake by plants or by direct ingestion, fall within this category.

The Dutch ABC values (Table 13.8) were introduced in 1983 as mandatory standards in support of the Soil Protection Guideline (Moen *et al.*, 1986):

- A values were 'target values' (remediation standards) and were set at either background levels or close to analytical detection levels
- B values represented that concentration above which additional investigation of the site was required
- C values represented concentrations above which remedial action was considered essential.

*Table 13.7. Trigger concentrations for contaminants associated with former coal carbonization sites**

Contaminant	Proposed site usage	Trigger concentrations[†] (mg/kg dry soil)	
		Threshold	Action
Polyaromatic hydrocarbons	Domestic gardens, play areas, allotments	50	500
	Landscaped areas, buildings, hard cover	1 000	10 000
Phenols	Domestic gardens, allotments	5	200
	Landscaped areas, buildings, hard cover	5	1 000
Free cyanide	Domestic gardens, allotments, Landscaped areas	25	500
	Buildings, hard cover	100	500
Complex cyanides	Domestic gardens, allotments	250	1 000
	Landscaped areas	250	5 000
	Buildings, hard cover	250	NL
Sulphates	Domestic gardens, allotments, Landscaped areas	2 000	10 000
	Buildings	2 000	50 000
	Hard cover	2 000	NL
Sulphide	All proposed uses	250	1 000
Sulphur	All proposed uses	5 000	20 000

NL, no limit set, the contaminant does not pose a hazard in this case.
[†] Values are for 'spot' concentrations, they do not apply to bulked or composite samples.
* From ICRCL (1987).

The most obvious difference between the UK and Dutch approaches was that there was a much greater range of contaminants quoted in the Dutch guidelines, and that in many cases the ICRCL did not always quote action values. However, there was an approximate relationship between the two sets of trigger values, with the Dutch B value corresponding to the ICRCL threshold trigger level and the C value to the action trigger value.

Environmental policy in The Netherlands has changed and the revised Dutch system comprises a 'target value' for a 'good quality' soil and an 'intervention value' which takes both human health and ecological risks into account (Table 13.9). The assessment process involves a preliminary investigation, comprising a desk study followed by a limited sampling programme, and if concentrations exceed the mean of the target and intervention values further investigation is necessary. If concentrations are below, or close to, target values the site is considered clean. The values given in the criteria index are based on the available knowledge of the substances involved (toxicity, vapour pressure, solubility, mobility, accumulation, corrosiveness, etc.), at the time of derivation.

As an alternative to assessment based on published standards, some countries, such as the USA and Denmark, have adopted a risk assessment approach. The aim of such an approach is to remove some of the subjectiveness inherent in the trigger value method. A number of factors are used (including geology, level of contamination, hydrogeology, proposed end-use and surrounding land use) to determine the levels of clean up required to reduce the environmental risk to an acceptable level. In the USA this process is highly developed and tends to be site specific, while in Denmark a set of national guidelines is used. Four key aspects have to be addressed for the effective risk assessment of contaminated sites:

*Table 13.8. Typical original Dutch (ABC) limits**

Substance	Concentration in soil (mg/kg dry weight)			Concentration in groundwater (μg/l)		
	A	B	C	A	B	C
Metals						
Chromium	100	250	800	20	50	200
Copper	50	100	500	20	50	200
Zinc	200	500	3000	50	200	800
Arsenic	20	30	50	10	30	100
Cadmium	1	5	20	1	2.5	10
Mercury	0.5	2	10	0.2	0.5	2
Lead	50	150	600	20	50	200
Inorganic pollutants						
Free cyanide	1	10	100	5	30	100
Total complex cyanide	5	50	500	10	50	200
Total sulphur	2	20	200	10	100	300
Bromine	20	50	300	100	500	2000
Aromatic compounds						
Benzene	0.01	0.5	5	0.2	1	5
Toluene	0.05	3	30	0.5	15	50
Phenols	0.02	1	10	0.5	15	50
Total aromatics	0.1	7	70	1	30	100
Chlorinated organics						
Total chlorobenzenes	0.05	2	20	0.02	1	5
Total chlorophenols	0.01	1	10	0.01	0.5	2
Total PCBs	0.05	1	10	0.01	0.2	1
Other pollutants						
Total pesticides	0.1	2	20	0.1	1	5
Fuel	20	100	800	10	40	150
Mineral oil	100	1000	5000	20	200	600

A, reference value (in the context of cleaning-up conditions); B, indicative value for further investigation; C, indicative value for cleaning up. PCB, polychlorinated biphenyl.
* After Moen *et al.* (1986).

- contamination
- hydrological (surface and groundwater) regime
- geotechnical properties and geological setting
- present and future targets and pathways.

Action values for the effect of sulphates on concrete can be found in BRE (1996). The question of sulphate and acid attack on building materials and the setting of appropriate limiting values illustrates the difficulty of drawing up guidelines. Although fairly simple chemical and physical properties are involved, the detrimental effects are likely to take a long time to manifest themselves and are difficult to observe because the concrete in question is buried.

An assessment of the results of a gas monitoring programme must usually address two main concerns: the possibility of immediate hazard (either within the site itself or to adjacent areas) and the propriety of building on the site. Information on the composition of the gas (e.g. concentrations of methane, carbon dioxide and oxygen) will often be sufficient to say whether or not a hazard exists. Information on the volume of gas being produced is essential to the design of measures to control the movement of gas or to the design of buildings to be erected on sites where gas is being evolved.

Table 13.9. Typical current Dutch limits

Substance	Concentration in soil (mg/kg dry weight)		Concentration in groundwater (μg/l)	
	Target (T)	Intervention (I)	Target (T)	Intervention (I)
Metals				
Chromium	100	380	1	30
Copper	36	190	15	75
Zinc	140	720	65	800
Arsenic	29	55	10	60
Cadmium	0.8	12	0.4	6
Mercury	0.3	10	0.05	0.3
Lead	85	530	15	75
Inorganic pollutants				
Free cyanide	1	20	5	1500
Total complex cyanide	5	650	10	1500
Aromatic compounds				
Benzene	0.05*	1	0.2	30
Toluene	0.05*	130	0.2	1000
Phenols	0.05*	40	0.2	2000
Xylene	0.05*	25	0.2	70
Chlorinated organics				
Total chlorobenzenes	—	30	—	—
Total chlorophenols	—	10	—	—
Total PCBs	0.02	1	0.01*	0.01
Other pollutants				
DDT	0.0025	4	—*	0.01
Aldrin	0.0025	—	—*	—
Dieldrin	0.0005	—	0.02	—
Mineral oil	50	5000	50	600

There is a correction for soil type to take account of clay content and organic content, provided these do not exceed 25% and 10%, respectively.
DDT, dichlorodiphenyltrichloroethane; PCB, polychlorinated biphenyl.
* Detection threshold.

When the hazard is an immediate one, such as the explosion of methane, it is not too difficult to decide whether or not a potential hazard exists, and it is relatively easy to arrive at control limits (e.g. a certain fraction of the lower explosive limit). When the question is of long-term exposure to some volatile organic compound, or more likely a cocktail of these, the setting of control limits is more difficult. It is customary to work on the basis of the upper limit being a small fraction of the occupational exposure limit (OEL) (e.g. not more than 2% or 3%).

13.3.3. Treatment selection

The aim is, almost invariably, to find a solution that is either permanent or which offers the best chance of long-term protection. Not only must there be a high degree of effectiveness at the time when the remedial measures are carried out, but they must remain effective in the long-term or be repairable or renewable in planned time spans (Stief, 1985). The long-term may range from 20 years to an indeterminate time-span. The effectiveness of the remediation measures must not be easily undermined by natural events (such as flooding, subsidence and vegetation growth) or unconscious intervention by man (such as excavation to install or repair services). Furthermore, it should be appreciated that there will always be flaws in execution; the theoretical effectiveness will never be achieved and environmental factors will generally serve to reduce effectiveness with time. Factors influencing the choice of remedial strategy and the actual design are summarized in Box 13.9.

Box 13.9. Factors influencing the choice of remedial strategy and design

Legal
- National, local legislation.
- Legal clean-up criteria.
- Contractual obligations.
- Requirements of water/disposal/health and safety authorities.

Political
- Type of development and end-use.
- Degree of clean up required.
- Timing and phasing with developments around the site.
- Speed of response necessary.
- Government policy.
- Requirements of planners and developers.
- Public profile.

Commercial
- Value of land (before and after).
- Specification of clean-up requirements.
- Volume and nature of materials to be treated.
- Clean-up targets and type of validation.
- Time available for reclamation.
- Space available.
- Transport and equipment costs.
- Commercial risk of a particular treatment strategy.
- Money available for treatment.
- Cost of monitoring and analysis.
- Capital outlay.
- Cost of disruption to the site.

Geographic
- Proximity of site to domestic dwellings, etc.
- Ease of access.
- Size of site and areas available for treatment.
- Availability of treatment facilities.

Environmental
- Proximity of the site to an aquifer.
- Hydrogeology of the site and its surroundings.
- Proximity of groundwater abstraction points.
- Local weather conditions.

Engineering
- Soil type (strength, permeability, homogeneity, etc.).
- Geotechnical requirements of treated materials.
- Depth of water levels.
- Volumes of material to be treated.
- Feasibility of the treatment system.
- Phasing of the reclamation.
- Availability of materials and equipment.
- Maintenance and operational requirements of the technology and ease of use.

Health and Safety
- Toxicity of the pollutants.
- Presence of underground or overhead services.
- Stability of the ground.
- Side-effects of treatment.
- Nuisance effects.
- Handling of materials.

Box 13.9. Continued

Managerial
- Availability of remediation companies.
- Availability of trained manpower.
- Degree of quality control and quality assurance required.
- Project management, relationships between participants.

Technical
- Limitations of treatment methods and application.
- Specification of the clean-up criteria.
- Availability of proven techniques and associated raw materials.
- Confidence level of definition of contamination problem.
- Ability to model contaminant profile/site characteristics/efficacy of treatment.
- Requirement and nature of feasibility or treatability studies.

Analysis techniques employing a Source–Path–Target (S–P–T) methodology may form a useful starting point for the rational selection of actual clean-up techniques. The objective is to relate an event (the source) through a chain of subevents (the path) to its effect at some sensitive point (the target) in the environment. In actual problems there may be multiple sources, targets and associated pathways. Having established the emission strength of a source, the position and sensitivity of the target envelope, and the transfer functions along the pathways it is possible to consider the influence that treatments will have. In general terms the remediation options available are:

- to avoid the problem
- to treat the pollution
- to interrupt the source–receiver path
- to harden/protect the target.

Remedial action can focus on either of the three elements of the contamination chain (source, pathway, target). The generic techniques are listed in Box 13.10.

Box 13.10. Treatment of the contaminant system

Source
- Source removal by excavation and off-site disposal.
- Source destruction.
- Reduction of contaminant mobility.
- Emission prevention by containment (to prevent both outwards and upwards movement).

Pathway
- Emission prevention by containment (to prevent both outwards and upwards movement).
- Direct path interception or the removal of a link in a food chain.
- Path redirection — the pollutant will still persist and so this method would usually involve the sacrifice of a less sensitive target.
- Changing path dynamics — subsequent slowing or delay of migration can increase the dilution and dispersion that take place.
- Enhanced (bio)chemical activity in the pathway — natural biochemical decay of substances can be very effective if allowed sufficient time. However, few inorganic and only some organic compounds can be dealt with in this way.
- Enhanced dilution and dispersal in the pathway — much less popular, and widely criticised because it contributes to the mass of contaminants in the environment.

Target
- Appropriate land usage.
- Move the target — politically unpopular when it involves the relocation of people. Viable for services, cabling and ducts. Does not deal with the contaminant itself.
- Harden the target — again unlikely for people, but a possibility for industrial or urban infrastructure.

Remedial action involving the source is aimed at reducing the capacity for pollution emission. The extremes of this approach can be regarded as excavation and off-site disposal and on-site encapsulation. Action associated with the pathway is intended to disrupt the direct line of communication between the source and the target. Remedial treatment involving the target is the most difficult and possibly the least satisfactory. It either involves a constraint on the location of the target or a 'fortification' of the target. Treatment technologies can be divided into three broad categories:

- ex situ — the soil is removed from the ground for treatment
- in situ — the contaminated soil is treated while it remains in place in the ground
- containment — the contamination is left untreated in the ground but is encapsulated within a containment system.

Within each category of treatment there are various techniques available (Table 13.10).

A combination of measures can be employed to many problems, to increase the effectiveness of treatment or to accomplish different aims. It is essential to keep track of the contributions from all the aspects that are involved in an S–P–T model to ensure that the contribution which each one makes to the whole is properly considered. Investigation (or monitoring) is necessary during and following remedial action. This is to demonstrate both the effective removal of excessive contamination and that all materials remaining in place comply with remediation values. Determination of the final cleanliness of the site is linked back to the assessment of the site during the initial site investigation stage, to the intended end-use and to current and anticipated legislation.

The likely cost of the remedial works is a major factor in the selection of an appropriate methodology. Costs of on-site treatment are strongly dependent on:

- the type of contaminants and type of soil
- the concentration of contaminants and the extent (spread and depth) of contamination
- the site situation and the presence of services, ground strata, etc.
- the need to take special measures
- the method of remedial action chosen
- the degree of removal of contaminants required.

Typical site clean-up costs are of the order of £300 000 per hectare. A typical site investigation cost (assuming general former industrial use, commercial industrial

	Method	Aims	Techniques
Table 13.10. General aspects of treatment technologies	Ex situ	To remove the pollution from the site	Excavation and disposal Washing Thermal stripping Bioremediation Separation
	Containment	To prevent the pollution from leaving the site	Barriers Covers
	In situ	To eliminate the potential for the pollution to leave the site by removing the contamination or by fixing it in place	Extraction Bioremediation Fixation Electrokinesis Hydraulic measures

Table 13.11. Summary of remedial treatments

Treatment	Suitability	Limitations	Cost (£/t)
Bioremediation	Versatile, applicable to organic contaminants, particularly water-soluble pollutants. Useful as a second- or third-stage option	Generally requires a long treatment time	20–60
Soil washing	Can treat organic and inorganic pollution, relatively quick	The extracted contaminants have subsequently to be treated	50–250
Thermal	Widely proven technique	High capital and energy costs. The costs for off-site incineration are particularly high	70–400
Stabilization	Particularly applicable to inorganic contamination	Treatment may miss some areas	30–50
Vitrification	Long-lasting treatment	Limited use to date. High energy and capital costs	≈ 200
Landfilling	Contamination completely removed from site	Requires a licensed landfill site	20–45
Capping	Straightforward to install	No treatment of the pollution, can be subsequently damaged	15–35
Vapour extraction	Probably the best in situ treatment for volatile and semivolatile pollutants	Cannot treat non-volatile contaminants	5–40
Fixation	Low cost, relatively simple process for shallow depths of treatment	Contamination not eliminated, pollution may become mobile again	5–30
Electrokinesis	Minimal disturbance to the ground, may also provide some ground improvement at the same time	Unproven for real sites, does not treat organic pollutants, expensive	25–200

after-use, and no gaseous or hydrogeological assessment) is likely to cost around £21 000 per hectare (based on Haines, 1991).

The overall treatment selection procedure consists of:

- identifying general response actions that, when taken singly or in combination, meet the requirements of the remedial action objectives
- determining volumes or areas of contaminated media to which general response actions can be applied
- identifying methods, and associated process options, appropriate to each general response action identified
- assembling selected methods into a range of alternative remedial strategies for the treatment of the sites as a whole.

A summary of the range of remedial treatments available is given in Table 13.11.

13.4. Ex situ treatment

This process involves the treatment or disposal of materials excavated from the ground. This clean-up approach is important because it offers a once and for all ultimate solution, even though there will often be an associated waste stream to dispose of.

The broad spectrum of remediation technologies that have been developed can be categorized as physical, chemical, thermal, and biological (see Table 13.10). The processes can be carried out using mobile, or temporarily erected, plant on site, or at a central treatment facility. The treated soil may be returned to the excavation, disposed of as a waste, used as general fill, or used for some other beneficial purpose. A decision to adopt such a process will raise the question of how clean the soil or treated material must be before it can be replaced in the ground or put to some other use. One of the unfortunate side-effects of many treatment processes is that the soil is rendered inert and is unsuitable for use as a growing medium without amendment with fertilizers and organic matter. Many cleaning processes are very contaminant specific and work well when faced with a single contaminant or family of contaminants. However, on actual sites the contaminants will be mixed.

During the excavation stage soil disturbance can cause escape of contaminants into the atmosphere. This may necessitate the employment of major, on-site health and safety measures to protect workers and neighbours. In some extreme cases the risks associated with excavation and removal might be higher than those associated with leaving the soil in place.

13.4.1. Excavation and disposal

The 'simplest' ex situ process is to excavate the contaminated soil, take it to a suitably licensed tip and fill the void created with clean soil. The method can be used to remove all the contaminated soil, or just the soil from the worst hot spots. A variation on this technique involves forming a sealed tip within the confines of the site being treated. All the contaminated material from the site is collected in the tip, which is then landscaped. The excavation and disposal remediation method does not treat the contamination, but relocates it to a controlled disposal point. However, the method does eliminate all contaminants from the ground, for all time, at one go.

Excavation and disposal is a relatively low-cost, technically simple operation and requires little specialist knowledge, although the replacement clean fill must be placed in a properly engineered fashion. Replacement materials are usually required to have satisfactory geotechnical properties (e.g. suitable grading and compaction characteristics). It is essential that all imported material is free of chemical contaminants and degradable or combustible organic matter, so that future contamination, gas, combustion and subsidence problems are avoided. This requirement is often not specified in detail, and is usually covered by stating that replacement material should be clean and inert.

Excavation and disposal may be the most technically satisfactory remediation solution when the contamination is fairly shallow and the earthworks volumes are relatively small. The great advantage is that the facilities are carefully managed to give a high degree of control over the clean-up process and also to prevent the occurrence of further pollution events arising from the clean-up process itself. However, there are various drawbacks with the technique:

- There are risks associated with transporting contaminated material through towns and the countryside, with the potential for spillage and road accidents.
- If there is a large volume to be removed there will be a lot of additional traffic, with related environmental effects such as noise and fumes.
- Although the quantity of the contaminated soil that must be removed will have been determined before work commences, there is usually no sharp edge to a contamination zone. In fact some of the contamination may have moved beyond the site boundaries. Thus the actual volume of material to be removed, tipped and replaced may be significantly different from the design estimate.
- If the water-table is above the bottom of the excavation there may be a major influx of water into the site, and this will have to be contained, removed and treated.
- Groundwater itself may cause flow-induced instability (piping and quicksand conditions; see Chapter 7, Section 7.4.5).

- The presence of water makes it difficult to place the fill in a properly engineered way. Hence it may be necessary to provide barriers to prevent groundwater inflow or to lower groundwater levels temporarily (see Chapter 7, Section 7.5). This will involve additional measures to protect any nearby rivers, streams and other watercourses.
- Even if the excavation remains dry there will be the possibility of slope instability in the soil faces around the excavation. Support to adjacent buildings and other structures must be assured.

13.4.2. Soil washing

Strictly, 'soil washing' is a term applied to a number of treatment processes, but the most widely applied form is an ex situ, water-based process using particle-size separation. The process is not generally a complete treatment or contaminant destruction process. Rather it is a volume-reduction step intended to recover a substantial portion of the original ground material as clean soil (for re-use or easy low-cost disposal) with a greatly reduced volume of contaminated material still requiring further treatment or safe disposal. The underlying principle is the fact that contaminants have greatest affinity for the fine particles in a soil or sediment (because of their mineralogy and high surface area). This applies to a wide range of contaminant types from heavy metals to hydrocarbons and organic chlorides. Thus the technique has wide applicability across different types of site (e.g. gas works, fuel storage depots, chemical plants, metal finishing works). The process has been applied successfully to real sites where the contamination is complex and a cocktail of different contaminants is encountered.

The washing process (Figure 13.13) has two cleaning aspects to it: the actual removal of contaminants from the surface of particles (for the coarse grains) and the removal of the fines (with their associated contaminants) from the soil. This work is achieved by physical means (sieving, screening, scrubbing, froth flotation) and by chemical processes involving solvents and chelating agents. This is usually carried out on site, using mobile plant, customized to suit the contamination and soil types. The soil-washing operation can be broken down into three stages: feed preparation, actual washing and dewatering.

In the *feed preparation* stage large clods of soil are broken down. Large pieces of non-soil debris (wood, metal, etc.) are removed by screening, and very coarse particles (which normally only have superficial contamination) are subject to high-pressure washing on screens and recovered as clean.

In the *washing* stage the soil is delivered to a multistage attrition scrubber and mixed intensively with an extracting agent (usually water). Contaminated silts and clays are detached from the coarser particles and transferred into the wash water, by a combination of mechanical dispersal and shear stress from vigorous rotating impellers and the addition of appropriate chemical aids. The fine particles are separated from the scrubbed granular material using hydrocyclone separators. It is possible to remove soluble contaminants and those that form stable colloidal suspensions in the extracting agent.

Dewatering is intended to separate the fine solid particles and minimize the volume of the resulting sludge. A combination of gravity sedimentation and dissolved air flotation commonly used to isolate the fines and any precipitation products. Coagulants and flocculants may be used to enhance these separation processes. The resulting sludge represents the bulk of the residual contamination from the plant.

As a generalization, coarse soils (sandy soils, made-ground with gravel, ash and clinker, etc.) show much greater potential for remediation through washing than silt or clay soils. The particle-size characteristics of the soil or sediment largely determines the relative proportion of cleaned coarse product and contaminated fines. For typical sites, the upper limit to recovery of clean

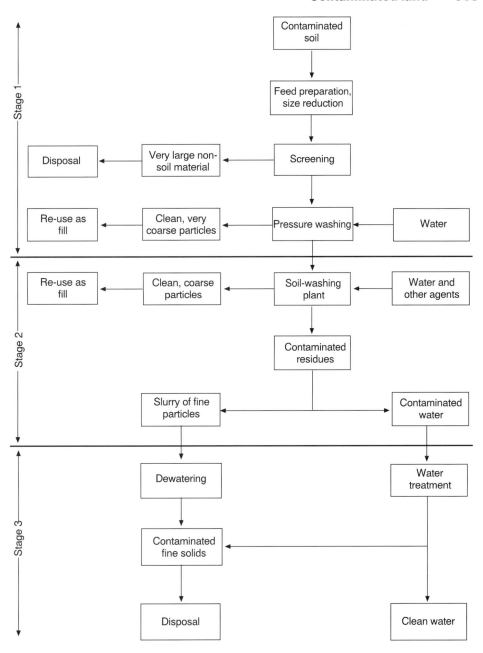

Figure 13.13. The soil washing process

material is about 70% (of the original volume of soil), while a value of 30% is about the lower limit for the process to be worthwhile. The presence of organic, humus-like substances creates difficulties in that some heavy metal compounds (and also organic contaminants) are preferentially adsorbed or absorbed by humus-like substances (Kuziemska & Quant, 1998). It is extremely difficult to separate the contaminated compounds from these humus-like substances. Water is the most commonly used extracting agent, but chemical aids may be used to enhance the separation and removal of specific contaminants. Typical aids include pH adjustment, detergents, surfactants, chelating agents, oxidizing agents, coagulants and flocculants. Soil-washing processes have proven to be effective in treating soils, sediments and oily sludges contaminated with a wide range of toxic organics, including herbicides, pesticides, polycyclic aromatic hydrocarbons (PAHs) and polychlorinated biphenyls (PCBs) (Mugglestone & Hughes, 1994). The removal of heavy metals is largely dependent on the form of the metals present in the soil (dissolved ions, organic and inorganic complexes, physically or chemically absorbed, etc.), but it is feasible.

13.4.3. Thermal treatment

Incineration has long been recognized as an applicable ex situ technology for permanently and completely removing toxic organic compounds and it is particularly suitable for eliminating hydrocarbons. Thermal processes (Figure 13.14) can be divided into two main categories: one-stage and two-stage systems (Pratt, 1994). One-stage systems destroy the toxic organic compounds directly within the contaminated soil or substance. End-products of this incineration are bottom ash (normally returned to the original site for backfilling or discharged to landfill) and gaseous emissions, which are usually subjected to a scrubbing treatment to remove fly ash, residual organic compounds and acidic gases. Two-stage systems use volatization and/or pyrolysis to convert the toxic compounds into the gaseous phase, with subsequent thermal combustion of the gaseous products or condensation of the gaseous inorganic compounds. The incineration can take place at either a licensed facility site or, less commonly, on-site through the use of a mobile incinerator.

Low-temperature thermal stripping is typically accomplished by disturbing the soil and then applying sufficient heat to the soil to drive off volatile and selected semivolatile organic contaminants. Soil temperatures below 600°C will often be sufficient to remove the contaminants. The method uses either direct heat transfer (convection or radiation) from heated air or an open flame to the soil or indirect heat transfer (conduction) to the soil. With steam stripping, volatile compounds are removed from a solid or liquid phase by passing steam through the contaminated material. The process can be applied to soil if the contaminants, which may be either water insoluble or soluble, are relatively volatile. Indirect heating (conduction) is realized by the use of heat-transfer pipes in a rotary kiln. Only a small gas stream is necessary for the transport of the evaporated soil water out of the kiln. The gas from the process may be subject to some form of emissions control (such as incineration) or may be discharged directly to the atmosphere if no treatment is required to meet emissions standards. Either way the gas will pass through a cyclone for separation of the fine material particles entrained in the exhaust flow. One of the main limitations is removal of the final traces of contaminants, because at low concentrations the contaminants are strongly adsorbed by the soil particles, which are covered with a water film.

Direct incineration can be achieved in a rotary kiln, but other options include multiple hearth and fluidized bed incineration. Temperatures of between 800°C and 1200°C are used and, to prevent vitrification or sintering of the mineral part of the soil, the particles are kept in continuous motion. Thermal systems using elevated temperatures (800–2500°C) induce physical and chemical processes, such as volatilization and combustion, for eliminating toxic substances (particularly organic compounds) from the polluted soil. Volatile heavy metals, such as mercury, can also be removed from soils by thermal processes, although

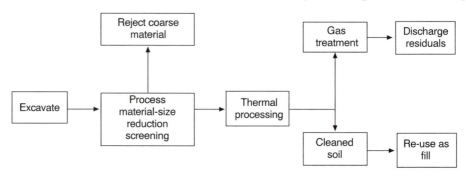

Figure 13.14. Thermal cleaning of soil

they are not destroyed and have to be condensed downstream of the process. Emitted gases can be processed in three ways: incineration at high temperatures; treatment at moderate temperatures using catalysts; and low-temperature wet scrubbing and purification of the washing liquid.

Vitrification involves excavating the contaminated soil and heating it in a furnace until molten, and then quenching it to produce a variety of glass products. Depending on the output product required it may be necessary to mix the contaminated soil with normal glass-making materials in order to control the quality of the product. Vitrification does not destroy the contaminants (although many organic compounds may volatilize or burn in the high temperatures), but instead locks them into the glass structure, rendering them virtually immobile and leaching-proof. With this technique, as for incineration, emission controls are required, both of the gases and vapours produced by the furnace and of the quenching water which will itself become contaminated.

Thermal systems can treat almost any type of contaminated soil, although soils with high clay and moisture contents require high-energy input and long residence times and are difficult to handle. Sandy, silty, loamy and peaty soils respond best to thermal processes. Clay-rich soil forms clods or nodules in rotary kilns. Incineration conditions need careful control, as incomplete combustion can lead to the formation of equally or more hazardous compounds (such as dioxins). However, it is the only known method of destroying certain contaminants such as PCBs. Thermal processes are most appropriate for soils contaminated with organic substances, although some of these chemicals are difficult to combust, and often the process in incomplete. Some inorganic chemicals (e.g. mercury, cyanides), can be combusted, but may leave toxic residues in the ash, thereby posing further disposal problems.

13.4.4. Biological treatment

Bioremediation is the process (Figure 13.15) of using micro-organisms to transform hazardous chemical compounds in a contaminated soil to non-hazardous end-products such as water and carbon dioxide. The system is relatively ineffective with inorganic compounds and metals, and the microbial degradation can be inhibited by heavy metals and complex organic compounds.

Almost all organic compounds and some inorganic compounds can be degraded biologically if sufficient time and proper physical and chemical

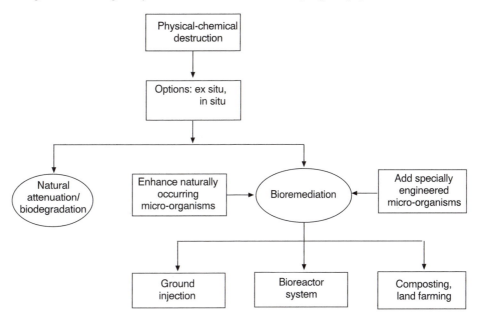

Figure 13.15. Biological treatment options

conditions are provided. The process has been shown to be effective in soils contaminated with wood-preserving chemicals, PAHs and chlorinated organic compounds (Holroyd & Caunt, 1994). The micro-organism population is obtained either by adding adapted toxicant-degrading micro-organisms to the soil or by injecting nutrients into the soil to stimulate the growth of the indigenous micro-organisms. To achieve biological degradation a number of conditions have to be fulfilled:

- presence of sufficient oxygen for aerobic processes and full depletion of oxygen for anaerobic processes
- presence of nutrients (mainly phosphorus and nitrogen)
- presence of sufficient water
- concentrations must be non-toxic to the micro-organisms
- sufficient availability of the contaminants to the micro-organisms
- presence of the appropriate organisms
- favourable temperature.

Small quantities of material may be treated by drum-mixing the soil with nutrients, suitable micro-organisms and water, and then placing the mixture in an aerated reactor (a simple container or suitable pit). Alternatively, wastes may be placed in a tank containing water, vigorously mixed and aerated to encourage contact between the microbes and the organic matter and to provide oxygen so that the biological reaction can take place. A prerequisite for an on-site mobile installation is that the microbiological conversion rate is fast.

An alternative, technologically simple, method for dealing with large quantities of materials involves constructing windrows (long, low heaps of materials, typically 1.5–2 m high and about 3 m wide at the base) of the contaminated soil (Jarvis, 1998). Microbes and nutrients are added and mixed in at appropriate intervals, often using equipment adapted from agriculture. This ensures adequate oxygenation and nutrient addition for the bacteria to perform aerobic decomposition and gives control over the rate of contaminant degeneration. Amendment materials (e.g. wood bark, straw) may be added to improve the physical conditions of the matrix by breaking up lumps, promoting drying through exposure to the air, etc. This will facilitate the access of air, nutrients and microbes to the contaminants. Any leachate from the process may by recirculated back into the contaminated soil, treated or discharged as appropriate. Treatment cycles can take up to 8–10 months, although enclosed greenhouse-type structures can be used to increase ambient temperatures and enhance degradation. The method is relatively inexpensive (although it requires a large area for operation) and is able to handle large amounts of soil over a long period of time. Conversion rates are strongly dependent on prevailing conditions (e.g. temperature, moisture).

13.4.5. Physical separation

The physical separation technique utilizes mechanical and physical differences between the contaminants, or the contaminant-rich soil, and the clean soil (CIRIA, 1995–98, Volume VII).

In separation based on *differences in specific gravity*, the contaminated soil and a suitable liquid (usually water) are thoroughly mixed and the soil phase is then separated from the liquid. The liquid contains the contaminants, usually as a floating layer, but in some cases as an emulsion or suspension. The process is particularly applicable to sandy soil contaminated with substances that have a lower density than water and are not soluble or miscible (e.g. oil).

Flotation is a widely used technique. The soil is converted to a slurry, which is dosed with a chemical that adheres selectively to the contaminated particles to give hydrophobic properties to the surface of the contaminants. Small air bubbles are then formed in the soil suspension and these bubbles adhere to the hydrophobic particles of contaminants and transport them to the surface of the suspension where they are removed as a foam. The foam is treated to produce a concentrated sludge of the contaminants, which is then taken away for treatment. The main difficulty with the flotation process is to obtain such selectivity that only the contaminants are removed and not the small soil particles themselves.

Another method is based on the *settling velocity* of a material in a suitable fluid. The process consists of a wet or dry classification step, where the soil particles and the contaminants are separated using apparatus such as an upflow column. The technique is applicable to sandy or slightly loamy soils with the contaminants present as particles or adsorbed to certain soil particles (the fines), which have settling velocities different from the settling velocity of the clean soil particles.

Separation based on *magnetism* may also be used as a means of extracting metallic contaminants. The contaminated soil is slurried with water and then passed through a magnetic field. However, the applicability of this technique is very limited. The most important group of soils that can be treated are those containing iron or magnetite particles 'clayed' with hazardous materials.

The most important general limitation with separation processes is the difficulty of separating clay particles from the water phase. The presence of humus or humic acid can also hinder particle settling and separation.

13.4.6. Fixation

Fixation techniques are not intended to remove the contaminants from the soil, but rather to eliminate physically or chemically the toxicity and/or mobility of the hazardous substances within the soil, which can then be left in place or handled in a safe way. Minimization of mobility or leachability can be achieved by the following mechanisms:

- a chemical reaction with the contaminants to form insoluble compounds
- isolating the contaminants from infiltrating water by adding hydrophobic compounds or chemicals to form hydrophobic compounds
- adding chemicals able to fixate water and thus influence the micro-leachability
- influencing the pH value and the redox potential to achieve minimum solubility of the contaminants.

Solidification is the formation of a dense, inert mass by mixing materials, whereas *stabilization* concerns wastes that are chemically converted or bonded to an inert, stable form. Both methods produce a tightly formed solid matrix that resists degradation and leaching. Most on-site stabilization methods use relatively simple techniques for mixing contaminated soil and suitable chemicals or other additives, such as:

- The soil is spread in thin layers, the top of each layer is dosed with chemicals and the soil and chemicals are mixed by rotovating or ploughing.
- The additive and soil (usually in slurry form) are mixed using rotating mixers or screw conveyors. The treated soil is then tankered to its final destination or pumped into a mould to solidify on-site.

Fixation techniques are applicable to heavy metals and organic contaminants, in either solid, sludge or liquid forms (USEPA, 1993a). The binding agents are typically inorganic materials such as cement, fly ash, pozzolans and silicates, although organic materials such as asphalt, epoxy resins and polyesters have been used. In the latter process the contaminated material is dried, heated and mixed with the thermoplastic material. After cooling and solidification, the final product can be dumped. The use of lime has been successful, although care must be taken that future acidification does not progressively occur since this would reverse the process. Stabilization using cement has been used to a lesser extent (Perry *et al.*,

1996 and New Civil Engineer, 1999c), and additives have been developed that overcome the inhibiting effects of organic compounds on cement hydration. Vitrification (see Section 13.4.3) is related to solidification in that the contaminated materials are encapsulated within glass. The most important chemical treatment methods are listed in Table 13.12 and potential treatment methods for specific contaminants are indicated in Table 13.13.

The provision of a sufficient degree of mixing or contact between the contaminants and the treatment agent presents a fundamental and difficult problem in the general case of multiple-constituent wastes intermixed with soil of different textures and permeabilities. Furthermore, because the contaminants are not removed, the treated soil can still form a potential source of hazard. Weathering and ageing may reduce the stabilizing effect, and leachability may be increased by freeze–thaw cycles. The presence of combinations of different compounds can sometimes mean that no adequate overall fixation technique can be found.

*Table 13.12. Chemical treatment methods**

Treatment method	Process	Direct applicability
Electrolysis	Reactions of oxidation or reduction take place at the surface of conductive electrodes (with an applied voltage) immersed in an electrolyte	Low
Neutralization	Elimination of acidity or alkalinity by the addition of appropriate chemicals	High
Hydrolysis	Generally refers to double decomposition reactions with water, usually carried out at elevated temperatures and pressures with appropriate catalysts	Moderate
Chemical oxidation	A reaction during which the proportion of the electronegative constituent in a substance is increased (e.g. by combining with oxygen or by losing hydrogen)	Moderate
Chemical reduction	The opposite chemical reaction to oxidation	Moderate
Ozonation	Ozone gas is a powerful oxidizing agent. It is led through a suspension of soil in water or a packed soil bed	Low
Photolysis	Chemical bonds are broken under the influence of ultraviolet or visible light	Low

* After Rulkens *et al.* (1985).

*Table 13.13. Specific applications of chemicals and electrolysis**

Contaminant	Treatment method	Usable chemicals
Cyanide	Oxidation to CNO^- Hydrolysis to CO_2 and NH_3 or N_2	$NaClO$, $Ca(ClO)_2$, $O_3 + UV$, Cl_2, alkali
Cyanide complexes	Oxidation	$O_3 + UV$
Heavy metals	Oxidation to change leachability, (e.g. Cr^{3+} to Cr^{6+}, or to enhance precipitation) Reduction (e.g. Cr^{6+} to Cr^{3+}) Electrolysis	$NaClO$, $KMnO_4$, O_3, H_2O_2 SO_3, SO_3^-, Fe^{2+}, Al —
Halogenated hydrocarbons	Hydrolysis Oxidation Electrolysis	Aqueous acids, alkali ClO_2, O_3, H_2O_2, $KMnO_4$ —
Organic compounds in general	Oxidation (e.g. phenolics) Hydrolysis Electrolysis	H_2O_2, $O_3 + UV$ Acids, alkali —

* After Rulkens *et al.* (1985).

13.5. Containment

Containment (encapsulation) is currently one of the most widely used pollution control strategies. In essence, it involves 'putting a box' around the contaminated ground. Consequently, components of containment systems are horizontal cover layers (over the pollution), vertical barriers (around the pollution) and horizontal bases (beneath the pollution). It may not be necessary to use all components in treating a particular site. Containment may be undertaken for several reasons:

- there is no proven or environmentally acceptable clean-up technique available
- it is an essential element of the clean-up process (e.g. to control ground-water ingress)
- the available clean-up techniques are so slow that the pollution would spread by an unacceptable amount during a clean-up process
- it is the most cost-effective form of treatment
- it is necessary to prevent the spread of pollution during a site clean-up phase
- it may enable natural biodegration (attenuation) to take place
- to modify the local environment (e.g. to divert or separate a contaminant plume).

13.5.1. General aspects of containment

Encapsulation offers no treatment to the contamination but isolates it from the end-user. One argument in favour of this treatment is that, with time, many contaminants, especially the organic ones, will decay naturally under biochemical processes to harmless constituents. Barrier systems provide relatively cheap solutions to immediate problems, but their long-term stability and resistance to chemical and biological attack is often questionable. Forces generated by expansion, contraction, land slippage and changing groundwater pressures also limit their long-term effectiveness.

The likelihood of achieving a high degree of separation between the source and the surrounding environment is a function of the quality of each component in placement. The potential efficiency of top sealing (specially designed covers) is very good. The potential efficiency of vertical barriers is good to poor depending upon site conditions, barrier type and workmanship. In-ground horizontal barriers are still in the development stage and unproven for this purpose, hence they are likely to remain of limited value due to the fact that it will be difficult to ensure reliability. The main problems with barrier techniques occur in forming joints and connections between horizontal (top or bottom) and vertical barriers as well as between adjoining sections of walls.

For lateral containment of contamination the most common solution is some form of vertical barrier wall taken down to a natural geological aquiclude. Alternatively, containment may be achieved by constructing vertical barriers so deep that contamination is unlikely to migrate beneath them. These barriers are usually long-term permanent features and they can be installed in limited physical space where the principal concern is to protect existing sensitive targets. The construction of a vertical barrier will normally influence the groundwater flow patterns and the actual level of the water-table may be affected in the vicinity of the barrier. The selection of a barrier system will depend primarily on:

- the site ground conditions
- the degree of integrity needed
- the depth of barrier required
- the aggressiveness of the ground conditions and the barrier durability
- the cost and ease of installation.

Vertical containment methods can be divided into:

- low-permeability soil zones (grouted barriers)

Barrier category 1
- Grout curtains
- Jet mix walls
- Ground freezing

Barrier category 2
- Driven pile wall
- Vibrated beam wall
- Vibwall

Barrier category 3
- Clay trench
- Slurry wall
- Bentonite–cement wall
- Narrow trench with membrane
- Cast in situ pile wall

- systems in which the ground is left relatively undisturbed (displacement barriers)
- constructions in which material is excavated from the ground (trench barriers)

A list of the most common types of vertical barrier is given in Box 13.11.

13.5.2. Low-permeability soil zones

These are created in situ, usually using some form of grouting method. Traditional *grout curtains* (see Chapter 14, Section 14.1.2) can be used to form cut-offs, but will tend to be used only in areas of restricted space. The most commonly used chemical grouts are composed of a sodium silicate base, a reactant, an accelerator and water. The actual composition of the grout will depend on the environment that is to be injected — some grouts are not acceptable because of their possible pollution effect on groundwater.

The *jet mix/jet grouting* process is based on the formation of columns or panels of cement grout mixed in situ with soil (to form a material such as 'soilcrete'), as illustrated in Figure 13.16. The columns or panels are formed by high-pressure jets, which emanate from rods lowered into a predrilled borehole. The jet erodes

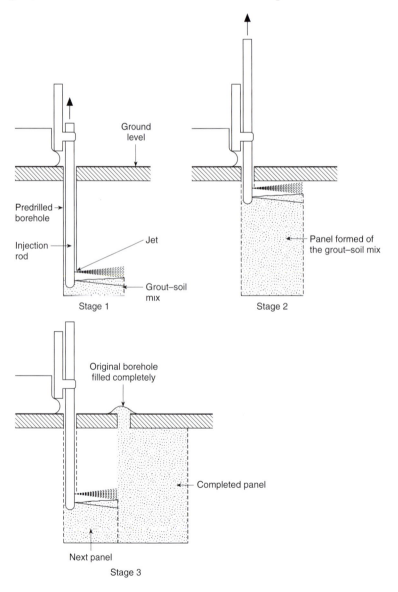

Figure 13.16. The jetmix grouting process

the existing soil, fine particles of soil are removed to the surface via an uplift pressure, and the remaining soil is mixed with grout. Soilcrete can be formed in a wider range of soils than is normally possible with more conventional grouting systems. Walls can be formed as a series of interlocking columns, or alternatively as a single or multiple row of soilcrete panels. Drilling and jet grouting can be via either vertical or inclined holes, although the presence of obstacles can frustrate attempts to achieve a complete seal. The maximum practical depth of installation is approximately 15 m.

13.5.3. Displacement barriers

These can only be used when the in situ material will allow penetration by the wall material or a void-former. The presence of boulders, rock, bulky wastes and strong large metal objects can easily preclude the use of these systems. The barriers may be formed by inserting preformed elements or by inserting a 'tool' to create a void, which is then filled with low-permeability material.

Driven pile walls may be formed by driving sheet-piles (see Chapter 10, Section 10.4.1) or membrane elements into the ground. Elements must be durable and joinable to form a suitable 'impermeable' wall. Steel sheet-piles may have particular potential in situations where barriers are required both to control pollution migration and to provide some ground support. Membrane walls use materials such as high-density polyethylene (HDPE), with the material that forms the screen being injected into the ground while coupled to a special injection plank or form (CIRIA, 1995–98, Volume VI). Steel sheet-piling is probably the least expensive method, is relatively reliable and can be installed to depths of 20–30 m. The junction between sheet sections is usually the main point of leakage, although flow through the joints may reduce over a period of time as they become clogged with fines.

To create a *vibrated beam wall* an I-section former is driven to the required depth, withdrawn and the void filled with a slurry mix (Figure 13.17). Each section is driven to overlap the preceding one, to ensure complete coverage of the area. The average thickness of the vibrated beam slurry wall is approximately 100 mm, and construction to depths of 25 m is feasible. The slurries used have usually been a mixture of soil plus cement and bentonite. However, mixtures are available that use additives such as asphalt emulsions.

'*Vibwalls*' are constructed in a similar manner, but using a heavy H-shaped lance which is driven into the ground. As the lance is withdrawn, grout is placed under pressure to leave an H-shaped grout column. The lance is then redriven so that flange positions coincide to give a continuous cut-off. The thickness of the resulting wall is generally 60–80 mm. Such a thin wall is not ideal for pollution control.

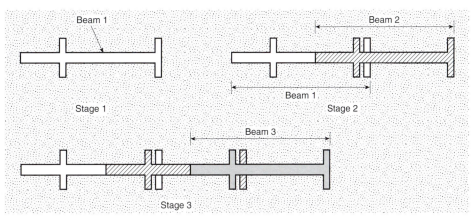

Figure 13.17. Formation of a vibrated beam wall

Plan view of wall formation

13.5.4. Trench barriers

This type of barrier is probably the most reliable system because the preparatory work can be undertaken with greater ease and certainty (i.e. the excavation is continuous and not as adversely affected by obstacles). The simplest form of excavated barrier is a *trench backfilled with clay* so that it has a lower permeability than the adjacent ground. The excavation should be sufficiently deep to reach an aquiclude so that the clay cut-off is keyed into it (Figure 13.18). In general, most barriers must be taken below the groundwater level and this may lead to instability of the trench (even if the sides are battered back) and/or difficulty with compaction of the backfill.

Diaphragm or secant walls (see Chapter 10, Section 10.4.1) can effectively contain a site, but such methods are only likely to be economic (for containment of contaminated land) if there is a structural aspect to the containment problem. The wall consists of a series of interlocking bored piles, drilled sequentially with a spacing that ensures that each pile penetrates its neighbours. There is concern about the leak-tightness of the joints because of the presence of 'foreign' matter.

Slurry trenching (see Chapter 7, Section 7.5.1 and Chapter 10, Section 10.4.1) is the most widely used form of trench barrier and involves excavating a trench through a dense slurry and then backfilling the trench (Figure 13.19). Most commonly the trench is excavated down to, and often into, an essentially impermeable stratum, depending on the application. The width of the trench can

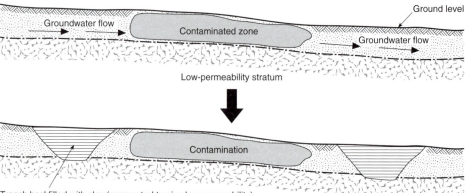

Figure 13.18. A shallow trench cut-off

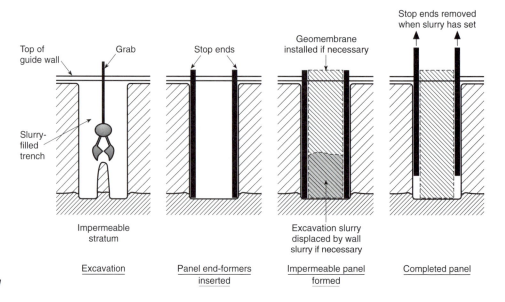

Figure 13.19. Slurry trenching

Figure 13.20. The formation of a bentonite–cement slurry trench

vary, but is typically 0.6–1.5 m and construction to great depths (around 80 m) is feasible. Once a trench has been dug the low-permeability element (a flexible wall) may be formed from a bentonite–cement mix, a clay–cement–aggregate plastic concrete or a soil–bentonite mix (with or without an included vertical geomembrane). Fluid backfills must displace fully the excavation slurry. This should not be a problem with non-setting bentonite or polymer slurries, but with clay–cement stiffening will occur and backfill should follow excavation as soon as possible. Alternatively, the trench may be excavated through a bentonite–cement slurry (Figure 13.20), which is designed to remain fluid during the excavation phase but which will set, when left in the trench, to form a material with permeability and strength properties similar to those of a stiff clay. A permeability requirement of 10^{-9} m/s or less is typical, although to control pollution a permeability of 10^{-8} m/s (easily achieved) should ensure minimal leakage.

Bentonite–cement slurries are the most widely used form of pollution cut-off in Britain. Typical mix proportions are given in Table 13.14. Very often some of the cement will be replaced by pulverized fuel ash or ground, granulated, blast-furnace slag (see Chapter 16, Section 16.4.3). Clay–cement grouts may exhibit severe cracking when exposed to sulphate solutions if the material is not maintained under a confining pressure by the surrounding ground. Generally, a slurry cut-off wall is required to possess a shear strength roughly equivalent to the surrounding soils, but, more importantly, to be of sufficient plasticity that cracks do not develop if the wall is subject to a large strain under confined stress conditions. Thus great emphasis is placed on the stress–strain behaviour of containment walls in order to avoid loss of cut-off performance.

Typically a wall specification will call for a sample of the cement–bentonite slurry to accommodate a strain of at least 5% without cracking failure. The drained triaxial test (see Chapter 6, Section 6.3.5) is used to test the material — the cell pressure conforming to the in situ confining stress. In practice, the ability of a cement–bentonite slurry to withstand these strains is rarely a problem, with values often in excess of 10% being achieved. Permeability tests should be carried out using a triaxial cell (see Chapter 7, Section 7.2.3). In general, a hydraulic gradient in the range 10–20 is satisfactory. Normal practice is to sample from at least two depths, and for deeper walls from three depths (the top 1 m, the middle and the bottom 1 m), during cut-off wall construction. Coring after the slurry has hardened may damage the wall and the resulting samples are likely to be cracked or remoulded.

The 'impermeability' of clay trenches and slurry walls can be ensured (and structural flexibility provided) by the addition of a synthetic membrane during construction (CIRIA, 1995–98, Volume VI). The use of such membranes may be

Table 13.14. Typical mix proportions for bentonite–cement slurry*

Component	Quantity (kg)
Bentonite	20–60
Cement	100–350
Water	1000

* From Jefferis (1990b).

appropriate where high levels of pollution exist or very aggressive chemicals are present, as membranes can be obtained to resist a wide spectrum of chemicals. For gas migration control the inclusion of a membrane within the trench is normally considered essential, as most slurry based cut-off materials have a relatively high water content and may become gas permeable if the water is lost. The supplementary membrane, usually HDPE at the present time, would be placed integral with the wall or incorporated into the clay fill to provide an extra continuous barrier. Barriers around 35 m deep have been installed. The main problems with membranes relate to the sealing of the joint between the membrane and the base of the excavation and the joints between membrane panels.

13.5.5. Covers and horizontal barriers

Covering systems may be used on their own or in conjunction with vertical and horizontal in-ground barriers or cut-offs. A total containment process is extremely costly to complete and therefore it will often be considered that simply placing a covering layer over the site is sufficient to isolate the underlying material. In fact in many instances this will be an adequate way of separating contaminant source and potential targets. Capping is thus one of the more commonly applied measures for dealing with soil (and groundwater) contamination. It normally insulates future site occupiers from contaminated deposits, while at the same time minimizing surface water infiltration (which could cause leaching of contaminants). The functions of covering systems are summarized in Box 13.12.

At its simplest a cover is simply a blanket of clean soil, modified soil or synthetic material. If clean soil is readily available this option can be very cost-effective. However, because of the number of functions that a cover generally has to perform (which sometimes lead to conflicting material requirements), a layered capping system is a better solution (in a similar way to caps for landfills; see Chapter 12, Section 12.7.2) and appropriate components would be as indicated in Table 13.15.

It must be remembered that the cover-up option does not offer any reduction in contamination levels, and therefore does not give any protection against lateral

Box 13.12. Functions of covering systems

> - **Control** of gas movement, leachate production, soil fluid movement, erosion.
> - **Prevention** of capillary movement of contaminants, windborne movement of pollution, biological translocation (via roots, etc.) of potentially harmful chemicals, root penetration through cover system.
> - **Reduction** of fire hazard, toxicity at the site surface.
> - **Improvement** of structural properties of site, aesthetic appearance, support for vegetative growth.

Table 13.15. Components of a covering system

Component	Functions
Top-soil layer	To support vegetation, or to provide hardstanding for vehicles
Barrier or membrane layer	To prevent the passage of water, gas or volatile compounds
Buffer layer	To protect the barrier/membrane by providing a firm, smooth base
Drainage channel or layer	To prevent accumulation of excess water
Filter layer	To prevent excessive movement of fines
Gas collection layer	To control gas movement and to act as a capillary break

migration or against the threat of pollution of percolating groundwater. A closure cap must be inspected regularly for signs of erosion, settlement or subsidence, but its useful life (if properly designed) can be 50 years or more.

Grouting involves pressurized injection of fluid materials into the ground in order to fill and seal voids and fissures and produce a low-permeability zone (see Chapter 14, Section 14.1.2). A variety of materials (clay particles, pulverized fuel ash, cement, chemical solutions) are used as the sealing agent within the fluid. Grouting can be used to create a bottom seal to contaminated land by:

- injection into the contaminated soil or waste
- injection into the natural soil if it is sufficiently permeable
- creation of a 'slab' or 'floor' in the waste
- creation of a 'slab' or 'floor' in the natural soil.

Two methods of grouting are available for bottom sealing by grouting of in-place material: permeation and claquage (hydraulic fracturing). With permeation grouting the grout is injected under pressure to make it travel the desired distance through the medium to be sealed. The method is intended to achieve a uniform distribution of grout to fill the existing voids in a block of soil or waste. In claquage, 'designed' fracture is caused by cutting horizontal slits in a borehole and then applying high grout pressures. The grout moves laterally outwards either through the fractures in the borehole wall, or, initially, along the horizontal slits, until it meets the grout from an adjacent hole.

Bottom sealing by means of a slurry 'floor' or 'slab' involves the creation of intersecting impermeable zones beneath the area of concern. The voids are created by a 'kerfing system', which is essentially a high-pressure, fluid-jet, cutting tool. At normal operating pressures it will cut a slit 1–3 m long in most soils. The high-pressure jet is oriented to cut horizontally from the bottom of a previously drilled borehole and the cutter is rotated without raising, so that a thin disc cavity is produced. The barrier material, usually in the form of a bentonite slurry, is then injected into intersecting cavities to form a continuous floor (Figure 13.21). Alternatively, the cutting jet fluid may be grout which displaces, and mixes with, the original material as it cuts. The soil or waste to be cut must be of a type that will allow the floor to be constructed without the 'roof' collapsing. As with permeation or fracture grouting, the grout needs to be injected from a grid of holes to form overlapping and continuous horizontal discs, which make up the complete barrier. Jet grouting may also be used to form a vertical barrier by producing a series of contiguous vertical columns composed of a remoulded mixture of soil and cement grout (see Section 13.5.2).

Grouting approaches have several advantages over other ground treatment methods:

- the techniques have been established and proven for many years
- construction is relatively straightforward and can be performed at any time of year
- there is minimal disturbance of the ground and the environmental and aesthetic impact should be negligible.

The major problem with forming low-permeability barriers using a grouting technique is the difficulty of ensuring that complete containment is achieved. Not only must the individual grouting 'sheets' be contiguous, but also their interfaces must not be contaminated with any permeable material. Controlling the distance and direction of grout flow is not straightforward in heterogeneous ground, and the flow and sealing properties may also be adversely affected by the substances contained in contaminated land. Several stages of grouting have to be undertaken to progressively infill between primary and secondary injection points. Unfortunately, there are no totally reliable methods for checking the quality of

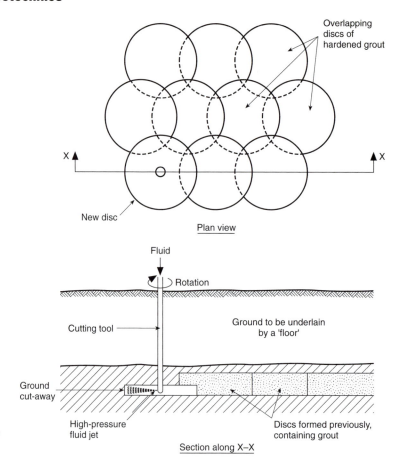

Figure 13.21. The formation of a 'floor' beneath contaminated land

the final horizontal barrier. In addition, drilling through contaminated soil may be difficult, and even dangerous, because of unknown materials.

13.6. In situ treatment

In situ treatment of contaminated ground means that the remediation is applied without excavation of the material to be treated, but some modification of the contaminant is undertaken. The potential methods of in situ treatment can be classified in terms of their objectives:

- physical removal of pollution (e.g. vapour extraction, electrokinesis)
- rendering the contamination harmless (e.g. microbial and chemical treatment)
- stabilization (e.g. render the contaminants insoluble)
- permanent fixation (e.g. production of a solid mass)
- control of groundwater (e.g. modification of seepage patterns).

Many of the treatments necessitate the injection of fluids into the ground in one way or another. Some systems have the added potential advantage of improving the engineering characteristics of the ground.

13.6.1. Extraction

Soil vapour extraction (SVE), also known as soil venting, is designed to physically remove volatile compounds from the vadose (unsaturated) zone within the ground. It is widely accepted as the best in situ on-site treatment for volatile and semivolatile contaminants such as hydrocarbons and chlorinated solvents (Licence, 1993). The system consists of a network of air-withdrawal or vacuum wells installed throughout the area of contaminated soil (Figure 13.22). The wells

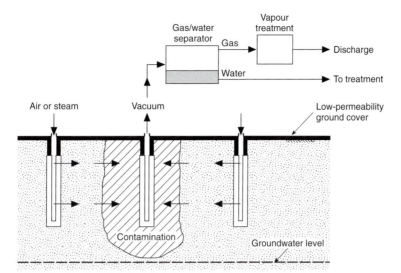

Figure 13.22. Soil vapour extraction

are connected to a vacuum pump system, which provides continuous air flow through the soil. The removed air is vented directly to the atmosphere or is passed through an activated carbon filtration unit or into an incineration unit. Usually some form of low-permeability surface cover is used to ensure that airflow pathways are nearly horizontal. The process continues until all the condensed-phase volatile organics are removed from the higher permeability soils. Contaminants in lower permeability zones are removed by advection. Contaminants from these zones, or in zones away from the direct air flow paths, must first diffuse to the air stream. In these cases the rate of diffusion becomes the controlling factor for the SVE process.

In some systems injection wells are used at the perimeter of the contaminated zone to increase airflow through the soil and to aid contaminant stripping. In situ thermal desorption can be carried out by injecting steam and/or hot air into the undisturbed subsurface. In situ steam injection facilitates the removal of moderately volatile residual organics, including non-aqueous phase liquids (NAPLs), principally from the vadose zone (although treatment within the saturated zone has also been carried out). Injected steam raises the temperature of in situ volatiles, increasing their rate of evaporation from the soil. Upward movement through the soil is enhanced by vacuum extraction at the surface. Some biodegradation usually occurs during soil vapour extraction/soil venting, but the primary purpose is usually to maximize the volatilization of lighter weight compounds.

Bioventing is a biological method for removing volatile and semivolatile organic compounds from soil and groundwater in situ, and is a variation of the vacuum extraction technique. It uses a combination of vapour extraction and air stripping (sparging) to create an air flow pattern through the contaminated soil and groundwater (Figure 13.23). As air is sparged into the saturated zone it strips volatile and semivolatile organic compounds from the groundwater and soil. After the air and organic compound bubbles have migrated to the groundwater surface they are drawn to the air-extraction line. As the organics pass through the oxygen-enriched and moisture-laden unsaturated zone, indigenous microbes absorb the organic compounds and biodegrade them. As the indigenous microbial population increases and more organic compounds are absorbed and digested, the organic concentration in the vacuum-extracted gas decreases. The system is designed and operated in such a manner that biological degradation is the main removal process for organic compounds. The method requires little or no gas treatment and no ex situ groundwater treatment.

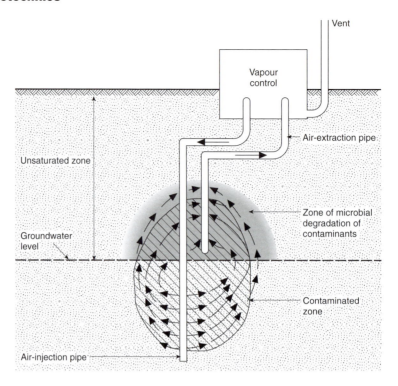

Figure 13.23. Bioventing

In recent years investigations have shown how plants can take up pollution from the ground (Daniel *et al.*, 1998) and studies have been made of the feasibility of using *vegetation* to extract contaminants from the ground (McGrath *et al.*, 1992). Plants draw in nutrients (and also some contaminants) from the ground via their roots, and these substances then pass into the stem, branches and leaves of the plant. The concept is to achieve site remediation by 'harvesting' shrubs and bushes, which have been planted on contaminated land for the purpose of contaminant extraction, and then process this vegetation (typically by incineration). To date the research has concentrated on the performance of plants which are fast-growing and which extract large quantities of water from the ground (e.g. willow), and on means of increasing extraction rates (e.g. by 'coppicing' plants to enhance foliage production). Unfortunately, this treatment is a rather slow process and it requires the contaminants to be waterborne.

13.6.2. Biological and fungal treatment

The general principles of this method of treatment are essentially the same as those for ex situ biological treatment (see Section 13.4.4). However, there is not the same degree of control as with the ex situ treatment, and therefore a significant degree of overdesign is required. Microbes tend to be selected from those types occurring naturally on the site and which are therefore tolerant to the prevailing conditions. Surface treatment is applied using conventional agricultural equipment (ploughs, rotovators). The application of in situ bioremediation methods at depth typically involves the use of a water recirculation system. Water is conditioned with nutrients (nitrogen, phosphorus), oxygen and biological agents and introduced into the contaminated zone at a central location. Water containing degradation products and residual contamination is extracted at the perimeter of the contaminated zone. For successful application, the subsurface strata must be sufficiently permeable to permit water infiltration. Formations with permeabilities of 10^{-2} m/s or more are most amenable to biorestoration.

Factors influencing the decision of whether or not to use in situ remediation are:

- soil structure and hydrogeology — heterogeneity of the subsoil, permeabilities (horizontal and vertical), organic carbon content
- microbiology — bacterial counts, enrichment cultures, biodegradation tests either in batches and/or soil columns, pilot investigation (if needed at complex sites)
- contamination — type of contaminants and concentration levels, free-floating layers.

In situ bioremediation is generally less expensive and less disruptive than other options. This technology is potentially applicable to a wide range of organic contaminants, soil types and contaminated water. However, the process can take a considerable period of time (typically 6 months to 2 years) and would not be suitable for land that has to be developed rapidly (Taylor & McLean, 1992). Because of its biological nature, bioremediation is rarely suitable for the treatment of inorganic contamination, and experience has shown that it does not readily work for heavy tars and other large, stable molecules. Recent research work, however, has identified some success in using the 'white-rot' fungus in reducing PAHs, PCBs and some pesticides to easily reducible, simple hydrocarbons (*New Nordic Technology*, 1995).

13.6.3. Fixation

In situ fixation treatments are intended to eliminate the mobility and/or toxicity of contaminants within the ground (see Section 13.4.6). In addition to treating contamination, the process may also improve the engineering properties of the ground.

Stabilization/solidification reagents may be introduced into the ground by using:

- soil-mixing equipment (e.g. agricultural rotovators (shallow treatment zone), hollow-stem augers through which treatment agents are injected)
- pressure injection techniques analogous to conventional grouting.

Only the first method has been developed on a commercial scale. With the latter approach it is difficult to ensure even permeation of the treatment agent throughout the ground. Also, the treatment cannot be applied at shallow depths (less than 2 m approximately) because overburden pressure is needed to withstand the injection pressures.

In situ *vitrification* uses electricity to literally melt inorganic materials into a glass-like inert product (Figure 13.24). Electrodes are placed in the ground in a box-shaped pattern. A starter path of flaked graphite and glass frit (the basic materials used in glass-making) is placed at the surface between the electrodes. As an electric potential is applied between the electrodes, current flows through

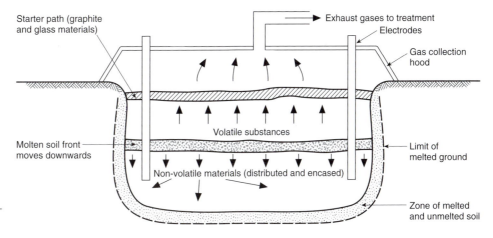

Figure 13.24. In situ vitrification (after CIRIA, 1995–98, Volume IX)

the starter path, heating it and the adjacent soil to its melting temperature (typically 1600–2000 °C). Once molten, typical soils become electrically conducting. The molten mass becomes the primary conductor and heat-transfer medium, allowing the process to continue. The molten mass grows downwards (at 25–30 mm/h) and horizontally as long as power is applied. Volume reductions of 20–40% have been achieved.

The system works primarily on solids such as soil and sand, it will not work on water or oily sludge (USEPA, 1994). Organic compounds are destroyed by being broken down by the intense heat. Volatile contaminants driven off as gases are collected by a vent hood placed over the area being treated, and are routed through a treatment system before release. Questions about this relatively unproven technology relate to its effectiveness, the need for large amounts of technical investment and the high electrical input.

13.6.4. Electrokinesis

A major limitation of the most successful decontamination technologies is that they are restricted to soils with relatively high permeabilities and hence are not particularly effective with fine-grained deposits. Furthermore, they are not particularly effective in removing contaminants adsorbed onto the soil particles. Electrokinetic soil processing utilizes the effects of a flow of electricity through a porous medium and is a promising, emerging technology for removing adsorbed toxic ionic species from fine-grained deposits and sludges. The process involves contaminant desorption, transport, capture and removal using an electric current. The basic process has been used for many years for dewatering (see Chapter 7, Section 7.5.2) and consolidating soils in foundation engineering applications. Electro-osmosis results in the movement of soil water towards the cathode. This causes the movement of any dissolved contaminants. As a bulk movement of water occurs by electro-osmosis, any dissolved organic compounds should also be removed.

Typically, a series of anodes and cathodes is installed (Figure 13.25) and a low level d.c. electrical current (a few amperes per square metre) is passed between these electrodes, which allow egress and ingress of pore fluid. Upon application

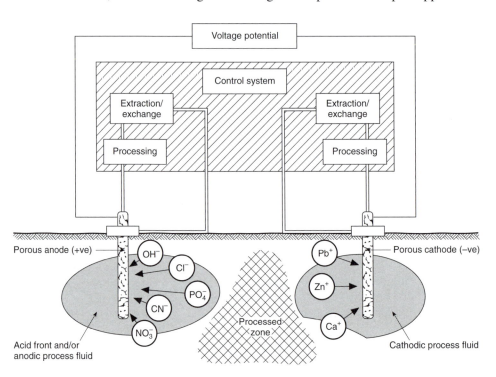

Figure 13.25. Electrokinetic decontamination

of the current the soil–water system undergoes physico-chemical, hydrological and mechanical changes, leading to contaminant transport and removal. Depending on the charge of the ions contained within the contaminated material, contaminants (contained in soil water) will either move to the cathode or the anode. In addition, the movement of soil water to the cathode results in the transport of dissolved ions and organic contaminants out of the soil. Dissolved contaminants can then be pumped from the electrodes to an effluent collector where they are removed from the solution using standard treatment systems. The composition of the treated water is adjusted and returned to the electrodes to 'condition' them (e.g. to maintain the pH between acceptable limits).

Power consumption is related to the concentration of contaminants and the time required to achieve remediation standards. Spacing between the electrodes is typically 1–2 m. Energy requirements tend to be in the range 30–300 kWh depending on the scale of operations and site conditions.

Laboratory and pilot-scale studies (Acar *et al.*, 1994) have shown the feasibility of extracting inorganic contaminants such as copper, zinc and cadmium. Field studies of soil decontamination by electrokinesis are limited, although trials in the USA and The Netherlands have shown high cleaning rates for heavy metals (Lageman *et al.*, 1993).

13.6.5. Hydraulic measures

Groundwater modification systems can be employed to achieve a number of objectives:

- to lower the groundwater table in the vicinity of the contaminated sites to create a separation between the contaminants and the groundwater
- to contain or isolate a contaminant plume
- to supplement a barrier system
- to facilitate treatment of the contaminated water in a post-extraction system.

The components in a hydraulic groundwater management system include extraction and infiltration wells/trenches (Figure 13.26). The extraction wells are used either singly or in groups to withdraw groundwater from the aquifer. The infiltration wells inject water into the ground after treatment or as part of a hydraulic system for plume management or water balance. Infiltration trenches are used primarily as a route for recharging the aquifer after interception and/or treatment.

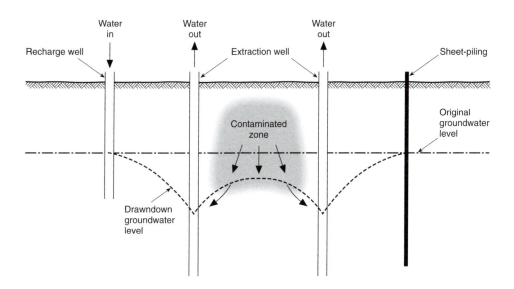

Figure 13.26. Groundwater management systems

In situ treatment of groundwater involves the passage of water through a treatment zone. One process involves the injection of agents (in solution or gaseous form) to cause or promote the natural degradation of contaminants. Alternatively, water is passed through a solid substrate that will support physical separation, chemical or biological degradation, or a reduction in toxicity. The two processes may be used in combination.

14 Derelict land

14.1. Introduction

Designations of land, such as vacant, derelict or damaged, can be confused with evidence for contamination. However, the two categorizations are independent. 'Contaminated land' is a designation indicative of the presence of an introduced substance which is harmful, whereas 'derelict land' is land damaged by past activities so that it is incapable of beneficial use without treatment. Contaminated land poses a potential threat to the environment, but derelict land is a threat to future development, although it may be aesthetically unattractive as well.

The Black Country Spine Road crosses derelict land polluted by waste tips, sewage and gasworks, chemical and metal works, brick factories, etc. As if this were not enough, further instability comes from deep fill, redundant canals and mine-workings. The region known as the Black Country is rich in mineral resources (coal, ironstone, fireclay, limestone, brick-making clay) and thus the area developed as a thriving centre of industry, thereby producing problems from a combination of factors:

- the vast amount of heavy industry
- the presence of mineral resources and disposal of wastes
- a legacy of mining, which developed in an ad hoc manner
- an extensive transportation system.

Mines pepper the area because people used to set up small family businesses with leases from the local gentry. Opencast mining was not a real option without today's bulk earth-moving equipment. Underground fires were frequent — oxygen and friction fuelled the small pieces of coal left behind, and fires could burn for years. The mines were inundated with water, and pumping from one frequently caused problems for its neighbours, until a coordinated approach was developed that reduced overall pumping. This revolutionized the mining, but left behind a network of culverts. Other old disused culverts and underground pipelines contain highly toxic residues (e.g. Blue Billy, containing cyanide and other contaminants).

Much of the Black Country is artificial made-ground, from 1–15 m deep, comprising 'Traditional Black Country fill' (silty material, with a mix of ash, clinker and colliery spoil) often found in clay or marl pits. When the Black Country Development Corporation undertook the clean up of Swan Village (the gasworks that had once supplied gas for the whole of Birmingham) they found 2–4 m of fill containing the gasworks infrastructure overlaying silty clay (*New Civil Engineer*, 1995e and 1995f). Removal of the existing foundation slabs uncovered many underground brickwork and concrete structures.

General domestic waste is also found, particularly in mined-out clay and gravel pits, and former sewage works have left a legacy of old sludge lagoons (1–4 m deep). Even though a crust has formed over the coagulated sludge, and trees are growing in the crust, the underlying sludge is still in a liquid state and the whole surface will bounce up and down under load.

Canals formed the original infrastructure of the region, and at its peak the canal network comprised around 260 km of waterways. The canals were built shortly before intensive mining in the region, which caused them to subside. Bed levels dropped by up to 9 m, but the canals were maintained by building up the sidewalls and relining with puddled clay. Redundant stretches of waterway have been filled in with debris, so that the lining clays have become significantly contaminated, and have been built over (as occurred at Love Canal, see Chapter 13, Section 13.1). One dramatic example of the endemic subsidence in the region is a major distortion in a tunnel which joins two sections of the Dudley Canal. The 200-year-old brick arch structure is 2.7 km long and 2.8 m wide — one section of wall has shifted 800 mm from its original position and at a dog-leg bend the canal has wandered laterally by half the tunnel diameter.

As the demand for developable sites becomes more intense, and the desire to conserve agricultural land and green-belt areas grows, methods of reclaiming land that has fallen into dereliction due to its past use are becoming increasingly important

14.1.1. Land dereliction and treatment

Many items in constant daily use originated in the ground, and mineral extraction is an integral part of civilization. However, this has not been without cost to the environment. There are many problems associated with mineral extraction, including:

- contamination by chemical pollution of the soil
- pollution of water sources
- change in groundwater regimes
- disfigurement of the landscape
- construction of tips
- increased sediment loads in rivers
- soil erosion
- loss of flora and fauna
- vegetation and soil degradation
- ground instability and subsidence
- creation of dangerous voids.

As early as the Neolithic period of prehistory, humans were modifying the Earth's surface in search of its mineral wealth. Many excavations in the chalk hills of southern England represent Neolithic pits dug to extract good-quality flint for tool-making. With time, the extraction of metallic ores became important to make metal tools and jewellery, and there are many examples of Roman mines still visible in the landscape throughout Europe. Fossil fuels have also been extracted from early times. Excavations for peat were widespread, and it is believed that the lakes and waterways of the Norfolk Broads in eastern England owe their origin to the removal of more than 25 Mm^3 of peat prior to the 14th century (Pickering & Owen, 1994).

The coming of the Industrial Revolution saw a major growth in the extraction of iron ore and coal. As a consequence, large opencast pits and mineshafts were constructed, some to great depths. Britain has over 100 000 old coalmines and at least as many more mines for other ores. Oil-shale, gannister (rock used for lining furnaces), fire-clay, brick-clay, limestone, ironstone, sandstone, flagstone, zinc, lead, copper, tin, silver, gold, barite and fluorspar have also been mined (Littlejohn, 1979; Waltham, 1989). Most mines are small, abandoned and shallow, and pose the maximum surface hazard (Figure 14.1). Substantial areas of several British towns and cities stand on ground that is riddled with disused untreated mine workings as little as 3 m below the surface (Box 14.1). The major problem with disused mine shafts is that the location of the workings is often unknown. Few plans survive, and even if they do they are often very inaccurate. The land overlying mine workings often cannot be redeveloped without treatment

Box 14.1. Examples of land blight due to Britain's industrial legacy

Building-stone mines (Bath) — subsidence

Two stone mines, which have been disused for more than 100 years constitute a vast subterranean warren of holes below part of Bath. These mines result from the rise of Bath as an architecturally important town and much of the characteristic stone used to build the town was hewn from the mines. The workings extend for 16 ha and the volume of voids created by the workings is around 400 000 m³. The workings are shallow (4–8 m below ground level), and during excavation hundreds of pillars of stone were left to support the roof. Over the years these pillars have been robbed of material and they have lost structural strength — roof collapses have occurred in around one-third of the the mines area. A plan to stabilize the area involved completely filling the voids using PFA grout. However, remediation was constrained by two environmental issues: since the abandonment of the mines, they have become one of the most important hibernation sites in Europe for Greater Horseshoe bats (classed as a priority species by the European Community); and there is concern that heavy metals will be leached out of the PFA and into the local water supply (*Construction News*, 1994).

Chalk-mining (Bury St Edmunds) — subsidence

Chalk-mining started at this location in the mid-19th century. The chalk was burnt, both for agriculture and for use as lime mortar in construction. The resultant caverns were very large, with underground corridors up to 50 m long. As Bury St Edmunds grew, houses were built over the workings, but problems were soon evident, with the first collapse recorded just 2 years after work started. Subsequently, the site was declared dangerous and houses were bought by the council and demolished. The 2 ha site has lain derelict since the early 1980s. A study to identify remedial work to stabilize the caverns identified several treatment options: to fill or support all the works (likely to be so expensive as to preclude making the site safe enough for housing); to prevent catastrophic collapse and use the site for recreation; to leave the tunnels as a colony for bats; and to convert the caverns into a tourist attraction.

Tin and copper mining (Cornwall) — mineshaft and adit collapse

Across Cornwall there are some 15 000 abandoned mineshafts (from tin and copper extraction). Some mines were worked into the 1920s and 1930s, but most were abandoned a hundred years ago. Many old shafts are still open or only lightly timbered over, and these are usually surrounded only by a circular stone wall. Collapses of these abandoned mineshafts are not unusual, and there have been instances of serious, occasionally fatal, injuries resulting from people falling into holes that have appeared overnight. The majority of the entrances to the adits, which are often found at the base of cliffs, have been blocked, but a few still remain open. These adits are now in a very poor state of repair in most cases, and may have sections where only rotting timbers form the floor or roof.

Coal and steel industry (Ebbw Vale) — uncapped shafts and waste tips

At Ebbw Vale an area of steep-sided valley (80 ha, 2.5 km long) was rendered derelict through decades of use by the coal and steel industries. Two collieries had stood on the site, but all the headworks have been demolished. Adjacent was a former steelworks area. There were two 40 m high spoil heaps, with steelworks slag to the west and colliery shale to the east. As the tips grew, the river Ebbw Fawr, which runs through the valley, had been progressively culverted to make more space for spoil. Subsequent reclamation work involved capping 11 mineshafts, removing toxic waste, moving 2 Mm³ of shale and slag, and re-laying 12 km of railway sidings.

PFA, pulverized fuel ash.

of the workings and it becomes derelict. However, shortage of building land (particularly for industrial concerns) means that developers are re-using more and more land in areas with a history of mining.

Extraction and processing of minerals has created huge quantities of waste materials, which have been disposed of by infilling depressions in the ground, by tipping to form spoil heaps, etc., usually in a non-engineered manner. Wastes

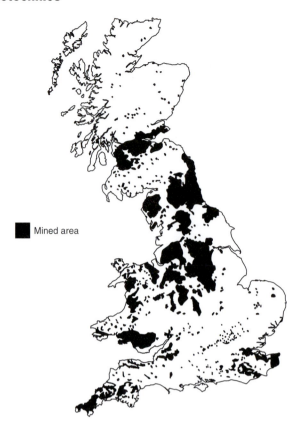

Figure 14.1. Mining in Britain

Mined area

produced directly by society (e.g. refuse, sewage), also have a history of indiscriminate uncontrolled, or semi-engineered, dumping in landfills or lagoons. All made-ground should be thoroughly investigated prior to development because of the likelihood of extreme variability in the ground conditions. Figure 14.2 shows a house where the foundations have failed because they were constructed partly on natural ground and partly on old backfill (note the abrupt change in the level of the brickcourse adjacent to the downpipe). The long-term stability of tips, infilled ground and lagoons cannot be assumed simply because at the present time they do not exhibit any outward signs of distress. The internal structure of many waste dumps, and the waste materials themselves, are metastable, and they are likely to degrade or breakdown with time.

Incidents involving landfill gas in buildings have increased in recent years (see Chapter 12, Section 12.1) and people are well aware that decomposing refuse generates methane and carbon dioxide. However, these gases also originate from

Figure 14.2. Settlement of a house founded partly on fill

other sources of land dereliction (e.g. coal strata, dumped river dredgings, sewage lagoons). Gas from the ground can enter buildings through gaps around service pipes, cracks in walls below the ground, etc. It can also accumulate in voids created by settlement beneath floor slabs. The force driving gas entry is principally the positive pressure arising from the conditions under which it is generated and the slight negative pressure (relative to atmosphere) that exists in buildings.

Ground treatment techniques for derelict land are basically of two types: those that stabilize the ground and eliminate the possibility of collapse; and those that reduce compressibility (usually by increasing the bulk density of the ground materials). Ground treatment can be used for both natural soils and fill material and general methods employed include:

- excavation and refilling in thin layers, with adequate compaction (large-scale earth-moving is a routine construction operation and all fill material can be checked)
- grouting (injection of stabilizing material into the ground)
- dynamic compaction (dropping a large weight onto the ground)
- 'vibro' techniques (penetration of a vibrating poker into the ground)
- preloading with a surcharge (as loose waste compresses immediately under load the surcharge need not be left in position for an extended period)
- soil stabilization by mixing in appropriate additions.

Box 14.2 outlines factors which affect the selection of a ground improvement technique.

Box 14.2. Factors affecting the selection of a ground improvement process*

Ground conditions
- Types of material present.
- Depth requiring treatment.
- Groundwater level.
- Permeability (particularly relevant for grouting operations).
- Presence of subterranean obstructions and voids.
- Stability of the ground (e.g. susceptibility to liquefaction, potential for collapse settlement).
- Presence of contamination.

Objectives of treatment
- To provide appropriate support for building foundations.
- To enhance the strength of the ground to facilitate construction.
- To restrict/reduce groundwater flow (temporarily or permanently).

General
- Establishment costs are often high and so large areas may have to be treated for a treatment process to be economic.
- Time available for undertaking the treatment.
- Local availabilty of appropriate construction materials.

Location of the site
- Proximity of adjacent structures and buried services.
- Possibility of ground vibrations and noise affecting adjacent structures and inhabitants.
- Ease of access to the site.

Performance requirements
- Magnitude of acceptable post-construction settlements and angular distortion of structures occupying the site.
- Amount of soil strength improvement required (improvement in clay soils is unlikely to be greater than twice the allowable pressure of the untreated ground).

---

* Adapted from Greenwood & Thomson (1984) and Leach & Goodger (1991).

14.1.2. Grouting

Grouting is the injection of a stabilizing medium, via pipes inserted into the ground, into large subterranean voids or channels or fissures within soils and rocks. The latter process may be achieved through permeation (grout flows through natural passages) or by fracture (whereby artificial passages are created within the ground). The injected material subsequently sets, and modifies the properties of the ground so that it becomes less compressible and/or stronger. Grouts usually consist of aqueous suspensions of cement or other pozzolanic materials (for large voids and also for the penetration of fissures in rocks and pores in sandy soil) or of chemical solutions (able to penetrate fine sands and sandy silts). Grouting penetration depends on the interaction of ground properties (permeability, porosity, pore size and shape, effective stresses, pore fluid properties) and those of the fluid grout (viscosity, shear strength, particle content, particle sizes). The applicability of *permeation grouting* techniques is indicated in Figure 14.3.

Fracture grouting (claquage) is frequently used to develop a network of grouted fractures in fine-grained soil types (silts and clays), which are not amenable to permeation grouting. The fractures are created by raising the pressure in the grout hole so that it exceeds the strength of the soil and overburden pressure. This produces a network of essentially horizontal fissures, which then fill with grout. If the grout has high shear strength, bulbs of grout paste can be expanded in almost spherical form. This radial expansion is at the expense of the total disturbance of the soil structure and leads to compaction of soil with high porosity — this is known as compaction grouting.

The most widely used form of grout is a mixture of ordinary Portland cement and water because of reproducible high performance, its high yield strength and its cheapness. Cementitious grouts may also consist of water, cement and other materials that combine chemically with the cement for special purposes or that serve as bulking agents. Pulverized fuel ash (PFA) (see Chapter 16, Section 16.4) is often used in grouts as a cement replacement because it has good flow properties. When grouting gravels, some sand or silt may be included in the mix. Cement–clay grouts are most suitable where the main purpose of grouting is to arrest water movement. Figure 14.4 shows a PFA–cement grout being made in a paddle mixer prior to being gravity fed (through the pipe on the right of the picture) into collapsed shallow mineworkings.

The principal factor affecting the properties of cement grouts is the water/cement ratio, with the amount of water governing the rate of bleeding (settling out of solids) and the plasticity and ultimate strength of the set grout. Pure cement grouts are unstable, and particles settle out as the grout travels from its pumping

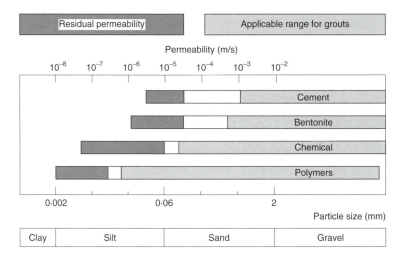

Figure 14.3. Applicability of permeation grouting

Figure 14.4. Void filling using a PFA grout

point. Hence the initially injected grout should be fairly thin (water/cement ratio 10 : 1 to 15 : 1) in order to minimize the chances of prematurely plugging the hole. Cement grouts are durable under most normal conditions. This quality may be increased by the incorporation of blast furnace or sulphate-resisting cement.

Chemical grouting of granular soils increases their cohesion, but the angle of internal friction remains more or less the same. Nevertheless, most chemical grouts set to form weak solids. Chemical grouts are solutions, with a viscosity only 2–5 times that of water, which contain no suspended solid particles (unless deliberately added for some specific purpose), so they are able to penetrate fine sands and sandy silts. However, they are expensive and so they may be used as the final seal after a preliminary stage grouting has been completed with a cheaper grout. Usually the reactive components are combined into two solutions which, when mixed together, produce a weak jelly-like material at some time after mixing. This time is determined by the temperature and the chemical concentration. Gel times are usually comparatively short (10–90 minutes) in practical applications. The resultant gel usually has very low permeability (about 10^{-10} m/s or less). Soils containing less than 10% fines can usually be permeated with chemical grouts. If the fines content exceeds 15%, effective grouting is likely to prove difficult.

A grouting programme requires the selection of suitable grout material and injection procedures and grout-hole patterns. Despite considerable practical and theoretical development, grouting remains largely an empirical art because it is used with natural materials that have a wide, natural variation in properties. Each grout injection must be carefully monitored and the results assessed to improve the technique for subsequent injections.

Economic spacings of injection holes are 2–3 m for cement grouts in open, fine gravels and 0.5–1.5 m for chemical grouts in sands. If, however, the soil properties are anisotropic or irregular, grout may travel considerably greater distances than that predicted theoretically. Hence grout should be injected in stages, in predetermined small quantities, with each injection being made after the prior infill has set so that continuous zones of treatment are gradually built up. Grout pressures should not normally exceed 10 kN/m^2 per metre of depth. Except for filling of mine workings and other large caverns, the volume flow rate is restricted by ground resistance to injection, and flow rates are not usually very high. Grouting continues until a specified pressure is reached. The grout tube is then partially withdrawn until the pressure drops and further grout can be injected. Records should be kept of the grout take at each drill hole to assess the extent of void filling by the process. It is more important to ensure that the full design soil volume is permeated with grout when the objective is water cut-off than when the objective is to improve mechanical properties.

When particulate grouts are injected into porous soil, filtering may occur, whereby the larger particles in the suspension tend to separate out at the entrances to pores. The ability of particulate grouts to penetrate a formation depends on the particle size of the suspended material. This ability has been quantified in terms of a groutability ratio ($d_{15(soil)}/d_{85(soil)}$). This ratio should exceed 25 if a grout is successfully to penetrate the formation concerned. Grouting is likely to be impossible if the ratio is less than 11 (Bell, 1993). Alternatively, the limits for particulate grouts may be taken as a 10:1 size factor between the d_{15} of the grout and the d_{15} size of the granular soil concerned.

As grout is pumped down the injection pipe it invades the ground with a steadily advancing front in three dimensions and in uniform soils the grouted zone assumes a spherical shape (Greenwood & Thomson, 1984). The distance travelled and the thickness of fissures induced by high pumping pressures can be controlled by varying the rheological properties of the grout itself. The time to penetrate to radius R (spherical flow) is given by:

$$t = \frac{na^2}{kh}\left[0.333\eta\left(\frac{R^3}{a^3}-1\right)-0.5(\eta-1)\left(\frac{R^2}{a^2}-1\right)\right] \tag{14.1}$$

where n is the porosity, a is the radius of the injection pipe, k is the ground permeability, h is the applied head and η is the grout viscosity. The differences between theoretical relationships for spherical and cylindrical flow are not large, considering the unknown flow patterns in real ground.

14.1.3. Dynamic compaction

Dynamic compaction is essentially a method of deep compaction of fill, with soil improvement resulting primarily from an increase in the density of the fill. The major use of the technique in Britain has been the deep compaction of loose, partially saturated fills. The process was devised by Louis Menard, and consists of dropping heavy blocks (5–20 t) onto the ground from great heights (up to about 30 m) (Figure 14.5). In most instances the applied energy is between 100 and 400 tm/m^2, and the high-energy drops may form imprints that are 1–2 m deep. The greatest increase in crater depth occurs in the first half a dozen drops. At the end of each pass the imprints are filled and the ground is levelled using the material heaved around them. The area to be treated is divided into grid patterns and each grid point receives several blows in a given pass. Several passes may be necessary to obtain the desired results. The final pass comprises overlapping drops from low height to recompact the soil at shallow depth which has been sheared and loosened by the high energy blows. To evaluate the extent of ground improvements, measurements are taken of the drop-weight imprint, ground heave, average settlement, changes in soil properties, porewater pressure, ground vibration, etc.

Figure 14.5. Dynamic compaction

Each compaction blow affects a zone of the ground, and the blows are arranged so that their influence zones overlap. The maximum depth to which a soil is influenced D_{max} (m) may be estimated from the equation:

$$D_{max} = I(WH)^{0.5} \qquad (14.2)$$

where I is the influence factor, W is the weight of the block (t) and H is the height (m) through which it is dropped.

Factors influencing I are:

- soil type
- initial soil density
- depth to groundwater table
- area of falling block
- grid spacing
- number and sequence of drops
- number of passes.

Suggested values for I range from 0.3 to 1.0, although a value of 0.5 appears to be the most widely accepted at this time (Slocombe, 1993). Alternatively, Figure 14.6 may be used to assess the depth of treatment.

The materials treated by dynamic compaction exhibit a high bearing capacity and low post-construction settlement. It works well with loose coarse-grained soils, rubble fills and non-hazardous landfills. The major use of this technique in Britain has been the deep compaction of loose, partially saturated fills. With unsaturated or highly permeable saturated granular materials the response to tamping (i.e. densification) is immediate. However, as the soil mass permeability decreases, so does the effectiveness of the treatment. Saturated in situ clays are unlikely to show any improvement, and may show considerable strength reduction if overtampered so that remoulding occurs. 'Dynamic consolidation' is used to improve the properties of natural, soft fine-grained soils. In this system the dynamic loading is used to produce a local pore pressure increase, which then dissipates by consolidation so that a denser mass results.

Dynamic compaction and vibrocompaction (Section 14.1.4) normally show considerable saving in foundation costs compared with pile foundations or deep piers. It is not unusual for the cost saving to be 30–50% of the substructure costs (Greenwood & Thomson, 1984).

It must be remembered that the impact from the falling weight causes transient vibrations in the ground (Figure 14.7). In built-up areas dynamic compaction may be prohibited by the possibility of damage being caused to nearby buildings by vibrations produced by impacts (see Chapter 17, Section 17.5). It has been suggested that the minimum safe distance between the point of impact of the falling weight and a building is about 30 m (BRE, 1997).

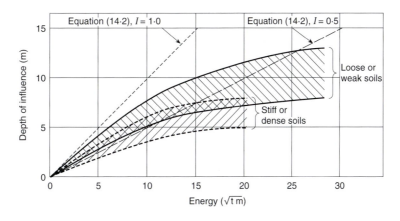

Figure 14.6. Dynamic compaction treatment depth (after Slocombe, 1993)

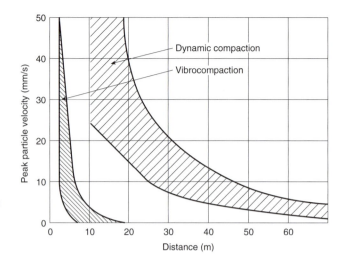

Figure 14.7. Ground vibrations due to ground compaction processes (after Slocombe, 1993)

The rapid impact compactor, which was originally developed for the rapid repair of military airfield runways, comprises a modified hydraulic piling hammer acting on an articulated compacting foot at about 40 blows per minute (BRE, 1997). Treatment is carried out by hammering the 1.5 m diameter compacting foot into the ground at closely spaced treatment points. After levelling the ground surface, further compaction treatment may be carried out and the process is completed with a general tamping of the whole area. Rapid impact compaction is similar in principle to dynamic compaction.

14.1.4. Vibrocompaction

This type of process, widely known as 'vibro', was developed to compact loose, naturally occurring sands. The compaction method relies on the fact that particles of non-cohesive soil can be rearranged into a denser condition under the action of oscillatory shear. The basic tool is a cylindrical poker containing in its bottom section an eccentric weight that rotates and generates vibrations which are transmitted to the soil as the poker penetrates the ground. The action of the vibrator is usually accompanied by water jetting. At the required depth of penetration the vibrator is surged up and down to agitate the soil, remove fines and form an annular gap around the vibrator. The water pressure is then reduced and, with the vibrator still in the ground, sand infill is poured into the void (which has been created by the vibrator) and compacted at the base of the vibrator. When the required compaction resistance has been achieved, the vibrator is raised and the infill and compaction process repeated until the hole is filled to ground level. The process is illustrated in Figure 14.8.

Compaction is usually undertaken on a regular grid pattern and a spacing between insertion points of around 1.8–3 m should produce a final relative density of the ground in the region of 70–80%. Vibration compaction works well with granular materials and filled ground, but there is a limiting silt or clay content of around 15–20% for efficient vibrocompaction of sand. In cohesive soils little compaction is achieved with this method, but the long cylindrical hole produced by the poker vibrator is backfilled with stone (usually coarse gravel-sized material) so that the cohesive soil is stiffened by stone columns.

Vibroreplacement consists of constructing stone columns (through fill materials and weak soils, although a soil shear strength of at least $20 \, kN/m^2$ is needed for stone columns to be effective) to improve the load bearing and settlement characteristics. To construct the stone columns the vibrator is allowed to penetrate to the design depth (usually to a firm stratum) and the resulting cavity is filled with hard inert stone, free of clay and silt fines (Figure 14.9). To develop the required interaction between the stone columns and the surrounding

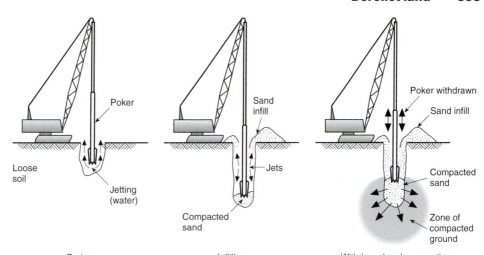

Figure 14.8. The vibrocompaction process

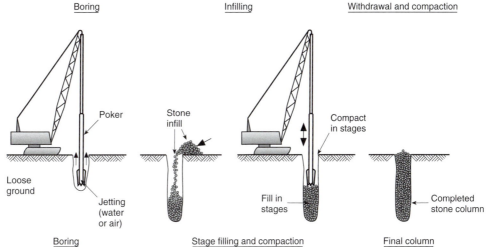

Figure 14.9. Stone columns

soils the stone infill is introduced and compacted in stages, each charge of stone being thoroughly compacted. The general spacing between columns is in the range 1.2–3.0 m.

As a guideline it can be assumed that, for isolated shallow footings, settlements will be reduced by around 50%. Furthermore, in cohesive soils, excess pore pressures are readily dissipated by the stone columns (because of shorter drainage path lengths), and for this reason settlements occur at a faster rate than is normally the case with cohesive soils.

14.1.5. Preloading and surcharging

The settlement characteristics of fills can be improved by densification through preloading (i.e. the placement of a temporary surcharge on the ground surface). This surcharge should produce an increment of effective vertical stress which is greater (typically by about 20%) than that induced by the weight of the development to be built on the compressible or filled ground. The imposition of the surcharge promotes settlement (and sometimes consolidation), thereby reducing the amount of movement that will occur due to the ground stresses imposed by the development (see Chapter 8, Section 8.5.3). With preloading, the short-term settlements of compressible deposits of waste can be cut by about half, compared with untreated refuse (Oweis & Khera, 1990). The secondary settlement (creep) may be reduced, depending on the length of time for which the preload is maintained.

The economics of the method are related to the costs of earth-moving and, in particular, the transport distance for surcharge material. The time required for the surcharge to be in position is also important economically. With most loose unsaturated fill it is unnecessary to leave the surcharge in place for any length of time. A large area can therefore be preloaded by moving a surcharge around the site in a continuous earth-moving operation.

14.1.6. Lime stabilization

Lime addition has been popular in America since the 1950s as a way of turning clay into a capping layer, or for solving sub-base problems in areas lacking in crushed rock. The addition of lime (typically 1–2% by weight) to a soil has several effects:

- *Drying*. Quicklime takes water from a soil in a reaction that produces heat and causes further moisture loss through evaporation, thereby producing immediate improvement of the soil.
- *Modification*. This is a subsequent reaction between the lime and clay minerals, which changes the properties of the soil, increasing the plastic limit. This improves the workability and the trafficability of the soil.
- *Stabilization*. A further reaction can happen between clay minerals and lime, which produces a cement-like paste, but this depends on the amount of lime added and the reactivity of the clay. Such stabilization can take a number of years. The best long-term results are obtained by adding enough water to increase the moisture content of the clay to around 1.2 times the natural plastic limit when the lime is added to the soil.

Two types of lime can be used in lime treatment work: calcium oxide (quick lime or burnt lime) and calcium hydroxide (slaked or hydrated lime). The ability of quicklime to reduce the moisture content of a soil is greater than that of slaked lime.

In essence, the process of lime treatment involves sprinkling a layer of lime onto the soil, mixing it in and compacting the ground using a roller (*Construction News*, 1999). Two pulverization stages may be applied after spreading. In the first stage the ground is rotovated to ensure intimate contact between lime and soil. Sufficient time has to be allowed before starting the second pulverization stage (which puts the stabilized soil into a form suitable for compaction), to enable the lime to react with the soil. This maturing period ranges between 24 and 72 h, depending on the plasticity of the original soil. Using this process, 100 000 m^3 of silt (which resulted from dredging operations in the Thames Estuary) was combined with quicklime and PFA to provide a lightweight fill material that met the requirements of the Highways Agency (Nettleton *et al.*, 1996).

When using lime stabilization personnel must be aware of various technical and health and safety aspects:

- lime is highly corrosive
- overcompaction
- rain can cause leaching out of the lime
- the occurrence of non-plastic material
- the problem of lime carbonation.

14.2. Underground voids

Almost all artificial cavities are formed as a result of mining operations. Mining operations were undertaken at a very early stage in Britain on a small scale, but the advent of the Industrial Revolution caused mining operations to become widespread. Mining instability occurs as a result of collapse of the underground workings, or from the collapse of shafts or adits that lead to the workings (Richards *et al.*, 1993). This gives rise to general or localized subsidence of the ground surface. Subsidence poses a particular threat in areas where mining was not intense, because there are often no physical signs that any activity ever occurred.

14.2.1. Mine workings

Most early workings for both coal and iron ore were carried out at surface outcrops. Drift mining was carried out in situations where seams had a shallow dip by progressively excavating the seam from an outcrop. Adits (horizontal or slightly inclined passages) provide access to underground workings from valley faces. A portal (brick, stone or timber) supports the weak surface materials and the adit is lined until competent rock is reached.

The earliest form of coal and iron ore mining operation on a wide scale in Britain employed bell pits (up to 12 m deep with diameters of 8–20 m at seam level). The technique involved sinking a shallow shaft (about 1 m diameter) down into the seam (Figure 14.10). Mining proceeded radially from the base of the shaft until the area became too large to be naturally or artificially supported, or problems of ventilation and water ingress occurred. The pit was then abandoned and a new pit started nearby. Areas containing bell pits present a risk of continued mining subsidence due to further collapse of voids, or the low bearing capacity of the loose fill within a collapsed bell. The collapse of a bell pit generally leads to the formation of a localized crater at ground surface. If the ground has been subsequently regraded the signs of former shallow mining activity may no longer be visible.

Room-and-pillar (or pillar-and-stall) extraction methods were introduced in the 17th century and involved leaving pillars of intact seam material to support the roof while removing the mineral from in between, thereby forming 'rooms'

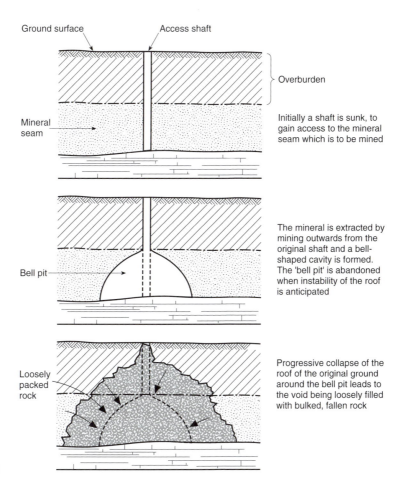

Figure 14.10. Bell-pit formation

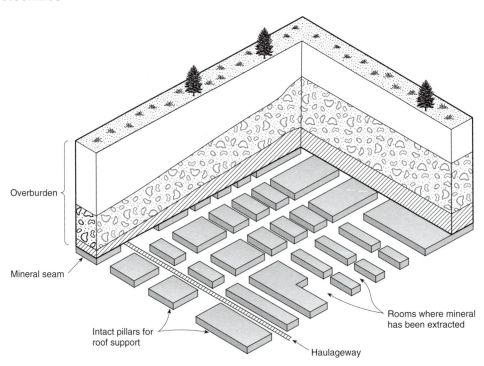

Overburden

Mineral seam

Intact pillars for
roof support

Haulageway

Rooms where mineral
has been extracted

*Figure 14.11. Room-and-pillar
mineral extraction*

(Figure 14.11). For deep workings pillar widths were normally in excess of 10% of the working depth to ensure pillar stability. After initial excavation pillars were often reduced in size or removed completely as the workings were abandoned. With time the pillars deteriorate — there is progressive reduction of pillar width due to spalling, slaking or erosion (Waltham, 1989). Areas of room-and-pillar workings present a risk of mining subsidence due to roof failure, pillar failure or floor heave.

Two common forms of surface subsidence arise: a sink-hole type resulting from collapsed mine junctions, and a widely spread saucer-shaped depression pillar failure (Whittaker & Reddish, 1989). In both situations, surface subsidence can occur many years after abandonment of the mining operations. Crown holes (localized surface failures) occur where progressive roof collapse causes a void to migrate upwards from the old mine workings until it reaches the ground surface. The vertical extent of a collapse before 'bulking up' of the collapsed material causes closure of the void is a function of the void shape, the mined-out thickness and the bulking factor (ranging from about 25% in some shales to 50% in stronger rocks). The limit of upward progression is in the region of 2–12 times the mined thickness.

Modern mining techniques generally involve total extraction by longwall methods. This involves working a single face, with access from roadways at right angles to the face at each end of the seam. The working face is temporarily supported and the support is moved forward as the face is worked, thereby allowing the roof to collapse into the space behind the works. Ground subsidence is inevitable and extensive, but it is also predictable and rapid, in contrast to the random, lingering hazard of many old pillar-and-stall mines.

14.2.2. Subsidence

In order for surface subsidence to occur due to mining operations some form of underground collapse or abstraction process must precede such an event. For example, below the Cheshire plain the extraction of salt has resulted in severe local subsidence and damage to buildings (Figure 14.12). The form of mining, the degree and extent of activity and the geological nature of the ground are prime factors in the development of subsidence.

Figure 14.12. Subsidence due to salt extraction

Subsidence is usually considered as the vertical displacement of the ground, but it also implies a measure of horizontal movement of the adjacent area by virtue of the lateral shift of ground generated by the downward movement (Whittaker & Reddish, 1989). Vertical displacement may have little effect on low-rise structures, providing it is uniform, but it may have a significant effect on drainage, tall structures and roads (Box 14.3). Tilt or differential subsidence will have a significant effect on buildings — subsidence damage most commonly arises as a result of horizontal extension and compression. Figure 14.13 shows a severe ramp in a trunk road resulting from subsidence due to coal-mining. Figure 14.14 shows a railway line that was abandoned because of damage caused by subterranean salt extraction.

Ground strain is developed as tension over the crest of the subsidence wave and as compression over the trough, with horizontal displacement reaching a maximum in the wave centre (Figure 14.15). Subsidence from longwall mining occurs in the form of a wave moving parallel to and at the same rate as the line of the advancing coal face. Most of the subsidence is transmitted to the surface in a

Box 14.3. Damage effects due to subsidence

Buildings
Pulling open of joints in brickwork or fractures within masonry are common characteristics of tension. Squeezing in of voids, such as doors and windows, and the occurrence of horizontal movement along well-defined lines of thrust are characteristics of compression.

Roads, railways
Distortion of horizontal and vertical alignment, disruption of drainage, displacement of kerbs and channels.

Watercourses
Canals, drains, etc., may suffer leaks, detrimental changes in gradient and loss of lateral support/confinement.

Figure 14.13. A road affected by mining subsidence

Figure 14.14. A railway affected by salt extraction

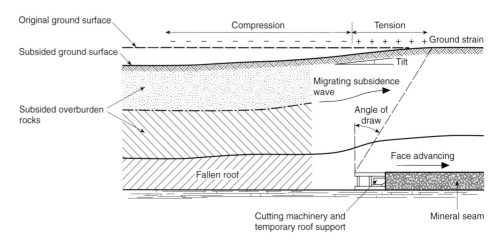

Figure 14.15. Ground motion due to subsidence

relatively short period of time, but residual subsidence may occur up to 2 years after mining has taken place. The subsidence produced is fairly regular and can be accurately predicted.

The principal methods for predicting mining subsidence can be divided into the following groups (Whittaker & Reddish, 1989):

- *Empirically derived relationships.* A number of formulae have been established from observed surface subsidence behaviour. When applied to the general region and mining conditions that correspond to those prevailing where the observations were made, such relationships can give a fairly accurate assessment of mining subsidence.
- *Influence functions.* It is assumed that there is an area of influence around a point (due to extraction of a small element within the sphere of influence of that surface point). A typical approach would be to consider annular zones around the point of interest and relate the percentage extraction in areas of each annulus to the resulting subsidence at the surface.
- *Analytical models.* Models have been derived using various representations of the ground (elastic, plastic and viscoelastic). The elastic treatment of ground movement has attracted considerable attention in the past and a marked degree of success has been achieved. To derive the relevant equations, the geometry of the problem is generally represented in a simple

two-dimensional form. The mass behavioural properties of the overburden are frequently determined by back-analysis of previous subsidence events.

The response of building structures to mining subsidence is difficult to predict with accuracy, but important factors are:

- size
- shape and orientation in relation to the workings
- the foundation design and type,
- age of the superstructure.

There are no definitive values for acceptable differential movements of buildings. However, recommended movement limits for a building 'panel' are usually in the region of:

- a maximum overall rotation of between 1/300 and 1/500 of the length of the panel
- a maximum deflection ratio (the maximum vertical displacement relative to a straight line joining the ends of a panel divided by the length of the panel) of 1/5000 for hogging and 1/2000 for sagging.

14.2.3. Mineshafts

Shafts provide vertical access to mine workings and, while their sizes and depths vary, typically they are in the region of 3–9 m diameter and 300–1000 m deep. Shafts are often circular, but can also be oval or square. Linings of timber, brick, stone or concrete have generally been provided, except in strong rock where a lining may not have been considered necessary.

Old shafts will be encountered in three conditions of abandonment: completely filled, partially filled or open. Many shafts have been only partly filled, with the fill having been placed on a staging (usually wood) within the shaft. The level for the staging varies, but it is often found just below ground level or in the vicinity of rock head. It seems that in many shafts large objects, such as mine cars or trees, were thrown in to form an obstruction on which to start filling, and these may have bridged a shaft so that its lower part is completely open. Until recently, it was not standard practice to close off galleries and drifts. Hence fill could migrate from the shaft, thereby creating voids within the shaft. Figure 14.16 shows a hole that developed in a car park due to a loss of infilling from an old mineshaft. Typical fill materials included colliery spoil, building waste, general debris, old iron castings, pit wood, etc. Obviously this type of fill was not compacted when placed, and any consolidation was due to self-weight alone.

Figure 14.16. Void resulting from the loss of mineshaft infill

Engineered infilling of shafts utilizes good-quality, free-draining, coarse granular material to achieve a dense stable infill.

When abandoned shafts collapse there is likely to be loss of support from a large area of the surface. Where rock head is near the surface and the collapse arises from the disintegration of staging, which supports infill, then the subsidence will be severe but localized. In the case of a breakdown of the shaft lining, or where a thick deposit of unconsolidated material overlies the shaft, a 'tulip' failure may occur (i.e. severe surface cratering, wherein soil surrounding the shaft rushes into it to form a depression several times the shaft diameter). In April 1945 an old, unfilled shaft suddenly collapsed and cratered in a railway sidings at Wigan and a whole train of 13 laden wagons and locomotive promptly disappeared down the shaft (Whittaker & Reddish, 1989). The size of the crater depends on the diameter of the shaft, the depth to the bedrock and the angle of repose of the surrounding material. Causes of shaft collapse are:

- deterioration and breakdown of the shaft lining
- settlement or degradation of the fill material within the shaft
- the collapse of staging (as the wood decomposes) within a partially filled shaft.

In some cases collapse may occur slowly, leading to a gradual loss of support and worsening damage to nearby structures.

14.2.4. Gas and water effects

Decades of pumping will have artificially lowered water-tables in the vicinity of mineworkings, and after abandonment it may be many years before the natural water level of the area is restored and porewater pressure stabilizes within the surrounding rock. Figure 14.17 shows water flowing over the floor of a factory that had been built over a capped mineshaft. After the cessation of mining in the area the groundwater level rose, progressively each year, until it was above ground level at this location. In wet mines water will flood the workings and shafts, and any fill material will become saturated. This is likely to cause fill in shafts to behave as a column of dense fluid, which will flow into the old workings if they have not been thoroughly filled or plugged (see Chapter 7, Section 7.1). If a platform has been erected across a shaft and fill has been placed on top of the platform subsequent flooding can undermine the platform and result in the entire column of fill falling to the shaft bottom.

Mining also causes serious problems because the extracted waste materials or overburden are often rich in toxic minerals. Acid mine drainage (i.e. flow of acidic groundwater into streams) was first recognized as a problem in

Figure 14.17. Flooding due to rising groundwater travelling up an abandoned mineshaft

underground coal mines. Many coal seams are bounded above and below by shales that are rich in pyrites. Groundwater entering such a mine, through rock fractures and mined-out areas, reacts with the pyrites and oxygen in the air to form sulphuric acid. The increasing acidity of the groundwater enables it to leach metals from rocks, so the acid drainage may become enriched in metals as it migrates. When tin production ceased in 1991 at the Wheal Jane mine (Cornwall), groundwater pumping was also stopped. Consequently, the water-table began to rise back to its original level, and eventually the water head had risen sufficiently to burst a plug in an underground adit. This caused a flood of over 55 Ml of water contaminated with iron, zinc and cadmium to sweep into the Carnon River, Falmouth Bay and the sea, causing a major pollution alert. When mine drainage water reaches external streams, dissolved iron becomes exposed to oxygen and turns formerly clear streamwater a rusty or bright orange colour.

The most common gas encountered in mines and mining areas is methane (fire damp). Carbon dioxide (black damp), resulting from the oxidation of methane, may also be present in coal mine workings. Geological methane is usually associated with coal-bearing carboniferous strata, and is produced by the anaerobic decomposition of ancient vegetation trapped within the rock. Methane is a buoyant gas, having a density about two-thirds that of air, and will migrate from a site as a result of diffusion and flow due to pressure differences — fissures, voids, pipelines and tunnels provide ideal pathways for the gas. Increasing groundwater levels and flooding of underground workings after abandonment can result in the release or expulsion of methane from old workings because of a pumping action by the moving water. In 1995 the mining town of Arkwright (near Chesterfield) was evacuated due to methane seepage, from coal mines, into houses (*Construction News*, 1996). More than 400 residents were moved to a new neighbouring development.

14.2.5. Investigation

A ground investigation will be required to obtain detailed information on:

- ground strata and relevant geotechnical properties
- the presence, or otherwise, of shallow workings and their condition
- the location and condition of shafts or adits
- the groundwater regime and how it is changing with time.

Non-invasive geophysical survey techniques would appear to offer a good means to conduct searches for old shallow mine workings and buried shafts. However, successful application of these methods requires that a distinct physical contrast exists between the target and the overlying or adjoining materials, and unfortunately there is usually a lot of background 'interference' on derelict urban sites:

- Ground resistivity surveys are believed by many investigators to be ambiguous and extremely difficult to interpret, and to have an unacceptably low recognition rate.
- Ground-probing radar suffers from a limitation to the possible depth of penetration (only a few metres, especially in wet clays).
- Gravity and microgravity surveys are of no value, as the effects of adjacent structures and topography mask the very small anomaly of a shaft.
- Seismic surveys fail as the target is too small.
- Magnetic surveys (comprising a walk-over survey of the site with a hand-held proton magnetometer) have been recommended as the most useful geophysical survey technique (Waltham, 1989). The magnetic anomalies resulting from a subterranean void are much larger than the void itself so that a clearly recognizable pattern is observed.

Figure 14.18. Drill rods that were bent when trying to penetrate old shaft infill

Direct investigation methods (trial pits and boreholes) are more expensive than indirect methods, but generally provide more reliable information. The use of trenches and trial pits is limited to searches for very shallow features. Old workings are normally located by drilling, when penetration rates, fluid loss and the rock quality (on cored holes) provide the most useful indications. The recommended minimum hole depth is 10 times the seam thickness or mine height. A typical specification for a building development in an area of shallow workings would be a hole depth of 30 m and a staggered grid spacing of 5 m or less.

Old mine shafts should be drilled to the bottom in order to determine the quality and extent of the filling employed. This can prove very difficult or even impossible because of the nature of the fill within a shaft. Metal and timber are extremely difficult materials to penetrate, and the loose nature of the fill means that hard materials rotate with the drill bit. Figure 14.18 shows drill rods that were bent (during drilling operations when they were confined within a 300 mm diameter casing) while attempting to pass through old fill in a mineshaft. Very great care is required when investigating concealed shafts, as these can be subject to sudden collapse when the fill or support staging is disturbed. Equipment needs to be founded on wide support beams or rafts (Figure 14.19). All personnel involved in the investigation should be equipped with harnesses and safety lines secured to anchor points outside the collapse zone (assumed to be a cone projecting upwards with sides at an angle of 45° from the point where the walls of the shaft intersect rock head). Closed circuit down-the-hole television cameras can be used within boreholes to check the condition of underground voids.

14.2.6. Remediation

The redevelopment of a site over shallow workings (less than about 30 m deep) generally requires some form of treatment to consolidate the workings followed by the use of appropriate foundations to support structures. The treatment is generally intended to fill open or partly filled underground cavities and to prevent upwards migration of voids. The main treatment options are:

- excavation and backfilling with material compacted in layers (only suitable for very shallow depths)
- partial grouting (to reduce the risk of void migration)
- full grouting (to improve the overall ground-bearing characteristics).

The grout typically consists of a mix of cement, sand and PFA (with proportions in the region of 1 : 1.5 : 12–20, respectively), injected through boreholes 50–75 mm in diameter. For collapsed workings, the void locations are difficult to predict and grouting is first carried out from primary holes (typically at about 6 m centres) and secondary infill holes may be drilled if necessary (at

Figure 14.19. Drilling to define the position and condition of an old mineshaft

about 3 m centres). Redrilling and further grouting will be necessary in areas where the specified pressure cannot be reached or the grout take is excessive.

The methods commonly used to treat old shafts are outlined in Box 14.4. Treatment selection depends on the:

- depth to rock head
- shaft size

Box 14.4. Methods for treating old shafts

Capping using reinforced concrete
The cap should ideally be located at or below rockhead, with the cap width being at least twice the diameter of the shaft.

Plugging
The plug would normally be installed at rock head and the rock should be trimmed back to provide a stable keyed surface.

Filling and plugging with grout
The shaft is filled with granular material and then grouted from it, top down to just below rock head.

Drilling and grouting
The existing shaft filling should be investigated by drilling prior to grouting operations.

Steel or concrete covers
These are appropriate for existing shafts which are to remain open and where bed rock is close to the surface, or where the existing lining is in a stable condition.

Filling and security fencing
The shaft may not be properly stabilized by filling only. However, this form of treatment is acceptable in open country away from populated areas.

Security fencing and warning signs
This is only acceptable as a short-term safety measure.

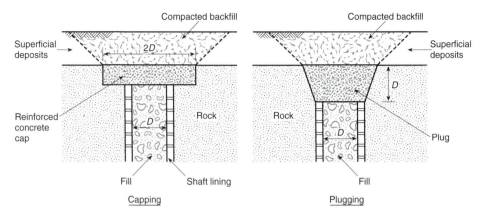

Figure 14.20. Capping and plugging of shafts

- condition of lining
- presence of infill within the shaft
- proximity to existing structures and urban areas
- need for future access
- available funding.

Capping with reinforced concrete is the most widely used solution (Figure 14.20). If gas could accumulate within the shaft, then venting should also be provided (Figure 14.21). Where the shaft cannot be capped at bed rock (due to excessive depth or limited access) it is acceptable to cap the shaft in competent ground at a higher level, provided the existing shaft lining below cap level is in a stable condition. If the superficial deposits are insufficiently consolidated there is a possibility that collapse may occur due to gradual deterioration of the ground because of lack of lateral restraint within the shaft or as a result of imposed loading due to new surface structures. Plugging, using reinforced concrete or clay, may be appropriate for small shafts (up to about 3 m diameter) where access is limited (see Figure 14.20). Clay plugs are often more economic than an equivalent mass concrete alternative.

When the depth to rock head is excessive, or when access is limited, shaft filling and grouting is an appropriate stabilization method. Free-draining single-

Figure 14.21. A gas vent through a cap being formed over an old shaft

size fill materials (such as pea gravel) should be used to allow subsequent penetration by grout. For shafts that are already filled, unsuitable fill should be removed down to the rock head and replaced with granular material prior to grouting. Even where an old abandoned shaft has been proved to be filled and the fill to be well consolidated, it is prudent to employ an amply large reinforced concrete capping before building over, or near to, it.

Adits are normally treated by stopping (blocking using brickwork, concrete or grouted fill, with filling and/or grouting behind). The fill may be progressively grouted either from within the adit or from the surface. Provision should be made for drainage or gas venting as necessary. Shallow lengths of adit (less than 5 m below ground level) may be treated by excavation and backfilling. Drilling and grouting may be used to stabilize lengths of open or collapsed adits in situations where access is difficult or dangerous (e.g. due to the proximity of nearby structures).

14.3. Filled ground

In urban areas the construction of buildings on fill or made-up ground has become increasingly common. A wide variety of materials (domestic refuse, ashes, slag, clinker, building waste, chemical waste, quarry waste, various soils, etc.) has been deposited over natural ground to infill topographical depressions and/or to raise the general ground level at sites. In urban areas it is not unusual to find:

- shallow fills where ponds and marshes have been reclaimed
- new buildings that have been erected on the rubble of older demolished buildings
- old basements (both filled and empty)
- disused clay pits and quarries that have been used for refuse disposal
- redevelopment of docklands
- infilled opencast mining areas.

The infilling processes may have been carefully controlled construction or it could have been simple dumping of a variety of materials. The extent to which an old fill area will be stable or suitable as a foundation depends on its composition, uniformity and age. These factors give some indication of the compressibility of the fill (both under its own weight and under superimposed loads) and its amenability to improvement.

14.3.1. Opencast mining

Opencast coal is relatively cheap to mine, and of a quality which is needed to blend with much deep-mined coal that would otherwise be unusable in many key markets. It is the scale of the operation that distinguishes opencast sites from other civil engineering schemes, and in the last 30 years very large open-pit mining operations have been developed (Figure 14.22). These operations, which have been made possible by the development of very large and efficient earth- and rock-moving equipment, produce huge volumes of spoil and filled ground.

The working of an opencast site begins with the stripping of topsoil and subsoil. On large sites the stripping is undertaken progressively in advance of mineral extraction. Scrapers are generally used for this operation, but considerable variation may occur in the methods of working opencast sites and in the type and sizes of machinery used. With relatively shallow seams it is usual to open an initial cut, usually along the outcrop, with the overburden going to dump. Work proceeds with successive parallel cuts, the overburden from each new cut being cast into the void of the previously completed cut. Because the work involves continuous shallow excavation over a long period of time, environmental noise pollution can limit working procedures (see Chapter 17, Section 17.2.2). On a large site with no

Figure 14.22. Opencast strip mining of coal

really hard rock, dragline excavation may be undertaken, but with multiseam sites face shovels and dump trucks are more usual.

In contrast to the creation of derelict land by the deep mining of coal, surface or opencast mining only leads to temporary land disturbance, since currently there is an obligation on operators to restore the land when mining ceases. Wherever possible, restoration is progressive, with the land being returned to agricultural use or being designated as public open space. However, the soil is substantially disturbed in the process, and the restored land requires prolonged and careful treatment to bring it back to its former state (older sites are unlikely to have been restored in a systematic way). Traditionally, backfill placement was simply carried out by the machinery used for the excavation of the overburden, e.g. draglines casting overburden directly into the dump area, truck-transported spoil was end tipped, scraper-transported spoil being placed in layers at various levels. The result of this approach is large volumes of backfill having varying degrees of compaction (Hills & Denby, 1996). Opencast extraction of minerals other than coal is undertaken to great depths, and has led to the creation of large craters and high heaps of soil and rock overburden and waste materials. For instance, the Bingham Canyon Copper Mine in Utah covers an area of over $7 \, km^2$ and extends to a depth of more than $700 \, m$ (Pickering & Owen, 1994).

The creation of a large, below-ground void can have a significant effect on the local groundwater regime. Drawdown associated with opencast workings may affect local ponds, streams and wetlands. On the other hand, the voids may fill with water to form lakes and meres. Parts of quarries and old workings that are left for some years may achieve Site of Special Scientific Interest (SSSI) status or similar, and particular parts of the site may require protection during any subsequent restoration.

14.3.2. Geotechnical problems

The construction of roads, housing or light structural development on deep fill involves several geotechnical problems, but the major one is that of settlement. Opencast backfills are typically prone to settlement by two mechanisms: creep and collapse. Following placement of a body of opencast backfill, significant long-term settlement is observed. This occurs under conditions of constant stress and moisture content. It can be assumed that the backfilled mass has sufficiently high permeability to prevent the build-up of porewater pressures, and thus the long-term settlements are a result of creep and not due to dissipation of porewater pressures. The mechanism for creep settlement is one of gradual rearrangement of the material fragments, due to crushing of highly stressed contact points, which results in a reduction in the voids ratio.

Table 14.1. Settlement of fills

Material	Settlement (% depth of fill)
Well-compacted shale and rock fills	0.5 (from self-weight)
Lightly compacted clay placed in deep layers	1–2 (from self-weight)
Nominally compacted opencast backfill	1.5 (from self-weight)
Opencast backfill compacted by scrapers	0.6–0.8 (from self-weight)
Backfill from opencast mining of ironstone: Untreated	1.5–3.0 (from self-weight) 1.0–2.5 (houses founded on the fill) 1.6–4.8 (due to inundation)

After ground treatment
    surcharge
    dynamic compaction
    inundation
	0.3–1.3 (houses founded on the fill)
	1.3–3.8 (houses founded on the fill)
	0.4–4.8 (houses founded on the fill)

Compacted mixed refuse	15–30 (from self-weight and decomposition)
Refuse placed in layers and well compacted	10–20 (from self-weight and decomposition)
Backfill of opencast site with sandstone and mudstone fragments	1–2 (collapse settlement due to saturation)
Non-engineered colliery spoil fill	7 (collapse settlement due to saturation)

The behaviour of waste fill is not, in general, amenable to quantification by normal soil mechanics theory, and thus the prediction of the settlement of a fill has a high margin of uncertainty (possibly of the order of $\pm 50\%$). However, an initial appraisal of likely settlement may be made by characterizing the fill material and using past experience of the behaviour of similar materials (e.g. from the data contained in Table 14.1).

Collapse settlement occurs when the fill is inundated by ground- or surface-water or when the surface load is increased (Figure 14.23). In the case of opencast backfill this effect can occur at depth due to cessation of pumping operations, or close to ground level as surface water penetrates the fill. For areas of old urban fill this type of settlement may occur when redevelopment of the site is commenced.

Restored excavations may represent a localized increase or decrease in the permeability of the ground. Bulking of excavated spoil, which is subsequently used as infill, may also increase the porosity of the ground. This local increase in

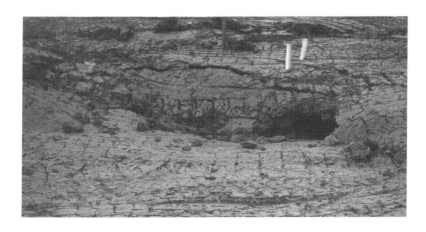

Figure 14.23. A void formed by the collapse settlement of underlying loose backfill

Table 14.2. Coefficient of secondary compression for fills

Material	C_α (= strain/log time)
Uncompacted opencast mining backfill	0.005–0.01
Landfill refuse	0.01–0.5
Heavily compacted sandstone/mudstone backfill	0.002

water storage will lower the groundwater table, on a temporary or a permanent basis. Argillaceous material will degrade with time to form a clay matrix, so that both the porosity and permeability of the fill reduce. Such a reduction in permeability can have significant effects on regional groundwater drainage, as the restored site is likely to act as an aquiclude.

With ground that contains landfilled refuse the worst effects result from the loose nature of the fill and the presence of matter which will decompose. The major geotechnical problem is likely to be associated with long-term settlement (occurring over decades) of the waste fill. Construction on old landfill sites also poses problems of gas and odour control and the durability of construction materials in the contaminated soil environment. Minor geotechnical aspects include bearing capacity (ultimate bearing capacities of around $200 \, kN/m^2$ can be achieved) and slope stability at the edge of the deposit (slopes of 1 vertical to 3 or 4 horizontal seem satisfactory). Even after stabilization treated refuse is only suitable for supporting light structures (one or two storeys). The ongoing long-term settlement due to decomposition cannot be totally eliminated by stabilization. For many waste fills it has been found that creep compression shows an essentially linear relationship with the logarithm of time that has elapsed since the waste was deposited (see Chapter 12, Section 12.2.4). Values of C_α (the coefficient of secondary compression; see Chapter 8, Section 8.4.3) for sites where full compaction has been carried out are typically in the range 0.001–0.003, whereas values in excess of 0.01 are likely for uncompacted fills (Hills & Denby, 1996). Typical values of C_α are given in Table 14.2.

14.3.3. Remediation

The most profitable form of reclamation and restoration of land from which mineral has been excavated is that which can be achieved following infilling of the excavation with imported wastes. The overburden and inherent quarry wastes can then be used as cover and capping materials. A pit backfilled to the original topographic levels, or deliberately contoured to another landform, may then be the site of a wide range of after-uses, including:

- leisure activities (sports grounds, golf courses)
- parkland
- urban development (for both domestic housing and commercial use)
- agriculture
- forestry.

There are three general ways of dealing with ground that has been backfilled in a non-engineered way, in order to eliminate settlement problems:

- excavate the existing fill and fill the void with suitable material placed in a controlled manner
- treat the fill to improve its stability and decrease its compressibility, as described in Sections 14.1.3 to 14.1.6
- pile through the fill so that surface loads are carried down to a stronger stratum.

The depth of the fill significantly influences the means of remediating a site. For shallow fills (up to 4 m deep approximately) remedial measures involving earth-moving may be the most economic, although large-scale removal is not desirable because of the inherent environmental problems. For deep fills (greater than about 15 m) the site treatment is dependent on the depth of the zone affected by the load from the proposed development. Hence high-rise buildings and heavy structures will need conventional piled foundations, but low-rise structures will require only a relatively shallow treatment of the fill.

A major concern regarding the erection of buildings on fill is gaseous emissions from the ground. Vibrocompaction and piling may provide pathways for the migration of gases or contaminated liquids. Dynamic compaction and surcharging may force gas out of voids within the fill. Methods that involve penetration of the ground could promote decomposition of the fill. Measures that can be adopted to ensure that gases do not accumulate in buildings include (Figure 14.24):

- Designing the foundation to avoid unventilated spaces within or beneath the structures and leaving an air space between the floor and the top of a granular layer overlying the fill. The air space is vented through screened openings on all sides.
- Providing ventilation for both below-ground and above-ground parts of the buildings, with monitoring alarm devices to respond to any potentially hazardous build up of flammable gases.

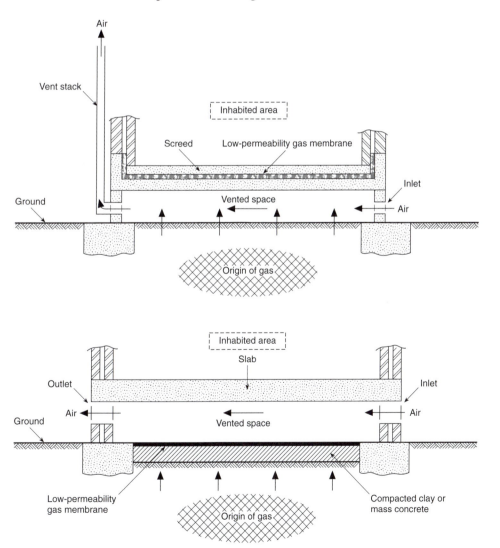

Figure 14.24. Gas protection systems for property

- Including a gas protection scheme, beneath the building, consisting of an impermeable sheet over a high-permeability layer. The gas can be vented to the atmosphere by means of a granular trench around the perimeter of the house, with regularly spaced pipes or airbricks providing a pathway through foundation walls. Alternatively, slotted gas-collection pipes can be inserted within the granular layer, and the pipes connected to a vertical riser.

14.4. Colliery spoil heaps

The creation of deep mines has led to the production of much rock waste from the sinking of shafts and the digging of underground passages. The waste brought to the surface has usually been dumped near to the pit head. With time, high heaps of waste material, which occupy large areas of land (often close to inhabited buildings), have been formed (Figure 14.25). Other industries, such as quarrying and china clay production, also produce large quantities of waste in their locality. However, tips from the latter industries do not present the same threat as colliery spoil heaps because:

- there are a lot less of them
- the dumps are usually relatively permeable
- they are not adjacent to residential areas.

14.4.1. Spoil heap formation

Spoil heaps are composed mainly of the rock types found adjacent to the mineral seams. For coal-mining this means shale and, to a lesser extent, sandstone, together with some coal that was not separated during processing (see Chapter 16, Section 16.2.1). The majority of tip material is essentially granular. Often most of it falls within the sand range, but significant proportions of the gravel and cobble ranges may also be present. Apart from those changes brought about by combustion, alteration of the material stored in the waste heaps will generally be quite small, and effects due to physical and chemical weathering processes will be restricted to the top metre. Physical weathering will be most effective on shales, which readily break down and release clay- and silt-sized material. There is considerable heterogeneity within colliery spoil heaps, as well as differences between them.

Waste materials were commonly deposited by discharge from a conveyor-belt system or by placement in layers. With the former method the tip slopes formed at the angle of repose for loose materials. Figure 14.26 illustrates the steep slopes achievable by this method. Since the material was placed dry it formed loose layers, and when it became wet slumping was likely to occur. Stratification also developed in the older tips, due to trafficking of the spoil during the winter months, leading to perched water-tables.

Modern practice is to place material in layers about 300 mm thick and compact it using either the construction traffic or rollers. Tips of this type are not liable to

Figure 14.25. A large spoil heap close to inhabited properties

Figure 14.26. A typical colliery spoil heap

spontaneous combustion. The moisture content of coal-mining waste increases with increasing content of fines, and generally falls within the range 5–15%. The water-table within such tips is generally low or absent, except where a tip has been built over springs or where surface water is allowed to flow into the tip. Nowadays tips are formed with slope angles less than the angle of shearing resistance of the tip material — coarse discards are cohesionless, with ϕ' usually varying between 25° and 45°. However, they may degrade, due to weathering, to a cohesive material with a plasticity index of around 15 and a residual ϕ' of 15–20°.

14.4.2. Stability considerations

Slope failures in spoil tips and lagoon banks can occur in the same way as for soil materials. Factors that reduce stability are the same as those for conventional slopes: additional loading at the top of the slope, increased steepness or height of the slope, changes in the water level or pore pressures, and sudden disturbances such as vibration or slippage of an adjacent part of the tip. Most failures in spoil heaps have been associated with high moisture contents or water seepage. The effects of such failures have been compounded by the looseness (high porosity) of the older tips, leading to the formation of flow slides, wherein the spoil mass is transformed into a liquefied state. Following disturbance of the material, due to rotational failure, mining subsidence or vibrations (from earthquakes or blasting), collapse of the loose soil structure takes place. However, closer packing of the grains is not possible immediately because of the water in the voids. Hence the effective stress falls to around zero and liquefaction occurs, thereby allowing the spoil to flow downhill as a slurry. Flow slides typically occur in saturated deposits of sand- or silt-sized material, and are a common failure mode in lagoon material following breaching of the lagoon bank. Once failure has been initiated flow slides develop rapidly and the liquefied material can travel very large distances (see Chapter 15, Table 15.1).

The problem of colliery tip instability was brought into sharp focus by the disaster at Aberfan in 1966, when a slide involving some 100 000 m^3 of colliery waste resulted in the death of 144 people (116 of whom were children). When the major slide occurred the height of the tip was approximately 66 m. The flowslide travelled down a 12.5° slope for a distance of about 480 m to the junior school, which it largely destroyed. Data for the discards showed that both the initial density and the volume decrease on subsequent saturation were very sensitive to placement moisture content. At Aberfan the material of the tip was loose colliery discards, tipped over the face of the spoil heap. The hydrogeology of the site was the triggering mechanism for the failure — after heavy rainfall artesian pore pressure developed in the sandstone beneath a relatively impermeable cover of boulder clay (Bishop, 1973).

Flowslides resulting from waste heap failure are not confined to tips of colliery spoil. Notable flowslides have also occurred within limestone waste, China clay spoil, fly ash, and gold and copper tailings. In all cases the general causes have been material dumped in a loose state (to come to rest at its angle of repose), which has subsequently been rapidly saturated. After the Aberfan disaster the stability of numerous waste tips was investigated and remedial works were undertaken as necessary. Works that may be used to increase or restore stability include:

- removal of soil from the upper part of a slope by constructing a berm (a horizontal ledge) part way up a slope, to reduce the overall disturbing moment
- reduction of the slope height and/or gradient, to reduce the disturbing moment
- maintenance of low groundwater by effective drainage, to maximize shearing resistance
- placing free-draining fill on the toe of the slope, to reduce the overall disturbing moment and increase the shearing resistance.

It is unlikely that new tips, where modern engineering and management practices have been introduced, will become unstable. Typical minimum factors of safety for the design of tip slopes are:

- 1.3–1.5 if using peak shear strength parameters
- 1.2 or 1.3 if using residual strength parameters
- 1.1 or 1.2 if using appropriate strength, with seismic effects being taken into account.

Higher or lower values may be appropriate according to the adequacy and reliability of the input data and design assumptions.

14.4.3. Spoil heap restoration

Colliery spoil heaps form prominent and obtrusive features and one of the objectives of reclamation is to make them blend in with their surroundings. Most reclamation schemes require an element of revegetation, even though in some cases this may be relatively limited (such as low-maintenance grass cover on sports fields, golf courses and open spaces). In other cases (such as agriculture, forestry and nature reserves) there is a need to provide a much better medium for plant-growth in order to maximize productivity or to foster plant growth characteristics for specific objectives. Detailed information on this subject can be found in Moffatt & McNeill (1994).

Table 14.3. Possible grass-based seed mixtures (%) for site revegetation*

Species	Site prone to waterlogging	Site prone to drought	
		Acid conditions	Alkaline conditions
Timothy	10		
Smooth meadow grass	15		20
Rough meadow grass	10		
Flattened meadow grass	5	30	10
Creeping red fescue	40		55
Sheep's fescue		30	
Browntop bentgrass	5	10	
Creeping bentgrass	10		10
Wavy hair grass		25	
White clover	5	2.5	2.5
Birdsfoot trefoil		2.5	2.5

* From Coppin & Richards (1990).

Figure 14.27. Lack of plant growth on a spoil heap

Spoil heaps are often slow to revegetate naturally. The bare surfaces dry out in the summer and water flows rapidly down them in winter. The essence of the problem is that the spoil weathers very slowly, so there is a limited supply of nutrients and plants are dependent on leaf litter being imported from elsewhere. However, given enough time and the absence of interference by man, the progress of succession from bare ground to vegetative cover is: herbs and grasses, then scrub, then trees. For Britain the 'theoretical' climax of potential natural vegetation is woodland (mainly oak, with some beech, birch and pine). On a lot of tips a grass sward can be established by following normal agricultural practices:

- cultivation of the ground with shallow tines to create a seedbed
- application of nitrogen and phosphorus fertilizers to the 'soil'
- harrowing to mix the fertilizers and the 'soil'
- spreading an appropriate seed mixture over the ground
- rolling the ground to provide a firm bed for the roots.

Table 14.4. Matters for inclusion in a specification for revegetation works

Aspect	Component	Specific items
Earthworks	Soil handling	Methods, working season, moisture effects
	Compaction	Methods, density and aeration requirements in upper layer
	Ground preparation	Site clearance (removal of weeds and excessive stone material, etc.), cultivation depth and intensity, timing, fertilizing, manures and composts
Seeding	Seeds	Species, mixes
	Method	Method of sowing (drilling, hydroseeding, etc.), storage of seeds
	Timing	Appropriate sowing season, suitable climatic conditions
	After-care	Fertilizing, weed control, mowing, assessment of performance, additional seeding if necessary, long-term management
Planting	Plants	Species, size, method of growing, pretreatment
	Method	Method of planting, provision of support, storage of plants, provision of ground cover, watering
	Timing	Appropriate planting season
	After-care	Fertilizing, weed control, pruning, thinning, coppicing/trimming, assessment of performance, additional planting if necessary, long-term management

The seed mixture is applied at a rate of around 60 kg/ha and usually contains perennial ryegrass, one or more varieties of clover and other grasses such as fescue, timothy and cocksfoot. Suitable grass seed mixtures are given in Table 14.3. The establishment of vegetative cover is usually successful, except on areas affected by acid production where pH values may fall to 4 or less. Conversely, the presence of bare patches on a revegetated site is usually a good indication of acid conditions (Figure 14.27). The specification for works associated with the establishment of vegetation should identify:

- materials to be employed
- performance targets
- means of assessing performance of the works
- special requirements of seed and plant materials that may affect construction operations.

Table 14.4 contains some items that should be considered for inclusion in the specification for revegetation works.

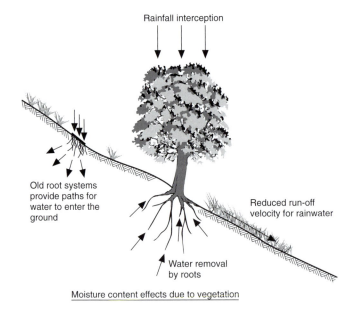

Figure 14.28. Slope stability aspects of vegetation

Box 14.5. Vegetation and the ground moisture regime

Rainfall interception
The volume of rainfall reaching the ground, and the rate at which it does so, is reduced so that saturation of the upper soil layer is likely to be prevented.

Surface water run-off
Vegetation reduces the velocity of flow so that the potential for surface erosion is less.

Infiltration
Vegetation increases the permeability and infiltration of the upper soil because of the voids formed by roots and the loosening of the surface soil that is caused by their growth. Potentially these effects could lead to an increase of the moisture content of the upper soil, but the effects are usually offset by increases in interception, transpiration and the ability to use steeper slopes.

Soil-moisture depletion
Plant roots extract water from the soil and the effect can extend well beyond the physical extent of the roots (zones of greatest moisture deficiency of 2–4 m below the ground surface are quoted by Coppin & Richards (1990)).

Adapted from Coppin & Richards (1990).

Apart from improving the appearance of a spoil heap vegetation also affects the stability of the slope by influencing the moisture regime of the upper layer of soil and by providing mechanical enhancement of the strength of the soil (Figure 14.28). The effects on the moisture regime and the modifications to the mechanical properties of the soil are summarized in Boxes 14.5 and 14.6, respectively.

Slate spoil heaps are particularly difficult to develop as landscaping projects. Because slate is durable and impervious to water, slate rubble is free-draining and will not retain the moisture that could support plant life. Even so, some natural colonization occurs on more sheltered slate waste tips. The usual cycle of colonization begins with saxifrage and foxgloves, followed by broom and heathers. Then comes birch and goat willow. It is likely to take hundreds of years for the climax species of oak to become established on slate spoil heaps.

Box 14.6. Effects of vegetation on the mechanical properties of the ground

Root reinforcement
The roots (which have relatively high tensile strength) and the soil form a composite material akin to reinforced soil (see Chapter 10, Section 10.3.1). The soil friction angle is unchanged but there is an apparent increase in cohesion of around $3–10 \, kN/m^2$ (Coppin & Richards, 1990), which can extend for several metres in depth in the case of tree roots. Thus the presence of vegetation, and its type, can be crucial in determining the surface stability of both natural and man-made slopes (see Chapter 9, Section 9.2.2).

Soil anchorage
The tap roots of many trees penetrate to great depths and thus into soil layers that are considerably stronger than the surface material. The trunks and principal roots then act as individual cantilever pile walls, providing a buttressing effect to up-slope soil masses. If the tress are sufficiently closely spaced, arching may occur between the up-slope 'buttresses'.

Surcharging
This effect is only relevant to trees because the weight of grass and shrub cover is insignificant. On a slope the surcharge can increase the effective stress and hence the frictional shearing resistance. Trees apply a surcharge, which also provides a down the slope force to destabilize or stabilize a slope, according to the type of failure surface considered and the actual position of the trees. The overall effect is beneficial, particularly if the trees are growing over the bottom portion of a slope.

Adapted from Coppin & Richards (1990).

14.5. Sewage sludge

In February 1992 the sludge lagoon adjacent to Deighton sewage works suddenly failed, engulfing it in 200 000 m³ of sewage sludge. A rotational shear prompted liquefaction of the sludge and the consequent major movement. The slide destroyed trees and retaining bunds, blocked a 30 m wide river, smashed through the walls of the reinforced concrete settlement tanks and destroyed a pump house. Fortunately, no one was injured. Sludge had been dumped on the site since 1906, up to a height of 25 m (Claydon *et al.*, 1997). The investigation of the failure revealed that the sludge was still digesting and a build up of gas and water pressure within the tip led to failure of a granular ash bund, which had been tipped around the edge of the sludge to retain it. Over the years high-density sludge had been dumped onto a horizontal waste layer with high moisture content and low shear strength. There was a build up of methane gas because it was trapped by a low-permeability 'filter cake' on the enclosing ash. The major restoration programme essentially involved scooping up the waste and dumping it behind a purpose-built, fully engineered rock dam.

14.5.1. Sludge production

Sewage generally consists of domestic wastewater and industrial (trade) effluents and contains approximately 1000 mg/l of impurities, of which about 67–75% is organic and degradable material. Sewage treatment usually involves the following sequence of processes:

- preliminary screening to remove large suspended and floating material
- primary sedimentation to separate out solids
- secondary, biological treatment of the sludge (digestion), commonly associated with an anaerobic stage in a digester
- aerobic digestion stage in settlement tanks and/or lagoons.

Sewage sludge contains a variety of pathogens, including bacteria, viruses, parasites, and fungi, reflecting the presence of the agents in the human and animal population contributing to the sewage. Consequently, further treatment may by required prior to ultimate disposal.

After purification the sewage sludge has a dry solids content of about 4–5% (Gay, 1991). An important part of the treatment of sewage sludge prior to ultimate disposal is thus the reduction of the water content (dewatering) to produced 'thickened sludge'. This facilitates handling and reduces transport costs, since the volume of sludge is much less. Traditionally, drying or disposal involved dumping the sludge in thin layers in open beds or in deep lagoons. Storage is a common feature of normal disposal operations, in which sludge is held in lagoons or in stockpiles until it can be transported away from the works. Nowadays sewage sludge is normally treated to reduce its water content by methods such as vacuuming, belt press filtration and centrifuging. A sludge cake containing 30–40% solids can be achieved by these methods. However, the gel-like nature of sewage sludge often makes tipping it as a thickened slurry more practical than dewatering it. The decomposition of organic matter in dumped sewage sludge occurs over several years, the major part taking place in the first 8–10 years after dumping (compare with domestic and commercial wastes in landfills; see Chapter 12, Section 12.2.1).

In Britain about 1 Mt (dry weight) of sewage sludge ('biosolids') arises each year. Prior to 1998, British sewage sludge was disposed of in the ways indicated in Table 14.5. The sewage production rates for some other countries in the early 1990s are given in Table 14.6. The potentially polluted nature of sewage sludges is indicated in Table 14.7. The organic substances present include carbohydrates, lignins (complex compounds of carbon, hydrogen and oxygen), fats, soaps and synthetic detergents, and proteins. The solid portion contains paper fibres, fine

Table 14.5. Disposal outlets for sewage sludge in Britain in 1998

Outlet	%	Comments
Spread on farmland	51	Sewage sludge has only a modest fertilizer content. If all the sludge in Britain was used for its nitrogen and phosphate content only about 5% and 10%, respectively, of national needs would be supplied
Dumped at sea	22	Sludge dumping at sea ceased in 1998 so that other forms of disposal, such as incineration, have subsequently become more important
Placed in landfill	11	The landfill option is limited because of the lack of availability of local sites and because some sludges are heavily contaminated. There is a limit to the quantity of sludge that is acceptable at any particular landfill due to difficulties in handling such material as well as the problems of odour
Incinerated	7	This produces maximum volume reduction. The ash produced by incineration represents about 30% of the original dry mass of the sewage sludge or 1–2% of the original sludge volume as produced at the sewage works. The ash is likely to go to landfill since sludges, especially sludges from heavily industrialized areas, often contain heavy metals and industrial chemicals (Lutzic et al., 1995).
Other	9	

Table 14.6. The annual sewage production of various countries*

Country	Sludge production (×10³ t)	Disposal method (×10³ t)			
		Landfill	Agriculture	Incineration	Other
USA	7000	1100	2900	1900	1100
Germany	2700	1600	800	300	0
Japan	1370	400	150	800	20
France	850	400	400	0	50
Netherlands	320	170	140	10	0
Switzerland	170	0	120	50	0
Denmark	130	40	50	40	0

* After Hall (1995) and Agg et al. (1992).

Table 14.7. Contaminants in sewage sludge*

Contaminant	Concentration (mg/kg)	
	Range	Median
Cadmium	0.4–3300	~20
Chromium	10–30000	~700
Copper	50–15000	~700
Iron	15–20000	~550
Lead	80–30000	~500
Zinc	50–30000	~2000
Benzene	≤600	—
Toluene	≤130	—
Xylene	≤1000	—
Phenol	≤100	—

* Sources: Campbell et al. (1988), Mott & Romanow (1992), Sincero (1996).

soil particles, glass and metal fragments. Generally, the more industrialized a community is, the greater the probability that heavy metals and persistent organic contaminants will be present in significant quantities.

Other principal sources of high-water-content sludges are:

- Water-treatment works: These sludges (approximately 20 000 t/year) are characterized by a low suspended solids concentration, a high resistance to mechanical or gravity dewatering, difficulties with dewatering, and problems associated with their handling and ultimate disposal (Knocke & Wakeland, 1983).
- Industrial manufacturing processes: Industrial sludges are difficult to quantify, but almost all go to landfill (either as co-disposal with other general wastes or in special or toxic waste sites). The paper industry alone generates approximately 0.8 Mt of paper sludge per year (which has generally been mechanically or chemically dewatered, but still has water contents up to 300%).
- Dredging activities in harbours and navigable waterways: In Britain dredged spoil amounts to around 1.5 Mm^3/year (over 90% is dumped at sea), the most common material being organic silty clay of high plasticity (Hall, 1995).

14.5.2. Disposal facilities

Old lagoons were constructed by excavating a suitable void space and using the excavated material to build up banks surrounding the lagoon to increase its volume. The older lagoons were constructed (in a semi-engineered way at best) within the curtilage of a treatment works and form an integral part of the works operation. While the old lagoons were not engineered for stability or containment, new lagoons have to be sealed as far as is practicable to prevent groundwater pollution. Depending on the permeability of the underlying soil, the lagoon may need to be lined with clay or an artificial liner (as for an engineered landfill site).

In the long term the stability of the thickened sludge mass in the lagoons must be a cause for concern. With time evaporation, settlement of solids and decanting surface water can lead to substantial thickening of the stored material. The extent to which a sludge will consolidate depends on its physical properties, such as density and particle size. However, the consolidation properties of sludges are also affected by the presence of industrial effluent and pollutants. Organic matter present in sludge can undergo anaerobic decomposition, producing biogas which tends to buoy the solids in sludge. Thus there is a limit to the thickening achieved by this 'natural' process and sludge may retain a 'thixotropic' nature indefinitely, especially if the deposit is several metres deep (Figure 14.29). Although surface drying may proceed to the point where a dry crust forms, which is sufficiently strong to support vegetation, the mass of waste may remain unstable and pose a danger to both humans and animals venturing onto the surface. The lagoon area is likely to remain completely unsuitable for reclamation for any useful purpose, other than perhaps as a wetlands nature reserve.

Because of their function, most sludge lagoons are in close proximity to watercourses and residential areas (Figure 14.30). Stability of the facility is thus important, but so is the potential for seepage flow out of the lagoon (either through the ground or the surrounding embankment) and the resultant pollution of land and water. Even after dewatering, sewage sludge is still very soft, and so when it is placed in a landfill, on-site plant will experience difficulties in travelling on the surface of the sludge and compaction of the waste will not occur. In addition, the sludge tends to block the pore spaces in household and industrial waste and hence reduce the overall permeability. A further problem is that, due to

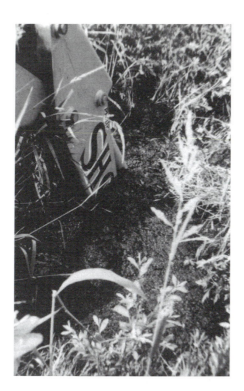

Figure 14.29. Liquid lagooned sewage sludge beneath a thin crust

Figure 14.30. An aerial view of an old sewage sludge lagoon

the fine nature of the sludge, it fills the voids within the refuse inside a landfill, and the overall shear strength of the waste mass is reduced. Hence, in Germany, landfills filled almost entirely with sewage sludge ('monofills') have been created, with typical treatment requirements for landfilled sludge being a shear strength of $10–20\,kN/m^2$, a solids content of more than 30% and a maximum water content of 70% (Voss, 1993).

Increasing amounts of sewage and dredged sludges are stored in lagoons or as mounds (in landraise and as embankments). This has the following geotechnical implications:

- For analysis of basal failure and slope instability, the shear strength of the sludge after deposition must be known. Stability investigations have to be done for all important phases of construction and for subsequent times. Shear strength parameters for undrained and drained conditions have to be distinguished.
- Consolidation characteristics have to be quantified because these influence the strength (and the rate at which it changes) of the sludge and also the reduction in volume of the sludge (which increases the capacity of a disposal facility).

14.5.3. Material properties

Despite the nature of sewage sludge the only means currently available for analysing its engineering behaviour is a geotechnical approach, with the sludge being considered as a fine-grained soil with a high organic matter content. Values of basic properties are indicated in Table 14.8. The shear strength of sewage sludge has been investigated using the laboratory vane shear (see Chapter 6, Section 6.5), to test disturbed samples consolidated in oedometers or using the triaxial apparatus (for undisturbed samples taken from consolidated sludge within landfills).

An empirical relationship between undrained cohesion (ϕ_u would be expected to be zero because of the saturated nature of the sludge) and bulk density has been proposed by Voss (1993) following field vane tests on sewage sludges placed in monodisposal landfill sites. The resultant empirical relationship is:

$$c_u = 2.09 \times 10^{-8} \times e^{17.5\rho} \tag{14.3}$$

where e is the voids ratio and ρ is the bulk density (Mg/m^3).

An alternative empirical relationship (Koenig & Kay, 1996) between undrained shear strength and the proportion of total solids S in a sewage sludge (derived on the basis of sludges which were collected after dewatering and then consolidated in oedometers) is:

$$S_u = A\,e^{-m/S} \tag{14.4}$$

where A and m are sludge-specific 'constants'. Most data fitted an m value of around 0.5, although A was found to be very sludge specific and ranged from 35.6 to 119.8. Equation (14.3) predicts shear strength values that are in keeping with the characteristics of lagooned sewage sludge (Figure 14.31) — for example, the undrained shear strength increases from 1 to $50\,kN/m^2$ as the bulk density increases from 1.0 to $1.25\,Mg/m^3$ (Klein & Sarsby, 1999). Furthermore, bulk

Table 14.8. Basic properties of common sludges*

Source of sludge	Moisture content (%)	Liquid limit	Plastic limit	Organic content (%)	Dry solids content (%)
Sewage:					
dewatered	120–160	130–175	50–65	20–40	10–20
lagooned	400–900	130–175	50–65	20–40	~5
Dredging	70–130	~150	~50	~20	10–35
Papermills	60–300	150–170	40–240	40–80	5–40
Water treatment	~700	~600	~200	—	~10

* Sources: Spinosa (1985), Blümel & von Bloh (1991), Hall (1995), Wang *et al.* (1996), Wu (1999), Sarsby & Cooke (1989).

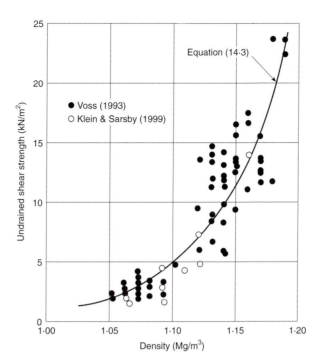

Figure 14.31. The undrained shear strength of sewage sludge

density is considered a better indicator of the likely shear strength behaviour of a sludge than just the water content or the fibre/solids ratio. This is because the bulk density encompasses the effects of both voids ratio (the pore fluid has zero shear strength) and fibre and solids content (the fibres will provide little frictional shear strength).

A large range of values (2.5–35°) has been quoted for the effective friction angle of sludge, although the effective cohesion is relatively low (0–10 kN/m^2). Even though sewage sludges are highly fibrous the fibres do not 'mat' together like peat to give tensile strength. This is because the sewage fibres are very short (unlike in organic soils or paper sludge) and are separated by a viscous pore fluid. The high values of the friction angle may be indicative of sludges with a significant grit content.

Settlement–time curves from consolidation tests of disturbed samples of lagooned sewage sludge (Sarsby & Cooke, 1989) tend to have the same form as curves for highly organic soils (Figure 14.32). The materials exhibit low values of C_v (0.1–0.7 m^2/year) even under low effective stress (1–5 kN/m^2), with a

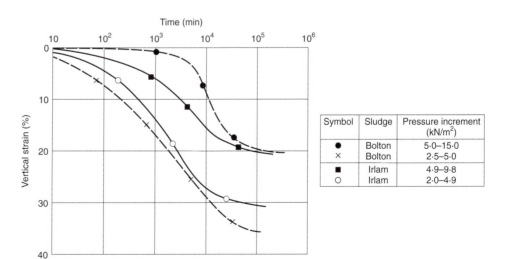

Figure 14.32. Consolidation curves for sewage sludge

noticeable decrease in C_v as the vertical effective stress increases (as low as $0.02\,\text{m}^2/\text{year}$ under $10\,\text{kN/m}^2$). Two factors seem to be responsible for the low consolidation rates: the composition and viscosity of the pore fluid (which is more akin to a gel than to water); and the ongoing digestion of the sludge and the generation of gas bubbles.

As would be expected, the sludge undergoes a large volume reduction (around 25% vertical strain for a pressure increment of $1–10\,\text{kN/m}^2$) during consolidation. This settlement at full consolidation can be represented by the form of equation used for a normally consolidated soil (see Chapter 8, Section 8.4.2), with C_c values in the range 10.0–1.5 (decreasing with effective stress level). For wastewater sludges, Koenig & Kay (1996) reported C_c values ranging from 1.8 to 2.4.

Despite the highly viscous nature of the pore fluid in these sludges they seem to exhibit little creep, with the coefficient of secondary compression C_α being around 0.02–0.04 (although, as stated in Chapter 8, Section 8.4.3, the low C_α value may be due to the very long time taken to complete primary consolidation).

Some work has been conducted on improving the shear strength characteristics by adding other materials (quicklime, fly ash, cement, sand) to sludge. The additives have reflected two approaches:

- the use of reactive materials that have good 'water-binding' properties to reduce the amount of 'free water'
- the creation of a frictional skeleton by adding inert hard particles.

With both methods substantial increases in shear strength can be achieved (Loll, 1990). However, this requires the addition of relatively large quantities of inert additive (a sand–sludge volumetric ratio of around $2:1$) although much less reactive agent (around 10–15% by weight) seems to be needed.

15 Tailings dams

15.1. Introduction

Tailings are a waste product of the mining industry. They consist of the ground-up rock that remains after the mineral values have been removed from the ore. Ore quality has deteriorated through the years as the best sources have become exhausted, causing a corresponding increase in the amount of tailings left after the extraction of each tonne of metal. The disposal of mine waste, chiefly tailings, has of late assumed an importance that transcends even the massive volumes of materials produced annually by mining operations.

Stava Creek is situated in a holiday-resort region of the Dolomite Mountains. In 1961, construction of a tailings dam (the lower dam), within the valley of the Stava Creek (northern Italy), was commenced to retain the waste from a nearby fluorite dam (Figure 15.1). After a low-height 'starter dam' (about 9 m high) had been constructed using locally won material, waste slurry was pumped into the lagoon behind the 'dam', where the fines settled out. As the lagoon filled up the starter dam was covered with sand (which was recovered by cyclone from the tailings) and was raised in steps of about 5–6 m. Hence, as the dam was raised, its crest moved in an upstream direction towards the lagoon. The final height of the

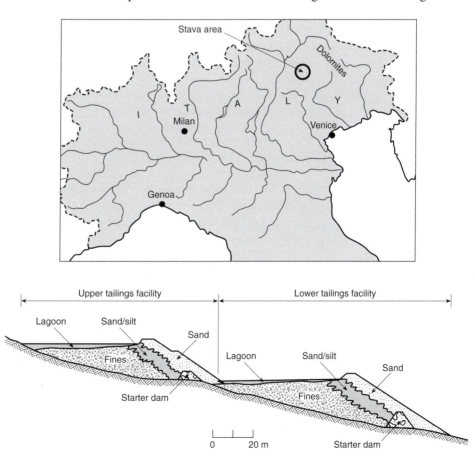

Figure 15.1. The Stava tailings dams

lower dam was about 26 m, with an average downstream slope angle of 32° (Berti *et al.*, 1988). In 1970 work commenced on the construction of a new impoundment (the upper dam) just upstream of the existing one on the same slope. The method of construction used was similar to that employed for the lower dam, and the downstream toe extended partly over the soft sediments impounded in the lower lagoon. The upper dam was raised progressively, and by 1985 it had a height of approximately 30 m with an average face slope of about 35°.

In July 1985 there was a disastrous failure of the downstream face of the upper dam which resulted in liquefaction of both dams. About 250 000 m^3 of liquefied tailings rushed down the narrow Stava Valley and the villages of Stava and Tesero, situated along the stream channel and at its end, respectively, were destroyed and buried under mud. At failure the debris travelled down the valley with a velocity of about 30 km/h as a consequence of liquefaction of the tailings. In the latter stages the speed rose to about 90 km/h as the tailings became more liquid due to water being released from the underlying sandstone (Chandler & Tosatti, 1995). The disaster claimed 269 lives and caused extensive property damage in the valley, making it the worst disaster of its type ever to have occurred in Europe.

The Stava tailings dams were built without hydrogeological or geotechnical investigations — the tailings were not even analysed for grain size distribution. There was no prior attempt to apply rational design and stability analysis procedures, and construction and operation was purely empirical. Post-failure stability analyses (undertaken using the Bishop simplified method) showed that the two dams were inherently unstable, and the calculated Factors of Safety were in the range 1.35–0.76 (Berti *et al.*, 1988). Dam failure was probably triggered by local liquefaction of the loose silty sands of the dams as a result of build-up of excess pore pressures due to additional loading (by raising the dam) on soft sediments.

The quantities of tailings that are transported and disposed of on the ground surface are enormous and the annual volume of tailings produced by industry throughout the world (of the order of 6000 Mt) far exceeds the volumes of materials handled in other branches of civil engineering, including the construction of embankment dams and road embankments. In addition, severe environmental pollution can result from tailings dumps, depending on the nature of the spoil and the dump's proximity to rivers and streams. Unfortunately, tailings dams have a long history of failure resulting in widespread death and destruction, as indicated in Table 15.1.

15.1.1. General aspects of tailings disposal

Mineral extraction processes often involve crushing mined ores down to silt size to expose the metal contents. The process leaves large quantities of saturated silt-sized 'valueless' material to be disposed of at minimum cost. Tailings dams are the most common disposal system for storing, or disposing of, the solid by-product of the mining industry. Historically, tailings disposal began as the practice of dumping tailings in nearby streams, and progressed to empirical design of impoundments, by mine operators, based on principles of trial and error. Since the disposal of tailings adds to the cost of mineral extraction, mine operators aim to dispose of the waste as cheaply as possible. This has resulted in the evolution of the tailings dam and, as mining operations have become bigger and bigger, these structures have become major works of construction. Tailings facilities are usually constructed by mine operators, with the height of the dam being increased as required to provide the needed waste storage. This has the advantage of slow construction, but construction control is usually very poor. To minimize tailings disposal costs the coarser fraction of the waste is used in the construction of most tailings facilities instead of importing more costly suitable material.

Table 15.1. Examples of tailings dams failures

Year	Outcome	Quantity released	Failure type	Dam location
1938	80 people killed	7 Mm^3 of hydraulic fill travelled 420 m	Foundation failure, liquefaction of fill	Fort Peck, USA
1964	6 ha of agricultural land lost, lead poisoning in cattle	13 000 t of lead and zinc contaminated tailings	Erosion	Pare Min, Wales
1965	300 people killed	About 2 Mt of slurry and flowslide	Liquefaction due to earthquake	El Cobre, Chile
1972	125 deaths, about 4000 people made homeless	About 70 000 m^3 of water and tailings (3 dams breached)	Overtopping of dam	Buffalo Creek, USA
1974	Groundwater pollution, damage to property	About 5000 m^3 of clay slimes	Overtopping of the dam	Berrion, France
1978	Severe pollution of 30 km of a major river and marine bay	100 000 m^3 of tailings contaminated with sodium cyanide	Liquefaction due to earthquake	Mochikoshi, Japan
1978	10 people killed, extensive property damage	About 40 000 m^3 of slurry	Piping through the embankment	Bafokeng, South Africa
1985	269 people killed, 2 villages destroyed	250 000 m^3 of slurry flowed down a narrow valley	Liquefaction induced by slope failure	Stava, Italy
1994	47 people killed, 200 houses swamped	2.5 Mt of acidic, radioactive slurry flowed for nearly 2 km	Internal erosion and possible overtopping	Merriespruit, South Africa
1997	Approximately 150 km^2 of a national wildlife park polluted	5 Mm^3 of acidic sludge containing heavy metals travelled 60 km	Internal erosion	Aznacollar, Spain

Design and construction procedures that were developed in the past for relatively small tailings dams, located in isolated areas where failure did not constitute a major threat to life or property, do not meet present-day standards of safety or pollution control. Most previous mine waste impoundments have not been subjected to formal engineering design. It is only within the last 25 years that serious attempts have been made to apply the principles of geotechnical engineering to tailings embankments.

Today there is a trend of building fewer, but significantly larger, tailings disposal facilities as a consequence of environmental pressures, decreasing availability of land and increasing disposal costs. Nowadays the largest dams in the world (in terms of volume) are composed of tailings. In 1989 an ICOLD world register of tailings dams included eight higher than 150 m and 115 higher than 50 m. A typical example is an oil sands project in Canada where tailings are stored in a pond which can accommodate approximately 270 Mm^3 of sand and 320 Mm^3 of thick sludge. To accommodate the waste the facility has approximately 18.5 km of hydraulic fill dykes (ranging from 27 to 82 m high) and the tailings pond has a surface area of around 17 km^2 (Morgenstern & Kupper, 1988). It has become increasingly common for the safety requirements for tailings dams to be the same as for their water-storing counterparts. However, tailings facilities differ from conventional water dams in several important aspects, as indicated in Box 15.1.

The advantages of progressively raised embankment dams are significant:

- construction expenditure is distributed over the life of the impoundment
- the total volume of fill for the ultimate embankment is not needed initially
- there can be much more flexibility in the selection of materials for embankment construction.

Construction materials
- Materials for constructing tailings dams are not selected on the basis of suitability criteria and there is no quality control of materials.
- The major part of the dam is usually constructed using the coarser fraction of the tailings — if loose and saturated this material may liquefy under seismic shock.
- Tailings dams are often constructed of cohesionless, highly erodible tailings sand, and therefore they do not have stabilized crests or downstream faces.

Construction procedure
- A tailings dam is not constructed to an engineering design.
- The tailings facility is built in a series of short stages and impoundment behind the dam commences as soon as part of the face has been raised.
- Most of the dam construction is carried out by the mining operators so that quality control of the product will be low.
- The dam is constructed slowly over a period of many years, so the design can be modified as necessary.

Stored material
- The bulk of material stored behind the dam is soft, loose, relatively impermeable tailings rather than water — all saturated tailings are likely to liquefy under seismic shock.
- When the tailings liquefy they become a liquid of high unit weight, thereby placing additional loading on the dams.
- There is a significant amount of lateral support for the upstream face of the dam from the retained saturated fines when they are stable.

Facility operation
- Modest overtopping can rapidly cause gullying and complete breaching of the cohesionless materials forming the dam face.
- Tailings dams cannot be breached at the end of their useful service life because they usually contain highly toxic fluids and solids.
- Tailings dams are normally subjected to only a nominal amount of drawdown of the free water in the retained fines.

Tailings facilities can be assigned to one of two general classes of impounding structures, depending on whether the embankment dam blocks a valley (cross-valley impoundments) or whether it encloses the impoundment partially or wholly. The latter class includes:

- ring dykes (best suited for flat terrain, but requiring a large volume of good embankment fill)
- sidehill impoundments (using embankments on three sides)
- valley bottom impoundments, using two embankments (where the drainage catchment area would be too large for cross-valley damming).

15.1.2. Hazards from tailings disposal
Mining activities produce environmental disturbance during all phases of the operations, as indicated in Table 15.2. With tailings disposal there are two general aspects of concern:

Table 15.2. Environmental impact of mining activities

Work phase	Cause of environmental disturbance
Exploration	Borehole drilling, access to site
Development	Open pits, mills, waste dumps, traffic
Extraction	Surface subsidence, mine drainage
Ore concentration	Wastes from crushing/grinding (tailings), contaminated waters
Processing	Pollutants, fine tailings, contaminated waters
Abandonment	Untreated and unstabilized waste dumps

- the structural stability of waste dumps and the possible release of very large volumes of semi-fluid tailings
- the possibility of pollution of the ground and groundwater due to egress of polluted liquid from stable dumps.

The difficulty associated with specifying reasonable short-, medium- and long-term stability requirements are compounded when environmental considerations are taken into account. Land disturbance does not necessarily produce a major long-term environmental problem because land used as a disposal area may be reclaimed after closure of the works. However, a structure may be designed and operated safely from engineering and other technical considerations, but may still result in an unacceptably high environmental impact. The tailings produced by most industrial and chemical operations and by some mineral beneficiation processes are fine-grained, often fluid or semi-fluid, and in many instances contain highly toxic materials. A common potential long-term impact on land associated with mining is the increased heavy metal concentrations in the tailings disposal area and the consequent effects on the aquatic environment. In general, the characteristics of contaminants of most concern are: high acidity, high total dissolved solids, heavy metals content, and, in the case of uranium mines, radioactivity. However, the potentially most harmful tailings component does not necessarily produce the greatest environmental impact. For instance, investigations of uranium tailings deposits produced over 30 years ago in Ontario (Canada) indicate that non-radioactive, non-reactive constituents (such as sulphates and calcium) have travelled hundreds of metres through underlying sand and gravel, while radium has travelled less than a few tens of metres away from the tailings (Parsons, 1981).

Currently the main thrust of legislation is directed towards ensuring that no catastrophic embankment failure occurs during the operational life of the dam. An important element in the development of a strategy to provide a common approach to safe design, construction, operation and rehabilitation has been the development of hazard rating systems (Table 15.3). The hazard rating is not only the basis for classifying and registering an individual tailings storage, but also for defining the on-going design and operating standards required to provide acceptable levels of safety, minimal environmental impact and adequate rehabilitation. The environmental threat posed by tailings can only be judged meaningful in relation to levels that are harmful to humans, plants and animals. The potential pollution hazards associated with the storage of the tailings slurry vary with different mining operations, and range from very severe to none (for mining processes that merely grind up an inert ore without the addition of toxic chemicals during processing). In between these two extremes is a wide range of conditions that represent either short- or long-term potential pollution problems. Most mine tailings fall into this intermediate category.

After back-analysing more than 100 tailings dams failures, Ivanov *et al.* (1989) defined the following principal reasons for failure:

Table 15.3. A tentative hazard rating system for tailings dams*

Hazard potential rating[†]	Potential loss of life	Potential economic loss	Size class[†]	Maximum wall height (m)
Low	None	Minimal	Small	>5 but <12
Significant	Not more than ten	Significant	Medium	≥ 12 but <30
High	More than ten	Great	Large	≥ 30

* After Jones *et al.* (1993)
[†] Hazard classification $\equiv$ hazard potential rating $\times$ size class.

- faults in the operation of discharge structures, causing excessive seepage through the dam (41%)
- overfilling of the mud-filled pond behind the dam so that the latter was overtopped (25%)
- faults in the construction of the starter dam (21%)
- incorrect foundation soil preparation (6%)
- increasing the dam steepness while the facility was in use (2%).

Design of tailings disposal facilities must include considerations of safe storage both during and after construction, and safety in terms of environmental concern. In fact, improved design methods have resulted from past failures. Careful site investigation must be carried out to detect adverse hydraulic conditions and the presence of weak layers or materials liable to deteriorate due to the tailings or its pore fluid. Knowledge is required of the geological history of the site in order to assess the probability of old shear surfaces being present. It is not sufficient for the geotechnical engineer to consider simply earth dam criteria and concepts when undertaking design of tailings facilities, the engineer also needs to have a thorough understanding of the materials being retained (in terms of the processes by which tailings are generated and their nature).

15.2. Tailings

The term 'tailings' is sometimes used in a generic sense to refer to any mine or mill waste in solid form, including rock from the initial development of the mine or pit, underground mine waste, etc. However, because of the divergent characteristics of these various materials, a more narrow definition is advantageous wherein tailings are defined as 'crushed rock particles that are either produced or deposited in slurry form' (Vick, 1983). This definition encompasses the vast majority of finely ground mill or mineral processing wastes remaining after extraction of mineral values, but it excludes pit-development waste and process wastes, such as fly ash, that are produced and handled in dry form. The grain size of tailings depends on the characteristics of the ore and the mill processes used to concentrate and extract the metal values. A wide range of grading curves exists for the wastes from various mining operations, and tailings may range in size from sands to clay-sized particles. The minerals which give rise to significant quantities of tailings (on a world-wide basis) are indicated in Box 15.2. Process industries that produce large quantities of mineral-based wastes in the UK are shown in Table 15.4.

*Box 15.2. Specific minerals giving rise to tailings**

Coal
- Processing is done to remove fine coal and rock (usually shale) that would produce excessive ash or interfere with burning. Processing relies to a major extent on gravity separation and froth flotation. Fine coal refuse consists principally of coal particles and silt–clay particles from shale.

Tar sands
- The production of crude oil from tar or oil sands requires large-scale mining and processing operations. The ore requires essentially no crushing or grinding. Petroleum values are recovered by aerating the ore slurry in hot water and steam. The tailings are relatively clean and coarse, with a significant slimes fraction (a mixture of silt, clay and oil residue).

Lead and zinc
- These metals commonly occur in association and are often extracted together. Concentration is by froth flotation. Hard, angular sands are typical. The tailings are generally of low plasticity and clay content, even for the slimes fraction.

Box 15.2. Continued

Gold and silver

- These metals are frequently associated in ore bodies and are often extracted in combination. Concentration is by froth flotation, and the product may be further processed by sodium cyanide leaching.

Copper

- The ore is mined from hard rock and is concentrated by froth flotation. The whole tailings are relatively coarse, although finer grinds may be produced by mills that extract other materials.

Molybdenum

- This metal is extracted chiefly as a by-product of copper generation. Concentration is by froth flotation. The tailings are relatively coarse and are similar to copper tailings.

Nickel

- Concentration is by flotation, sometimes in conjunction with magnetic separation.

Taconite

- This is the main ore in the USA for the production of iron. Concentration is usually by gravity, often followed by magnetic separation. Concentration results in relatively coarse tailings almost exclusively in the sand-size range. Sometimes further concentration by flotation methods is performed, producing slimes.

Phosphate

- The crushed ore is washed to remove the fines, which are discharged as slimes or phosphatic clays. Phosphate values are retained in the coarser particles, which are usually concentrated by flotation. The phosphate slimes are among the finest tailings produced by any type of processing operation (predominantly silt and clay particles). The phosphate concentrate produces gypsum tailings, which fall almost exclusively within the silt size range.

Bauxite

- Processing of the mineral to yield aluminium oxide produces a type of tailings known as 'red muds'. The common extraction process involves leaching by washing the ground ore in hot caustic soda solution. Most of the tailings are in the silt size range. Despite their low plasticity and low clay content, bauxite tailings share some of the properties associated with clayey tailings.

Uranium

- Tailings have radioactive properties. The ore is processed using leaching techniques with either acid or alkaline reagents. Large quantities of water are discharged into the tailings impoundment.

Trona

- This is mined for conversion to soda ash. Trona occurs in seams between shale layers. The crushed ore is beneficated by hot water solutioning, which leaves behind the insoluble tailings (principally shale particles).

Potash

- This is mined from sedimentary deposits. Tailings comprise a coarse sandy fraction containing a large proportion of silt, and a clay fraction.

---

* Adapted from Vick (1983).

15.2.1. Production

In the mining and extractive metallurgical industries the primary emphasis is on the extraction of mineral values from the parent ore. Tailings are simply a waste product and have seldom been treated as a separate entity, even though they are produced in vast quantities and they are the most important potential source of environmental impact for many mining projects. The extraction of mineral values requires procedures as diverse as the ores processed, but the fundamental steps in the processes are common to most ores (Figure 15.2):

Table 15.4. Major producers of mineral wastes in the UK

Source	Material	Examples
Coal-mining	Fine discard from washing	Lagoons retained by dams of coarse discard
Thermal power stations	PFA	Lagoons and tips
China clay processing	Micaceous residues/fines	Dams and tips
Sand and gravel pits	Silt and clay from screening plant	Low bunds around worked-out areas
Salt-based chemical industry	Hot liquids/sludges	Lagoons with dams
Dredging	Clay, silt, fine sand	Sludge holding lagoons

PFA, pulverized fuel ash.

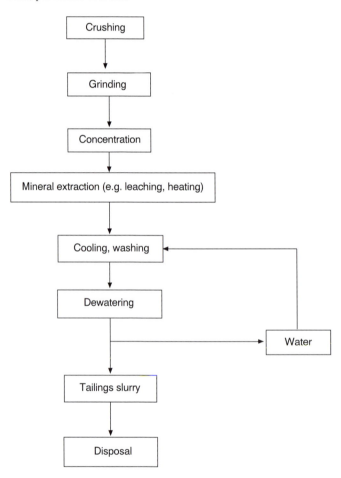

Figure 15.2. The elements of minerals processing

1.   Central to the processes are the initial steps of crushing and grinding. The purpose of crushing is to reduce the rock fragments to a suitable size for feeding into grinding equipment. The crushing is done in stages and is followed by grinding, which produces a further reduction in the size of the fragments by passing them through mills. This is the final stage in the physical reduction of the ore to tailings size. The gradation of the tailings depends on the degree of particle breakdown due to grinding as well as the clay content in the original ore. The mineral content of the individual particles produced by grinding varies.

2.   The purpose of concentration is to separate those particles with high values (concentrate) from those with lower values (tailings). Physical methods of concentration are:

- gravity separation (requires that the mineral and the host rock have different specific gravities)
- settlement (usually performed with water)
- magnetic separation (most useful for iron extraction)
- flotation (froth flotation is the most common, in which individual mineral-bearing particles in water suspension are made water-repellent and receptive to attachment to air bubbles; these particles thus rise to the surface of a froth which is skimmed off).

3.   Processes such as leaching and heating may be used to complete the process of concentration of the desired minerals. In leaching the mineral values are removed by direct contact with a solvent, usually a strong acid or alkaline solution. Sulphuric acid is the most common acid reagent, and hydroxides and carbonates of sodium or ammonium are the most common alkaline reagents. Leaching may change the physical characteristics of the tailings. Heating can be applied to either the ground ore or to a particle–slurry suspension. Separation and removal of mineral values from the concentrate leaves the remaining barren particles as tailings.

4.   Dewatering is the final process stage. It is not intended to produce complete drying of the tailings, but it removes some of the water in the tailings–water slurry (following concentration). The recovered water and reagents are recirculated where possible. The most common means of dewatering is by pumping into thickeners, which are essentially continuous-feed settling tanks.

15.2.2. Disposal

Many mine and mill operations produce a wet waste stream and, for all but the shortest distances, it is substantially more economical to transport solids hydraulically through pipes than by truck, conveyor or other dry handling methods. High-velocity, continuous flow is necessary if the pipes are not to clog due to the tailings settling from suspension. The tailings disposal area is usually a 'lagoon' that has been created by the construction of dykes or dams to retain the tailings slurry. Tailings materials are transported in slurry form (at concentrations varying from 20% to 35% by weight of solids to total weight of slurry) to hydraulic-fill tailings dams or impoundments. They are then discharged along the length of the impoundment wall (the dam), thus allowing the mixture of water and solids to flow towards a decant pool (Figure 15.3). In some systems a 'thickened' slurry is deposited some distance away from the dam and is allowed

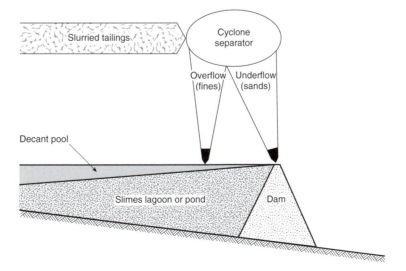

Figure 15.3. Tailings disposal

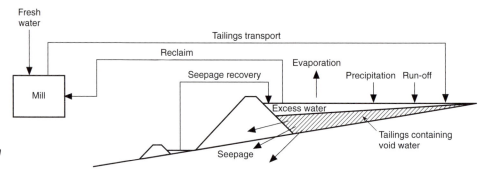

Figure 15.4. The water cycle in tailings disposal

to flow towards the dam. The material comes to rest at a flat slope to form a roughly conical surface.

The amount of free water that is stored in the tailings disposal area together with the tailings varies from very little at mining operations where most of the water is decanted and returned to the mill, to a very large amount for disposal areas that are used for water storage as well as tailings disposal. Where possible, water is decanted from the lagoon and returned to the mill for re-use. The complete cycle is illustrated in Figure 15.4.

The mineral tailings are dumped in an impoundment, but they are also used in the construction of that disposal area. This is the crucial factor in understanding tailings 'dams' and why the topic is important from the engineering point of view. The coarser fraction of the tailings is used to construct embankments (wholly or in conjunction with natural soils), while the finer grained portion is deposited behind these retaining dams to form the tailings lagoon. There are two very important items relating to grain-size distribution:

- The percentage of material coarser than $75\,\mu$m (i.e. the 'sands'), since this determines the amount of dam-building material that can be obtained from the total tailings by cycloning.
- The relative proportions of clay and silt in the material finer than $75\,\mu$m (i.e. the 'slimes'), because a high clay content makes it difficult to produce clean sand efficiently. High clay contents can also produce storage problems because the particles settle out in the decant lagoon in a very loose state and at a slow rate.

Nearly all tailings contain sands and slimes. For example, copper mining normally produces tailings having 40–60% of particles finer than $75\,\mu$m. On-site separation (using cyclones) of coarse and fine components of tailings may be undertaken to enhance the quantity and quality of material for dam building. Typical grading curves for whole and cycloned tailings are shown in Figure 15.5.

15.3. Tailings facilities

In general, a tailings disposal system consists of two parts: i.e. a dam and a slurry lagoon behind it (see Figure 15.3). However, the construction methods employed usually mean that there is no clear distinction between these two components, and the dam utilises available waste material separated from the tailings stream. The old, traditional method of tailings dam formation involved construction over previously placed hydraulic fill and well-known examples of tailings dams failure (El Cobre, Chile, 1965; Mochikoshi, Japan, 1978; Stava, Italy, 1985) were all constructed by this method. In recent years there has been a trend in major tailings projects to move closer to conventional dam design and construction procedures, and to date there have been no cases of collapse of tailings dams constructed by modern methods.

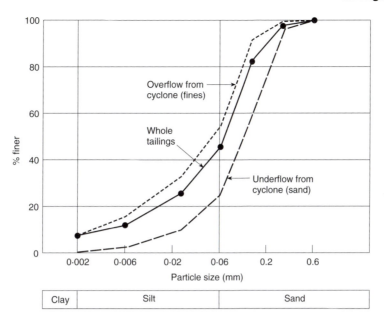

Figure 15.5. Typical grading curves for whole and cycloned tailings

15.3.1. The 'dam'

The 'tailings dam' is not really a dam in the conventional water-retaining sense (because it does not contain an engineered water barrier). It is more akin to a thick facing or veneer at the front of a waste dump. Typically, the construction of a tailings dam begins with the creation of a starter dyke using natural, permeable materials. Tailings slurry is then discharged from this dam crest so that the coarser particles settle first, while the fines are transported as a stream to form a lagoon or pond where they settle and from which water can be decanted. In most cases tailings dams are constructed from the sandy fraction of the mining wastes, and the finer fractions (the slimes) are stored behind the dam together with the ponded transport water. As the depth of slurry in the lagoon increases the tailings dam is raised, progressively, using the coarser portion of the slurry. Depending on whether the dam crest moves upstream or downstream, or remains fixed during construction, the embankment is constructed following the so-called upstream, downstream or centreline methods, respectively.

Segregation processing of fine-grained and coarse-grained tailings is one of the key aspects for safe containment of the tailings. Sand tailings generally share the engineering behaviour of loose to medium-dense deposits of natural sand. The coarse materials can also be separated by means of hydrocyclones and deposited close to the embankment, while the finer portion is discharged upstream to form the settling pond. Cyclones are simple devices that function on a centrifugal separation principle with no moving parts. Whole tailings slurry under pressure enters a cylindrical chamber and is spun round because of the input momentum. Coarser particles in the slurry spiral down (due to the force of gravity on the particles), while the finer fraction and most of the slurry water rise to the overflow outlet.

The *upstream method* of dam construction is the oldest and simplest method. It is particularly cost-effective since the volume of the containment embankment is kept to a minimum and the downstream face of the dam is often at or near the angle of repose (Figure 15.6). The normal construction process involves hydraulic deposition of tailings through a spigot in an upstream direction off a low starter dyke (Figure 15.7). The dam is raised by dragging coarse material from previously deposited tailings in the vicinity of the dam (Figure 15.8). In arid regions, particularly where there was a low water-table and relatively permeable ground, dams have been built by hand using only light tools and a low initial

Figure 15.6. The face of a tailings dam formed by the upstream method

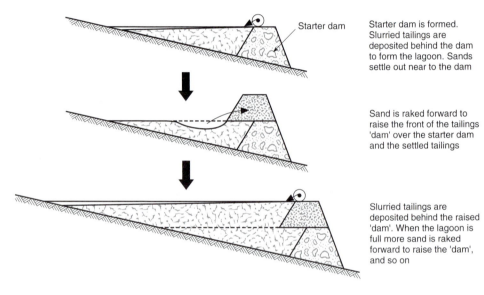

Figure 15.7. The upstream method of tailings dam construction

Starter dam

Starter dam is formed. Slurried tailings are deposited behind the dam to form the lagoon. Sands settle out near to the dam

Sand is raked forward to raise the front of the tailings 'dam' over the starter dam and the settled tailings

Slurried tailings are deposited behind the raised 'dam'. When the lagoon is full more sand is raked forward to raise the 'dam', and so on

Figure 15.8. The upstream method of constructing a tailings dam

earth bund ('starter dam') about 1 m high. The tailings slurry would be spread onto the land upstream of the bund and, periodically, coarse material would be raked up (to raise the bund by about 0.3 m) so as to form the downstream slope of the 'dam'. Success depended on the natural drying of the tailings shovelled up to maintain the height so that it developed sufficient strength to retain the rising lagoon (see Figure 15.8).

In tailings dams built by the upstream method the front is composed of cycloned sand and behind are the slimes. In reality this method does not provide a dam but a shell on the face of the waste mass (Figure 15.9). Between the dam and the slimes lagoon there is a material transition zone. The sands and transition-zone materials are generally permeable soils that consolidate rapidly and may be compacted. The primary difference between the slimes and the materials in the transition zone is particle size. The permeability of slimes is so low that long periods (many years) are needed for significant consolidation to occur. Central to the application of the upstream method is that the tailings form a reasonably competent material for support of the perimeter dykes. However, the dried pond surface and the base of each lift represent potential planes of weakness and areas of concentrated seepage. The dam is built on top of previously deposited unconsolidated tailings, and it is likely to be prone to liquefaction due to earthquake loading.

Factors controlling the upstream method include:

- the phreatic surface (location is critically important)
- the water storage capacity
- the seismic liquefaction susceptibility.

The phreatic surface must be kept below the downstream slope by a combination of downward drainage into the ground and evaporation from the exposed downstream face of the dam itself. Disruption of the fine balance of water gain and loss can lead to a rise of the phreatic level towards the downstream face and the risk of erosion, piping and instability. Furthermore, the rate at which upstream embankments can be safely raised is limited by excess pore pressure development within the deposit, particularly in the slimes zones (because of the low C_v value; see Section 15.4.5). This fact is believed to have contributed to the Stava failure.

The upstream-type dam does not meet conventional dam requirements of slope stability, seepage control (internal drainage) and resistance to earthquake shocks. Saturation of the outer face of the dam is likely to lead to piping or sliding failure, and the catastrophic failures of several upstream-constructed tailings embankments in Chile in 1965 highlighted the limited earthquake resistance of this form of construction. Hence the upstream method of construction is generally unsuitable for all but very minor tailings dams.

The *downstream method* of construction is a relatively new development which has evolved for constructing larger and safer tailings dams. The dam is raised in a downstream direction and is not underlain by previously deposited tailings (Figure 15.10). Also, the slimes layers are not intermixed with the sandy

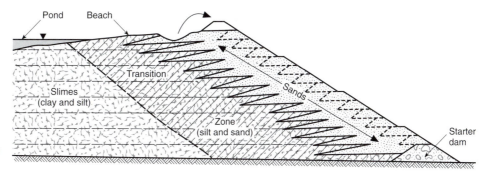

Figure 15.9. A section through a typical upstream dam

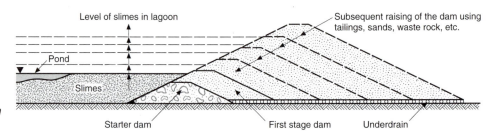

Figure 15.10. The downstream method of dam construction

tailings (which could form planes of weakness), which allows safe dams to be built to appropriate heights, provided adequate compaction and seepage control are provided (this can be easily incorporated). However, each successive lift requires more material for construction than the previous one if the coarse tailings assume their natural angle of repose. The volume of fill required often increases exponentially with height, so there is corresponding high cost. Because the phreatic surface can be maintained at low levels within the embankment and because the entire body of fill can be compacted, downstream construction is liquefaction-resistant and can be used in areas of high seismicity. Downstream methods are essentially equivalent in structural soundness and behaviour to water-retaining dams. Hence they require design and careful advance planning. Sufficient space must be left downstream of the starter dyke.

Significant embankment heights can be achieved safely with the *centreline method*, which is a variation on the downstream method in which the crest is raised vertically with the centrelines of raises coinciding as the embankment gains height (Figure 15.11). The part of the embankment corresponding to a dam is not built on previously deposited fine tailings, and internal and underlying drainage systems can be installed. The main body of the embankment fill can be placed and compacted in an organized manner and saturation can be controlled so that the method produces good seismic resistance. However, the embankment should not be raised too quickly, so that the possibility of failure of the upstream slope is minimized. The dam can be designed so that it can usually be raised above its original design height if need be, and the construction process is relatively rapid. Although less sand is required than with the downstream method, large volumes of sand are still needed to raise the dam, and the early stages of mining operations may produce insufficient coarse material to raise the dam.

15.3.2. The lagoon

The tailings lagoon (sometimes called the 'slimes pond') is formed by the hydraulic deposition of the tailings. The lagoon extends backwards from the dam. There is no clear boundary, just a transition zone from predominantly sands to primarily slimes. The supernatant water that appears as the fines settle out is

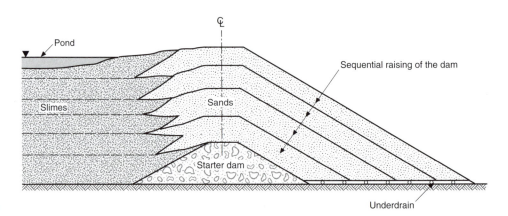

Figure 15.11. The centreline method of dam construction

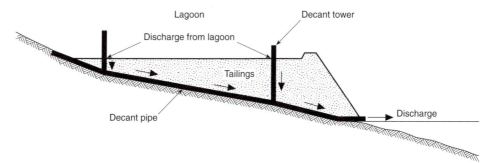

Figure 15.12. The old system of drawing off water from the slimes pond

usually drawn off to be recycled into the ore treatment process. A decant draw-off is placed in the lagoon to keep the pool of water well away from the dam. However, several failures of upstream embankments can be attributed to an inadequate separation distance between the decant pond and the embankment crest. Furthermore, most of the older dams used decant towers for reclaiming water from the slimes pond, with discharge lines running through the base of the dam to a downstream pumphouse (Figure 15.12). These latter systems represent a major potential risk as a source of seepage and piping problems.

Slimes are complex materials that may exhibit geotechnical properties similar to natural sands or natural clays or a combination of both. The slimes in the tailings pond are extremely fine grained and saturated, and may never consolidate completely during the life of the mine.

Hydraulic fill placed under water behaves differently from fill placed above water — it tends to assume flatter slopes, be more sensitive to the presence of fines and ends up in a looser state on average than hydraulic fill placed on land. For most types of tailings the beach slopes downward to the decant pond, with an average gradient of 0.5–2.0%. The deposition process produces a highly heterogeneous beach deposit, which is frequently layered in the vertical and horizontal directions. The proportion of fines may vary by as much as 10–20% over a depth of several centimetres. The coarser particles are found near the dam, with the material increasing in fineness towards the decant pond. Furthermore, the deposition of tailings slurry is an intermittent process. Material deposited on one section is usually left to drain and dry out for a few days before the next layer is deposited. The drying process has the beneficial effect of considerably reducing the voids ratio, but it causes the formation of a network of shrinkage cracks which are subsequently filled when the next layer of slurry is deposited. During subsequent drying cycles the slurry in the shrinkage cracks also shrinks so that a network of open subsurface cracks may form in the tailings deposit. This network constitutes potential channels for piping erosion, either when fresh slurry is deposited or when rain water accumulates on the surface of the tailings.

15.4. Engineering properties of tailings

It must be remembered that there is no real quality control of the materials in the tailings system and inherent variability is quite significant. In terms of geotechnical engineering, the properties that bear on both the design and the performance of tailings dams are:

- the density state of the sands in the dam
- the permeability of the dam material
- the consolidation characteristics of the slimes
- the shear strength properties of both the dam material and the slimes
- the earthquake resistance and liquefaction potential of the dam and the slimes (particularly important).

15.4.1. Physical properties

Tailings are usually angular to very angular (except for oil sands which are usually subgranular to subrounded), bulky grained sand- and silt-size particles. The particle sizes and particle-size distributions of tailings materials and industrial wastes cover wide ranges (Figure 15.13) and are influenced by the parent ore (as indicated in Table 15.5). Soft-rock tailings are derived principally from shale ores. Ordinarily they contain some sand-sized materials, but the clayey nature of the slimes significantly influences the behaviour of the material as a whole. Fine tailings, having little or no sand, include phosphatic clays, bauxite red muds, fine taconite tailings and slimes from tar sands. The characteristics of slimes predominate and render the tailings incompetent from a structural viewpoint. Hard-rock tailings are dominated by sands (primarily finely crushed silicate particles), and slimes are derived from the crushed host rock rather than clay. Coarse tailings are those whose characteristics are determined, on the whole, by a sizeable coarse sand fraction.

The grain size and clay content control in situ voids ratios, which for sands generally range from about 0.6 to 1.1 (typical data are given in Table 15.6). The in situ density exhibits considerable scatter within any single tailings deposit, and it depends on the parent ore, but typically it increases by 0.09–0.17 Mg/m^3 per 30 m depth for tailings sands. Although the relative density can vary considerably

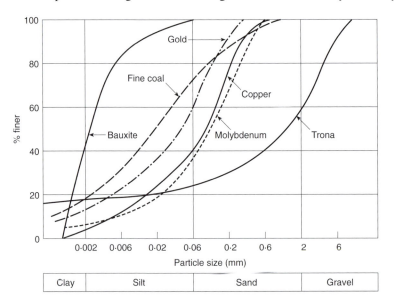

Figure 15.13. Grading curves for tailings

Table 15.5. General characteristics of tailings from different ores

Category	Examples	General character
Soft rock	Fine coal, potash, trona (for soda ash)	Produce both sand and slimes fractions. Slimes may dominate the overall properties because of the high clay content
Hard rock	Lead, zinc, copper, gold, silver, molybdenum, nickel	Produce both sand and slimes fractions. Slimes are usually of low plasticity to non-plastic. Sands usually control the overall properties for engineering purposes
Fine tailings	Phosphatic clays, bauxite muds, tar sands slimes	Generally little or no sand fraction. Behaviour is dominated by silt- or clay-sized particles, which may pose significant disposal problems
Coarse tailings	Tar sands, uranium, gypsum	Produce mainly sand or non-plastic silts. Usually have favourable engineering properties

Table 15.6. Typical state parameters for tailings*

Ore type	Tailings type	Specific gravity	Voids ratio	Dry density (Mg/m³)	Bulk density (Mg/m³)
Coal	Slimes	1.4–2.1	0.5–1.1	0.7–1.4	0.8–1.7
Lead	Slimes	2.6–3.0	0.6–1.1	1.3–1.8	1.7–2.2
Molybdenum	Sands	2.7–2.8	0.7–0.9	1.5–1.6	1.6–1.8
Copper	Sands	2.6–2.8	0.6–0.8	1.5–1.8	1.8–1.9
	Slimes	2.6–2.8	0.9–1.4	1.1–1.4	1.5–1.9
Taconite	Sands	3.0	0.7	1.8	1.9
	Slimes	3.1–3.3	0.9–1.2	1.5–1.7	1.9–2.2
Bauxite	Slimes	2.8–3.3	8.0	0.4	1.2
Trona	Sands	2.3–2.4	0.7	1.5	1.7
	Slimes	2.4–2.5	1.2	1.1	1.6
Phosphate	Slimes	2.5–2.8	11.0	0.22	1.1

* After Mittal & Morgenstern (1975) and Vick (1983).

from point to point within a deposit, it appears that many beach sand tailings attain average relative densities in the range 30–50% from spigotting or similar procedures. Relative densities in this range can be achieved only by relatively clean sands. Slimes that are of low to moderate plasticity show high in situ voids ratios, ranging from about 0.7 to 1.3. Slimes of highly plastic clay or unusual composition (notably phosphatic clays, bauxite and oil sands tailings) are likely to have very high in situ voids ratios, ranging from about 4 to 10. Slimes can exhibit low in situ bulk densities (see Table 15.6), but typically they show an average increase of about $0.17 \, Mg/m^3$ for every 30 m depth.

15.4.2. Permeability

The coefficient of permeability of tailings materials can range widely because of the variability of grading and in situ density. The average permeability of tailings spans five or more orders of magnitude, from about 10^{-4} m/s for clean, coarse sand tailings to 10^{-9} m/s for well-consolidated slimes (Table 15.7). The actual permeability varies as a function of:

- grain size
- plasticity
- depositional mode
- depth within the deposit.

However, for a preliminary estimate it has been demonstrated (Mittal & Morgenstern, 1975) that the average vertical permeability for sand tailings is predicted reasonably accurately by the 'loose sand' form of Hazen's equation (see Chapter 7, Section 7.1.2):

Table 15.7. Permeability values for tailings

Material	Source	k_v (m/s)	k_h (m/s)
Sands	Mittal & Morgenstern (1975)	2×10^{-4} to 9×10^{-6}	—
	Genevois & Tecca (1993)	4×10^{-5} to 5×10^{-6}	—
	Vick (1983)	10^{-4} to 10^{-5}	—
Slimes	Genevois & Tecca (1993)	10^{-8} to 5×10^{-9}	—
	Routh (1984) — (China clay)	5×10^{-6} to 5×10^{-7}	2×10^{-5} to 5×10^{-5}
	Routh (1984) — (Tungsten)	2×10^{-7}	10^{-6}
	Blight (1994)	10^{-7} to 10^{-9}	—

$$k = 0.01 \, d_{10}{}^2 \tag{15.1}$$

where d_{10} is the 10% particle grain size (mm) and k is the coefficient of permeability (m/s). It has been suggested that Hazen's formula can be extended to non-plastic slimes tailings and cycloned sands, although the permeability of slimes is significantly affected, because of their compressibility, by changes in the voids ratio.

Although absolute permeabilities vary greatly, the change in k with decreasing voids ratio is reasonably consistent for most tailings sands. For the range of voids ratios encountered in most tailings sand deposits a decrease in k (with depth) by a factor of up to 5 is typical. Because of their layered nature, tailings deposits exhibit considerable variation in permeability between the horizontal and vertical directions – the ratio k_h/k_v is generally in the range 2–10 for reasonably uniform beach sand deposits. Transition beach zones between the areas of relatively clean sands and slimes are likely to have high anisotropy ratios (for the coefficient of permeability) due to interlayering of fine and coarse particles, and values of k_h/k_v of 100 or more are likely. Although the deposition process produces segregation or sorting of the particles within the transition zone, the overall variation in the average permeability with distance from the discharge point is not very significant. Typically the range is only an order of magnitude.

15.4.3. Shear strength

When tailings materials are deformed slowly under fully drained conditions their strength and deformation characteristics are usually unexceptional and their shear strength properties may be determined by conventional soils tests (direct shear box and triaxial tests, as described in Chapter 6, Sections 6.2 and 6.3, respectively). However, the behaviour of tailings during undrained loading is more problematical. The stress–strain characteristics of tailings in triaxial shear are generally similar to those of loose to medium-dense natural soils of similar grading. At typical in situ densities dilatancy seldom occurs. Normally there are high strains to failure and little or no reduction in post-failure strength at large strains (Figure 15.14). Stress–strain curves typically rise continuously throughout a test, often without showing any well-defined peak. For undrained loading of slimes the pore pressures rise with strain to reach a peak value (at around 5–10% axial strain) and then remain constant or decrease slightly.

Notwithstanding their generally loose depositional state, tailings have high drained strength owing primarily to their high degree of particle angularity. It is

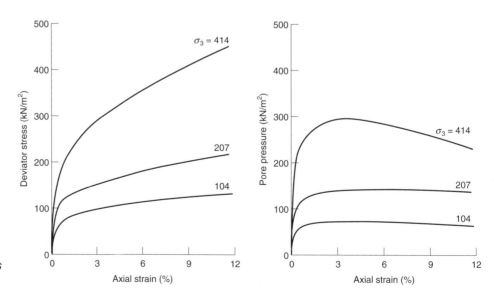

Figure 15.14. Stress–strain characteristics of fine tailings (copper)

Table 15.8. Shear strength parameters for tailings

Source	S_u/σ_v'	c'	ϕ'	Comments
Street (1987)	—	0	25–33	China clay sands, relative density 50–60%
Genevois & Tecca (1993)	—	0	42.3	Sands — fluorite tailings
	—	0	35–38	Slimes — fluorite tailings
	0.15–0.22	—	—	Slimes — fluorite tailings
Routh (1984)	—	0	28–38	Sands — tungsten
	—	0	32	Fines — tungsten
	—	0	25–36	Fines — china clay
Wroth & Hughes (1973)	—	0	34	China clay slimes
Vick (1983)	—	0	31–32	Gypsum slimes
Jennings (1979)	—	0	36	Gold slimes
Mittal & Morgenstern (1975)	—	0	33.5	Sands — copper
Charles (1986)	0.2–0.4	—	—	Hydraulic fills — slimes
Volpe (1979)	—	0	33–37	Sands — copper

not uncommon for tailings to exhibit an effective friction angle ϕ' that is 3–5° higher than that of similar natural soils at the same density and stress level. With rare exceptions, tailings are cohesionless and c' is zero. The effective cohesion intercepts for tailings that have been reported usually result from testing errors or erroneous linear extrapolation of curved failure envelopes. For most materials ϕ' generally falls in the range 25–40° over a wide stress range (Table 15.8). There is usually little variation between ϕ' for sands and slimes tailings (except for fine coal refuse, which may contain layers of highly plastic clay). This is because the slimes are generally composed of ground-up hard minerals (unlike normal clays) and so they are angular not flat.

The effect of the voids ratio on the effective-stress strength of tailings is small. The value ϕ' seldom varies by more than 3–5° for sand tailings for the range of densities commonly encountered. Similarly, overconsolidation, if present, has relatively little effect on the effective friction angle for slimes tailings. The most important factor influencing ϕ' for tailings is the stress level. Even at relatively low stress levels the contact stresses of the angular grains are very high, producing particle crushing. This usually results in curvature of the strength envelope, especially at low applied stress. The combined effects of particle crushing and dilatancy are most pronounced at stresses up to about $300\,\text{kN/m}^2$. At higher stresses ϕ' becomes essentially constant.

Field vane tests suggest that in situ tailings slimes exhibit undrained strength profiles similar to soft, normally consolidated deposits. Thus they can be characterized by the ratio S_u/σ_v', where S_u is the undrained shear strength and σ_v' is the vertical effective stress. Although the strength ratio S_u/σ_v' varies as a function of the overconsolidation ratio, slimes are mainly normally consolidated material with S_u/σ_v' in the range of 0.15–0.40. Undrained triaxial tests are not normally undertaken because of the problem of obtaining undisturbed samples. Since for fully saturated fines ϕ_u is essentially zero, it is better to estimate the undrained shear strength using an appropriate value of S_u/σ_v'. Published data for ϕ_u for saturated slimes (values of 14–24° have been proposed) are not reliable. Alternatively, an effective stress analysis can be undertaken.

15.4.4. Liquefaction potential

Liquefaction of a soil is a temporary state in which the structure of the soil is disturbed, causing the particles to lose contact. Fine sand and silt particles are susceptible to rapid and large reduction in strength due to very minor disturbances if deposited in a loose, saturated condition. Liquefaction can be caused by changes in static stress conditions, particularly in very loose soils. However, the more general cause of structural disturbance leading to liquefaction is dynamic loading such as can occur during an earthquake. The disturbance tends to cause a decrease in volume if the sand is not initially in a dense state. If drainage is prevented, or if it cannot occur rapidly, the volume change cannot occur and the relaxing sand skeleton transfers some of its load to the water, thereby causing a rise in the pore pressure. In general, the higher the intensity of the shaking and the longer its duration, the greater the increase in the porewater pressure. For liquefaction to occur the soil must have a sufficiently low permeability that no significant migration of water (and consequent pore pressure relief) is possible within the time frame of the earthquake. Under appropriate conditions the porewater pressure may rise sufficiently to cause the effective stress to become zero, so that the soil shear strength falls to zero — this is liquefaction. The soil behaves like a dense fluid and is unable to support loads or to resist significant shear forces.

The response of a soil to cyclic stress depends on a number of factors, but the soil type, relative density, grain characteristics and method of deposition are fundamental parameters. In general, clayey soils and coarse permeable cohesionless materials do not liquefy. Also, for dense sands, liquefaction is not a problem, provided that a minimum relative density of 75% is achieved. The most problematic materials are loose saturated silts and fine sands, typically at relative densities of 30–50%, and these are very prone to liquefaction as a result of vibration. In contrast to most natural soil deposits, tailings are usually highly angular, have been deposited by very specific hydraulic mechanisms, and have been recently deposited and are unlikely to have experienced previous seismic shaking. Relative densities of tailings are lower than those for many natural soil deposits, and therefore they may undergo large cyclic strains after a relatively few cycles of stress reversal. Attainment of approximately 10% cyclic strain, commonly used as a strain-related failure criterion, often more or less coincides with initial liquefaction (pore pressure equal to confining pressure) for many tailings samples. Slimes tailings overall generally have higher cyclic shear strengths than do sands or undifferentiated tailings.

Tailings with relative densities in excess of about 50–60% will not liquefy under accelerations of less than $0.1g$ approximately. For soils of the types and densities used in constructing tailings dams, liquefaction is assumed to occur only in saturated zones (i.e. below the phreatic surface). Partially saturated sand, such as may exist where water is percolating through the fill above the phreatic surface, will not liquefy because the air is highly compressible, thus preventing a significant rise in porewater pressure. However, even if the soil does not liquefy, the densification induced within a tailings dam is likely to reduce the effective stress to such an extent that large deformations occur.

15.4.5. Consolidation characteristics

Primary consolidation for tailings sands occurs so rapidly that it is difficult to measure. The few data available suggest that C_V varies from about 1000 to 300 000 m^2/year for beach sand deposits (Table 15.9). It is thus reasonable to assume that consolidation processes within the tailings dam take place rapidly, and that excess pore pressures are normally negligible (the steady-state pore pressures result from the seepage regime which is dominated by the liquid within the lagoon). For slimes C_V is generally about 0.3 to 300 m^2/year (see Table 15.9), which is in the same range as that exhibited by conventional clay soils.

While both sands and slimes tailings may appear to be more compressible than most natural soils of similar type, this is because of their loose depositional state

Table 15.9. Consolidation parameters for tailings

Source	Sands		Slimes		
	C_v (m²/year)	C_c	C_v (m²/year)	C_c	m_v (m²/MN)
Vick (1983)	—	0.05–0.07	0.3–30	0.19–0.35	—
Genevois & Tecca (1993)	—	—	60	0.15–0.37	—
Guerra (1972)	—	—	—	—	0.24
Volpe (1979)	1200	0.09	—	—	—
Mittal & Morgenstern (1975)	—	0.05–0.11	3–300	0.2–0.3	—
Nelson et al. (1977)	320 000	0.05–0.13	—	—	—
Chandler & Tosatti (1995)	—	—	95	—	—
Routh (1984)	39–142	—	11–43	—	0.1–0.9

and grading characteristics. If these materials are subjected to oedometer tests the compressibility values obtained are generally comparable to those of ordinary materials, provided that the density states are similar. Typical values of C_c (from one-dimensional consolidation) are included in Table 15.9. For sand tailings C_c usually ranges from 0.05 to 0.13. Most slimes of low plasticity show C_c ranging from 0.15 to 0.35. The looser or softer the initial state of the material, the higher the compression under loading.

Tailings do not always exhibit the well-defined break between recompression and 'virgin consolidation' portions of the consolidation curve shown by natural clays. In fact many sand tailings show broad curvature of the e–log(stress) relationship even after preconsolidation, so that compressibility coefficients often require specification of the stress range over which they apply. Slimes tailings usually exhibit preconsolidation effects similar to those shown by clay soils in that overconsolidation (by desiccation or capillary suction on exposed slimes deposits) produces a flatter initial recompression portion of the e–log(stress) curve.

For most types of tailings secondary compression produces continuing deformation under constant load, even after pore pressure dissipation is essentially complete. Secondary compression of tailings is, however, usually small and insignificant when compared to primary consolidation.

15.5. Engineered tailings disposal

Tailings dams may be considered to pass through three distinct stages: operation, rehabilitation and long-term. Tailings dams must be designed and constructed to meet adequate safety standards through all three phases of their life. Safe and economical tailings facilities can be built by applying the engineering knowledge and experience presently available from conventional water storage dam designs, suitably modified (as appropriate) to satisfy the special requirements of the mining industry. Properly engineered tailings dams are currently built to stringent requirements and design standards (e.g. to withstand the worst floods within a 100-year period), and regular checking and monitoring are standard.

Tailings structures pose similar dam safety problems to those posed by conventional water storage dams. However, important differences exist (see Box 15.1), which can be summarized as:

- the fluid retained is a pollutant
- a large part of the dam is constructed using waste materials
- tailings facilities are not constructed by civil or geotechnical engineers
- rapid drawdown of the lagoon is not possible.

15.5.1. Design aspects

There are several potential critical situations for analysis: the end of construction, during staged construction, and under long-term steady seepage. With staged construction there is the problem of whether the embankment and lagooned material is drained fully or partially. Within each raise and each time step, processes of excess pore pressure dissipation and generation occur simultaneously. Sources of pore pressure include seepage, initial excess pore pressure due to rapid uniform loading, and pore pressure changes due to shearing. In tailings embankments thin sand seams will normally be present and consolidation will be two-dimensional (outwards towards the face as well as vertically downwards) so that the estimation of pore pressure is usually conservative.

Long-term analyses are usually applied to raised embankments at their maximum height. It is normally assumed that excess pore pressures resulting from raising have dissipated (for steady-state seepage a reasonable initial approximation is that the static head corresponds to the depth below the phreatic surface). Other factors to be considered include:

- In addition to the usual data, the site investigation must cover site seismicity and the geochemical properties of the ground. Since seepage losses from a tailings lagoon could contaminate the surrounding groundwater, detailed information on existing groundwater conditions and the permeability of strata is vital.
- Internal filters and drains should be installed to prevent piping and to lower the phreatic surface within the dam. Special design features may be required at areas where cracking of the dam is considered a possibility. The design should include adequate monitoring of the phreatic surface and of the vertical and horizontal displacements.
- Under severe seismic shock all saturated tailings are likely to liquefy, becoming a fluid with a bulk unit weight significantly higher than that of water. Hence both static and dynamic analyses of the dam should be performed. Conventional pseudo-static analyses (which represent the effects of shaking by the use of a seismic coefficient or horizontal force) cannot predict failures due to high pore pressures or liquefaction.
- A large part of a tailings dam is constructed using the coarser fraction of the tailings. These building materials must be separated from the finer waste (slimes), and ideally strict quality control procedures should be employed to ensure compliance with grading and permeability specifications. Tailings are far from ideal dam-building materials in that they are highly susceptible to internal piping and they present highly erodible surfaces.
- Most of the dam construction is carried out by the mining operators, as part of the tailings disposal operation, and appropriate procedures are needed to ensure compliance with design assumptions.
- Because tailings dams are usually constructed slowly over a period of many years, modifications can be made as required throughout the long construction period. This is a critically important aspect of design as it allows far more flexibility than is available in the design of conventional water-retaining dams.
- The design should include a consideration of the very long-term situation in terms of abandonment, reclamation and rehabilitation of the site.

15.5.2. Slope stability

Pore pressure conditions within a tailings embankment exert a controlling influence on its stability. For embankments raised less than 5–10 m each year excess pore pressures may be assumed to dissipate as rapidly as the load is

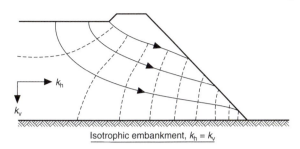

Isotrophic embankment, $k_h = k_v$

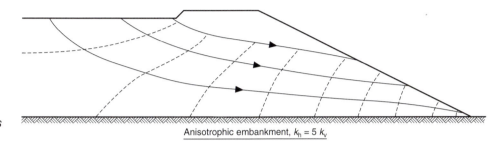

Figure 15.15. Typical flownets for tailings facilities

Anisotrophic embankment, $k_h = 5\,k_v$

applied (Vick, 1983). For faster rates of construction the embankment has to be modelled as a series of discrete raises and analysed at incremental time steps. Seepage in tailings embankments is commonly assumed to occur under gravity flow and is usually determined for steady-state conditions. Once the phreatic surface (or top flow line) within the tailings facility has been located, the pore pressures needed for stability analysis can be determined. The most common procedure for determining phreatic surface location is by drawing a flownet (see Chapter 7, Section 7.4.3). Typical examples are shown in Figure 15.15.

Slope stability analyses for tailings embankments concentrate on initial rotational-type slides incorporating the rigid-body assumptions of limiting-equilibrium analysis. These analyses only represent conditions of incipient failure and are not intended to describe the behaviour of the embankment after failure has been initiated. In fact, although most tailings embankment slides probably have a rotational-type slide as their trigger mechanism (except for cases of seismic liquefaction), it is the flow-type behaviour that is so destructive. The stability of tailings facilities is analysed using the method of slices for rotational failure (see Chapter 9, Section 9.3) or a wedge failure mechanism (see Chapter 9, Section 9.5). The sand and transition zone are regarded as free-draining sections, while for the slimes shearing will take place under undrained conditions. In many cases involving slopes in tailings, shallow critical failure surfaces within the dam are found to approach the infinite-slope condition (see Chapter 9, Section 9.2). In such cases the approximations given in Table 15.10 are often useful in conjunction with a simplified analysis for cohesionless materials ($c' = 0$).

Table 15.10. Approximations for the Factor of Safety (FOS) against shallow failure within tailings dams

Case	Relationship
Dry slope	$\mathrm{FOS} = \dfrac{\tan \phi'}{\tan i}$
Seepage parallel to the slope, phreatic surface at ground level	$\mathrm{FOS} = \left(\dfrac{\gamma - \gamma_w}{\gamma}\right)\dfrac{\tan \phi'}{\tan i} = \left(1 - \dfrac{\gamma_w}{\gamma}\right)\dfrac{\tan \phi'}{\tan i}$
Horizontal seepage, fully saturated soil	$\mathrm{FOS} = \left(\dfrac{\gamma - \gamma_w \sec^2 i}{\gamma}\right)\dfrac{\tan \phi'}{\tan i}$

The exact determination of the Factor of Safety for a tailings embankment generally requires an extensive physical property sampling programme and the use of a computer model. However, with the upstream method of construction the tailings 'dam' is in reality a facing or veneer to the slimes deposit. The same is also true, to a large extent, for the centreline method of construction. Hence a circular failure surface will pass mainly through the slimes, which have relatively little variation in effective shear strength parameters. Consequently, slope stability charts (see Chapter 9, Section 9.4.2) for homogeneous materials and uniform conditions are useful for rapid, preliminary assessment of the effect of such variables as phreatic surface location and strength parameters, even though they do not produce a precise answer for the real situation.

Tesarik & McWilliams (1981) have derived slope stability charts for tailings embankments using the Bishop simplified method of analysis, with the failure mass being divided into 80 slices. The failure surface assumed was a circular arc, and different trial circles were tried by using a grid of centres and several different radii. To eliminate the calculation of an average r_u value, the bulk unit weight was defined as a fixed parameter (15.7 kN/m^3); for small differences from this value the relevant parameters can be interpolated. The results of the analysis were assessed by comparing them with the data reported by Bishop & Morgenstern (1960) and O'Connor & Mitchell (1977). The differences between the three sets of values (where they could be compared) were insignificant.

A sequence of least-squares curve-fitting steps was used to prepare the final charts for embankments with slopes ranging from 1:1 to 1:5 (vertical to horizontal). A sample of the charts is presented in Figure 15.16. The charts cover

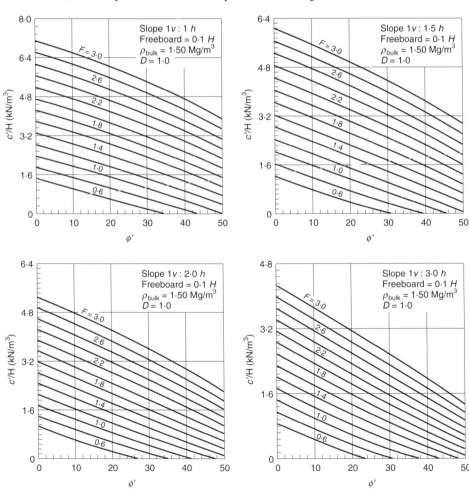

Figure 15.16. Slope stability charts for tailings facilities (after Tesarik & McWilliams, 1981)

two different groundwater conditions: no phreatic surface, and embankments with 10% freeboard, i.e. the free water surface behind the tailings embankment is at a depth of $0.1\,H$ below the surface of the lagooned material (where H is the height of the dam). The charts in Figure 15.16 are for a depth factor D of unity (i.e. the tailings dam/lagoon is sat on an effectively rigid stratum), since this is the most realistic situation in practice. Interpolation between charts (to account for the variation in the bulk density) is permissible.

15.5.3. Dynamic stability

With tailings dams constructed of saturated cohesionless soils a primary cause of slope failure is the build up of porewater pressure and the consequent reduction in shear strength resulting from strong shaking or vibration of the dam. Hence stability analyses of tailings embankments are commonly performed using either a pseudo-static or deformation approach.

The *pseudo-static analysis* differs little from conventional static methods, and in effect the embankment is treated as a static structure but with the incorporation of horizontal and vertical forces which are induced by vibration or an earthquake. The value of this force is expressed as the product of a seismic coefficient (typically $0.1-0.2$) and the weight of the potential sliding mass. This approach does not accurately model true embankment behaviour under major earthquake shaking (i.e. acceleration is not constant over the surface of the embankment and the approach does not account for cyclic liquefaction or major pore pressure build-up). These factors suggest that the use of a simple pseudo-static approach should be restricted to embankments of low height where the seismic risk is low.

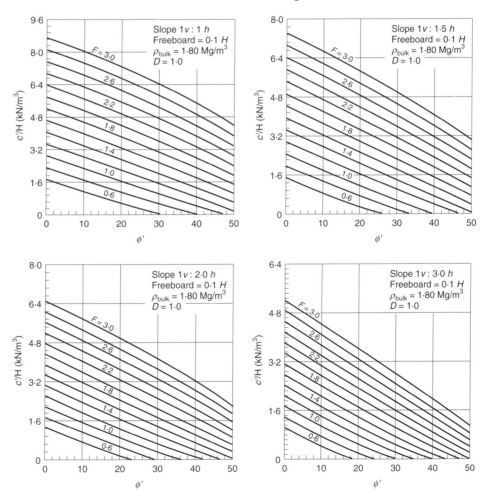

Figure 15.16. Continued

Furthermore, it is recommended that undrained strength parameters are used for saturated uncompacted materials of low to moderate permeability to account for pore pressures generated (due to shear) by the rapidly applied earthquake loading.

Modified pseudo-static methods incorporate relevant effects. The shear strength envelope for embankment materials is obtained from cyclic shear triaxial tests (thus pore pressure generation is accounted for), and in the analysis it is assumed that slurries liquefy and produce a rapidly applied load on the embankment itself.

Deformation approaches may be applied to embankment materials that do not experience severe strength reduction during the application of cyclic stresses. The method assumes that the embankment slope will not undergo complete failure by sliding, but will experience some degree of permanent deformation (an assumption verified by the observed performance of compacted dams). The basis for evaluating the suitability of embankment behaviour under seismic shaking is whether or not the computed deformations lie within acceptable limits (e.g. crest deformations are not large enough to cause overtopping).

Complete dynamic analysis is complex and requires substantial information on the dynamic behaviour of the embankment and underlying ground. In general, the analysis incorporates the following steps:

- determination of the pre-earthquake static stresses in the embankment (typically performed using a static finite-element analysis)
- determination of the dynamic responses of the embankment and foundation (requires the establishment of a bedrock acceleration – time history)
- evaluation of the dynamic soil behaviour (ordinarily accomplished by performing cyclic triaxial tests under appropriate stress conditions)
- evaluation of embankment performance (in terms of stresses or strains occurring locally)
- comparison of the latter values with those required to cause liquefaction or to exceed prescribed limits of local strain.

Properly engineered dams constructed of clay on clay or rock foundations have withstood strong shaking under accelerations of $0.35–0.8g$ without significant damage. Upstream-constructed tailings embankments have experienced, and survived, accelerations up to about $0.15g$. It has been proposed (Seed *et al.*, 1977) that well-built hydraulic fill dams, with reasonable slopes and good foundations, should survive moderate earthquake shaking with peak accelerations up to $0.2g$ (resulting from earthquakes of magnitude 6.5–7.0 on the Richter scale). For a particular site the anticipated design acceleration level can be estimated from historical data or by using probabilistic methods.

15.5.4. The lagoon

The tailings pond should operate in such a manner that a substantial beach of slimes is maintained between the upstream face of the tailings dam and the free water surface in the pond. These slimes should act as an upstream low-permeability membrane and reduce seepage through the sand dam to a very small and acceptable quantity. The wider the beach the lower the seepage line through the dam and the smaller the danger of piping failure within the sand dam. Particular care should be taken to prevent the water from issuing from the outer face of the dam, otherwise erosion, pipe formation and collapse are likely to result.

The beach slope of the hydraulic fill is usually very flat. Typical beach gradients are of the order of 0.5–2.0% and they change with distance from the dam because of the particle-size separation that occurs after deposition. Although tailings slurries may be transported as reasonably homogeneous materials, this state ceases upon deposition. The coarse material in the slimes stream settles out

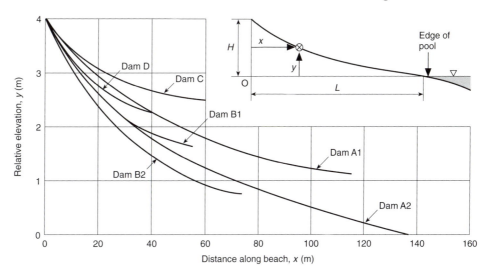

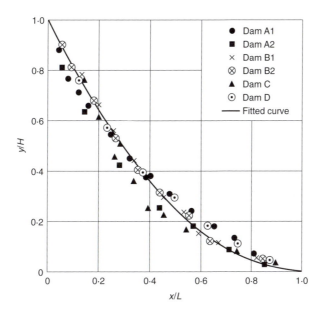

Figure 15.17. Beach profiles for tailings lagoons (after Blight & Bentel, 1983)

first. It also drains rapidly and gains strength more quickly than the finer material, and hence there will be a progressive flattening of the beach gradient with distance from the dam. Figure 15.17 shows beach profiles from six platinum tailings facilities.

Studies done in Russia in the early 1970s by Melent'ev (1973) on the form taken by hydraulic fill beaches for water-impounding dams built by hydraulic filling showed that beaches constructed of a specific type of particulate material, with similar solids content at placing and a similar rate of placement, have a similar geometric form (Figure 15.18). A two-dimensional profile of the beach can be expressed as a dimensionless master profile (or Melent'ev profile) by the equation:

$$\frac{y}{H} = \left(1 - \frac{x}{L}\right)^n \tag{15.2}$$

where y, H, x and L are as defined in Figure 15.17. Typical values of n are indicated in Figure 15.19.

In addition, the trend line for the variation of particle size with distance x along a beach can be characterized as:

Figure 15.18. Master profile for tailings beaches (after Blight & Bentel, 1983)

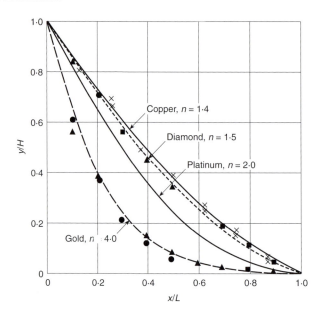

Figure 15.19. The effect of tailings material on the n *value of Melent'ev*

$$\left(\frac{d_{50\ \text{at}\ x}}{d_{50\ \text{max}}}\right) = e^{-Bx/L} \tag{15.3}$$

A typical value of B would be 2.5 (Blight, 1994). Since the permeability of a granular material has been shown (Hazen, 1892) to be related in some way to particle size, the permeability of hydraulic tailings fill will decrease continuously from the point of deposition up to the decant pool, according to a relationship of the form:

$$k = a\,e^{-bx} \tag{15.4}$$

15.5.5. Hydraulic aspects

The location of the phreatic surface (internal water level) within an embankment exerts a fundamental influence on its behaviour, and control of the phreatic surface is of primary importance in tailings embankment design. The prime objective is to keep the phreatic surface in the vicinity of the embankment face as low as possible. The primary factors influencing phreatic surface location are:

- the location of the ponded water relative to the embankment crest (often the most important item)
- the degree of grain-size segregation
- lateral variation in permeability
- the permeability of the foundation relative to that of the tailings.

For an upstream tailings dam the initial seepage entry equipotential is essentially horizontal (corresponding to the flat bottom of the decant pond), and as a result the flow lines are initially directed vertically downward (see Figure 15.15). Where a slimes zone is present, considerable head loss occurs within these low-permeability tailings. As seepage flow becomes directed horizontally outwards towards the embankment face, the resulting phreatic surface may be considerably lower than that which would result from conventional water dam boundary conditions. For downstream and centreline tailings embankments, internal zoning and boundary conditions are often sufficiently simple and similar to those for conventional water dams that published flownets can be adapted to determine the phreatic surface. Figure 15.20 shows the influence of the relative permeabilities of the embankment and lagooned material on the location of the phreatic surface for an isotropic downstream embankment.

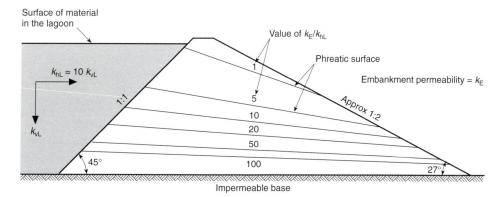

Figure 15.20. The location of the phreatic surface within a downstream embankment

Due to the method of their deposition, tailings deposits have marked anisotropy of permeability and this has a greater influence on the seepage pattern than does the actual value of the coefficient of permeability. Figure 15.21 illustrates phreatic surfaces for several homogeneous (but anisotropic) embankments with different beach widths L. It is important to note the dramatic influence of the beach width on the phreatic surface directly beneath the embankment sludge. A value of L/H less than 5 would be undesirable.

Perhaps the second most important factor influencing the location of the phreatic surface for upstream embankments is the degree of lateral permeability variation (Figure 15.22). If the variation in the beach permeability is large (e.g. $k_c = 100 k_L$), even a narrow beach may produce acceptable phreatic conditions. The whole phreatic surface will, of course, be considerably lowered by a permeable foundation to the tailings facility.

For upstream construction it is important to provide a starter dyke that is more permeable than the tailings. However, problems related to piping and improper filter zones in tailings embankments are most often encountered where coarse mine waste is used as a construction material directly in contact with the tailings, especially where end-dumping of the mine waste in high lifts has produced severe size segregation. Thus the arrangement and types of materials within the embankment are governed not only by the control of the phreatic surface, but also by filter requirements to prevent migration of soil or tailings into adjacent coarser zones. A general principle is that the permeability of any internal zones should increase in the direction of seepage flow. Filter requirements to prevent piping are well established from conventional dam design practice (Cedergren, 1967):

$$\frac{d_{15 \text{ filter}}}{d_{85 \text{ protected soil}}} < 5, \qquad \phi \frac{d_{50 \text{ filter}}}{d_{50 \text{ protected soil}}} < 25 \qquad (15.5)$$

The use of synthetic filter fabrics to replace conventional graded sand filters may be attractive for new tailings constructions if natural materials are difficult or expensive to obtain.

If the stability of a tailings facility is reasonably assured it is essential to evaluate the seepage in order to provide a tailings impoundment design that will

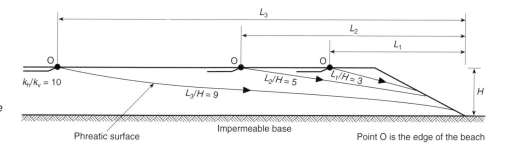

Figure 15.21. The effect of the beach width on the phreatic surface

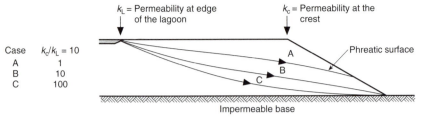

Figure 15.22. The effect of permeability variation on the phreatic surface (after Vick, 1983): k_c, permeability at the dam crest; k_L, permeability at the edge of the ponded water

minimize environmental damage. Permeability is the physical factor that most influences groundwater contamination potential. Tailings impoundments on naturally impermeable foundations or those that incorporate impermeable liners will have little or no effect on the underlying groundwater regime. Geological discontinuities and heterogeneity significantly affect aquifer behaviour, and evaluation of the degree of fracturing, jointing and faulting is necessary to assess the potential for concentrated seepage through discontinuities.

Seepage control systems range from unrestricted flow (which is collected), through partial retention, to total restriction of seepage by structural measures. Seepage return systems do not attempt to restrict seepage flow, but rather to collect it (thereby minimizing migration of pollutants) and return it to the source. The two basic systems are collector ditches or sumps and collector wells. Contaminant removal as seepage progresses through the subsurface is an important consideration for groundwater quality and protection of the environment. It is a result of dilution processes and chemical reactions. Removal by chemical reaction is often far more significant than dilution by volumetric mixing in reducing contaminant concentrations.

Seepage barriers are only fully effective when permeable foundation materials are underlain by a continuous impermeable stratum and the barrier penetrates the permeable layer completely. A barrier that penetrates 90% of the permeable stratum will reduce the seepage by less than two-thirds. The primary types of seepage barrier are cut-off trenches, slurry walls and grout curtains (see Chapter 7, Section 7.5.1 and Chapter 14, Section 14.1.2).

15.5.6. Reclamation of tailings facilities

Tailings management must continue until such time as the deposited tailings are assured to be permanently stable and environmentally innocuous. Subsequent reclamation of the site is a relatively new concept in the field of tailings disposal, and this may be a reflection of the likely cost (in the region of £1 million to £10 million). The fundamental, long-term objectives of reclamation are:

- impoundment stability
- erosion prevention
- avoidance of environmental contamination
- eventual return of the area to productive use.

The principal difference in stability between the operating and abandonment cases is in phreatic conditions. Ordinarily when tailings discharge ceases and a continuous source of water no longer supplies the decant pond, the phreatic surface within the embankment drops dramatically, resulting in greater embankment slope stability after reclamation than during operation. Immediately after tailings discharge ceases, the lagoon surface will consist of a relatively firm and dry sand beach, a soft and saturated slimes surface, and a submerged area covered by the decant pond. The decant pond may be dried by evaporation or by drainage — desiccation and consolidation of the slimes surface may take considerable time. Figure 15.23 shows the first stage of an attempt to improve a tailings lagoon, by covering it with a geotextile and then applying a surcharge of free-draining fill to cause consolidation. Chemical stabilizing agents have been

Figure 15.23. A geotextile cover being placed on top of slimes in a tailings lagoon

used for temporary tailings erosion control, but they cannot be considered as a permanent reclamation measure.

The anticipated partial saturation of the tailings following abandonment and reclamation usually precludes liquefaction (see Section 15.4.4), even under major seismic shock. However, this expectation is based on the assumption that the phreatic surface falls after the site is closed, but this situation may not occur if rainwater is not prevented from entering the wastes. Hydrologically induced failures are the major cause of mass instability of abandoned tailings deposits. Accumulation of run-off water in an impoundment, in addition to raising the possibility of slope or seismic instability, can cause direct failure by overtopping or by erosion of the embankment toe. Provided effective measures are taken to prevent long-term accumulation of water in the impoundment, any tailings embankment slope that was stable during operations should maintain its overall integrity after operations cease.

Wind erosion is primarily a factor on flat, unbroken surfaces, while water erosion is most often a problem on embankment slopes (tailings are notoriously susceptible to gullying by water run-off erosion). Experience indicates that embankment slopes flatter than 1:3 will usually provide reasonable erosion resistance. The use of riprap for erosion stabilization follows from its conventional use for engineering purposes as channel protection and slope protection to prevent water erosion. Riprap includes not only conventionally sized rock fragments but also gravels.

While seepage ordinarily diminishes and eventually ceases after termination of tailings discharge, special reclamation-related precautions may be necessary in some cases. Partial saturation may enhance oxidation, thereby reducing the pH and increasing the liberation of metallic contaminants. These oxidation-produced contaminants may be potentially much more noxious than those present during impoundment operations. To prevent leaching of contaminants in such cases, a clay cap is often required over the impoundment surface, in conjunction with grading to prevent ponding of run-off. When mining operations cease the tailings pond passes through a rehabilitation phase. This often involves draining the pond and diverting surface water, sealing the surface of the tailings pond, providing vegetative cover, and providing whatever permanent spillway facilities may be required to handle surface runoff.

While the return of the land to productive use is certainly a desirable goal, the term 'productive use' needs to be clarified. It could be defined in the context of land-use patterns that existed prior to development of the site and reinstatement of the area to its former state. However, some treatments may not return the area

Table 15.11. Examples of plants for the reclamation of tailings facilities

Species	Varieties	Erosion control	Deep reinforcement	Water removal	Comments
Grasses and sedges	Common bent	X			Wide soil tolerance, some varieties tolerant of heavy metals
	Couch grass	X			Not tolerant of extremes
	Creeping red fescue	X			Very wide soil tolerance, some varieties tolerant of heavy metals and salt
	Flattened meadow grass	X			Tolerant of very infertile conditions
Legumes	Crown vetch	X			Wide tolerance, slow to establish
	Birds-foot trefoil	X	X		Wide soil tolerance, salt tolerant
	Alsike clover	X	X		Tolerates waterlogging
Trees	Common alder		X	X	Suitable for wet sites and derelict land
	Birches	X	X		Tolerant of infertile conditions, good pioneer species
	Hawthorn		X	X	Wide soil tolerance, salt tolerant

to its original state, but they may still make the land fit for use in some productive way.

Vegetation is by far the most common and usually the preferred stabilization option. If a self-perpetuating vegetative cover can be established (Table 15.11), not only can wind and water erosion be minimized, but also the impoundment can be returned to some semblance of its original appearance. Vegetation growth is dependent on two principal factors: climate characteristics and the nature of the growth medium. Many tailings facilities are located in climates that are arid, cold or otherwise inhospitable to vegetative growth. Soil characteristics that strongly influence vegetative growth are texture, fertility and toxicity. The texture will affect the degree of aggregation of individual particles (and hence the aeration potential of the ground) and the water-retention capability. Tailings sands have poor moisture-retention characteristics, and slimes are poorly aerated and become compacted on drying (both factors inhibit root penetration). Soil fertility is reflected by the availability of nutrients required for plant growth (including nitrogen, potassium and phosphorous), as well as the necessary bacteria and fungi (many tailings are virtually sterile in this sense because of their mode of generation). High levels of toxicity stunt or kill developing plants, although some heavy metals are necessary for healthy plant growth (but only in very small quantities). Acidic water (due to pyrites oxidation) and excessive salinity may kill plants. Species chosen for revegetation should be matched to the soil characteristics, erosion potential and microclimate of the tailings impoundment. Examples of suitable grass-seed mixtures are given in Chapter 14, Table 14.3.

16 Waste materials in geotechnical construction

16.1. Introduction

The advancement of society and infrastructure development is supported by the work of the construction industry, and this results in increasing demands for materials to produce aggregates, cement, concrete, bulk fill, asphalt, etc. Over a 35-year period from 1960, the total annual production of aggregates (sand, gravel and crushed rock) within the UK increased from about 100 Mt to over 300 Mt. Construction materials have usually been obtained from primary sources, but there are many millions of tonnes of waste material and industrial by-products that are potentially capable of satisfying a substantial proportion of industry's demands for aggregates, fill materials and other construction materials.

Mining spoil has been used extensively and successfully as a bulk fill material in the construction of motorway and rail embankments, river and sea defences, etc. In 1926, minestone (the waste from coal-mining) was used as the inner body of a river embankment in the Ruhr Coal Region (the minestone was overlain by layers of lower permeability materials). Minestone has also been used in the Silesian Region of Poland for the construction of large embankments, dykes for reservoirs, and settling lagoons for slurry, tailings and fly ash since the early 1960s. In the 1970s, when the motorway building programme was at its height in the UK, about 8 Mt of mining waste was being used in engineered construction each year. This represented the biggest single useful commercial outlet for colliery spoil.

During the construction of the Spalding bypass (Lincolnshire) local absence of competitively priced traditional granular fill led the contractor to look for alternative materials (*New Civil Engineer*, 1993). The nearby Peterborough brickfields turned out to be a good source of material (i.e. rejected house bricks). The contractor's proposal to use crushed and graded house brick for the drainage blanket ($90\,000\,m^3$) and whole and broken bricks for the main embankment fill ($170\,000\,m^3$) was approved. Bricks for the drainage blanket were crushed and uniformly graded to a maximum size of 125 mm, and the moisture content of the aggregate adjusted at the source of supply. The graded mixture complied with the Specification for Highway works requirements in terms of fines value and strength. The highest (5 m) embankment sections are around 42 m across at the toe and 12 m across at the crest, with 1 : 3 side slopes.

The five main sources of construction aggregates in Britain are land-won sand and gravel, crushed limestone, crushed igneous rock, crushed sandstone or gritstone, and marine-dredged material. The environmental impacts of the extraction and transport of geotechnical materials are a source of significant concern and include:

- the loss of mature countryside
- visual intrusion
- heavy lorry traffic on unsuitable roads
- noise

- dust and blasting vibration
- effects on flora and fauna
- pollution of surface and groundwater.

The extraction of aggregates involves the loss of two finite natural resources: the aggregates themselves, and the unspoilt countryside from which they are extracted.

Any construction project should be planned to minimize resource depletion, environmental degradation, environmental impact and energy consumption. The use of alternative or waste materials in place of naturally occurring materials can make a major contribution to these objectives.

16.1.1. Waste materials

Industrial development and economic growth have led inevitably to the production of large quantities of waste materials and by-products. There are various sources of solid waste:

- quarrying overburden and unwanted rock during the extraction of minerals (e.g. colliery spoil)
- wastage during the preparation/processing of quarried materials (e.g. slate waste)
- heat and electricity generating industries (producing pulverized fuel ash (PFA) and furnace bottom ash (FBA))
- process industries (e.g. the production of blast-furnace slag).

Coal extraction, whether by opencast or deep-mining, produces large quantities of waste rock and stone because of the intimate intermingling of coal and rock layers. The China clay industry in south-west England produces vast quantities of waste, and other industrial activity has left legacies in Wales, the Midlands, Scotland, etc. Although there are many mineral wastes in Britain, six are by far the most important by virtue of the scale of their stockpiles or their annual production or their potential for utilization, as indicated in Table 16.1.

Table 16.1. Sources of mineral wastes in Britain*

Material	Quantity	Location
Coal-mining spoil	Approximately 2000 Mt in tips and 40 Mt produced each year	Numerous locations in the UK, but the largest volumes are found in tips in Yorkshire, Nottinghamshire and South Wales
China clay spoil	Approximately 600 Mt in stockpiles and 27 Mt produced each year	South-west England
Slate waste	Between 400 and 500 Mt in tips and 6 Mt produced each year	The largest proportion is in north Wales, but there are also tips in the Lake District, the west of Scotland and the West Country
Power station ashes	Around 13 Mt of ash is produced each year, of which about 6 Mt is dumped annually in lagoons, tips, etc.	Throughout Britain, but concentrated in coal-producing areas
Demolition and construction waste	About 24 Mt is produced each year, about 12 Mt is re-used and the rest goes to landfill sites	Throughout Britain, but mainly close to large conurbations
Blast-furnace and steel slags	All blast-furnace slag (about 4 Mt per year) is fully utilized, little steel slag (about 2 Mt per year) is used	Old tips are to be found in the Midlands, Wales, the north-east of England and Scotland

* After Whitbread *et al*. (1991).

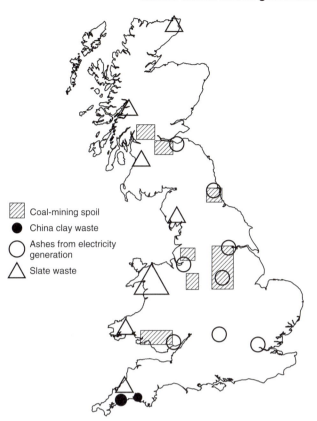

Coal-mining spoil

China clay waste

Ashes from electricity generation

Slate waste

Figure 16.1. The location of the major stocks of solid waste materials in Britain

Figure 16.1 shows the primary locations of waste production areas and by-product stockpiles.

Within construction, more mineral waste utilization takes place than is commonly realized. At the beginning of the 1990s in Britain, of the total consumption of aggregates of around 332 Mt per year, about 32 Mt was derived from secondary materials (Whitbread *et al.*, 1991). Major secondary aggregates are power station ash and blast-furnace slag and these represent some of the best examples of productive use of waste materials. Other widely used materials are demolition and construction wastes and asphalt road planings. While most of this sort of usage is not as graded aggregates, it still represents a provision that would otherwise have to be met from primary materials.

In terms of potential greater re-use of waste materials in construction, studies have identified that:

- Colliery spoil is of variable quality and its low value means that there are only a limited number of situations where the benefits of using it are not outweighed by transport costs.
- Because of its geographical location, China clay waste would need considerable financial support, to offset transport costs, for it to be used.
- Only a relatively small cost incentive might be required to initiate bulk shipping of slate aggregate to the south-east of England because of the high amenity value of this region.
- Future production of power station ash is likely to decrease and locations will continue to be concentrated more unfavourably. Increased sales could be encouraged if more high-value, non-geotechnical end-uses (such as cement substitution) could be generated.
- There appears to be some scope for greater use and recycling of demolition wastes since landfilling costs are rising due to the Environmental Protection Act.

16.1.2. Environmental aspects

Construction materials and products range from naturally occurring materials to complex elements comprising many components and a variety of processed materials. All such materials affect our environment to a greater or lesser extent, whether in terms of extracting a finite source, or in terms of the energy consumed by a product during its use. However, the construction industry's consumption of materials is not solely an issue of depletion of natural resources, since the winning, processing, use and disposal stages all have the potential to create impacts relating to the landscape, noise, air quality, water and land pollution, local and regional ecology, transport, etc. At any stage in the construction process, from the extraction of materials to the disposal of demolition waste, industry creates (directly or indirectly) various forms of pollution, which are discharged to the atmosphere, the land or various water systems. Thus the construction industry has a substantial impact on the environment by virtue of both the large quantities of resources that it uses and the wastes that it produces. Table 16.2 contains an outline of some of the materials implications of construction. It is difficult to quantify accurately the environmental impact of construction materials because of:

- the paucity of available data and uncertainty about the reliability, accuracy and consistency of these data
- the complexity of the life cycles associated with the winning, manufacture, use and disposal of construction materials
- the difficulties in balancing environmental, social, health and economic uses.

With the exception of timber and timber-derived products, few raw materials used in the construction industry can be replaced. From environmental considerations it is desirable that more waste materials should be used instead of natural aggregates. The adoption of waste minimization and recycling techniques will help to extend the reserves of non-renewable materials.

Due to an increased awareness of the environment, planning approvals to develop new quarries in the UK are currently running at about half the rate of aggregate extraction (O'Mahony, 1990). The use of secondary materials may not completely remove the problem of a potential shortage of aggregates, but it could

Table 16.2. Materials used in construction

Works	Specific items	Materials produced	Materials needed
Site preparation	Demolition Access Decontamination	Hard rubble Used timber Inert spoil Contaminated soil	Temporary access
Earthworks	Excavation Filling Drainage provision	Spoil (made ground, soil, rock)	Suitable fill (clay to gravel size) Aggregates
Structures	Concrete Steelwork Timberwork	Quarry waste Slags Ashes	Aggregates Cement Timber Steel
Drainage and water	Excavation Pipes Backfill	Spoil Quarry waste Slags Ashes	Aggregates Cement Steel
Roads	Flexible pavement Rigid pavement Supporting layers	Planings Quarry waste Ashes	Aggregates

Figure 16.2. Quarry waste (more than 100 years old)

alleviate it significantly. Long-term projections indicate that over the next 20 years the British construction industry may need as much as 7000 Mt of aggregates. The raw materials for civil engineering construction are obtained by opencast methods, which scar the landscape, while mineral wastes, for which there is little demand, are usually stockpiled in spoil tips. All countries with an industrial heritage have large areas of derelict land containing pits and waste heaps (Figure 16.2). Habitat reinstatement, substitution and translocation are often accepted as good practice for certain types of extractive process. In some cases, the land can be put to a beneficial use after its extraction phase has been completed. However, most extractive industries produce irreversible effects on the natural environment, involving the destruction of habitats and resulting in barren land that is difficult to reclaim or recultivate, or producing vast quantities of spoil. Industrial re-use of waste produces benefits by reducing industrial dereliction, making land at present covered with unsightly spoil tips available for use, conserving supplies of natural aggregates and reducing the need to open new quarries and pits.

If there is a convenient source of waste that is suitable for construction usage, the decision to use it rests partly on the economic factors and partly on the environmental benefits to be gained by using the waste or by-product. It should be appreciated that there are both benefits and disadvantages to the re-use of stockpiled or dumped waste materials (Box 16.1). A dominant environmental factor is the extent to which working a waste deposit will assist the eventual restoration of the land. Limited removal of material from a tip may hinder rather than help eventual restoration and, if only partial removal of a deposit is planned, then the stability and landscaping of the remainder will require significant consideration in order to avoid aggravating dereliction.

Care should be taken in the use of wastes from the mining or quarrying of metal ores, as these often have high concentrations of metallic contaminants, which may pose hazards to public health or the growth of plants if used as fill or cover materials on residential sites. Many waste materials and by-products contain traces of toxic elements that, given the right conditions, could become mobile due to tip instability (Figure 16.3) or due to leaching. There is also the question of worker safety during exploitation of waste materials, because of the effects of contact with the materials, and the nature (potentially dangerous due to

Box 16.1. Environmental aspects of the re-use of waste materials in construction

- **Partial removal of waste tips.** It is seldom feasible to use the whole tip, and this may hinder eventual restoration. However, recontouring of the remaining material may give visual improvement
- **Increased haulage costs.** The waste tip will almost invariably be further away from the site than a borrow pit
- **Avoidance of borrow pits.** Although a conventional borrow pit can be located close to the site, thus avoiding the need for haulage, it leaves a hole that needs careful restoration
- **Conservation of natural aggregates.** The best materials are saved for situations where they have to be used
- **Disturbance caused by haulage.** There are numerous disbenefits from haulage using public highways (e.g. congestion, dust, noise)
- **Variability of the waste materials.** Variations in material chemical composition, stability, etc., can have a serious effect on the final construction works
- **Possibility of long-term pollution problems.** Leaching of soluble minerals is possible in fill situations, but most suitable wastes are chemically inert

Figure 16.3. Instability of an old industrial tip

non-engineered construction) of spoil tips and deposits. Caution should also be exercised in working spoil tips in case excavation exposes a hazard (e.g. a zone of burning shale, or a cavity due to previous combustion). There may also be high local levels of gases such as carbon monoxide, sulphur dioxide and hydrogen sulphide.

16.1.3. Use of waste materials in construction

Recycling of solid wastes has steadily progressed and there has been a greater recognition of the inherent value of many recovered materials, as well as the need to conserve more of our limited natural resources. In order to complete the loop for recycling of solid wastes uses must be found for many recovered materials. However, the use of 'cheap' waste material can rapidly become uneconomic as haulage distances increase and if the material needs 'processing' to make it comply with the specifications and standards. The main constraints to re-use of waste materials can be summarized as:

- transport costs due to stockpiles being distant from major areas of use
- variability of the properties of a 'natural' material
- variable rate of production of waste and uncertainty of availability when required
- client reluctance to using a waste material in a manufactured product
- the need to comply with existing, established material specifications
- the costs of processing the waste to convert it to a suitable form for usage.

There is little doubt that waste materials (and by-products) have been used more in highway engineering than in other branches of construction, because of the opportunities to use large quantities of bulk fill (Figure 16.4), although wastes can be used at various elevations within road construction. The bottom of the road structure on which the pavement layers are constructed (the subgrade) may be in situ material (usually soil) or fill material that has been imported. If the load-carrying capacity of the subgrade is low, it may be advantageous to replace the upper part of the subgrade by a material that has better strength characteristics. Fill used in this way is often referred to as *selected fill*, and the layer formed by it is termed a *capping layer*. The thickness of a road pavement is not just determined by the strength of the underlying subgrade, but also by its stiffness (i.e. its deformation under load). This has led to subgrade strength being expressed in terms of the resistance developed when the subgrade is penetrated by a certain amount. An empirical test has been derived wherein the resistance to penetration by a standard plunger (50 mm diameter) is expressed as a percentage of the known resistance of the plunger to two different penetrations (25 and 50 mm) of an ideal aggregate (i.e. crushed limestone). The higher of these two percentages is called the Californian Bearing Ratio (CBR). The test results are plotted in the form of a load–penetration diagram by drawing a curve through the experimental points.

Many waste materials have good load-carrying capacity when compacted. Indeed some compacted wastes may comply with specifications for the road pavement layer above the subgrade (i.e. the sub-base). In addition to providing strength, the sub-base can also act as an insulating layer against freezing where the subgrade material is likely to be weakened by the action of frost.

Considerations about the stability of earthworks constructed from waste materials are the same as those for situations where naturally occurring materials are used. The methods for controlling the sequence of operations in the construction of an embankment are also basically the same. The tests for determining whether or not a waste material is suitable for use in embankment construction are therefore the same as those used for other materials. The major requirements of imported fill are that it should be relatively easy to transport, place and compact to form a stable bed, which is strong enough to support the

Figure 16.4. Reinforced earth using ash from power generation

layer above it. These conditions are relatively easy to meet, and for this reason bulk fill provides by far the biggest potential volumetric outlet for the use of alternative materials.

Waste materials and industrial by-products can be used in building in processed and unprocessed form. In an unprocessed form (e.g. as fill), waste materials may be used to bring a site up to a required level or they may be placed within the foundations as support for a concrete floor slab. The materials may be processed to a limited degree for use as aggregates in concrete, or they may be used as the raw material for manufactured building products such as bricks or aerated concrete blocks. The consequences of a serious failure in a building are severe, so it is particularly important that the use of a waste material or industrial by-product is based on adequate knowledge, is well controlled and is in accordance with any appropriate standard or specification. Compared with the use of wastes for the construction of road embankments, the use of wastes in building is much more restricted (this is partly because the amounts of fill in a building project are likely to be small).

The feasibility of using waste materials for land reclamation depends on the economics of transportation. Where the waste is produced locally the economics are highly favourable, particularly when the waste can be transported by pipeline. Appropriate land reclamation schemes would be the formation of new land on coastal fringes and tidal estuaries and the rehabilitation of land damaged by industrial activities and mineral workings.

Waste materials are often used in void filling. The purpose of the waste is to provide bulk (and hence save more valuable, quality material) and form a stable structural matrix, which is often fixed rigidly by filling the voids with a cementitious paste.

16.1.4. Material characterization

If alternative materials are to be used in construction they have to be classified and meet specification requirements in the same way that conventional building materials do, for instance:

- The requirements of the *Specification for Highway Works* (Highways Agency, 1998) are mandatory for trunk roads and motorways and almost invariably they form the basis of the specification for other roads constructed in Britain. The specification will usually set grading limits, define minimum compressive and shearing strengths, and specify limiting values of classification indices to ensure that the material behaves as a generic 'type' (e.g. granular drainage material, cohesive fill).
- Building Regulations state that 'no hard core laid under a floor shall contain water soluble sulphates or other deleterious matter in such quantities as to be liable to cause damage to any part of the floor.' Concrete components exposed on one side only, with evaporation taking place on the other side, are particularly vulnerable to attack, as the sulphate solution will be continually drawn into the concrete. Concrete floor slabs are a particular case of such a vulnerable component, and where there are significant amounts of sulphate in the fill an impermeable separation membrane should be used in addition to the use of sulphate-resisting material.
- Aggregates for concrete should be sufficiently hard and strong for the grade of concrete required, they should not react adversely with the cement nor contain impurities that do so, and they should remain reasonably stable when subjected to changes in moisture content. The shape and grading of the aggregates have important effects on the properties of concrete, and in particular should be such as to allow adequate workability of the concrete.

Box 16.2. Material characteristics affecting usage in engineered construction

- **Compaction.** To define the optimum conditions for compaction (i.e. the maximum dry density and optimum moisture content) and set targets and limiting field values
- **Moisture content.** To assess compliance with the specification for placement and to check the end result
- **Stiffness (CBR).** To assess the adequacy of the fill to support an imposed 'structure'. The thickness of road pavements is decided on the basis of the CBR value of the subgrade
- **Frost susceptibility.** Some materials exhibit significant volume changes during freeze–thaw cycles. However, this is unlikely to be a problem if the fill is more than 450 mm from the surface of the finished road. Materials most likely to be frost susceptible are PFA and burnt colliery spoil
- **Chemistry.** Concrete can be attacked if the sulphate content is high. This should not be a problem with China clay wastes, slate waste and most quarry wastes, but burnt colliery spoil and blast-furnace slag may contain soluble sulphates in harmful concentrations
- **Strength.** The fines content of material when crushed will indicate its ability to perform as free-draining fill. The shear strength dictates the final design
- **Permeability.** To classify the material as free- or slow-draining

CBR, Californian bearing ratio; PFA, pulverized fuel ash.

The basic properties required for technical effectiveness and acceptance of industrial by-products and waste materials are that they should perform their intended functions throughout their particular design lives without exhibiting effects deleterious to the environment or associated constructional features. There is a primary requirement that the use of the waste, when used in preference to other materials, must not shorten the life of the completed works. Hence the use of waste materials and by-products must be based on an adequate knowledge of their properties and performance (on both a short-term and long-term basis). Relevant materials parameters, and their general purpose, are indicated in Box 16.2. It is important to recognize that most deposits of waste materials are very large and many contain a variety of materials. The selection of an appropriate sample for testing to ensure that representative values of parameters are obtained may be a problem. Methods of sampling that are suitable for soils and aggregates may not be appropriate, and test data may require statistical analysis.

The evaluation of a material for any particular application is most readily done by comparing its properties with those of materials known to be satisfactory. At present the technology of the use and behaviour of waste materials and knowledge of their properties is in advance of the ability to use them economically.

16.2. Minestone

Minestone (colliery spoil) is the waste from the mining of coal and consists of rock fragments brought to the surface during the development of mine shafts and roadways or during the extraction of coal. Minestone invariably contains a proportion of coal. Older tips often ignited, and the materials in these are generally a mixture of unburnt, partially burnt and well-burnt spoil. Tips formed more recently comprise predominantly unburnt, well-compacted material. Table 16.3 contains data on the accumulated minestone quantities in various countries.

Underground coal-mining produces huge quantities of waste materials, which for many years have been stockpiled close to the collieries or deposited in lagoons of various types. By far the greatest proportion of spoil has been tipped on land (Figure 16.5). This has an unfavourable influence on the environment and may also cause problems with spoil-heap stability (see Chapter 14, Section 14.4.2). Although coal production in Britain has gone into decline, the amounts produced annually are still likely to mean that colliery spoil will remain a major

Table 16.3. Indicative world-wide accumulations of minestone*

Country	Accumulation (Mt)
USA	3000
UK	2000
China	1200
South Africa	1000
Poland	700
Japan	600
France	200

* From Skarzynska (1995a).

Figure 16.5. An old colliery spoil heap

source of waste for a long time, especially with the huge stockpiles available from past production (see Table 16.1).

Colliery spoil is by far the most significant of the mineral waste materials available for use in construction, and its use has important and unquestionable national implications with respect to economic and environmental impacts. The high economic and social costs of minestone storage make virtually all methods of using it as a raw mineral material attractive propositions.

16.2.1. Production

Coal resources were formed in zones of vast and waterlogged morphological depressions with very rich vegetation and also in regions of subsidence resulting from mountain formation. In periods of relative stability organic material accumulated, whereas in times of intensive movements coal deposits were covered up, mainly by material eroded at the peripheries of the subsidence regions. Hence, coal resources contain alternate layers and lenses of clayey shales, mudstones and sandstones, as well as occasional fine-grained conglomerates. Regardless of whether coal is mined from considerable depth or from an open pit (Figure 16.6), the seams are usually relatively thin and are often separated by layers of different rock types, which are frequently interrupted by numerous faults. Thus, since the advent of mechanized excavation, large quantities of mining waste have been extracted along with the coal.

In the late 1940s, when coal production was largely unmechanized, the ratio between coal and spoil production was 10 : 1. High mechanization of the coal-mining industry has changed this to about 2 : 1 (i.e. much more spoil is now produced) (Sleeman, 1990). This colliery spoil is composed of claystones and clay shales, mudstones, coal shales, sandstones, sporadic conglomerates and chips of coal. The material that results from the development of a mine and sinking of shafts is usually predominantly sandstone and differs greatly from that obtained by the exploitation of coal seams or from preparation plants where the

Figure 16.6. Deep open-cast mining of a thin coal seam

clay-type rocks predominate. The physical and mechanical properties of the rocks present in minestone vary considerably (Table 16.4).

The spoil can be divided into two main groups:

- Rock removed to gain and maintain access to the coal faces and from routine coal preparation processes. These materials are usually coarse-grained (up to 500 mm diameter approximately).
- The rock that is unavoidably brought out of the pit (with the coal) and which is collected during the coal separation process. These wastes can be separated into three size categories: coarse-grained (produced from suspension plants; mainly 5–150 mm diameter), fine-grained (produced by sedimentation during washing; predominantly 0.5–30 mm diameter), and slurries or tailings (from flotation procedures; mainly finer than 1 mm diameter). The precise range of the particle sizes present in freshly deposited spoil depends on the technology of the specific coal production operation.

Table 16.4. Rock materials in minestone

Property	Claystone and clay shales	Mudstones	Coal shales	Sandstones
Specific gravity	2.4–2.8	2.3–2.8	1.6–2.8	2.5–2.8
Porosity (%)	3–14	4–10	—	1–10
Absorption (%)	0–7	0–4	0–3	0–8
Unconfined compressive strength (MN/m^2)	8–60	9–110	10–60	6–170
Deformation modulus (MN/m^2)	70–180	100–180	20–80	70–300
Friction angle (°)	23–27	26–29	17–26	30–38

The amount and size of fine material (infilling the space between large particles) depends on the petrographic composition and degree of weathering. Mudstones are not eroded by water and show considerable mechanical resistance. However, when exposed to air, mudstone easily disintegrates into sharp-edged lumps and small slabs. The mineral composition of mudstones is similar to fine-grained sandstone with a clay binder. Coal shales belong to the carbonaceous rock group and are composed of alternate thin layers of coal and sterile rock, and they tend to weather along intercalations, forming thin laminae. Sandstones occurring in the carboniferous series usually show high mechanical resistance. Minestone contains varying amounts of coal (overall 3–30%, but more usually in the region of 8–10%). The content of coaly material is high in fine-grained spoils, and hence these discards are of doubtful value for direct use as construction materials.

16.2.2. Disposal

Most of the spoil has been tipped on land, with different particle sizes being deposited together (apart from the fine 'tailings' from coal recovery plants, which are usually placed in lagoons). Spoils originating directly from mining or sorting plants undergo grain segregation when loosely tipped. Until the late 1960s, the only compaction that was performed at most colliery spoil tips was from the plant used to transport and place the waste (i.e. in most tips the spoil was in a loose state). Minestone is readily combustible when the coal content is over 25%, and fires occurred in the older colliery spoil tips because of external heating (such as the emptying of steam locomotive boilers) or from self-heating of the waste resulting in spontaneous ignition. The self-ignition process is probably due to the chemical reaction of pyrites, water and oxygen, and for it to occur three main factors are needed:

- easy access of air or oxygen into the interior of the tip
- the presence of sufficient material that is prone to oxidation
- the potential for heat accumulation in the tip.

Once the temperature rises above 50–70°C the rate of reaction increases rapidly and spontaneous ignition is likely to occur. When the coal and other carbonaceous material begin to burn there is a rapid rise in temperature in the tip (even up to 1200°C). At the same time the oxygen content of the material is rapidly reduced, and carbon monoxide is produced, which reacts with other materials to produce oxides of sulphur. As a consequence, old tips often ignited, which resulted in a mixture of unburnt, partially burnt and well-burnt materials being present in a tip.

When first deposited, coal-mining spoil is usually grey to black in colour and has a significant clay mineral content. When burnt it usually becomes reddish in appearance (although lack of oxygen during cooling may produce a blackish colour) and is mechanically stronger. Newer tips consist mainly of unburnt material because they are constructed in layers that have been compacted to increase their stability and their resistance to spontaneous combustion. The mine wastes in newer tips are characterized by gradual weathering, which depends on the material and the time of its dumping.

In compacted tipped material the weathering process is slow and the intensity of the weathering depends mainly on depth. Investigation has shown it to take place within 0.5–1.0 m of the surface of the material where warming–cooling, wetting–drying and possibly freezing–thawing processes cause many of the larger particles of the more susceptible rock types to break down. The surface layer (0.1–0.3 m deep) weathers the most extensively. Chemical weathering activity is very slow and does not seem to extend deeper than about 3 m below ground level.

16.2.3. Properties

The properties of colliery spoil can vary considerably, both within a tip and from tip to tip, and this should be borne in mind when sampling and using the material. Minestone is susceptible to disintegration during transportation, handling and compaction, and by weathering, with consequent changes in its physical and mechanical properties. Hence the characteristics of the original tip material are not necessarily directly indicative of the properties and behaviour of minestone after incorporation into a structure. This potential variability of properties is the major reason for the reluctance of many engineers to specify minestone as a construction material. On the other hand, material taken from colliery spoil heaps constructed during the last 20 years tends to be homogeneous because in almost all cases it has been spread by bulldozers or scrapers in layers from 300 mm to 1.5 m deep.

The mineralogical composition of minestone is important when considering its potential use. Clay minerals are the major component of the material. Typical spoil contains about 50–70% clay minerals (micaceous minerals are the principal clay mineral types present), 20–30% quartz and 10–20% other minerals and carbonaceous matter. The presence of this latter material is a characteristic feature of minestone. However, the organic carbon content usually does not exceed 2.5% by weight. On the other hand, the inorganic carbon content is typically in the range 18–25%, and influences the properties of the minestone, as well as its long-term chemical and thermal behaviour.

Since colliery spoil is prone to breakdown by weathering, its chemical composition is important because of the formation of secondary products and the possibility of environmental pollution resulting from washing out of soluble salts such as sulphates and chlorides. The variation in chemical composition of minestone from different countries is shown in Table 16.5. Silica (SiO_2) is the most abundant component and accounts for 30–70% of the mass, with aluminium oxide (Al_2O_3) being the second largest component (around 15–30%). Sulphur is often present in minestone, and ferrous sulphate and sulphuric acid may result from exposure to atmospheric conditions. Freshly wrought minestone is normally neutral or slightly alkaline, but as weathering proceeds the spoil may become increasingly acidic, although the pH value seldom falls below 5. Sulphates arising from minestone are very important when considering the use of the spoil in conjunction with concrete (specific information is given in BRE (1996)). Figure 16.7 contains data on the sulphate content of minestone. Water seepage through the tips and the body of dams or embankments constructed from minestone may result in the leaching out of sulphates, chlorides and some heavy metals (manganese, zinc, iron). However, chloride levels rarely exceed 0.1% (Skarzynska, 1995a). Coals are generally less radioactive than other rocks and pose no radiation risk.

Table 16.5. The chemical composition of colliery spoil

Chemical component	Brazil	Spain	Poland	Britain
SiO_2	43–59	38–67	35–60	30–67
Al_2O_3	30–34	8–30	17–28	15–27
Fe_2O_3	3–13	2–10	2–6	3–10
SO_3	0.8–3.0	0.6–1.0	0–1.8	0.2–7.5
K_2O	—	2–4	0–6	1–5
CaO	0–4	1–5	0–2	0–4
MgO	—	1–2	0–2	1–4
P_2O_5	0.1–0.3	0.2–0.3	0.1–0.3	0.1–0.3
Cl^-	0–0.1	0–0.1	0–0.1	0–0.1
pH	5–9	5–10	4–8	3–10

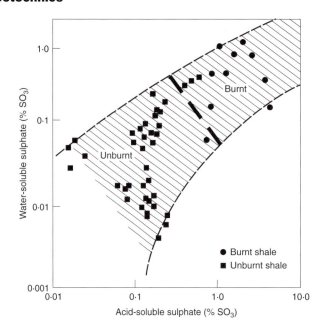

Figure 16.7. The sulphate content of British minestone

It is well established that minestone is a material in a transitional phase which has generally not reached its residual state. The final product will be an inert soil. Particle size distribution within minestone depends on the manner in which seams were worked, the coal-processing procedures used, and the physical and chemical weathering experienced. Typical grading curves are given in Figure 16.8. With the increased use of mechanical mining technology, the proportion of the smaller size of rocks has increased greatly due to rock breakage by coal-cutting equipment.

In 'fresh' minestone, coarse fractions (cobbles and gravel) predominate and there is a significant shortage of fine fractions. Hence the moisture content of freshly wrought minestone is usually low (4–7%), although it may be slightly higher after prolonged periods of precipitation. The moisture content of the spoil increases (typically up to about 15%) with the degree of weathering. In general, coal-mining wastes are comparable to boulder clays or clay–gravel mix soils. Because of the lack of natural selection, fragments of soft, easily disintegrating rocks are mixed with hard ones. However, the length of time for which minestone is exposed to atmospheric agents is extremely short compared with the exposure time for natural soils, and so the rock fragments in colliery spoil are essentially angular in shape.

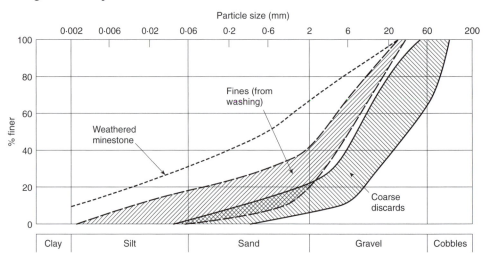

Figure 16.8. Typical particle-size ranges for coal-mining waste

The porosity of bulk samples of fresh minestone may reach 30% because of the predominance of coarse material and a shortage of fines. In situ bulk densities of dumped minestone vary considerably, although most lie in the range 1.4–1.9 Mg/m^3 (the value is not significantly affected by weathering). The specific gravity of minestone depends on the quantities of the various rock types present, and can range between 1.5 and 2.8, with most spoils having an average of 2.3–2.5. The lowest values correspond to waste having the highest coal content.

The Atterberg limits are only determined for minestone when the percentage of fine material exceeds 15%. In such cases the liquid limit is typically in the range 8–50, with the plasticity index varying between 5 and 25.

It is important to recognize that particle crushing is likely to occur in laboratory compaction tests (Sarsby, 1987b) and a fresh sample should be used for each stage of a compaction test (BSI, 1990). Heavy (modified Proctor) compaction results in a significantly higher maximum dry density and a lower optimum moisture content than does standard compaction. This is due to the mechanical disintegration of the rock fragments produced by high impact energy; this is also true for field compaction when heavy rollers are used. The actual optimum moisture content and maximum dry density depend on the grain size distribution, the strength of the rock fragments and the applied compactive effort. Typical data are presented in Figure 16.9.

In Britain the shear strength of minestone is usually determined using the direct shear box (typically 300 mm × 300 mm in plan) and triaxial apparatus. Failure envelopes tend to be linear for strong spoil. However, curvature of the envelope can be marked (because of particle breakdown) when a large normal stress range is considered, and this feature is important in slope stability considerations for high embankments. The angle of internal friction of fresh minestone is relatively high because of the angular shape of the rock pieces and the roughness of their surface. Reported upper and lower values of ϕ' for 'fresh' minestone are approximately 50^0 and 23^0, respectively. The angle of friction falls with weathering, typically to about 30–35^0, although lower values (around 20^0) have been reported for well-weathered material. The effective cohesion for 'fresh' material is commonly zero, but it increases with weathering and values in excess of 50 kN/m^2 have been measured for highly weathered minestone (Skarzynska, 1995a). However, because weathering of typical spoil only occurs at shallow depths, care should be exercised with respect to the validity of a c' value

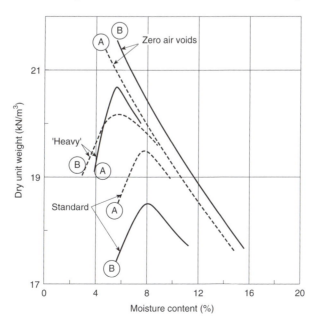

Figure 16.9. Compaction curves for colliery spoil

when considering shearing failure in minestone. Experience of using minestone in engineered construction suggests that for preliminary design purposes the average shear strength parameters of zero effective cohesion and ϕ' equal to 33° would be appropriate.

Laboratory vane tests on lagooned coal-mining sediments (i.e. minestone fines) suggest that the peak undrained shear strength is approximately equal to $0.3\sigma_v'$ (where σ_v' is the vertical effective stress), while the remoulded undrained strength may be reasonably represented by Skempton's empirical relationship: $S_u = \sigma_v' (0.11 + 0.0037 I_P)$ (Taylor & Cobb, 1977).

The permeability of colliery spoil depends not only on the grain-size distribution and the degree of compaction, but also on the intensity of the weathering process it has experienced. Permeabilities of the order of 10^{-4} to 10^{-5} m/s for newly tipped materials and 10^{-4} to 10^{-7} m/s for weathered material in older tips are common. Laboratory values for compacted material are generally several orders of magnitude lower (10^{-5} to 10^{-10} m/s) than in situ tipped material.

Compacted coarse discards are usually free-draining (C_v about 800 m²/year) with the finer discards consolidating somewhat more slowly (C_v values in the range 10–40 m²/year). In general, the compressibility of minestone is low, with compression moduli being close to those for cohesive soils of low plasticity; the actual value will depend significantly on the extent of weathering of the spoil. At low stress levels the compressibility is relatively insensitive to pressure changes, but at high stress levels changes in compressibility may become significant.

16.2.4. Utilization

Minestone is a material with potential for use as fill in civil and hydraulic engineering, and its greatest outlet would be in bulk earthworks. There are sometimes reservations about the use of fresh minestone because of the particle-size distribution of the material, but laboratory and field experience has shown that the majority of colliery wastes can be readily compacted into stable fills of high dry density (particularly weathered material which is usually well graded). Furthermore, the moisture content of material in a spoil heap more than a few months old has usually stabilized to a suitable value for compaction in road embankments. Even when coarse particles predominate, minestone can be used successfully because the mechanical crushing of the large particles, when subjected to the appropriate degree of compaction, produces sufficient infilling to give a well-graded end-product. On the other hand, excessive crushing of minestone during compaction of an embankment can produce a thin layer of highly degraded material having a low shear strength. Unless this layer is broken up (usually by scarifying) such material could provide a potential shear plane with considerably lower shear strength than the surrounding material.

Optimum compaction does not require intensive crushing of the material, but it should produce an end-product having sufficient fines to infill pore spaces significantly. This effect can be obtained for coarse-grained material when the proportion of small particles (those smaller than 7 mm) is greater than about 30%. A vibratory roller will usually give better compaction than a static one.

Unburnt colliery spoil is usually excluded from use as a selected granular fill material. This is not surprising because, although it has the appearance of a granular material, it does not possess the properties generally associated with such a medium. The coarse particles are not discrete and are usually aggregations of smaller particles, which means that the long-term stability of the aggregated particles is open to question. In evaluating the potential of a minestone for construction purposes it is necessary to determine how susceptible it is to weathering and how fast the process will take place. Burnt minestone is acceptable material for use as common fill. However, this use is somewhat extravagant (unburnt minestone will do just as well) and stocks of burnt material are conserved for more technically demanding uses (e.g. as a capping layer immediately below the formation level in roadworks) (Sleeman, 1990).

The multipurpose use of minestone in earth structures, as in the building of river and road embankments and for levelling ground for future housing or industrial developments, has drawn special attention to the problem of frost susceptibility of minestone and frost heave. It is a well-known fact that certain soils are susceptible to frost action due to water accumulated during freezing, and cyclical repetition of freezing temperatures causes substantial increases in heave. After thawing the original strength of the material can decrease, either for a certain period of time or permanently. In general, burnt shales seem to be frost susceptible, but for unburnt shales the behaviour may be summarized as: shales with more than 20% finer than 75 μm are unlikely to show substantial heave, shales with less than 20% finer than 75 μm are likely to be highly frost susceptible if the water absorption exceeds 6%. However, since permissible limits of soil frost heaving are 13 mm in England and Wales and 16 mm in Scotland, it may be concluded that most of the recorded values of the frost heave of minestone are within the permissible limits.

When minestone is used as a civil engineering material it is necessary to determine how soluble chlorides may affect steel and concrete incorporated within, or in close proximity to, minestone structures. Stringent specifications, with respect to the degree of compaction, are now imposed on minestone used in civil engineering projects, since high compaction usually prevents percolation of the water that could dissolve the chloride salts.

The use of untreated minestone as a sub-base is now widely accepted in many countries. At one stage approximately 10 Mm^3 of minestone was being used in Britain, each year, in this way. To provide strength and durability for heavy vehicular traffic, unbound minestone can be stabilized by means of a variety of binder materials, such as cement, lime, bitumen or tar (the most common binder is Portland cement). In a field trial of cement-stabilized unburnt shale (Kettle & Williams, 1978), the material was spread to a loose depth of 230 mm before being compacted to a nominal thickness of 150 mm. The field trial confirmed the suitability of the material for sub-base construction, and even after several years' exposure to the elements the material was weathering in a satisfactory manner (Kettle & Williams, 1978).

Minestone has also been used for:

- land reclamation and backfilling of opencast quarries — preloading techniques (see Chapter 14, Section 14.1.5) can be successfully applied to reclaim areas, which are filled with minestone, for housing and industrial uses
- mine-shaft filling and stabilization
- raw material for the manufacture of some building materials (e.g. lightweight aggregates, building blocks, bricks, etc.)
- formation of a final cap over landfilled refuse (Figure 16.10).

16.3. China clay

China clay (kaolin) is used in the paper and ceramic industries to provide whiteness and surface smoothness. The world's production of kaolin is concentrated in Britain and the USA (North Carolina and Georgia), with Britain being by far the largest producer. In the UK the commercial extraction of China clay is concentrated in the St Austell area, with subsidiary workings in the nearby Bodmin and Lee Moor areas (Figure 16.11). There is an estimated accumulated stockpile of waste of about 600 Mt. Nearly all of the 5000 ha of land covered by spoil heaps in the south-west of England can be attributed to China clay

Figure 16.10.   Minestone used to form a final cap over landfilled refuse

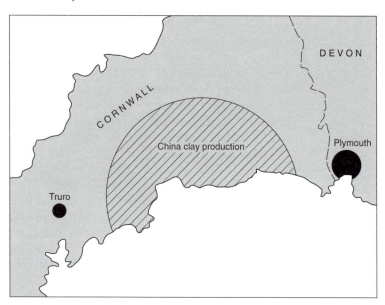

*Figure 16.11.   China clay
extraction in Britain*

workings. Most (97%) of this land is in a small area of Cornwall and it represents the most concentrated stockpiling of wastes in Britain. The stockpile continues to grow, with the current production of wastes being estimated as 27 Mt. The demand for the wastes is insignificant compared with the annual output — 30–40% of Cornwall County Council's aggregate requirements are easily met by using China clay wastes (Sherwood, 1995). The China clay workings are particularly intrusive in that they occur in an otherwise attractive area with an important tourist industry.

16.3.1. Waste production
China clay is directly associated with the large granite intrusions that occur in the south-west of Britain. The China clay has been formed by hydrothermal decomposition of feldspar minerals within the original granite. China clay was first discovered in Britain in the mid-18th century and the industry has for many years been the mainstay of the economy in the south-west of England. Around 50% of the China clay is used in the paper-making industry, and about 15% is used for pottery.

Mining of China clay is only technically and economically feasible by means of an open-pit operation (Street, 1987). The China clay is extracted from steep-sided open pits by subjecting the face to high-pressure jets of water. The broken-up rock flows in a slurry to the pit bottom from where it is pumped to a separating plant. Here the bigger grains, which are predominantly quartz with small but variable amounts of other minerals (including a few flakes of mica), are separated from the sand waste. The China clay slurry produced undergoes three refining stages: removal of coarse micaceous residue using hydrocyclones, removal of fine micaceous residue in gravitational settling tanks, and separation of coarse clay using centrifuge techniques.

Where suitable, the pits formed by mineral extraction (up to 60 m deep) are used for storing water, which is required in large quantities for extracting the China clay. The wastes are usually tipped on land less suitable for China clay working. In recent years there has been concern over the stability of the waste tips and some slips have occurred (*New Civil Engineer*, 1990a and 1990b).

For each tonne of China clay produced, about 9 t of waste is produced (Sherwood, 1987). The approximate composition of this waste material is:

- 2 t of overburden
- 2 t of waste rock (Stent)
- 3.7 t of coarse sand waste
- 0.7 t of micaceous residue.

16.3.2. Waste properties

Of the wastes produced by the extraction of China clay, the coarse sand is not only the largest component, but also that with the most desirable engineering properties. The particle-size distributions of typical sands are given in Figure 16.12. It is a quartzitic sand, which is chemically inert, is free from sulphate and does not swell upon wetting. China clay sand is a good fill material which compacts well, and the small quantity of mica present has no detrimental effects on compaction characteristics (Tubey, 1978). Compaction test results (for equivalent heavy compaction) gave maximum dry densities of 1.9–2.2 Mg/m^3 with corresponding air voids of 6% or less (Figure 16.13).

The waste rock (Stent), which can vary in size from less than 100 mm to greater than 2 m in diameter, essentially consists of massive quartz, quartz–tourmaline and partially kaolinized granite. It exhibits large variability in composition.

16.3.3. Waste utilization

The waste sand is a good-quality aggregate which needs only the same basic grading and washing processes that are applied to other natural aggregates before

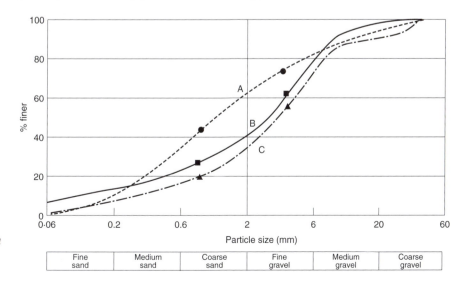

Figure 16.12. Particle-size distribution of sand from China clay extraction (after Tubey, 1978)

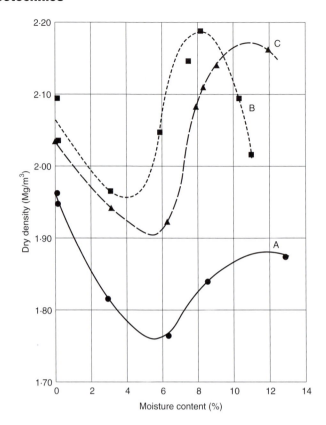

Figure 16.13. Compaction of China clay sand (after Tubey, 1978)

it can be used. It is possible to choose sands that meet the specifications for many road-making purposes, and China clay sand has, in fact, been used extensively in road construction in south-west England. China clay sand was used satisfactorily for sub-base construction on the Camborne bypass (A30 trunk road). The CBR values were well in excess of that required (i.e. a minimum value of 30), for use as sub-base. On the basis of this trial it was concluded that China clay sand was a valuable material for bulk filling and road sub-base construction (Tubey, 1978). However, the markets for the wastes available in this area are considerably less than their annual production. Since the cost of this bulk fill at source is so low, the cost of haulage is the most significant factor in deciding which material to use, with hauls of more than 10 km usually being prohibitively expensive.

There is a reluctance to use China clay sand as a bulk fill (although it would appear to be an excellent material) because the presence of mica in the sand has given rise to suspicions that compaction difficulties may result. When dry the material requires wetting, and when excessively wet the residual clay and mica content give the fill an apparent thixotropy, which eventually disappears with drainage. Once compacted the material develops a natural 'set' due to the cementing action of the clay–mica, resulting in a marked increase in stability.

The sand has been used with considerable success as bulk fill for earthworks where a strict moisture content control was exercised to enable adequate compaction to be achieved. Samples that have been examined have satisfied the requirements for granular sub-base materials. Although some particle breakdown occurred during compaction, this was small and would not pose problems in use (Tubey, 1978).

China clay sand, with a suitably adjusted grading, will pass the requirements for its use as an aggregate for concrete, although it does reduce the workability of the mix slightly (Collins & Atkinson, 1995).

16.4. Residues from coal-fired power stations

The production of electricity from coal results in the generation of ashes as a consequence of combustion of the coal. The current UK production of power station ashes is around 13 Mt per year. The development of an integrated power supply system in the UK resulted in a network of power stations built as close to coalfields and water supply as possible, and hence relatively close to centres of population. The ash is therefore seldom more than 50 km from its market and generally it is considerably less (Cripwell, 1992). Most of the large modern power stations are in central England (Figure 16.14) and this area produces more than half of the available ash supplies. Availability of ash is therefore rarely constrained by transport distribution factors, as virtually all power stations have rapid, and immediate, access to both road and rail networks. Availability of ash is, however, dependent upon electricity generation, as power stations only produce ash when they produce electricity. Although most base-load power stations generate most of the time there are occasions whey they do not work at full capacity. The amount of ash produced is expected to decline with the run-down of coal-fired power stations and the increasing proportion of electricity produced by gas-fired and nuclear power stations

16.4.1. Waste generation

In a modern coal-fired power station boiler, the coal is pulverized into a fine powder (such that typically 80% will pass a 75 μm mesh) before it is injected into the boiler in a stream of hot air and combusted at a temperature of 1000–1700 °C (Figure 16.15). This is in excess of the melting temperature of many of the minerals contained within the coal, and physical and chemical changes to the mineral elements result. The hot gas produced is then passed through heat exchangers and into exhaust flues.

Some of the incombustible mineral matter is deposited on the relatively cool boiler tubes, from where it periodically falls to the lower section of the furnace. This material comprises 20–30% of the ash from the pulverized fuel and it clinkers in the furnace to form conglomerates, which collect at the base of the furnace to be removed as furnace bottom ash (FBA). After quenching it is removed for storage.

The coarsest fraction of ash remaining in the gas stream is arrested by mechanical (cyclone) collectors, the fine material passing to electrostatic collectors for removal — pulverized fuel ash (PFA) is the fine powder removed from the exhaust. PFA varies in colour from light to dark grey, depending on the unburnt

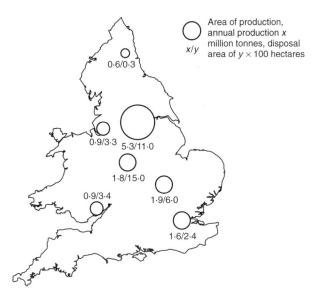

Figure 16.14. Ash production in the UK

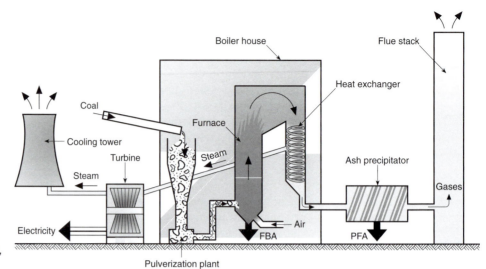

Figure 16.15. A coal-fired electricity generation station. FBA, furnace bottom ash; PFA, pulverized fuel ash

carbon content. Much of the PFA is deposited in stockpiles or lagoons (Figure 16.16), but it may also be stored in hoppers at the generating power station.

About one-third to one-half of ash production is currently utilized, principally in the construction industry, as:

- load-bearing structural fill
- aggregate for lightweight material or building blocks
- a pozzolan in the making of concrete.

A particular advantage of PFA (in the fill context) is that many ashes harden with time and the result is that the fill becomes self-supporting and settlement is less than with other materials. This makes it particularly useful as a selected fill behind bridge abutments. The major commercial outlets for ash are summarized in Box 16.3. Each PFA market has its own specific quality requirements. The use of PFA in concrete and purely structural materials is not covered in this text, which concentrates on the use of PFA in geotechnical engineering situations.

Figure 16.16. Dumped pulverized fuel ash

Box 16.3. Commercial outlets for ash

Fill

The relatively low bulk density of compacted PFA makes it particularly suitable as load-bearing fill. Particular advantages are:

- it undergoes little consolidation and settlement after placement
- it has a high immediate shear strength because of the high value of apparent cohesion, and it hardens with time
- its low bulk density and hardening behaviour mean that lateral earth pressures are low when ash is used as backfill to walls

In grout

Mixtures of PFA and cement are widely used to seal cracks and fissures and to fill cavities to improve the strength of ground. The advantages of using PFA are:

- improved pumpability due to the spherical nature of the ash
- a reduced water/solids ratio
- reduced shrinkage and enhanced resistance to chemical attack
- cost savings

In cement

The pozzolanic properties of PFA are well known. It can be used as a cement replacement, thereby reducing heat generation during setting. Furthermore, the addition of PFA results in less bleeding, improved workability, lower creep and better pumpability

Blocks and bricks

PFA is used as a partial replacement for cement in the manufacture of concrete building blocks. It is also used as a lightweight aggregate. Lightweight blocks are used internally and externally in domestic and industrial buildings, and one particular advantage of PFA blocks is that they can be readily sawn and cut. As PFA originates from the clay minerals present in coal, it has ceramic properties and can be used as a partial replacement for clay

Granular fill

Blended mixtures of FBA have been used beneath road construction when the performance of the ash has been improved by cement stabilization.

FBA, furnace bottom ash; PFA, pulverized fuel ash.

16.4.2. Ash characteristics

FBA consists of a well-graded mixture varying in particle size from 50 mm down to dust, with usually less than 10% of the material being finer than 75 μm (a typical grading curve is included in Figure 16.17). The particles in FBA range from a highly vitrified, glossy and heavy material, to a lightweight, open texture and more friable type. The precise nature will depend on the boiler plant and coal type.

PFA is a fine particulate material composed predominantly of rounded, well-fused, glassy spheres (with a very small number of irregular particles). Chemically, PFA is an alumina–silicate glass containing some iron, calcium, magnesium and alkali metals, together with carbonaceous particulates resulting from incomplete combustion (Table 16.6). The constituent chemistry and mineralogy of PFA is entirely dependent on the coal type from which it was produced and its properties are affected by the geological origins of the parent coal. As UK coal is predominantly bituminous in origin, most PFA produced in Britain is mineralogically consistent.

Ashes generally have particle sizes of medium–coarse silts, and lagoon ashes fall into the medium silt to medium sand size range, although individual particles range in size from less than 1 μm to greater than 300 μm diameter, with specific

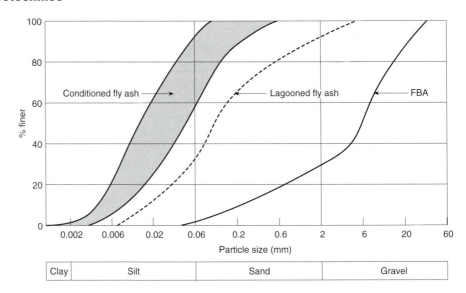

Figure 16.17. Particle size distributions for power station ashes. FBA, furnace bottom ash

gravities ranging from 0.3 to 5.0. The typical particle size range is indicated in Figure 16.17. Ash variability may be a significant problem unless small quantities of conditioned ash are required and they are delivered from one power station alone.

In lagoon ash the particle size is likely to vary within the lagoon, the finer material being furthest from the outfall (compare with tailings dams and lagoons; see Chapter 15, Section 15.5.2). The ash is likely to have a variable moisture content, which may be above the optimum for compaction of earthworks, and material may need spreading or mixing to allow it to dry or otherwise reduce its moisture content. Fly ash is also available from power stations, either as a free-flowing dry powder or 'conditioned' with water to a moist semi-damp mass. Loose ash may be picked up by light winds, and this can cause considerable nuisance, both on and off site, during bulk handling operations. This problem is avoided by keeping the ash at 18–20% moisture content during transport and by placing the ash at the optimum moisture content.

For compaction purposes fly ash lies within the 'uniformly graded' category and it should be supplied at a moisture content close to its own optimum for effective compaction. Standard (Proctor) compaction tests give optimum moisture contents typically in the range 18–30%, with maximum dry densities lying mainly in the range 1.1–1.5 Mg/m^3. Originally the low mean specific gravity of PFA particles (2.0–2.4) led to comparatively low bulk densities after compaction, and PFA gained wide utilization as a lightweight fill. However, the

Table 16.6. The chemical composition of pulverized fuel ash

Chemical component	% by weight	
	Range	Typical value
Silicon (as SiO$_2$)	48–52	50
Aluminium (as Al$_2$O$_3$)	24–32	28
Iron (as Fe$_2$O$_3$)	7–15	10
Calcium (as CaO)	1.8–5.3	2.3
Magnesium (as MgO)	1.2–2.1	1.6
Potassium (as K$_2$O)	2.3–4.5	3.6
Sodium (as Na$_2$O)	0.8–1.8	1.2
Total sulphur (as SO$_3$)	0.5–1.3	0.7
Chloride (as Cl)	0.1–0.2	0.1
Water-soluble sulphur (as SO$_3$)	1.3–4.0 g/l	2.0 g/l

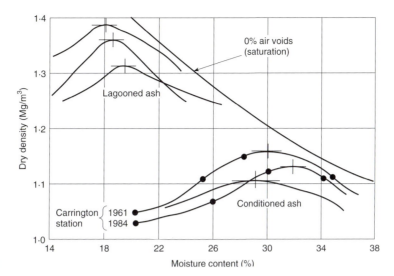

Figure 16.18. Compaction curves for pulverized fuel ash from Carrington power station

specific gravity has been gradually increasing (it is now typically 1.90–2.72, depending on the source) so that from some stations it can no longer be classed as a lightweight fill (Clarke, 1992).

The air voids content of compacted PFA is often much higher than that for other materials. For standard compaction the value may be as high as 18% at maximum dry density and optimum moisture content, although it may also be as low as 5%. Tests with many PFAs over several years have shown that the actual compaction characteristics of ash from a single source show little variation. Compaction curves from one power station, over a period of 23 years, are shown in Figure 16.18. It can be seen that the optimum moisture content has remained virtually unaltered despite considerable changes in the operation of the plant over this period. For some stockpile and lagoon materials, because of either their grading or a significant content of coarser material, the compaction curve is much flatter than for fresh, conditioned ash.

There have been no reported cases of significant sulphate attack of concrete structures adjacent to PFA fill (despite continuous exposure over many years). However, it is normal to have a layer of tar, bitumen or polythene between concrete and PFA as this prevents loss of moisture during curing of the concrete. PFA was accepted as a bulk fill in reinforced soil structures a number of years ago, provided that suitable non-metallic reinforcement elements are used (Department of Transport, 1987).

The effective shear strength parameters of fly ash are most easily determined from drained shear box tests on soaked samples consolidated under appropriate vertical stresses. The effective friction angle is typically in the narrow range 28–36°, but quoted values of initial cohesion cover a wide range (0–200 kN/m^2) approximately (Table 16.7). One of the reasons for this is that many ashes exhibit time-dependent hardening when compacted with water. The strength after 28 days is usually significantly greater than the immediate strength, although the rate of increases varies according to the ash source. When conditioned ash is

Table 16.7. Typical shear strength parameters for conditioned pulverized fuel ash

Age (days)	Effective cohesion (kN/m^2)	Effective friction angle (°)
0	0–20	28–36
7	5–60	32–37
14	10–80	34–38
21	15–95	34–39
28	20–110	34–40

compacted this damp material may gradually harden to produce material like soft sandstone. The hardening of PFA appears to be due to water-soluble components, and ash that has been recovered from a water-filled lagoon will exhibit only moderate hardening. A further factor affecting quoted cohesion values is that triaxial tests on ash have been frequently conducted on partially saturated samples, and so the shearing would not be under true undrained conditions. With most ashes it would appear that a minimum effective cohesion of $5–10 \, kN/m^2$ can be relied upon.

On the basis of its particle-size distribution, PFA might be regarded as a rapidly draining material. However, falling head tests on laboratory specimens of PFA indicate the coefficient of permeability to be in the range 10^{-5} to $10^{-9} \, m/s$ and field measurements have generated values of 2×10^{-5} to $3 \times 10^{-7} m/s$. The drainage behaviour of fly ash therefore ranges from poor to practically impermeable, despite air voids ratios of 5–15%. Furthermore, the permeability appears to decrease, by up to an order of magnitude, with hardening. However, this low permeability prevents leaching of soluble material from a mass of compacted PFA.

For compacted ash normal rainfall may be held in the air voids without saturation occurring, and this is confirmed by the quick recovery of PFA after a period of rain, which can allow work to resume sooner than with other fill materials. On the other hand, thin layers of PFA can become saturated and unstable because of the capillary rise of water from underlying strata. For bulk fill operations a drainage layer of coarse material (150–400 mm deep), placed before the PFA is laid, will eliminate capillary action from below.

16.4.3. Ash utilization

Over 30 Mt of PFA have been used in the UK as engineered fill for road and bridgeworks during a 30 year period (Clarke, 1992). The use of PFA for structural fill, ranging from fill beneath houses to embankments for motorways, effectively started in Britain around 1953 when work at the Road Research Laboratory concluded that PFA could be used for road embankments (Fox, 1984). For embankment construction on compressible soils the lightweight properties of PFA gave it an advantage over other materials.

PFA taken from a stockpile is generally close to the optimum moisture content and can normally be used without pretreatment. A range of compaction moisture contents of 0.8–1.2 times the optimum value has proved successful for many years. Relative densities of 90–95% and above are quite easy to achieve with modern compaction plant and an appropriate moisture content. Vibrating or rubber-tyred rollers provide good compaction of PFA, giving a closely knit compact surface free from shear planes.

Rain does not normally have an adverse effect on an ash embankment, but it can render the surface material unsuitable where it has been loosened. Once ash is mobilized by water erosion it will travel on shallow gradients, and once saturated it may not dry out easily. Although fresh ash is sterile and contains no organic matter, in time it will become completely vegetated.

PFA has also been used widely as backfill behind retaining structures (Figure 16.19) because the horizontal pressure exerted by this fill is reduced by the self-hardening properties of the ash. Furthermore, the load in the ground beneath the wall is less because of the lightweight nature of the fill.

The Elland Trial wall (Jones *et al.*, 1990) and the retaining walls on the Dewsbury Ring Road (Sarsby & Marshall, 1987) demonstrated that fly ash works well in reinforced soil walls, and it is now an accepted technique of construction. Usually the fill employed in reinforced soil is quality, free-draining granular material, but there are considerable benefits to be obtained by using PFA as a bulk fill:

Figure 16.19. Pulverized fuel ash backfill to a retaining wall

- it has a high effective friction angle and it exhibits significant effective cohesion
- it behaves as a drained material (unless saturated artificially)
- it provides good shearing interaction when in contact with soil reinforcement
- the particles form a densely packed mass that dilates during shear
- it hardens with time.

A grout (Section 14.1.2) may be defined as a fluid material that, after being poured or pumped into a void or fracture, sets and hardens to produce useful engineering or geotechnical properties. The void being filled may range in size from very small to very large. The particles of PFA are predominantly spherical in shape, and therefore the use of fly ash in grout mixtures imparts a lubricating action that facilitates pumping and injection. Furthermore, the rounded particles, particularly when mixed using a high-speed mixer, increase the stability of the grout and reduce segregation and bleeding (settling out of solids). Types of PFA grout are listed in Table 16.8.

Table 16.8. Types of pulverized fuel ash (PFA) grout and their applications

Type of grout	Typical applications
General purpose — mixes of PFA and cement	Structural and large void filling Gravel and coarse sand stabilization Annulus filling behind tunnel segments in dry conditions
PFA–cement–clay	Annulus filling behind tunnel segments in mild wet conditions Grouting of alluvium and fissured rocks Annulus grouting around pipes
PFA–cement–polymeric additive	Marine works in, or under, water Annulus grouting behind tunnel linings in wet conditions
Selected PFA–cement	Grouting of alluvium and/or fissured rock
PFA–cement accelerated grout	As for general purpose use, but with shorter setting times Grouting for non-bolted tunnel segments

Cement–PFA grout will seal some water-bearing fissures in rock strata but will not penetrate fine sand, which acts like a filter to the larger particles in the mix. PFA has approximately the same grain-size distribution as cement, and thus using PFA as a filler does not further limit the size of cracks or pores that can be injected. By varying the PFA/cement ratio, dense grouts having a wide range of strength can be obtained. Ratios of PFA/Portland cement in common use vary from 1 : 4 to 7 : 1 depending on the strength and elastic properties required.

Grouts containing PFA alone are cheap and have only low strength after setting, but they are useful for filling large cavities in the ground. PFA–water mixes of 50% flow easily and, provided that the excess water needed to pump the grout can subsequently drain away, it is possible to obtain fills with bulk densities very near to those obtained by compacting the PFA at its optimum moisture content. For void filling, such as in underground mineworkings (see Chapter 14, Section 14.2.6), where stability has to be enhanced significantly, it is normal practice to use a PFA/cement grout ratio of around 15 : 1, which is sufficient to stabilize the PFA. This mix is readily pumped into the ground under pressure (Figure 16.20), filling the voids and any smaller fissures resulting from the mineworking.

Principal advantages to be gained from the use of PFA grouts are:

- *Economy*. PFA acts as a filler replacing sand (thereby improving the flow properties and penetration) and also forms additional cementitious compounds that result in greater strength and more durable composition.
- *Reduced water/solids ratio*. It is important to keep the water/solids ratio as low as possible, as an excess of water results in increased bleed levels, lower strength and durability. Water/solids ratios for PFA–cement grouts range between 0.4 and 0.6 by weight.
- *Reduced permeability*. The precipitation of gel products (from the pozzolanic behaviour) acts as a blocking mechanism within the pore structure.
- *Improved pumpability*. Lower grouting pressures are needed and the grout will flow for greater distances.

'Stabilization' is defined as the treatment of a material to improve its strength and other physical properties. The techniques that have been evolved are based on those used for the stabilization of soils. The most commonly used additive stabilizer is probably Portland cement, and the resultant stabilized material, usually known as soil cement (see Chapter 14, Section 14.1.6), has given good results in many parts of the world, especially in road and aircraft runway bases. The various processes have one thing in common in that the stabilized material is

Figure 16.20. The preparation of a PFA–cement grout prior to injection

made by intimately mixing a predetermined amount of additive with the soil and then compacting it under as near optimum conditions as possible.

Most PFA stabilization in Britain has been carried out with cement as the binder. Compaction should be used so as to give the maximum dry density for the compactive effort employed, which means that the moisture content must be the optimum for the PFA–cement mixture. Stabilized PFA will not form a permanent wearing surface, but can be used for road bases, highway hard shoulders, footpath bases, hard standings, etc. Blended mixtures of FBA have been used as granular sub-bases for pavement structures and, while good performance has been achieved, the 'particle' strength is low due to the porous nature of the large 'particles', which are weakly fused conglomerates. The performance of this material has been enhanced by cement stabilization and 10% cement mixtures have given 28-day cube strengths in excess of $20\,MN/m^2$ (Kettle, 1984). At this strength the material will not be frost susceptible.

16.5. Slate waste

In terms of waste quantities immediately available for use, slate waste (total amount around 500 Mt) is second only to colliery spoil. The current production of slate waste is much less than in former years. Nevertheless, the legacy of waste in old tips throughout the world is enormous (Figure 16.21).

The term 'slate' has been applied generally to any rock that can be cleft or split into thin sheets. The majority of commercial slates are derived from fine material, formed by the weathering of rocks, that had originally been carried to the sea or into lakes by the drainage of surface water. Eventually the sedimentary deposits of this material became buried and were then subjected to movement and to heat and pressure beneath the surface of the Earth, and the consequent 'metamorphism' resulted in the formation of slate. The most significant property of slate, as far as its commercial development was concerned, is its cleavage, which enables blocks to be split into thin sheets. The way in which slate has been formed generally gives it low porosity and an imprevious nature, and the principal minerals that it contains are relatively stable and inert. Thus many slates are highly resistant to weathering. While clays, shales and slates have similar chemical compositions they have very different properties as a result of their different modes of formation.

Slate is well known for its use in the form of roofing slates, but it is found in many other building applications, such as dampcourses, walling, cladding, flooring and paving. Slate is also processed into granules and powder which find many applications in different industries (e.g. slate granules are used for surfacing roofing felts and the powder is used as an inert filler in bituminous compositions).

Figure 16.21. Slate waste

16.5.1. Waste generation

During the 19th century there was a dramatic growth in roofing slate production in Britain because of increased urbanization. By the end of the 19th century the annual production of slate had reached its peak at 650 000 t, and nearly 80% of this was produced in Wales. Other areas of slate production include the Lake District, north Devon and Cornwall, and the west of Scotland. In the 20th century the output of slate declined rapidly, and by the early 1970s it had fallen to around 60 000 t per year. This was due to the competition offered by alternative materials such as mass-produced tiles and the increasingly high cost of the craftsmen required to split the slate.

The method of extracting slate depends very much on the nature of the geological formation in which it is found. If the strata are steeply inclined then the slate bed can often be followed by deep, open quarries or by terraces. If the strata are less steeply inclined then, as the quarry is extended, progressively more overburden has to be removed and mining techniques become more profitable. However, in most cases, slate is extracted by means of open quarries. The Penrhyn Slate Quarry in north Wales, is the largest surface excavation in Britain, it is about 2 km long, the actual excavation covers an area of nearly 40 ha and the overall area, including the workings and waste tips, is over 200 ha.

For the production of roofing slates, the material from the quarry is first reduced to blocks of convenient size. These are then usually split by hand along their cleavage direction using mallets and chisels, to form separate slates of the correct thickness. After splitting, the edges are trimmed to the required dimensions, although sometimes the slates are split from blocks that have previously been sawn to the right size. A very high proportion of the slate extracted by quarrying and mining has to be discarded as waste — in Britain the overall ratio of waste production to finished product is about 20 : 1 (by weight).

16.5.2. Waste properties

Slate waste from tips is normally very coarse, with typically only about 15% passing a 10 mm sieve (Figure 16.22). In its 'natural' state the waste is uniformly graded with a low coefficient of uniformity. Grading is critical to the compactability of granular materials, as aggregates with a minimal fines content (such as slate waste) are difficult to densify. Densities are variable and are

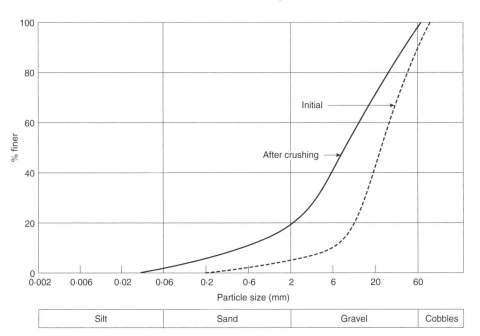

Figure 16.22. Grading curves for slate waste

dictated by the distribution of voids within the compacted mass. Slate is very platey in nature, with a predominant length/thickness ratio of 5–8, and particles are typically twice as long as they are wide (Neden, 1996). If the slate is crushed, finer particles are produced, but the length/thickness and length/breadth ratios are still maintained. As noted earlier, the slate waste is deficient in fines. Nevertheless, in some cases slate waste can comply with typical highway specifications for general fill and sub-base material.

The compaction curve for slate discards is normally more or less flat and the moisture content has no significant effect on compaction (Figure 16.23). After compaction the waste has a fairly high porosity (30–40%) due to the shortage of natural fines. The fines content may be increased by processing or crushing some slate waste or by the addition of other waste material such as fly ash or China clay sand. Compacted slate waste has a high optimum dry density (range 1.7–2.0 Mg/m^3) because of the high specific gravity (around 2.9) of the grains.

Slate discards are, of course, cohesionless. In the direct shear box test high effective friction angles are developed (50–55°), even though the slate grains tend to orientate themselves in a horizontal direction so that there is little dilation during shear, even for densely packed samples. Instability of slate waste tips is usually not a problem because of the coarse nature of the stone, which gives a good friction angle and a free-draining nature. However, the lack of particle interlock means that slate waste needs to be confined if high shearing resistance is to be achieved in engineered construction. Compacted waste gives CBR values of the order of 25–45% depending on the level of compaction. Soaking has a negligible effect on the material, as would be expected.

Problems of toxicity and spontaneous combustion do not arise with slate waste tips, and the factors that inhibit the growth of plant life and reclamation of these dumps tend to be physical. The surface of slate waste tips can be steep and is generally uneven and without fine soil-forming material. Furthermore, the material is deficient in nutrients as well as micro-organisms and beneficial fauna such as earthworms. The inhospitable and generally moisture-deficient conditions

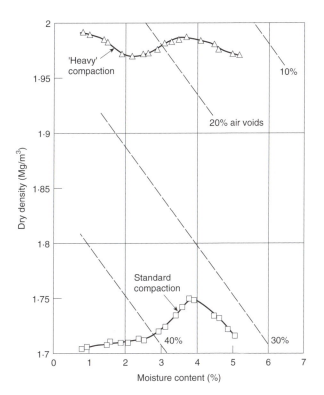

Figure 16.23. Compaction curves for slate waste

at the surface of tips inhibit the development of shallow-rooted plants such as grass, but deeper rooted plants can survive once they are established.

16.5.3. Waste utilization

The disposal of waste from quarries, mines and their associated processing facilities has always been a major problem for the slate industry. Many slate quarries are located in thinly populated, remote areas of poor agricultural value and there are sometimes few obvious incentives to clear the tips. An early example of utilization of slate discards for bulk fill was the construction of a railway embankment at the Cilgwyn Quarry in 1895–97 to reach a new tip. Since the turn of the 20th century there have been sporadic attempts to re-use slate waste, but few have been successful (except for its use as a powdered filler). In recent years serious attention has been paid to the use of slate waste in road construction. The largest use of slate waste is as sub-base material, which consumes on average about 200 000 t per year. Slate waste is crushed and graded for sale as construction material, and can be screened to produce particle-size ranges suitable for drainage blankets, pipe trench backfill, french drain fill, capping or blinding layers, and 'rip-rap' stone of 150 mm upwards (Richards *et al.*, 1993).

The major drawback with slate waste is the remoteness of available material (Figure 16.24). Like China clay sand, slate waste is distant from areas of high construction activity (although long-distance transport by sea may be a way to overcome this problem). The flakey nature of the slate particles has caused some problems in compaction, although grid rollers have been found to be effective as they break the elongated pieces of slate into shorter pieces. Using this technique slate waste has been successfully employed for sub-base construction in road works in north Wales. The degradation did not cause the material to become clayey (generally the fines were in the sandy range) and little trouble was experienced with compaction, even in very wet weather.

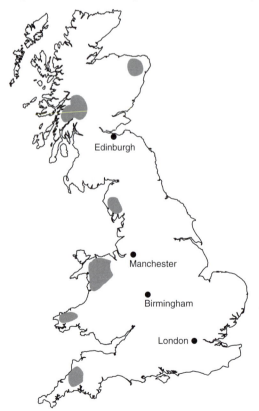

Figure 16.24. The location of slate waste in Britain

Slate granules have been used for surfacing bituminous roofing felt since the beginning of the 20th century. There is considerable scope for using slate powder in bituminous compositions (for road construction, roofing, waterproofing, etc.) and a large amount of development work has been done.

There has been significant use of slate in concrete, particularly in lightweight concrete (expanded slate is used as lightweight aggregate), and in the manufacture of autoclaved (high pressure steam curing) materials.

The primary benefits of slate waste use are the conservation of existing raw materials, aesthetic improvements in areas of waste tips and reduced energy consumption in the production of construction materials. The major factors that inhibit large-scale use of slate waste in construction are:

- the distance of tips from potential markets
- reticence by the construction industry to use a comparatively unknown material (in terms of its engineering properties)
- the inferior quality of slate waste by comparison to conventional materials
- the non-conformity of the waste with existing specifications for suitable materials, unless additional processing is applied.

16.6. Demolition and construction wastes

The construction industry is a major generator of waste, producing more than the household sector. Recycling of demolition materials is not a new idea and reported cases of recycling rubble date back to World War II. Putting construction waste to a positive use has the following benefits:

- it reduces waste transport costs
- it reduces waste disposal costs
- it reduces expenditure on new materials.

It has been proposed (CIRIA, 1993) that the construction industry could conserve natural resources and benefit the environment by considering various actions:

- the use of recycled materials derived from other industries
- the investigation of alternatives to traditional construction materials
- the use of recycled demolition waste
- recycling unavoidable site waste
- examining the scope for specifying recycled materials
- designing to facilitate recyclability of materials and components
- developing salvaging and redistribution of materials between sites
- minimizing waste through design and specification
- designing buildings to facilitate recycling activities during operation
- designing buildings to facilitate adaptation and re-use
- eliminating barriers to recycling (e.g. by revising specifications and regulations).

16.6.1. Waste generation

Construction and demolition waste represents about 16% of the total weight of waste generated in Britain annually, around 70 Mt (DOE, 1995b). The majority of this waste is bulky and inert, and it is not susceptible to treatment such as incineration or biodegradation. However, there is considerable potential for using recycled construction and demolition waste as a substitute for primary aggregates and other quarried building materials. Major components of demolition and construction wastes are:

- soils (often mixed with other materials)
- made ground and fills

- concrete (mass, reinforced with steel, precast)
- masonry and brickwork
- stone from bedrock excavation
- metal (mainly steel)
- timber
- plasterboard and internal finishes
- bituminous materials (including road planings).

In 1992 it was estimated that within Europe approximately 50 Mt of concrete materials were demolished each year, and of this about 11 Mt was dumped in landfill sites each year (Hansen, 1992). There is no evidence of a significant improvement in this situation. Currently around 24 Mt of 'hard' demolition (from demolition of buildings, road pavements and airfield runways) and construction waste is generated in Britain every year, although about half of it is re-used or recycled (around 11 Mt for levelling, 1 Mt for aggregate). Using construction and demolition waste in this way has a double benefit: it reduces both the amount of this waste that is landfilled, and it reduces the environmental impacts of quarrying primary minerals.

Successful cases of recycling construction wastes have occurred where the recycling plant has been located in a large city so that there was sufficient demolition to provide a consistent supply of rubble for fill (O'Mahony, 1990). Consequently, the remainder of this section concentrates on the use of demolition wastes in geotechnical engineering.

16.6.2. Demolition waste properties

Potentially, these wastes can have a wide range of sizes, but crushing and processing generate materials with particle sizes predominantly in the range 0.5–100 mm approximately (Figure 16.25). The specific gravity of recycled and processed demolition rubble is relatively low, generally being in the range 2.2–2.5. Hence, when compacted, these materials usually have relatively low dry density (1.5–1.8 Mg/m^3) which, like PFA, makes them useful as lightweight fills for applications such as backfill to structures. The results of compaction tests on demolition rubble and crushed concrete, conducted on both the 'complete' fill material and that portion smaller than 37.5 mm, are given in Figure 16.26. The porous nature of the particles means that they will absorb water when soaked, with a weight increase of around 8% being typical.

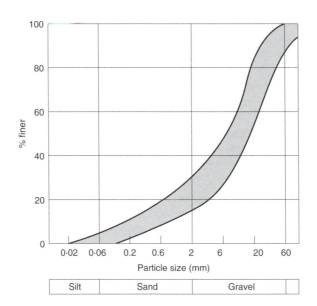

Figure 16.25. A typical particle-size range for demolition waste

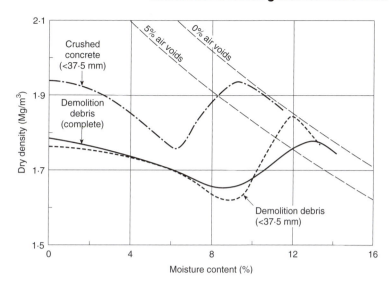

Figure 16.26. The compaction of demolition rubble

The shear strength parameters are best determined using a large (300 mm × 300mm) direct shear box and conducting drained tests (Figure 16.27). Both recycled concrete and demolition rubble have high effective friction angles and no effective cohesion. Typical ϕ' values are 35–55° for crushed concrete and 40–60° for demolition rubble. These values compare well with effective friction angle values for traditional high-quality aggregates such as limestone (ϕ' range 40–55°). Before testing the coarse rubble it is necessary to hand-pick it (to remove wood, plasterboard, old electrical wire, nails, reinforcement, etc.). However, despite the apparent variability of demolition rubble there is good consistency of essential properties such as maximum density, grading curve and friction angle (Table 16.9).

CBR values for both demolition debris and crushed concrete are high, being in the region of 40–150% and 100–500%, respectively (when compacted at

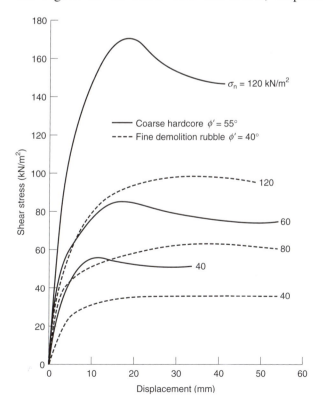

Figure 16.27. Shear box tests of demolition waste

Table 16.9. Parameters from tests on demolition wastes

Material	Natural moisture content (%)	Optimum conditions		Friction angle (°)
		Water content (%)	Dry density (Mg/m^3)	
Coarse hardcore	7–12	9–12	1.58–1.80	49–51
Coarse demolition fill (P)	4–10	8–12	1.57–1.70	49–57
Fine demolition fill (C)	4–10	10–14	1.70–1.78	44–49
Coarse demolition fill (O)	8–12	9–11	1.48–1.88	45–64
Crushed concrete (O)	—	8–10	1.52–1.97	38–55
Fine demolition fill (B)	9	7	1.53–1.67	37–39

optimum moisture content and peak dry density). These CBR values indicate that both materials have more than sufficient bearing capacity to be used as aggregates for granular and cement bound road sub-bases.

Recycled and processed aggregate made from mixed building rubble will usually contain less than 1% impurities. While this material may be acceptable for road construction, it is not suitable for concrete. However, when recycled aggregates are made from raw materials that contain more than 95% old concrete the end-product will usually be clean enough for concrete aggregate.

16.6.3. Waste utilization

The waste hierarchy (prevent waste if possible, if not recycle the waste, and only dispose of it as a last resort) could be applied to construction in the following ways:

- *Reduction (design).* Waste arising from unsuitability, temporary works or the design may be avoidable if considered sufficiently early and integrated into the works (e.g. use of lime stabilization to improve the load-bearing performance of soils that would otherwise be excavated and taken to tip).
- *Reuse.* Direct re-employment of materials after refurbishment or use in lower grade applications is simple and effective (e.g. the reinstatement of a slope failure using the original soil combined with reinforcement).
- *Recycling.* Recover the usable element from waste materials (clean, process and screen them as necessary) and return them to construction (e.g. cleaning of contaminated ground by soil washing and recycling the granular material to fill the excavated void).
- *Beneficial disposal.* Process materials (which are redundant or used) to convert them to a usable form (e.g. crush and screen demolition rubble on-site to produce granular bulk fill).
- *Disposal (landfill).* The final option, when it would be environmentally unsound to re-use the material in any form (e.g. the disposal of excavated contaminated, organic clay where decontamination and conversion would require more energy, and would produce a greater amount of waste, than simple disposal and replacement).

Re-use of construction and demolition waste is likely to necessitate some dismantling, rather than simple demolition, and this is labour intensive and hence costly. Expensive sorting, distribution, cleaning and testing operations might also be required. For these reasons the costs of reclaiming low-price products such as bricks, stone and tiles are usually much greater than the 'price' of new products. Around 50% of the construction and demolition waste produced annually in Britain is recycled. However, in practice, the great majority of this 'recycled' waste is in fact only roughly broken-up for low-grade uses, such as for the construction of access roads at landfill sites. Only some 4% of the waste is subjected to high-level processing to meet the standard required for use in place of primary aggregates in more demanding construction uses. Most of the remaining waste is dumped in landfills, and much of it is not even used as a low-grade construction material at these sites.

The major use of construction waste is as fill and hard-core. Demolition waste is frequently crushed on site and used at the same location as a base of new construction. There is also a long history of recycling concrete from road and airfield pavements for use as a road sub-base (i.e. in the unbound aggregate layer of a flexible road pavement). The sub-base performs three main functions: it acts as a structural component of the pavement, it insulates the sub-grade against freezing, and it provides a working platform for construction traffic. The sub-base material must be densely packed to produce a stiff layer, while having a high permeability and a low capillary rise.

Plant for the production of recycled aggregate can be mobile (Figure 16.28) or stationary, and is not much different from the machinery used for the production of crushed aggregate from other sources. It incorporates various types of crushers, screens, transfer equipment and devices for removing foreign matter. The basic method of recycling is one of crushing the debris to produce a granular product of a given particle size. The plant incorporates processes for removing larger pieces of foreign matter (mechanically or manually, before crushing) and for cleaning. Products from the primary crusher are screened and the final granular material is washed to eliminate fines. Coarse material is crushed down further. The input to centralized crushing facilities is potentially highly variable as there are no statutory requirements to separate waste before its arrival at the recycling facility. The plant operator must rely heavily on visual inspection of the waste and on his charging policy for the acceptance of materials.

Concrete and brick rubble from the demolition of buildings, and road and airfield pavements may contain potentially deleterious materials such as wood, glass, steel reinforcing bars, gypsum plaster and asphalt. The volume of wood in non-selected rubble can be around 2.5%. Timber must be removed from the waste because it rots and leaves cavities in the fill. Concrete from demolished structures may have various types of finishes, cladding materials, lumber, dust, steel, etc., attached to it.

Quality control and the availability of the material when needed are major problems that have to be addressed. It has to be recognized that designers will not risk using recycled material that is not covered by adequate testing and standards. In most cases material specifications are based on the use of natural materials, and opportunities for the use of waste materials are not necessarily clear. Specifications are needed that promote, rather than hinder, the appropriate use of waste materials. For instance, in building construction unbound aggregates are

Figure 16.28. Mobile plant for crushing and sorting demolition wastes

used as fill and hard-core. This is not covered by British Standards and the main source of advice is *BRE Digest 276* (BRE, 1983). This recommends the use of wastes such as colliery soil, clean demolition waste, modern blast-furnace slags, PFA and oil shale residue (subject in all cases to limits on the water-soluble sulphate content to prevent attack on concrete floor slabs, etc.), but warns against the use of steel slags, old blast-furnace slags, refractory bricks and gypsum mine waste, which could cause heave.

16.7. Other materials

Oil was extracted from shales in the Lothian region of Scotland for nearly 100 years until 1962 and the spent oil shale (i.e. shale after processing), was deposited in large tips (bings). Waste from the associated mining operations was usually deposited in separate tips. Spent oil shale is generally of a pinkish appearance and consists of particles generally smaller than 50 mm in size, but being fairly soft it crushes easily under compaction to give a finer grading. It exhibits similar properties to burnt colliery shale, so spent oil shale may be used in situations where burnt colliery spoil is permitted to be used. It has been widely used as a bulk fill material in central Scotland, with very good results (Fraser & Lake, 1967).

The main metallurgical slags in the UK are blast-furnace slags from the production of iron, and steel slag from the production of steel. Blast-furnace slag is a by-product of the manufacture of pig-iron in a blast furnace, and is formed by the combination of the earthy constituents of the iron ore with the limestone flux. The physical appearance and mineral structure of blast-furnace slag depends largely on the method by which it is cooled. Because iron production has fallen and imported iron ores are used (with a higher iron content), increasingly less slag is produced. Blast-furnace slags are used extensively:

- in road-making
- as aggregate
- in coated macadams and rolled asphalt
- as a cementitious binding agent (when granulated).

These slags are processed to provide dense or lightweight aggregates for concrete, and they also provide part of the raw material for blast-furnace slag cements. All blast-furnace slag currently produced in Britain is used either for aggregate or for cement. Steel slags may have quite distinct properties from blast-furnace slag and may contain residual iron, free lime or free magnesia. The latter may make them unstable and liable to expand. To avoid instability, steel slags should be allowed to weather before being used in roads and they should not be used where failures can be caused by their instability (e.g. as fill under buildings or as aggregate in concrete). Steel slags are usually denser, and may be mechanically stronger, than blast-furnace slags.

The direct incineration of domestic and trade refuse leaves a clinker material that contains iron and other metals, glass and cinders, together with smaller amounts of unburnt papers, rags and vegetable matter. In addition, a fine dust is extracted by precipitation from the combustion flue. Little information is yet available on the composition or usefulness of the fine dust, except that the composition is variable and, on occasion, a high proportion of heavy metals may be present. Some of the clinker may also contain appreciable quantities of heavy metals, and care should be taken in the use of these materials (e.g. as a fill under houses), because of their possible toxicity.

17 Noise and ground vibrations

17.1. Introduction

A certain level of sound in the environment is natural, but when noise intrudes on the normal activities of life it becomes a nuisance. The effects of noise include: disturbance of work, leisure or sleep; interference with communication; annoyance; and effects on mental and physical health.

In 1920, work started on the foundations for a new 11-storey building in London. Deep trenches were dug through the fill (about 5 m deep) and 300 mm square piles were driven from the base of the trenches into the underlying London clay. Initial driving was undertaken using a steam hammer, but later a heavy drop-hammer was employed. The stiffness of the London clay was such that as the piles approached their required installation depth each hammer blow was only producing about 5 mm penetration. Pile-driving began in November 1920 and continued until July 1921, by which time 670 piles had been driven. Opposite the piling site there was an old building which had been inspected by the District Surveyor in 1919. The inspector had noted that, although the property had undergone some settlement the building was stable and, apart from two iron pillars and a pier, the walls and columns were not out of plumb.

Shortly after the pile-driving commenced the owners of the old building claimed that ground vibrations from the piling (which at its closest was only 12 m from the property) had caused walls to start leaning and cracks to appear in external and internal walls and in ceilings. In July 1921 the City of London Corporation issued a dangerous structure notice, which required that the old building was 'shored up', and subsequently (January 1922) the Corporation served a notice requiring the structure to be demolished. The owners of the old building subsequently sued the foundation contractor for compensation. The final judgement was that there was no doubt that the vibration produced by the piling operations had caused serious structural damage to the building, and the owners of the building were entitled to compensation from the piling contractor (All England Reports, 1922).

Most construction processes, and many other human activities, cause detectable levels of motion in the ground (at least in the immediate vicinity of the activity). Within the construction industry, demolition, blasting, pile-driving and dynamic compaction produce the most severe ground vibrations. The end-terrace house in the middle of Figure 17.1 lost the top of its chimney due to the ground vibrations generated by driving a temporary sheet-pile wall close to the property.

17.1.1. Construction noise and ground vibrations

Increased mechanization in the construction industry has generally led to an increase in noise levels associated with construction and other open sites such as opencast mines. However, this increase has been offset to some degree by improvements in 'sound-proofing' and sound-screening measures. When a site is first being developed, adjacent properties may receive significant construction noise due to the relatively close proximity of the plant and the type of operations and machinery used (e.g. earthworks using bulldozers, excavators, etc.) and the

Figure 17.1. An end-terrace house damaged by ground vibrations

absence of any screening. However, one of the first site operations is likely to be topsoil stripping and stockpiling, and if this is used to form a bund at the site boundary then a noise screen will soon be created between the main construction operations and adjacent premises.

Almost all construction operations generate vibrations within the ground, and pile-driving, by either impactive or vibratory means, usually produces the highest vibration levels likely to emanate from a construction site. High levels of vibration can cause severe disturbance to people and damage to structures. However, the acceptable level of vibration is invariably determined by the response of the building occupants. Such acceptable levels are low (because ground-shaking is an uncommon phenomenon) and can cause major problems for works in urban environments because the piling works (particularly those for soil-retaining purposes) are, of necessity, carried out in close proximity to inhabited premises. Construction operations with particular potential for affecting the environment due to noise and vibrations are listed in Box 17.1.

All practicable means should be used to control construction-site noise at source and to limit the noise that spreads to noise-sensitive buildings in the immediate neighbourhood. Designers and contractors must determine how construction can be undertaken while complying with noise and vibration specifications and requirements. Such analyses may indicate that specialist equipment and techniques are necessary. However, minor adjustments to the proposed design or the selection of an alternative method of working, may obviate the need to use specialist services plant or processes that will produce significant noise pollution (Boxes 17.2 and 17.3).

It is not possible to provide detailed guidance for determining whether or not noise or ground vibrations from a site will constitute a problem. However, factors that are likely to affect considerations of acceptability are outlined in Box 17.4.

Box 17.1. Construction operations with particular potential to produce noise and vibration pollution

---

Extreme risk
- Blasting (explosive)
- Hammer-driven piling
- Concrete finishing (scabbling, etc.)
- Concrete, rock and road breaking and drilling.

Significant risk
- Non-hammer piling
- Compacting
- Excavation and earth-moving
- Dynamic compaction.

Box 17.2. An example of the development of an environmental problem due to construction

Project

A new road is to be taken through an urban area in a cutting. At a particular location this work will require the construction of a 5 m high retaining wall in silty clay.

The design

On the basis of the ground conditions and the height of soil to be retained, the designer has chosen to install a reinforced-concrete cantilever retaining wall at this location. The positions of 'services' and 'land-take' are designed on this basis.

Construction works

The contractor for the works decides to construct the wall by installing temporary sheet-piling (driven and then anchored) to provide temporary ground support, excavate the ground, build the wall and then backfill behind it. The phasing of these works is included in the contractor's programme of work for the whole scheme.

The problem

After the piling contractor has driven several sheet-piles some metres into the ground the work has to be halted because of claims of nuisance (due to noise and ground vibrations) from a local factory which makes precision balances and from a nearby library.

Possible solutions (at this stage)
• Change the wall design (e.g. use bored secant piles).
• Change the method of installing the ground support.

Outcomes
• Disruption of the contractor's programme of work and an associated delay.
• Additional expense due to changes to the works.

Possible solutions (at the design stage)
• Amend the road layout.
• Choose a different form of permanent ground support.

Box 17.3. The inclusion of noise and vibration considerations in the 'design' process

The site

The site is derelict land, which was previously occupied by a mill, lying at the heart of a residential area. The streets adjacent to the site are narrow and much of the property consists of terraced housing aged 70–100 years. The ground is contaminated to a depth of 2 m, and remediation options suitable for this site are excavate–dispose–replace (see Chapter 13, Section 13.4.1) and in situ bioremediation (see Chapter 13, Section 13.6.2).

Excavate–dispose–replace

Benefits:
• contamination will be eliminated once and for all
• the works are of relatively short duration
• backfilling of the excavation provides an engineered foundation.

Disadvantages:
• excavation results in site noise — this is probably acceptable to adjacent residents because the works will bring improvement of the site
• a large number of lorries will travel to and from the site and give rise to a significant noise increase in the narrow streets leading to the site — the relatively short-term nature of the traffic movements will probably mean that the temporary increase in noise will be acceptable.
• the movement of heavy lorries in narrow streets will induce ground vibrations which are received by the houses and there will be low-frequency vibration of windows — inhabitants are likely to imagine that their property is being damaged (cracks that have been in existence for many years may now be noticed), or genuine architectural damage may occur (often because of internal modifications to the property).

Box 17.3 Continued

In situ bioremediation

Benefits:

- no excavation or significant volume of material to be moved
- only a limited amount of site traffic, and thus no significant increases in noise and ground vibrations in the surrounding areas are anticipated

Disadvantages:

- the work will take a long time to complete
- some residual contamination may exist within the ground after treatment
- the ground may require modification or treatment to improve its geotechnical characteristics for redevelopment of the site

Box 17.4. Factors affecting the human response to noise and vibrations

Site location

The nearer a site is to sensitive premises, the more stringent should be restrictions on noise or vibration emanating from the site.

Existing noise and vibration levels

The likelihood of complaint increases as the difference between the industrial noise and vibration and the existing noise and vibration increases. However, a large difference may be tolerated when it is known that the operations are of short duration.

Duration of site operations

Good public relations are important. Local residents are likely to be willing to accept higher levels of noise or vibration if they know that such levels will only last for a short time.

Hours of work

For any noise-sensitive premises some periods of the day will be more sensitive than others and the times of site operations outside normal weekday working hours will need special consideration. Sensitivity to vibration is more complex than sensitivity to noise. In general, human sensitivity to vibrations is less when sitting or standing (usual daytime situation) than when lying down (usually the night-time situation).

Attitude to the site operator

The acceptability of the project itself, or the perception that the site operator is doing everything possible to minimize noise and vibration will influence the public reaction to the disturbance.

Noise and vibration characteristics

A particular characteristic (e.g. the presence of impulses or specific tones) may make the disturbance less acceptable than might be concluded from the overall or mean level.

Concern about the effect on buildings and/or contents

The effect that people believe the disturbance is having on their properties can significantly affect their reaction to the event.

17.1.2. Environmental protection

Noise pollution has increased dramatically over the last 20 years in developed countries, and complaints about excessive environmental noise have shown a corresponding significant increase (Figure 17.2). The noise levels experienced by most people are unlikely to produce hearing defects, with the exception of those working in close proximity to heavy unsilenced construction plant. However, noise can produce secondary health effects arising from stress due to annoyance and sleep disturbance, and the noise can create communication and task-performance difficulties. Furthermore, in any neighbourhood some persons will be more prone to disturbance than others (e.g. night-workers who sleep during the daytime) and some localities will contain 'sensitive' facilities (such as hospitals and libraries).

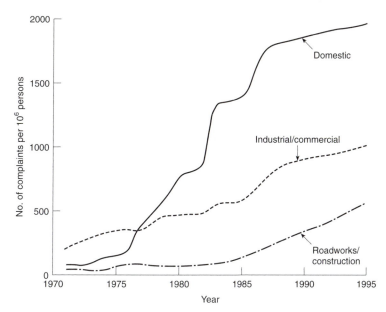

Figure 17.2. The increase in complaints about noise pollution (after Adams & McManus, 1994)

The Control of Pollution Act gave wide powers to local authorities for the control of noise from construction sites. Furthermore, it also defined 'noise' as including vibrations, and so the latter were subject to the same stringent controls and regulations as noise. The subsequent Environmental Protection Act defined noise (and vibration) that emanates from premises, and which is prejudicial to health, as a statutory nuisance. The actual level of noise or vibration that constitutes a nuisance depends on the specific situation being considered. A local authority has the power to serve a notice requiring abatement of the statutory nuisance and prohibiting or restricting its recurrence. Moreover, in stating its requirements for noise and vibration control a local authority may specify:

- acceptable and unacceptable plant or machinery
- permissible working hours (apart from emergency operations)
- maximum allowable noise and vibration levels at the boundary of the site.

The foregoing powers of control may be exercised before work has commenced or after construction operations have started.

However, contractors can take the initiative and ask local authorities to define their noise and vibration requirements prior to the commencement of site works. In the 'prior consent procedure' the contractor (or developer) submits details of the proposed works to the local authority to show that noise and vibration requirements will be satisfied in order to obtain 'prior consent' for the works. The applicant should give the authority as much detail as possible about the works and construction details, together with any noise or vibration control measures that have been incorporated. If the consent is granted (conditions may be attached) and if the contractor adheres to the terms, then the local authority cannot serve subsequent notices during the period of the consent. Compliance with a consent does not, however, exempt the contractor from action by a private individual over noise and vibration pollution or nuisance.

The emphasis should be on solving potential noise and vibration problems before work starts. Planners, developers, architects, engineers, environmental health officers, etc., can all play a part in avoiding potentially excessive noise and vibration levels. This can be achieved by giving careful consideration to the design of the proposed project, the processes and equipment implied by the design, and the phasing of operations (see Boxes 17.2 and 17.3). Thus an assessment can be made of the feasibility of undertaking the proposed construction, while complying with noise and vibration requirements. The intention at each stage of the project (i.e. initial concept, design, tender, construction) should be to minimize levels of noise and vibration while having due regard to the practicability and economic implications of any proposed control or mitigation measures. Because construction works are of limited duration (and sites have a relatively short lifetime), annoyance due to construction noise is generally short term, and there is evidence to suggest that people living near construction sites will accept higher levels of noise from these sites than they would from some fixed installation or permanent source of industrial noise.

Construction works pose different problems of noise and vibration management from other types of industrial activity, because:

- they are of temporary duration
- they are carried out mainly in the open
- noise and vibrations emanate from many different activities
- the 'pollutant' sources are often mobile, or at least they are likely to move during the lifetime of the works
- noise and vibration levels will vary according to the phase of the works
- it is very difficult to accurately predict ground vibration levels before any site operations commence
- sites cannot always be divorced from sensitive areas.

Detailed information on assessment procedures and a large body of data are contained in BS 5228 (BSI, 1996).

17.2. Sound

Sound may be defined as the transmission of energy through gaseous, liquid or solid media via rapid fluctuations in the pressure of the medium. The pressure fluctuations originate from some vibrating object (e.g. human voice, loudspeaker diaphragm, machine tool). A sound wave in air consists of a number of regions of increased and decreased pressure travelling longitudinally from the source to the receiver. The range of pressure that can be felt by the human ear is so large (from 20×10^{-9} to $0.2 \, kN/m^2$ approximately) that a logarithmic scale is used to define sound levels. The actual unit of measurement relates to the energy content of the sound. Each ten-fold increase in energy content is termed a 'bel'. Because the bel is such a large unit it is subdivided into 10 as decibels (dB).

17.2.1. Fundamentals of sound

The frequency (pitch) of a sound is the number of cycles of pressure fluctuation occurring per second and is measured in hertz ($1 \, Hz \equiv 1$ cycle per second). Wavelength is the distance between two successive pressure maxima or minima, (Figure 17.3), and the time interval between these peaks is called the period (of the wave). Frequency is therefore the reciprocal of period.

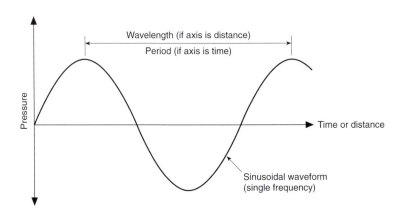

Figure 17.3. Sound wave characteristics

For a pure sinusoidal wave the ratio of the pressure p to the particle velocity v is termed the impedance z of the medium:

$$z = \frac{p}{v} \tag{17.1}$$

The impedance is also the product of the density ρ of the transmitting medium and the velocity of sound c, and thus

$$z = \rho c = \frac{p}{v} \quad \text{and} \quad v = \frac{p}{\rho c} \tag{17.2}$$

The intensity I of a sound wave is the rate of flow of energy (power per unit time W) per unit area:

$$I = \frac{W}{A} \tag{17.3}$$

The intensity is also equal to the product of the pressure and the particle velocity, so that

$$I = pv \equiv \frac{p^2}{\rho c} \tag{17.4}$$

Intensity (or, alternatively, energy content) is used as a scale for measuring sounds. The range of audible intensities, at a frequency of 1 kHz, varies from 10^{-12} W/m^2 at the threshold of audibility to 10^2 W/m^2 at the threshold of pain. Because the intensity range that can be experienced by humans is so large the sound intensity level (in bels) is defined on a logarithmic scale by comparison with the lowest detectable intensity level I_0:

$$L = \log\left(\frac{I}{I_0}\right) \text{bels} = 10\log\left(\frac{I}{I_0}\right) \text{decibels} \equiv 10\log\left(\frac{I}{I_0}\right) \text{dB} \tag{17.5}$$

Since the human ear is pressure sensitive, a pressure scale is generally a more useful parameter for noise considerations than is sound intensity level. Furthermore, microphones, which are the first stage of scientific sound level measurements, are pressure sensitive and hence the sound pressure level is the quantity that is actually measured when a microphone is placed in a sound field. The impedance ρc is constant for a given medium, so the sound intensity is proportional to the square of pressure. Consequently, the sound pressure level L_p, or alternatively the 'sound level' defined on the basis of pressure, is given by:

$$L_p = 10\log\left(\frac{I}{I_0}\right) = 10\log\left(\frac{p^2}{p_0^2}\right) \text{dB} = 20\log\left(\frac{p}{p_0}\right) \text{dB} \tag{17.6}$$

where p_0 is the reference sound pressure and is equal to 20×10^{-9} kN/m^2. A typical scale of sound levels for everyday events is given in Figure 17.4.

With sound levels being expressed on a logarithmic scale even a small numerical change in the decibel value represents a significant change in the loudness: a change of 3 dB corresponds to a doubling of the sound energy, and a change of 20 dB corresponds to a change of the sound pressure by an order of magnitude. Each increase of 10 dB(A) in sound level represents, subjectively, an approximate doubling of the loudness of most common sources. Furthermore, since decibel values are based on logarithmic scales, sound levels cannot be combined by simple addition of the numerical decibel values. The total sound pressure level due to individual sources with sound levels of L_1 to L_n is given by:

$$L_{\text{total}} = 10\log\lfloor 10^{L_1/10} + 10^{L_2/10} + 10^{L_3/10} + \ldots + 10^{L_n/10}\rfloor \text{dB} \tag{17.7}$$

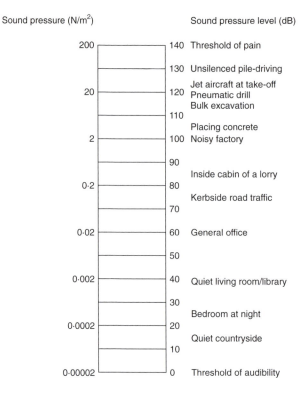

Figure 17.4. Sound levels of everyday events

If two equal sound pressure levels are added together the resultant total sound pressure level is 3 dB higher than an individual sound pressure, but the sound energy is doubled.

The sound power level L_W is a measure of the energy output of a noise source and is independent of the environment in which the source is measured. This allows direct rating of machines, machine tools, domestic appliances, etc., in terms of their sound output. The sound power level is defined as:

$$L_{\mathrm{w}} = 10 \log \left(\frac{W}{W_0} \right) \mathrm{dB} \tag{17.8}$$

$$L_{\mathrm{w}} = 10 \log \left(\frac{IA}{I_0 A_0} \right) \equiv 10 \log \left(\frac{I}{I_0} \right) + 10 \log \left(\frac{A}{A_0} \right) \tag{17.9}$$

$$L_{\mathrm{p}} = L_{\mathrm{w}} - 10 \log \left(\frac{A}{A_0} \right) \tag{17.10}$$

Sound intensity decays with distance from the source. For a point source located on a hard, reflective surface the sound power would be dissipated over the surface area of a hemisphere that expands with distance from the source. At distance R the area of the hemisphere will be $2\pi R^2$ and, since A_0 is equal to unit area, then

$$L_{\mathrm{p}} = L_{\mathrm{w}} 10 \log \left(2\pi R^2 \right) \mathrm{dB} \cong L_{\mathrm{w}} - 20 \log R - 8 \mathrm{\ dB} \tag{17.11}$$

Note that the units of the distance between the source and the receiver R are metres because of the definition of I_0.

The sound power level of a source is an absolute measure of the sound output, but it cannot be measured directly. Thus it is back-calculated by measuring the sound pressure level at various distances from the source and applying equation (17.11).

17.2.2. Human sensitivity to sound

The range of frequencies over which sound is audible varies from person to person, but normally it is in the region 20 Hz to 20 kHz. Sounds below 20 Hz are termed infrasonic and those above 20 kHz are termed ultrasonic. However, the response of the human ear to frequency is not linear — it perceives sounds in the middle range (1–5 kHz approximately) more strongly than at either end, and it is relatively unresponsive to low frequencies. Microphones, which are the first stage of sound level measurements, are pressure sensitive and hence the sound pressure level is the quantity that is actually measured when a microphone is placed in a sound field.

Since the same sound pressure level will produce different responses at different frequencies, it is necessary to consider the frequency content of a sound if an estimation of the subjective effect on the ear is to be made. Most everyday sounds are complex in nature in that they contain many frequencies generated simultaneously at different sound levels. A complex sound could be split into its component frequencies using frequency analysis, but this process is rendered unnecessary by equipping sound level meters with built-in filters to apply weightings to different frequency ranges (lower frequencies are reduced relative to others) so that the 'frequency sensitivity' of the meter approximates that of the human ear.

Initially, three weighting networks, A, B and C (Figure 17.5) were used. The A-weighting network was found to be the most applicable to human response, and hence this is now used for most types of industrial and environmental noise. Sound measurements made using this frequency spectrum are quoted in units of decibel(A-weighted), or dB(A). The D-weighting network has been designed specifically for assessing jet aircraft noise, for which the sensitivity of the ear to frequencies around 4 kHz is important.

Loudness is a function of the frequency content of a sound and the sound pressure level. Previously the loudness of a sound was defined in terms of phons (which are numerically equal to the sound pressure level of an equally loud 1 kHz pure tone). The phon scale of loudness has been largely superseded by the use of weighting networks.

When a peak noise level of about 90 dB(A) is reached, a person becomes physically affected by it. However, the levels of noise associated with most environmental pollution are unlikely to produce hearing defects (with the exception of amplified music and close proximity to construction plant and power

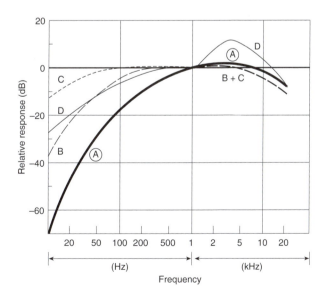

Figure 17.5. Sound meter weighting networks

tools), but they can have secondary health effects arising from stress due to annoyance, sleep disturbance, and communication and task performance difficulties. The relationship between the measured noise level and its effect is extremely complex, and varies greatly from person to person. A large number of factors can influence an individual's opinions or feelings about noise, such as their general state of health, their personality, the community (urban or rural) in which the person lives, and prejudices against the noise source.

It has been known since the early 1970s that a correlation exists between public dissatisfaction with excessive road traffic noise from freely flowing traffic and the hourly L_{A10} sound values, (the level exceeded for not more than 10% of the time), averaged arithmetically over an 18-hour period. However, for intermittent noise events the equivalent continuous sound pressure level L_{eq} is the preferred parameter for assessing the acceptability of noise. L_{eq} is defined as that sound level which, if generated continuously over a given time period T, would give the same energy content as that imparted by a fluctuating signal over the same time period:

$$L_{eq} = \frac{1}{T} \int_0^T L_p \, dt \qquad (17.12)$$

During construction operations the sound level from a particular source will invariably fluctuate due to changes in the rate of working, the position and orientation of the source, the stage of the work, etc. If the variation in the sound level with time can be approximated to a series of discrete sound values, $L_1, L_2, \ldots, L_n$ which occur over the time periods $t_1, t_2, \ldots, t_n$, respectively, then the equation for L_{eq} can be written as:

$$L_{eq} = 10 \log \left[\frac{1}{T} [t_1 \times 10^{L_1/10} + t_2 \times 10^{L_2/10} + \ldots + t_n \times 10^{L_n/10}] \right] \qquad (17.13)$$

Typically the period over which L_{eq} is assessed is 12 hours, to cover the time-span during which it is normal to expect external noise within a community. This period is frequently equated to 7a.m. to 7p.m.

Unfortunately, it is not possible to give precise guidance on whether or not a certain noise level from a site will constitute a nuisance and provoke a reaction from nearby residents. Factors that are likely to affect the acceptability of site noise are outlined in Box 17.5. Good liaison between the contractor and the

Box 17.5. Factors influencing the acceptability of construction noise

Site location
- Proximity to noise-sensitive premises and persons
- Whether within a rural or an urban setting
- Whether it is a large or small group of people that is affected.

Previous environment in the vicinity of the site
- Ambient noise level in the neighbourhood prior to the start of the work
- Magnitude of the noise level increase.

Nature of the works
- Total duration of the operations
- Hours of work
- The extent to which the finished works will benefit or disadvantage the affected community
- Whether some individuals are likely to be particularly affected by the noise.

Noise emission
- Characteristics of the sounds (frequency content, variability)
- Evidence of attempts to control the escape of noise from the site
- Whether the noise is going to be accompanied by ground vibrations.

community can significantly reduce the incidence of complaints and help to forestall problems.

The Environmental Protection Act does not specify permissible noise levels and the responsibility for setting such limits rests with the local authority. Consequently, the limiting value of L_{eq}, which must not be exceeded during construction works, is not a constant throughout Britain. However, an L_{eq} (12 hours) of more than 75 dB(A) is likely to prove annoying to most people, and limiting L_{eq} values of 75 dB(A) and 70 dB(A) for urban and rural settings, respectively, are commonly specified. An approximate guide to people's perception of sound level changes relative to the original ambient level (i.e. the total background sound), and their reaction to these changes, is given in Table 17.1.

Special consideration is needed for noisy operations outside the period Monday to Saturday 7a.m. to 7p.m. Work in the evening (7p.m. to 10p.m.) should be restricted so that the noise level is about 10 dB(A) below the daytime limit. The periods when people are getting to sleep or just before they wake appear to be particularly sensitive, and L_{eq} values may need to be as low as 40 dB(A) to avoid sleep disturbance.

For opencast works it has been suggested (DOE, 1991c) that the 1 hour L_{eq} values at the worst affected house (or school, hospital, open area used for relaxation by the public) should not exceed 55–60 dB(A) by day, 45–50 dB(A) in the evening or dawn (where considered appropriate) and 40–45 dB(A) at night (typically 10p.m. to 6a.m.).

17.2.3. Sound level data

Site noise is produced by many different activities and types of plant, the noise from which not only varies in intensity and character but also in location and over time. There may also be many combinations of these activities of both a static and a mobile nature. There are three general ways of obtaining sound levels for noise analysis:

- carry out noise measurements on a similar item of plant or process
- obtain the maximum permitted sound power level of the plant under current regulations
- use published values of the sound power level or overall sound level for a particular activity from sources such as BS 5228 (BSI, 1996).

Typical sound levels for items of plant and construction activities are given in Table 17.2.

In general, the highest construction noise levels are generated by piling operations, particularly when unsilenced, unscreened driving methods are used. Diesel hammers are usually significantly more noisy than any other type of hammer, but in some situations they may be the only way of driving the piles. All impact and vibratory systems of piling have the additional problem that the vibrations associated with them can be a source of nuisance. Pile-driving is a noisy operation when considered either as an isolated event (i.e. the actual driving

Table 17.1. Likely effects of sound level increases*

Increase in sound level above ambient level (dB(A))	Change in apparent loudness	Possible consequences
2.5	Just perceptible	No objections
5	Clearly noticeable	Objections possible
10	Twice as loud	Objections likely
20	Very much louder	Objections certain

* After Sarsby (1992b).

Table 17.2. Typical sound levels from construction operations

Plant or operation	Sound power level, L_W (dB(A))
Pneumatic breaker	110–119
Dragline excavator	106–111
Tracked excavator	96–111
Dozer	104–118
Diesel pile-driver	128–138
Air hammer (piling)	114–126
Drop hammer (piling)	94*–122
Hydraulic pile-driver	94–104
Bored piling	112–116
Pumping concrete	100–109
Compressor	90–105
Face shovel	106–115
Dynamic compaction	115–120

* Enclosed.

only), or as an activity that includes elements of driving, pitching, manoeuvring, servicing, etc.

Although earthworks and embankment construction involve a significant number of machines operating over a substantial time period, the works are not normally in close proximity to residential properties and maximum induced noise levels at the site boundary are relatively low (in the region of 75–80 dB(A)). However, in mineral working the construction of noise bunds (screens) and topsoil stores are noisy operations (from an environmental viewpoint) because this preliminary work is usually conducted close to the noise-sensitive dwellings to be protected. Bunds may also be visually intrusive, they may affect TV reception and dust may be created during their construction, and these effects may trigger a complaint against the level of noise from the site. While blasting gives rise to significant noise in the vicinity of the work, the number of noise events each day is very low so that the L_{eq} value is low. More serious environmental problems arising from blasting are vibrations transmitted through the ground, pressure waves through the air (overpressure) and flyrock (fragments of rock propelled into the air).

17.3. Noise assessment

Both local authorities and contractors need to know the sound levels that can be expected at occupied premises as a result of site operations — the former so that realistic noise limits can be set and the latter so that the works can be planned to meet these limits. Published sound data can be used to undertake preliminary calculations to ascertain rapidly whether site noise is likely to constitute a nuisance or not. There are three main methods of prediction:

- The *sound power level method* is used where sound data are not available for a specific construction operation as an entity. The overall noise level is built up by considering each individual item of plant and adjusting the power level for the proportion of on-time during the assessment period.
- The *activity method* can be used for stationary and quasi-stationary activities when these activities and their locations are clearly defined. This can be done provided that data are available for the appropriate activity (which may involve several items of plant or just one machine undertaking a specific job).
- For haul-road traffic (i.e. mobile plant using a regular route and passing at a more or less constant rate), there are *specific formulae*.

17.3.1. Sound power level method

The general procedure of the method is outlined in Figure 17.6. The noise sources are divided into categories (point, line, mobile) that enable their positions and noise level to be determined accurately. Sound power levels L_W for each source are obtained from appropriate data sources (e.g. BS 5228 (BSI, 1996)) or by measurement of similar machinery. If the plant moves about a limited area on site then the average distance from the source to the point of interest is determined.

The sound level L_p at a receiver placed at a distance R metres from an item of plant or operation is calculated using the appropriate form of equation (17.11). For propagation over hard ground (concrete, asphalt, etc.)

$$L_p = L_w - 20 \log R - 8 \tag{17.14}$$

and for propagation over soft ground (vegetation, etc);

$$L_p = L_w - 25 \log R - 1 \tag{17.15}$$

Soft ground attenuation does not apply for propagation distances less than 25 m. Also, for distances over 300 m caution should be exercised in applying the distance attenuation equations (especially that for propagation over soft ground) because of the increasing importance of meteorological effects.

While the period of assessment is T (usually 12 hours), it may be that a noise source is only active for a limited period t during the day. Thus each sound level received by the target has to be converted to an L_{eq} activity value $L_{A\,eq}$ and from equation (17.13),

$$L_{Aeq} = 10 \log\left(\frac{t}{T} 10^{L_p/10}\right) \equiv L_p + 10 \log \frac{N}{100} \tag{17.16}$$

where N is the percentage on-time (the working time as a proportion of the full monitoring period) for the noise source.

If there is a solid obstacle between the noise source and the receiver then part of the sound will be reflected or absorbed (i.e. the sound will be attenuated, as indicated in Figure 17.7). As a simple guide, the attenuation due to screening can be taken as at least 5 dB(A) when a line joining the source and the receiver just grazes the top of the screen. When the screen completely hides the source from the receiver then a minimum attenuation of 10 dB(A) can be assumed. These

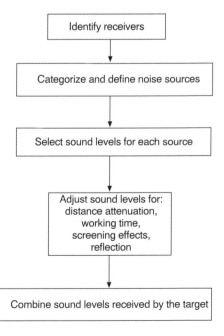

Figure 17.6. Flowchart for the calculation of the overall sound level

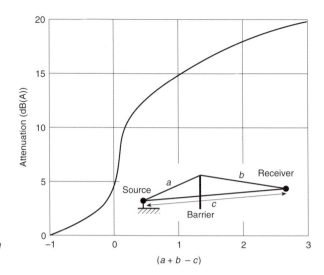

*Figure 17.7. Sound attenuation
due to screening*

reductions are applied to each noise source as appropriate. It is usual to take account of either the combined attenuation from the screening and the hard ground propagation, or the attenuation due to the propagation over soft ground alone.

To assess the effect of noise sources on inhabited buildings the total sound level at a point 1 m from the affected facade (and 1.5 m above ground level) is determined. In this case the target will receive sound directly from the source and also as a reflection from the facade. Thus 3 dB(A) should be added to each calculated sound level to allow for this reflection.

Because sound levels are defined on a logarithmic scale, the sound values received by the target cannot be added arithmetically. Instead, pairs of values are combined progressively, until a final, single value is obtained. The resultant sound level from each pair of values depends on the difference between the two sound levels. Table 17.3 gives the sound level increment (dB) that has to be added to the higher of the two values when adding a pair of noise levels.

17.3.2. Activity method

This method uses the same overall analytical components as does the sound power level method (see Figure 17.6 and Section 17.3.1). However, the sound level used in the analysis, the equivalent activity sound level L_{Aeq}, is for a whole construction operation, which typically involves several different items of plant, although it may be for just one item of machinery. Furthermore, this value is the equivalent sound level, over the assessment period, at a distance of 10 m from the noise source. This output is obtained by taking sound measurements, over the assessment period, for a similar construction operation or item of plant operating in the relevant mode and at the appropriate power level. Alternatively, published values of L_{Aeq} (as per BS 5228 (BSI, 1996)) may be used. The advantage of this method is that variations in the plant cycle time and interactions between various items of the plant during the activity, and the consequent overall variation in the

*Table 17.3. The addition of
sound levels*

Difference between sound levels (dB(A))	Addition to the higher sound level (dB(A))
0–1	3
2–3	2
4–9	1
>10	0

noise level with time, are automatically taken into account. However, the activity L_{Aeq} still has to be corrected for source–receiver distance, reflections and screening or soft ground attenuation.

If the distance from the receiver to the source (the geometric centre of the plant or activity) is other than $10\,\text{m}$, then the following corrections ΔL_{Aeq} must be subtracted from the L_{Aeq} value: for propagation over hard ground

$$\Delta L_{\text{Aeq}} = 20 \log \frac{R}{10} \tag{17.17}$$

and for propagation over soft ground

$$\Delta L_{\text{Aeq}} = 25 \log \frac{R}{10} - 2 \tag{17.18}$$

Corrections are made for screening and reflection effects (as explained in Section 17.3.1), and then the individual activity values are progressively combined in pairs to give one final value.

17.3.3. Plant using a well-defined route

For mobile items of plant that travel a specific route at regular intervals (such as earth-moving machinery passing along a haul road), it is possible to predict an equivalent continuous sound level using the following expression (Martin & Solaini, 1976):

$$L_{\text{eq}} = L_{\text{w}} - 33 + 10 \log R + 10 \log Q - 10 \log V - 10 \log d \tag{17.19}$$

where; L_{w} is the sound power level of each power unit, Q is the number of power units per hour travelling at an average speed of $V(\text{km/h})$, and d (m) is the distance from the receiving position to the centre of the path followed by the plant.

Where the angle of unobscured view α_v of the traffic path, at the receiver, is less than $180°$ an angle of view correction ΔL_{eq} is applied, wherein:

$$\Delta L_{\text{eq}} = 10 \log \frac{\alpha_v}{180} \tag{17.20}$$

If necessary, the equivalent sound level has to be adjusted for reflections, as outlined previously.

17.3.4. Noise control

Regular noise monitoring should be undertaken in the vicinity of the works. Sound levels are usually measured, or calculated, for a position 1 m in front of a relevant building facade (and 1.5 m above ground level). Since it is the effect of the sound levels on persons adjacent to the site that is important, the monitoring instruments (Figure 17.8) must record the A-weighted sound and must be capable of giving the L_{eq} value, either directly or indirectly. For rapidly fluctuating and

Figure 17.8. The measurement (evening) of sound levels from opencast operations

impulsive noise the preferred instrument would be an integrating sound level meter to give L_{eq} directly. If the noise consists of a number of clearly distinguishable values of sound level then the separate levels can be measured as steady noise and combined (taking account of the durations) using the previously given approximation for L_{eq} (equation 17.13). Precautions should be taken to ensure that measurements are not affected by the presence of personnel, by wind, or by other extraneous effects such as electric fields.

The general principles of noise control are to keep the source–receiver distance large and use screening where possible (this can be in the form of covers that are part of the equipment, or it may be buildings, stockpiles, etc., which shield affected areas). While these principles apply to all works, piling is the noisiest construction operation (in terms of the sound level and the proximity to dwellings), and so the common noise-reduction methods are illustrated by reference to this type of operation.

From the acoustic point of view a complete enclosure of the piling operation is the simplest and most effective method of controlling the propagation of the sound produced. However, such enclosures have many practical drawbacks, and in general there are four basic methods of managing the noise arising from piling works, as described below.

Selection of the appropriate construction process

Pile installation methods include driving, jacking and vibratory, and at most sites the ground conditions are likely to be such that at least two methods of pile installation (Table 17.4) would work (from an engineering viewpoint). In conventional pile-driving a hammer impacts on the top of a pile thereby generating high sound levels. In concept the Taywood Pilemaster (see 'hydraulic pile-driver' in Table 17.2) applies a large static force to the head of a pile and simply jacks it into the ground. This force is developed by using the reaction (of adjacent piles) to pull-out. There is no impact or vibration, and therefore no noise except for the electric motor and hydraulics and pitching of the piles. Jacking

Table 17.4. Alternative piling methods

Bearing piles

Bored replacement			Driven		Jacked	**Type**
Injection	Preformed, grouted in place	Bored, cast in situ	Small displacement	Large displacement	Preformed	**System**
Continuous flight auger, jet mixing	Concrete section, with or without temporary support	Straight underreamed	Steel sections (box, tube, H, screw)	Preformed (timber, concrete). Driven tubes (filled with concrete after driving, plugged during driving with void being filled afterwards)	Concrete or steel	**Forms**

Retaining piles

Replacement			Driven sheets	Jacked	**Type**
Contiguous bored piles	Diaphragm wall	Secant piles	Steel (mainly) concrete timber	Steel	**Options**

works with most cohesive soils and silty sands. Vibratory driving is similar to jacking in that a heavy weight is sat on top of the pile. However, to achieve penetration in this latter case the pile is subjected to vibration.

Reduction of the sound level at source

Impactive piling noise can be reduced by using different connecting media between the hammer and a pile (e.g. non-metallic, fluid). Significant reduction of noise emission can be achieved by shielding at source, and there are commercial systems available that employ partial enclosure of the hammer or complete enclosure of the hammer and the complete length of pile being driven (see 'enclosed drop hammer' in Table 17.2). Continuous sound generators such as engines should be fitted with effective exhaust silencers and acoustic engine covers.

Control of the propagation of generated noise

Screening by separate barriers (Figure 17.9) is less effective than enclosure, but it can be a useful contributor to other control measures, and the prevention of a direct line of sight between the recipient and the noise source usually has a psychological benefit. Hoardings around a site may be designed to act as a noise screen and can easily produce an attenuation of 5 dB(A). In most practical situations the overall attenuation will be limited by transmission over and around the barrier, provided that the barrier material has a mass per unit of surface area in excess of about 7 kg/m^2. Ideally, the screen should be as close as possible to either the source or the receiver. Some sound will pass round the ends of short straight barriers. As a rough guide, the length of a barrier should be at least five times greater than its height. A shorter barrier should be bent round the noise source. However, it should be remembered that a barrier may, by reflecting sound, simply transfer a problem from one receiving position to another.

Some screening may result from site topography and the position of on-site buildings. In urban situations buildings close to the piling operations can provide substantial protection for more distant properties (Figure 17.10) and noise reductions of 10 dB(A) or more are not uncommon.

Figure 17.9. Screening by a free-standing barrier

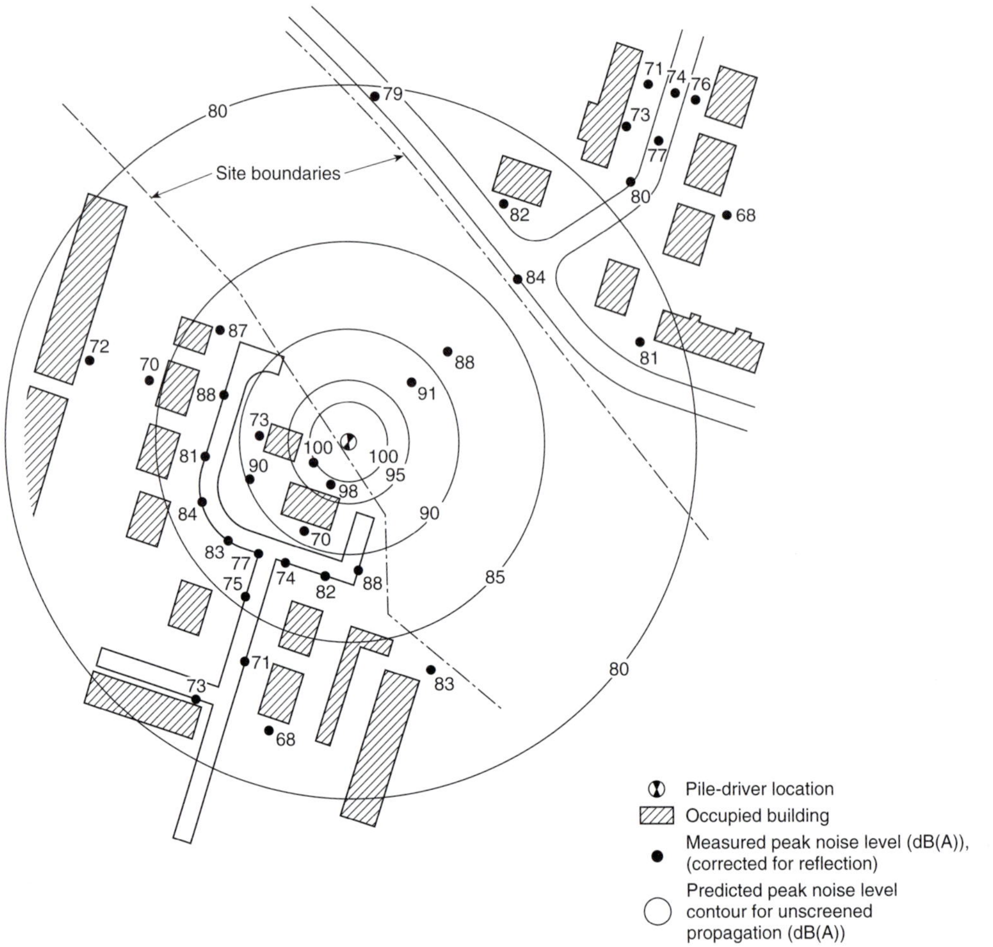

Figure 17.10. Screening of distant dwellings by other properties

Phasing of the piling and adjacent construction operations
Reduction of the length of the working day for noisy operations reduces the daily L_{eq} value. However, a halving of the working period will only reduce the value by 3 dB(A). Consequently to achieve large reductions the working day would have to be shortened to an unrealistic time.

17.4. Ground vibrations

The acceptability of construction methods is often critically dependent on interactions with nearby structures or people. Various geotechnical operations (e.g. driven and bored piles, earth-moving, compaction, quarrying) produce significant levels of ground vibration. The vibrations may be sufficiently intense to annoy occupants of nearby buildings, or they may have detrimental effects on structures or equipment.

Construction vibrations may be divided into three categories (Figure 17.11):

- *Steady-state or continuous*: an approximately constant level of vibration is maintained for a substantial time (e.g. vibrations from vibratory pile-drivers, large pumps and compressors, earthworks).
- *Transient or impact*: the vibration decreases from its peak level to nothing before a new quantum of energy is applied, which causes the vibration level to rise rapidly to a peak value again (e.g. dynamic compaction, pile-driving using a drop hammer).
- *Intermittent*: a string of vibration incidents, each of short duration, separated by intervals of much lower vibration magnitudes (e.g. ground motion arising

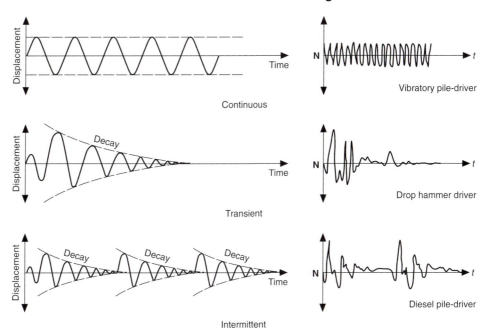

Continuous

Vibratory pile-driver

Transient

Drop hammer driver

Intermittent

Diesel pile-driver

Figure 17.11. Classes of ground vibration

from a diesel pile-driver, boring for piles cast in situ, vibrations resulting from blasting with a succession of charges).

17.4.1. Vibration characteristics

Vibrations are physically characterized as wave phenomena and they may be transmitted in a number of wave types. Each type of wave travels at a velocity that is characteristic of the material properties of the medium through which it passes. The wave velocity does not, however, determine the severity of the vibration at a receiving point. As the wave passes through this receiving point the particles of matter undergo an oscillatory motion and it is the intensity of these particle motions that defines the severity of the vibration.

A body is said to vibrate when it describes an oscillatory motion about a reference position. The motion can consist of a single component occurring at a single frequency, but in practice vibration signals usually consist of very many frequencies occurring simultaneously. The term 'frequency' has the same meaning as when it is applied to sound (see Section 17.2.1).

The motion of a discrete element of a vibrating material has its simplest form when it is sinusoidal (harmonic). Vibratory pile drivers often create this type of motion in the ground (see Figure 17.11). For sinusoidal motion (Figure 17.12),

$$Z = A \sin(\omega t - \theta) \tag{17.21}$$

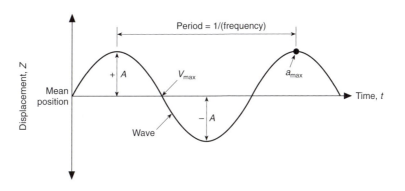

Figure 17.12. The motion of a vibrating particle

where Z is the displacement about a mean position, A is the maximum displacement (amplitude of vibration), ω is the circular frequency ($2\pi f$, where f is the frequency), t is the time and θ is the phase angle.

The peak values of velocity v and acceleration a are interrelated:

$$v_{\text{peak}} = 2\pi fA, \qquad a_{\text{peak}} = 2\pi f v_{\text{peak}} = 4\pi^2 f^2 A \qquad (17.22)$$

The severity of vibrations can be defined in terms of particle displacement, velocity or acceleration. Various investigations have reported good correlation between dynamic strains and peak particle velocity for buildings subjected to blast vibration (Wiss, 1967; Nicholls *et al.*, 1971). For assessing human tolerance to vibration the relevant parameter is often stated to be acceleration (see BSI, 1984). Human reaction to acceleration depends on the frequency of the vibration and, at their source, construction vibrations contain a wide spectrum of frequencies. However, the ground motion felt at some distance from the source will have only a very limited frequency range (due to the 'filtering' effect of the ground). Thus, over this frequency range there is a more or less direct correspondence between ground acceleration and peak particle velocity, and human sensitivity to peak particle velocity is essentially independent of the frequency of vibration (Figure 17.13). Consequently, the usual parameter for assessing the severity of ground vibrations from construction is the peak particle velocity, for both consideration of the risk of damage to structures and where human reaction is of prime concern.

In general, there will be vibration components in three orthogonal directions, and ideally the three components should be considered in any measurement or assessment process. Simultaneous ground measurements of peak particle velocities in the vertical, radial and tangential directions (relative to a straight line joining the source and the receiver) have shown the vertical values to be greater than, or equal to, the radial ones and significantly larger than those in the tangential direction. If the effects of ground vibration depend primarily on the

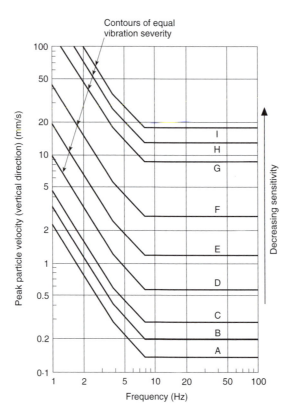

Figure 17.13. Vibration sensitivity curves for erect humans (after BSI, 1984)

Situation	Vibration type	
	Continuous	Intermittent
Critical	A	A
Home (night)	B	F
Home (day)	C – D	G – H
Office	D	I
Workshop	E	I

energy content of the disturbance, it would be appropriate to consider the vibration vector as the parameter for assessing vibration severity. However, most methods of analysing and assessing ground vibrations from construction are based on the peak particle velocity in the vertical direction, for the following reasons:

- An erect human is most sensitive to motion in the vertical direction.
- A prone or supine person would be most affected by horizontal vibrations, but construction-induced ground motion occurs at a time when only a very small proportion of the population (e.g. night-shift workers, hospital patients) is likely to be lying down.
- Human beings are much less tolerant of ground vibrations than are structures (possibly because of their much smaller inertial mass and their internal systems, which can be easily affected by motion and changes in acceleration). Hence, well before an occupied building suffers damaging ground vibrations, its residents will have found the motion intolerable.
- For well-constructed buildings in a good state of repair, damage from ground-borne vibrations results from flexure and cracking of building elements rather than from shear. Structural members that can be excited by vibration most easily, and possibly brought into a state of resonance, are those with a high length/depth ratio and a relatively low mass. Such elements tend to be aligned predominantly in a horizontal direction (thus they are 'plucked' by vertical motion).
- Proven cases of damage to buildings from blasting-induced vibrations have suggested that there is a correlation with the vertical intensity of motion rather than with movement in other directions.
- Most available information relates to vertical motion of the ground.

17.4.2. Generation of ground vibrations

The transfer of energy from construction sources to the ground mass and the subsequent propagation of the vibration involves very complex processes (as illustrated in Figure 17.14), particularly when the factors relating to the actual construction process are included. For instance, the general variables relating to

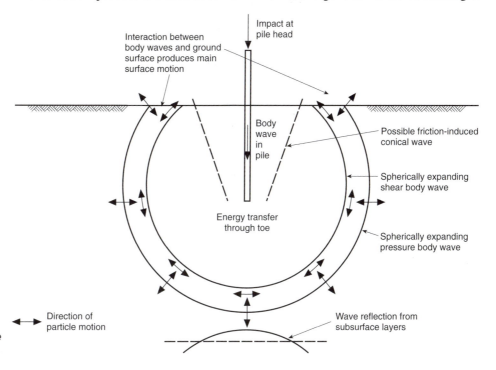

Figure 17.14. Wave components near a driven pile (from Martin, 1980)

the generation of ground vibrations by piling may be summarized as driver system, pile type and ground conditions. The energy imparted to the ground is dependent on:

- the characteristics of the hammer or vibrator
- the deformation of any helmet (a 'cap' sitting on top of a pile) or packing
- the elastic distortion of the pile
- the rate of penetration of the pile or casing.

In a soft clay most of the applied energy directly advances the pile, whereas in stiff clays, dense gravels and weak rock a significant proportion of the energy is transmitted to the ground. At refusal, all the energy is used in vibrating the soil.

The need for blasting varies significantly between the types of mineral being worked. For coal workings it is often, but not always, necessary to loosen or 'heave' the overburden. Most of the energy stays in the ground because the overburden is only loosened. For hard rock it is necessary not only to loosen the rock but also to fragment it and move it away from the face of the quarry. Because of this more energy is lost to the atmosphere than with coal, so the overpressure will probably be greater. The basis of good blasting design is to achieve the desired degree of fragmentation in the rock safely and economically.

Not only is the initial radiation pattern from excitation sources difficult to define, but it constantly changes with propagation through the ground mass (an elastoplastic, anisotropic medium) due to the following factors (New, 1986):

- geometrical spreading
- the progressive separation of compression, shear and surface wave types due to their different propagation velocities
- the presence of discontinuities, which cause reflection, refraction, diffraction and scattering
- internal friction causing frequency-dependent attenuation.

The response of the ground to continuous excitation is to vibrate in sympathy with the source (i.e. at the same frequency). However, soils have a frequency-dependent damping mechanism and they act as filters so that only frequencies within a certain band are readily transmitted (Table 17.5). Coincidentally, many structures, building elements, services and human beings are particularly sensitive to vibrations in the frequency range 5–50 Hz approximately.

17.4.3. Vibration data

The instrument that is more or less universally used for measuring ground-borne vibrations is the piezoelectric accelerometer. It works on the principle that the motion acts on a small mass, which then applies a force (which is proportional to the acceleration) on an adjacent piezoelectric element. When the slice of piezoelectric material is mechanically stressed it generates an electrical charge (which is proportional to the applied force), which is measured by a read-out unit. Numerical integration, of the acceleration indicated by the electrical charge, is undertaken within the read-out instrument to obtain the velocity of motion. Piezoelectric accelerometers have very wide frequency and dynamic ranges with

Table 17.5. Preferential frequencies for the transmission of ground vibrations

Ground type	Frequency range (Hz)
Very soft silts and clays	5–20
Soft clays and loose sands	10–25
Compact sandy gravels and stiff clay	15–40
Weak rocks	30–80
Strong rocks	>50

good linearity throughout the ranges. Velocity transducers (geophones) are also used, although they have smaller frequency and dynamic ranges and they are physically larger. These latter instruments operate on the principle that the vibrations cause a magnet and coil to be relatively displaced, and a voltage is generated that is proportional to the instantaneous relative velocity. An extensive collection of relevant vibration data is contained in BS 5228: Part 4 (BSI, 1996) and typical vibration levels are indicated in Figure 17.15.

Peak particle velocity values are often obtained for everyday situations that are of the same order as those induced (at the recording point) by piling operations (e.g. walking across a suspended wooden floor can produce a vertical velocity of 3 mm/s, a washing machine on the spin cycle can generate a vertical velocity of 5 mm/s in the floor). Demonstrating that vibrations from piling works are similar to those from everyday acceptable events may help in public relations efforts.

The design operating frequency of most vibratory pile-drivers is typically in the range 25–30 Hz, whereas resonant pile-drivers operate at higher frequencies (up to 135 Hz). Although the operating frequency of vibratory piling systems may be higher than the 'preferred' frequency of a particular ground-type, it must be remembered that whenever the vibrator is switched on or off it accelerates or decelerates through the 'preferred' frequency range.

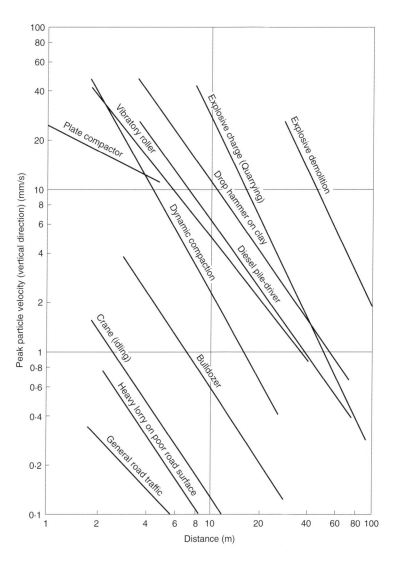

Figure 17.15. Vibrations from typical construction operations

17.5. Vibration assessment

The general assessment process involves:

- prediction of the vibration level at the point under consideration
- estimating the resultant vibration level induced in the target
- assessing the likely effects of these vibrations.

Prediction of ground-borne vibrations will usually be on the basis of some semi-empirical relationship that accounts for the attenuation of the vibratory energy (by the ground) as a result of geometrical spreading and dissipation (damping). Ideally, the limits for vibration should be set (by the local authorities and the specification for the works) after due consultation with the parties involved and taking into account the specific circumstances of the site. Ground-borne vibration propagation has to be assessed very carefully where sites are close to the categories of building or structure identified in Table 17.6.

Before starting the construction operations it is necessary to survey the most-affected structures. The survey should include a detailed record of:

- existing cracks and their size and orientation (photographic records are particularly useful)
- level and plumb survey (including any evidence of movement along the damp-proof course)
- measurements of tilting walls or any bulges
- any unevenness of paths
- cracked window panes and other existing damage, including loose or broken tiles, pipes, etc.
- incidence of jamming of windows and doors.

Vibration monitoring using static tell-tales is a useful adjunct to the measurement of the actual vibration levels. If the works are expected to generate significant motion within adjacent structures then a thorough survey of existing building defects in the area should be undertaken prior to commencement of the works. This survey should identify the locations at which tell-tales would serve a useful purpose (e.g. across cracks, between different structural elements).

17.5.1. Vibrations from impactive piling

Most of the energy from a pile-driver goes into temporary compression of the pile and penetration of piles through the ground. As the pile penetrates into the ground, the driving hammer generates a body wave within the pile, which travels along the shaft to the interface between the pile base and the soil. A portion of the wave energy is reflected within the pile, but most is transmitted to the soil. Body

Table 17.6. Vibration-sensitive buildings and structures

Building/structure category	Examples
Likely to contain people who are particularly sensitive to disturbance	Hospitals, nursing homes
Likely to contain an environment/ambience that is particularly sensitive to disturbance	Museums, art galleries, theatres, laboratories, libraries, places of worship
Likely to contain equipment and/or instruments that are particularly sensitive to disturbance	Precision machine workshops, laboratories, automatic telephone exchanges, fibre optic transmission systems, buildings containing relay systems and switchgear
Likely to be directly sensitive to disturbance	Historical monuments, ancient buildings, housing and buildings in poor condition, swimming pools

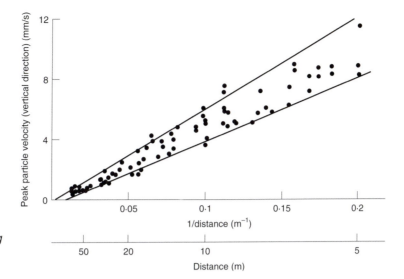

Figure 17.16. The variability of vibration values: a sheet-piling diesel hammer (one panel of piles)

waves are produced at the pile base and a shear wave reaches the ground surface, creating a head wave. This complex system of particle motions (see Figure 17.14), together with the inherent variability and inhomogeneity of even 'uniform' ground, renders precise prediction of ground vibration levels, from 'first principles', totally unrealistic. The data in Figure 17.16 refer to piling undertaken in essentially one spot, using the same driver, with the same type of pile throughout — the variability of values is readily apparent.

Theoretically, compression and shear waves attenuate at a rate inversely proportional to the separation distance, whereas surface waves attenuate at a rate inversely proportional to the square-root of the separation distance. Also, as would be expected intuitively, the magnitude of the induced ground vibrations increases with the energy applied to a pile. The foregoing factors have led to the development of a simple, usable, semi-empirical equation for the prediction of peak particle velocities (mm/s) within the ground, due to piling operations:

$$v = C \frac{\sqrt{W}}{r^x} \tag{17.23}$$

where W is the source energy (joules) per blow or per cycle and r is the horizontal radial distance (m) between the source and the receiver.

Most measured values of the power x have been in the range 0.8–1.5 and, while there have been suggestions to employ different values of x according to the type of piling process, a commonly assumed value is unity. Attewell & Farmer (1973) suggested that an upper bound to generated vibration levels was given by $C = 1.5$. On the basis of the author's experience, alternative values of the parameter C are proposed (Table 17.7). These values define upper-bound peak

Table 17.7. Values of the parameter C (equation 17.23)

Driving method	Ground conditions	C
Impact	Very stiff cohesive soils, dense granular media, rock, fill with large solid obstructions	1.5
	Stiff cohesive soils, medium-dense granular media, compact fill, well-compacted refuse	0.75
	Soft cohesive soils, loose granular media, loose fill, organic soils, loose refuse	0.25
Vibratory	All soil conditions	1.0

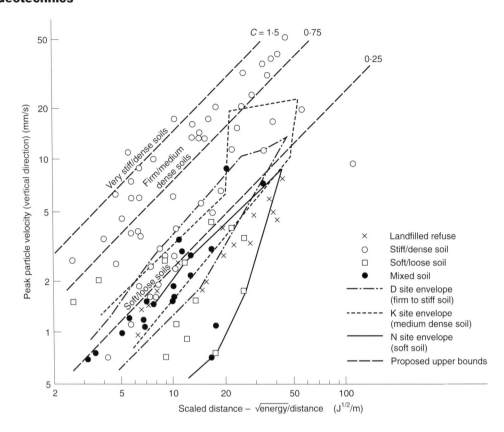

Figure 17.17. The bounds to ground vibration levels

particle velocities, which are unlikely to be exceeded, for the particular situations stated, in the ground, or on a load-bearing part of a structure at ground or foundation level (Figure 17.17). Stiff soil and intact rocks transmit vibrations more readily than do softer, more compressible materials. Thus the presence of unknown hard layers, through or into which the piling penetrates, can give rise to vibration intensities corresponding to the upper case values.

More accurate assessment can be achieved by 'calibration' of the actual site (i.e. determination of a site-specific formula). The necessary data can be obtained by conducting a trial drive or by dropping a large weight onto the ground surface and recording the vibration levels at various locations around the source. Vibration measurements can also be taken inside affected buildings.

17.5.2. Vibrations from quarrying

Quarrying will usually involve relatively large rounds of explosive at substantial distances from residential structures, so that the predominant ground motion will arise from surface waves with low frequencies. Vibrations from blasting on construction sites is likely to be induced by small charges at very close range, so that the body waves will dominate and much higher frequencies will result.

During blasting, the rock mass around a hole is subjected to repeated loading and unloading by the shock waves. The amount of deformation in a particular rock mass depends on the magnitude of stress developed by the shock waves which is, in turn, dependent on the charge size. The greater the deformation, the greater will be the reduction in shock energy. Hence the rate of ground attenuation of vibrations will be greater for large charges than for small charges.

Empirical relationships for the estimation of the peak particle velocity v (mm/s) are usually of the form:

$$v = \frac{KM^a}{R^b}$$

(17.24)

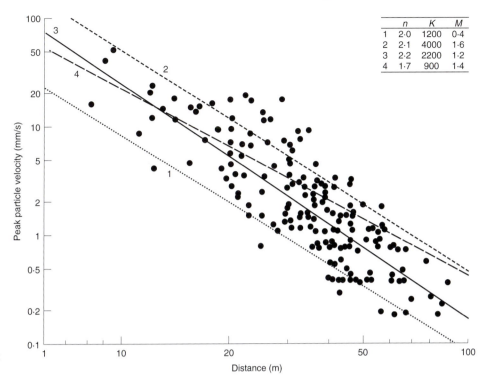

	n	K	M
1	2·0	1200	0·4
2	2·1	4000	1·6
3	2·2	2200	1·2
4	1·7	900	1·4

Figure 17.18. Ground vibrations from blasting (adapted from Chakraborty et al., 1996)

where M is the charge weight (kg) and R is the distance (metres) between the source and the receiver. A commonly used form of equation (17.24), which is very similar to other predictive equations for ground motions, is

$$v = K \left(\frac{\sqrt{M}}{R} \right)^n \tag{17.25}$$

Typical values of K and n lie in the range 900–4000 and 1.7–2.2, respectively (Figure 17.18).

Whenever explosives are detonated, airborne waves are generated. When a blast wave passes a given position, the pressure of the air at this point rises rapidly to a value above the ambient pressure, and then falls more slowly to a value below, before returning to the ambient value. The maximum excess pressure in this wave is known as the *overpressure* (air blast) and when this overpressure is excessive damage can occur.

17.5.3. Other construction operations

Although earth-moving plant and heavy lorries may produce measurable ground vibrations, these dissipate very rapidly with distance from the source and their effects on potential targets beyond the site boundary are generally negligible. Dynamic compaction (see Chapter 14, Section 14.1.3) may produce ground motion that affects recipients outside the site boundary, because of the rate at which energy is imparted to the ground. The variation in the peak particle velocity with horizontal distance from the point of compaction (Figure 17.19) is very similar to that observed for impactive piling. Equation (17.23) may be used to predict ground vibration levels, typically with x equal to unity and C around 0.5.

17.5.4. Effects of vibrations on humans

The human body is very sensitive to vibration, and when the induced peak particle velocity exceeds the threshold of perception (about 0.15 mm/s at frequencies between 8 and 80 Hz) complaints may arise. High levels of vibration

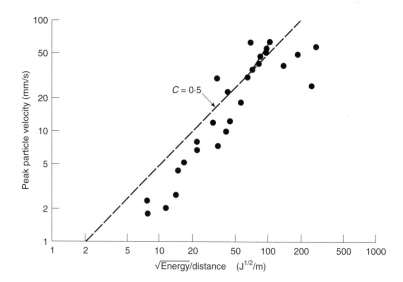

Figure 17.19. Ground vibrations from dynamic compaction

are unpleasant and even painful, and they can have physical effects on health. Such levels of motion are very unlikely to be induced outside the confines of a construction site, except by piling operations. With blasting, people seem to be more concerned about a few high-level blasts than a greater number of small ones.

Various scales of vibration perception criteria have been developed. Many of the scales were developed for permanent steady-state vibrations or transient events such as earthquakes. One of the most important contributions to the knowledge about human sensitivity was provided by the comprehensive investigations undertaken by Reiher & Meister (1931), wherein a representative group of people was subjected to vertical or horizontal vibrations. From their experimental data, Reiher & Meister (1931) classified vibrations into several broad bands or zones in terms of amplitude and frequency (Figure 17.20). The demarcation between 'just perceptible' and 'clearly perceptible' corresponds to a

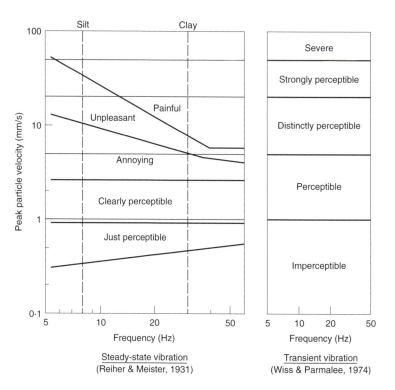

Figure 17.20. Criteria for the effect of ground vibrations

maximum velocity of about 1 mm/s and an annoying vibration would have a velocity of 2.5–7.5 mm/s. The results of later studies have broadly confirmed the applicability of the data to transient vibration situations (Wiss, 1981).

Magnitudes of vibration below which the probability of adverse comment (from the occupiers of affected buildings) is low are presented in BS 6472 (BSI, 1984). The base curves for assessment are given in terms of acceleration in three orthogonal directions (one vertical and two horizontal). For the predominant frequencies usually associated with piling vibrations (i.e. frequencies greater than 5 Hz), the human body is most sensitive to motion in the head–foot direction and the threshold values correspond to a unique value of the peak particle velocity (see Figure 17.13). In the case of intermittent and impulsive vibration, the limits given in BS 6472 (BSI, 1984) assume only several occurrences per day, and consequently they do not immediately correspond to the conventional piling situation.

The primary factors influencing the degree of annoyance to humans from ground vibrations are indicated in Table 17.8. The complaints arise for two reasons: disturbance and possible discomfort, and concern regarding damage to property and belongings. Even the slightest vibration may draw the attention of the occupier of a building to minor cracking that was pre-existing and that may otherwise have remained unnoticed. However, it is widely acknowledged that human tolerance will be greatly increased if the origin of the vibrations is known in advance and if occupants are assured that their property and possessions will not be damaged. If ground vibrations from construction operations are kept below the values of the peak particle velocity (in the vertical direction) indicated in Table 17.9 then the probability of complaints will be low. Table 17.9 has been drawn up on the basis of available information and past experience. A serious number of complaints is likely if peak particle velocities are greater than four times these values.

*Table 17.8. Primary factors affecting the response to ground vibrations**	For humans	For buildings
	Age and physical condition	Stiffness of building and elements, general condition of the building, internal modifications
		Damping characteristics of the building
	Vibration characteristics	Relationship between natural frequencies of the building and its elements, characteristic frequency of the ground vibration
	Duration and timing	Triggering of defects already present
	Proximity of source	Magnitude of vibration
	Quality of previous environment	—
	Accompanying noise	Overpressure effects
	Activities affected	Use of the buildings
	Mental condition and attitude	—
	—	Dimensions of the building, materials used in construction

* Adapted from Steffens (1966).

*Table 17.9. Suggested human tolerance boundaries**

Area	Peak particle velocity (mm/s)	
	Continuous vibration	Transient vibration
Sensitive locations (e.g. hospitals, libraries, precision laboratories)	0.15	0.15
Residential	0.3	1.0
Offices, shops	0.6	2.0
Workshops, factories	1.2	4.0

* After Sarsby (1992).

Blasting overpressure is of considerable importance because it is usually accompanied by noise, such as the rattling of windows, which is then attributed to ground vibration. Thus, air blast magnifies the perception of vibrations and incurs increased complaints concerning blast vibration.

17.5.5. Effects of vibrations on buildings

It is vitally important to distinguish between architectural and structural damage. Architectural damage (e.g. the initiation or propagation of cracks in plasterwork), is annoying and it begins at a much lower level of motion than does structural damage. Architectural defects, however, are easily repaired and they do not detrimentally affect the integrity of the property. Serviceability failure and structural damage include loss of weatherproofing, sticking of doors and windows, sloping floors and wall movements. Documented instances of vibrations (resulting from construction operations) causing such damage to buildings are very rare, and in those cases reported it is noted that the levels of vibration were unpleasant and painful to the occupier. Usually the harm has resulted from either amplification effects within dwellings, due to large floor spans and/or lightweight construction, or induced settlement of the ground beneath the foundations.

When considering the effects of piling vibrations on properties or their inhabitants it must be remembered that the form of building construction, or internal structural modifications, may cause amplification of incoming motion. For standard dwellings, vibration levels monitored during the driving of sheet-piles by drop hammers and diesel drivers have indicated amplification factors ranging between 0.3 and 0.9 in buildings with concrete floors and between 1.5 and 1.8 for buildings with wooden floors. Significant amplification can occur in upper storeys and in the middle of long-span suspended floors. Amplification factors vary according to circumstances, but values of between 1.5 and 2.5 have been recorded under the aforementioned situations. The situation is even worse for retaining walls and retained facades, which have little lateral restraint near their top, and vibrations can be amplified three-fold.

There are no universally agreed criteria for the assessment of the effects of construction-induced, ground-borne vibrations on buildings, but conservative threshold values of the peak particle velocity, for the onset of non-structural damage within soundly constructed, well-maintained properties, are given in Table 17.10.

- Damage is unlikely (even improbable) at true peak particle velocities of less than 2 mm/s measured on structural foundations. For a given value of the peak particle velocity the risk decreases for higher frequencies.
- For peak particle velocity values between 2 and 10 mm/s there is the increasing possibility of plaster cracking, of the enlargement of existing or visible defects and of cracking in finishes.

Table 17.10. Suggested vibration limits for buildings and vulnerable structures

Type of building/structure	Peak particle velocity (mm/s)	
	Continuous vibration	Transient vibration
Ruins, buildings of architectural merit, tailings and sludge lagoons	2	4
Residential property, tailings dams, waste tips	5	10
Offices, shops, contaminant barriers, backfilled shafts	10	20
Workshops, factories	15	30

- At peak particle velocity values between 10 and 50 mm/s the risk of the onset of damage is high, depending on the structural form, materials and the construction quality of the building. Undamaged buildings of modern construction, built to a high standard and well-maintained, should not be affected.
- For particle velocities greater than 50 mm/s some form of damage is likely, and this might include adverse effects on load-bearing units.
- Where buildings contain existing structural defects the limits in Table 17.10 may have to be reduced by up to 50%. For ruins, buildings of particular architectural interest and environmentally sensitive situations (e.g. schools, hospitals, computer installations) the limits may need to be even lower than those suggested previously.

Piling vibrations can induce densification of granular deposits (Figure 17.21) and can even liquefy saturated fine-grained soils. Damage is most pronounced where shallow foundations overlie loose or medium-dense uniform fine sands and silty sands below the groundwater table, and where ground vibration produces movements that are differential in nature. The resultant settlements or loss of bearing can cause greater distress to adjacent buildings than the vibrations transmitted directly to them. The effect of vibrations from vibratory pile-drivers and vibrocompactors becomes very marked as the characteristic frequency of the ground is approached. Consideration should also be given to the possibility of the vibrations 'triggering' distortion of dilapidated buildings, metastable slopes (Figure 17.22), aged sewers and culverts, etc.

17.5.6. Reduction of effects
It is impossible to obtain a precise and unique prediction of ground vibration levels prior to the commencement of the works. However, the preceding sections

Figure 17.21. Playground settlement due to soil densification resulting from piling vibrations

Figure 17.22. Vibration-induced slope failure

can be used, with a reasonable degree of confidence, to assign a site to one of the three categories shown in Table 17.11.

As with noise-control methods it may be possible to select a piling process or technique that generates sufficiently low levels of motion for it to be acceptable at a 'difficult' location. BS 5228 (BSI, 1996) contains a considerable amount of vibration data relating to various piling systems and ground conditions. The installation of piles by jacking produces little ground excitation. Bored and cast in situ piling is less disturbing than is impulsive driving. Preboring and excavation

Table 17.11. Categorization of predicted vibration levels and suggested outcomes*

Prediction	Vibration category	Actions/options
Predicted particle velocities less than the threshold values	No problem	Proceed with the work Monitor vibration levels during construction
Predicted particle velocities greater than the threshold values for human sensivity, but significantly less than the threshold values for architectural damage	Further investigation required	Obtain site-specific data Assess the sensitivity of the neighbourhood Identify whether there are any areas that are particularly vibration sensitive Undertake a public liaison awareness campaign Monitor vibration levels during construction
Predicted particle velocities close to, or in excess, of the threshold values for architectural damage	Normal piling likely to be unsuitable	Select a specialist, low-vibration installation system, if there is a suitable one Employ methods to control the propagation of vibrations Redesign the works

* After Sarsby (1992b).

of a 'starter' trench can be used when there is a likelihood of serious vibration during penetration of the upper ground stratum (as might occur when driving through fill containing large obstructions).

According to the proposed simple predictive analysis (equation 17.23) the peak particle velocity is dependent on the square root of the energy input from the driver. It has sometimes been noticed that reducing the driving energy has had an appreciable effect on the vibration close to the driver. However, reduction of the driving energy means that more 'blows' and a longer driving period will be needed to install the piles in order to meet the engineering requirements. This longer period may cause 'fatigue failure' of people and buildings (Crockett, 1979).

A cut-off trench (analogous to a noise screen) can be interposed between the source and the receiver to interrupt the direct transmission of vibrations. The trench needs to be as deep as possible and, with the depths feasible on site (typically 4 m or less), such trenches are only really effective when the source or the receiver is in the immediate vicinity of the trench. Factors such as the stability of the trench, the disruption of site traffic movements and the hazard to personnel mean that this method of vibration control should only be used after much careful consideration. Generally there is a consensus that the use of trenches or ditches (interposed between the source and the receiver) is of little value, particularly since open, deep (6–10 m) trenches would be required for any significant effect.

There is some possibility of limiting the need to blast by the use of modern earth-moving plant, but the main way of controlling vibration and overpressure from blasting is by good design and practice. Mitigation measures include:

- reducing the instantaneous charge weights
- using short delay detonators, decking of explosive charges within boreholes
- ensuring that the maximum free face (fragmented or open) is available for each shot
- using adequate stemming with an appropriate material such as gravel or stone chippings
- using greater 'Factors of Safety' when blasting in variable conditions or unusual situations.

Trial explosions can be performed to determine the size of the actual explosive charges and their disposition so as not to exceed appropriate vibration limits (Highways Agency, 1998).

18 Radioactive waste disposal

18.1. Introduction

In November 1983 Greenpeace informed the British public that the beach below the Sellafield nuclear plant in Cumbria had become highly contaminated by radioactive waste. The waste was washed up from the Irish Sea after being discharged from Sellafield. In December the Government issued a warning to the public that a 40 km stretch of beach had become contaminated to levels over one thousand times the normal background radiation. Ironically, independent studies of the same beach showed that the beach had begun to be contaminated as far back as 1979, but little was done to counteract the problem (Pickering & Owen, 1994).

The disposal of radioactive waste ('radwaste') is currently one of the most sensitive environmental issues throughout the world and is the subject of considerable scientific effort in many countries. Radioactive wastes have to be managed and disposed of in ways that protect the workforce handling them, the public and the environment. The principal problem is that radioactive waste can remain toxic for a very long time and this means that pollutants from the waste must not reach the environment for hundreds or thousands of years. In Britain, responsibility for radioactive waste management is shared between the Government, the regulators and the producers of the waste. NIREX (the Nuclear Industry Radioactive Waste Executive) was set up specifically to implement strategy for the disposal, on land and at sea, of solid low level and intermediate level radioactive waste. In Germany, underground disposal of radioactive wastes (low and medium level) commenced in 1967 in a former salt mine near Braunschweig. Disposal continued up to 1978, when the licence expired and was not renewed. At that time around 25 000 m^3 of waste had been placed in the mine (Wiedemann, 1990).

Since the early 1990s a multidisciplinary team, led by engineers and geologists, has been searching for the safest place to build an underground repository in which to entomb the tonnes of radioactive debris created as a by-product of the British nuclear power industry. At the favoured repository site in Cumbria some 50 boreholes, many 500 mm in diameter and nearly 2 km deep, have been sunk (*New Civil Engineer*, 1996b). Tests within boreholes have included those of; permeability, stress–strain characterization, geophysical identification of fissures, and faults and weakness zones. The investigation is needed to ensure that the various sandstone and limestone layers present and the underlying volcanic tuff (within which the repository would lie) will retain the decaying radioactivity effectively. Mathematical modelling is used to assess the long-term behaviour of such a repository.

Radioactive waste (which may be in gaseous, liquid or solid form) is very much a legacy of the 20th century. Figures 18.1 to 18.3 show the facilities at an abandoned nuclear submarine base (of the former Soviet Union) in Estonia. The major producers of radioactive wastes in Britain are outlined in Table 18.1.

Despite the extensive research that has been undertaken into the safe disposal of radioactive waste, there is still significant public concern, for several reasons:

Figure 18.1. Buildings in which
nuclear submarine engines
were serviced

Figure 18.2. The entrance to a
building which was formerly
the store for nuclear material

Figure 18.3. A repository
(reinforced concrete) for
nuclear material (the size of
the repository can be gauged
from the person's image on
the right-hand-side of the
picture)

Table 18.1. Major producers of radioactive waste*

Organization	Operations
British Nuclear Fuels Ltd (BNFL)	Reprocesses spent nuclear fuel from UK and foreign reactors; fabricates fuel and carries out enrichment
Nuclear Electric	Operates Magnox reactors and Advanced Gas-cooled Reactors
Scottish Nuclear Ltd	Operates Advanced Gas-cooled Reactors
United Kingdom Atomic Energy Authority (UKEA)	Together with AEA Technology, has a number of research establishments
Amersham International Plc	Supplies radioisotopes for research, medical and industrial uses
Ministry of Defence	Maintains the UK nuclear defence capability and operates nuclear-powered submarines

* From the Royal Society (1994).

- fears about detrimental health effects due to leaks and accidents
- the long time period which it is acknowledged is needed for the waste to become innocuous suggests to many people that total containment is impossible to achieve
- the radioactive waste will be left as a problem for future generations
- the disposal site will have negative effects on the economy and attractiveness of the surrounding region
- a belief that environmental damage will result from increased traffic and other disturbances
- a lack of awareness, or appreciation, of the quality of the scientific and engineering expertise being applied to the problem.

18.1.1. Sources of radioactivity

According to the Department of the Environment (DOE, 1992), only about 0.1% of the total annual radiation received by members of the public comes from the disposal or discharge of radioactive wastes from UK nuclear installations. By far the greatest proportion (over 87%) of the received radiation is from natural sources, while an average of 12% of the total originates from the beneficial use of radioactive substances in medical treatment (X-rays, radiotherapy, etc.). Mankind has always been exposed to naturally occurring sources of ionizing radiation, some of which has caused genetic mutations (e.g. the injuries to miners exposed to radon gas in the pitchblende mines of Bohemia during the 16th century) (Hughes, 1982).

In general, background radiation originates from space (cosmic radiation) and the ground. A small proportion of the radiation originating in space consists of electromagnetic radiation that reaches mankind directly. The major part of the radiation from space comprises nuclear particles of a very high energy content, which collide with matter in the atmosphere and so are annihilated. However, the annihilation process results in the formation of a large number of particles such as electrons, neutrons and protons, which impact on humans.

Radon, which is a natural radioactive gas formed by the decay of uranium and thorium in rocks is the biggest contributor to radiation exposure in Britain, constituting approximately 60% of the radiation from natural sources. There are traces of uranium in all soils and rocks across Britain, and consequently radon is being emitted continuously from the ground. The highest concentrations of the gas are associated with granite masses in the south-west of England and parts of the Pennines and Scotland. The rate at which radon passes through the ground is important in controlling its concentration at the ground surface or within a soil. Rocks with relatively high permeability, such as highly fissured and faulted rocks, allow radon, in the form of gas or dissolved in groundwater, to migrate at a fast rate into the human environment.

Although radon is easily dispersed by the wind when out in the open, it can enter and collect in buildings. Concentration of the gas within a building is aided by the indoor pressure being lower than the ambient atmospheric pressure (the pressure gradient is greatest when buildings have chimneys). Double-glazing and draught-proofing impede the escape of radon. Levels of radon vary annually, the highest concentrations being between November and March, when houses are kept warm and ventilation is reduced. The National Radiological Protection Board (NRPB) has estimated that up to 100 000 homes in Britain may contain radon concentrations in excess of recommended safety limits.

Tentative links have been drawn between high incidences of cancers (e.g. lung cancer, leukaemia), and high concentrations of radon in houses. As a result, the NRPB has estimated that as many as 2000 deaths per year in Britain may be the result of the radiation produced by radon, and under certain circumstances the increased risk of contracting lung cancer may be comparable to those risks faced by heavy smokers (Pickering & Owen, 1994).

The utilization of radioisotopes in industry, agriculture, education and medicine gives rise to substantial radioactive waste. Radioactive waste is also produced in varying quantities and levels of importance at virtually every stage of the nuclear fuel cycle:

- mining of the ore
- uranium extraction and enrichment
- fuel fabrication
- reactor operation
- spent fuel storage
- spent-fuel reprocessing
- waste management.

All nuclear reactors rely on uranium fuel. The uranium is fabricated into fuel rods, which are inserted into the nuclear reactor. Within the reactor they become intensely radioactive due to the production of fission products and the formation of transuranic elements such as plutonium and americium. The main long-lived fission products require isolation for about 1000 years, but some products (e.g. americium-241 and caesium-135), are estimated to remain biologically dangerous for in excess of 1 million years. Because of the very high energy density of nuclear fuel, the volume of solid radioactive waste is small in comparison with other wastes, accounting for only 0.02% of the total annual waste production in Britain. Nearly 80% of the radioactive waste that is produced contains only a relatively small amount of radioactivity. Estimates of radioactive waste quantities in Britain are given in Table 18.2.

Table 18.2. Estimated radioactive waste quantities in Britain

Waste type	Approximate quantities in some form of storage (m^3)	Approximate annual production (m^3)
Low level	Around 700 000 has already been disposed of	30 000–50 000
Intermediate level	61 000 (raw), 81 000 (conditioned)	3 000–5 000
High level	1 950 (raw), 720 (conditioned)	50–100

Table 18.3. Estimated nuclear reactor decommissioning*

Time period	No. of reactors	Cumulative no. of reactors
–2000	3	3
2001–2010	4	7
2011–2020	5	12
2021–2030	5	17
2031–2040	2	19
2041–2050	2	21

* Adapted from Openshaw *et al.* (1989).

In many countries nuclear power is an important energy resource. It is economically competitive with other forms of energy used in electricity production. The Chernobyl accident in 1986 (when operators lost control of a nuclear reactor and about 5% of the radioactive inventory escaped) has caused some reassessment of the commitment to nuclear energy. However, it seems certain that, in the medium-term and beyond, nuclear energy will make a growing contribution to national energy supplies.

A major problem for the future is the decommissioning of old nuclear power stations, because certain components of such facilities will have become highly radioactive through continuous exposure over the life-time of the reactor. Table 18.3 contains estimates of the numbers of nuclear reactors that may be decommissioned by AD 2050. The environmental and financial cost of decommissioning a nuclear reactor is dependent on the extent to which the site is required to be restored. Restoration to a greenfield site involves removal of all the contaminated components to a special disposal site. However, the bulk of the site buildings and generation equipment, which are never exposed to radiation, can be demolished using conventional techniques. On the other hand, the radioactive portions of the structure (the reactor vessel and primary heat exchangers) will have to be isolated for many years.

18.1.2. Ionizing radiation

Radioactive decay is a spontaneous nuclear transformation that is unaffected by pressure, temperature, chemical form, etc. Radioactivity tends to decay exponentially with time, with the half-life being the period required for the activity of a specified radionuclide to reduce by half through radioactive decay. Emitted nuclear radiation can be considered to consist of three components: alpha (α), beta (β) and gamma (γ) particles. α-Radiation has been shown to be identical to helium ions, whereas β-radiation is identical to electrons. γ-Radiation has the same electromagnetic nature as X-rays, but is of higher energy.

There are three naturally radioactive chains whereby the elements produced by decay themselves subsequently decay, producing new elements. Eventually the elements decay to a non-radioactive product, and in each case this is one of the isotopes of lead. The natural chains are for radium (or uranium-238), thorium and actinium (or uranium-235). The chain for radium is outlined in Figure 18.4. The first energy change in the transformation of the naturally radioactive uranium and thorium chains results in the release of an α particle. Subsequent transformations in these chains also involve the emission of β and γ particles.

The intensity or activity of any radioactive material is measured in terms of the number of disintegrations occurring in unit time. The unit originally used for activity was the Curie (Ci), such that a source of 1 Ci was one in which 3.7×10^{10} disintegrations occur per second. This is the rate of disintegration of nuclei in 1 g of radium. The current unit of activity is the Becquerel (Bq), whereby a source is of intensity 1 Bq if disintegrations occur at the rate of 1 per second. It therefore follows that 1 Bq is equivalent to 2.7×10^{-11} Ci.

Parent element	Nuclides: Parent		Nuclides: Daughter	Particle emitted*		Half-life
Uranium (U–I)	$^{238}_{92}U$	→	$^{234}_{90}Th$	$^{4}_{2}He$	(α)	$4{\cdot}51 \times 10^{9}$ years
Thorium (UX_1)	$^{234}_{90}Th$	(≥99%) →	$^{234m^{*}}_{91}Pa$	$^{0}_{-1}e$	(β)	24·1 days
		(0·63%) →	$^{234}_{91}Pa$	$^{0}_{-1}e$	(β)	
Protactinium (UX_2)	$^{234m^{*}}_{91}Pa$	→	$^{234}_{92}U$	$^{0}_{-1}e$	(β)	1·2 min
Protactinium (UZ)	$^{234}_{91}Pa$	→	$^{234}_{92}U$	$^{0}_{-1}e$	(β)	6·7 h
Uranium (U–II)	$^{234}_{92}U$	→	$^{230}_{90}Th$	$^{4}_{2}He$	(α)	$2{\cdot}47 \times 10^{5}$ years
Thorium (ionium)	$^{230}_{90}Th$	→	$^{226}_{88}Ra$	$^{4}_{2}He$	(α)	$8{\cdot}0 \times 10^{4}$ years
Radium	$^{226}_{88}Ra$	→	$^{222}_{86}Rn$	$^{4}_{2}He$	(α)	$1{\cdot}6 \times 10^{3}$ years
Radon (Ra-emanation)	$^{222}_{86}Rn$	→	$^{218}_{84}Po$	$^{4}_{2}He$	(α)	3·8 days
Polonium (Ra–A)	$^{218}_{84}Po$	(≥99%) →	$^{214}_{82}Pb$	$^{4}_{2}He$	(α)	3·0 min
		(0·02%) →	$^{218}_{85}At$	$^{0}_{-1}e$	(β)	
Lead (Ra–B)	$^{214}_{82}Pb$	→	$^{214}_{83}Bi$	$^{0}_{-1}e$	(β)	26·8 min
Astatine	$^{218}_{85}At$	→	$^{214}_{83}Bi$	$^{4}_{2}He$	(α)	2·0 s
Bismuth (Ra–C)	$^{214}_{83}Bi$	(≥99%) →	$^{214}_{84}Po$	$^{0}_{-1}e$	(β)	19·7 min
		(0·04%) →	$^{210}_{81}Tl$	$^{4}_{2}He$	(α)	
Polonium (Ra–C′)	$^{214}_{84}Po$	→	$^{210}_{82}Pb$	$^{4}_{2}He$	(α)	$1{\cdot}6 \times 10^{-4}$ s
Thallium (Ra–C″)	$^{210}_{81}Tl$	→	$^{210}_{82}Pb$	$^{0}_{-1}He$	(β)	1·3 min
Lead (Ra–D)	$^{210}_{82}Pb$	→	$^{210}_{83}Bi$	$^{0}_{-1}e$	(β)	22 years
Bismuth (Ra–E)	$^{210}_{83}Bi$	(≥99%) →	$^{210}_{84}Po$	$^{0}_{-1}e$	(β)	5·0 days
		(2×10^{-4}%) →	$^{206}_{81}Tl$	$^{4}_{2}He$	(α)	
Polonium (Ra–F)	$^{210}_{84}Po$	→	$^{206}_{82}Pb$	$^{4}_{2}He$	(α)	138·4 min
Thallium	$^{206}_{81}Tl$	→	$^{206}_{82}Pb$	$^{0}_{-1}e$	(β)	4·2 min
Lead (Ra–G)	$^{206}_{82}Pb$	→	Stable	—		Infinite

Figure 18.4. The decay chain for radium

The amount of absorbed radiation energy per unit mass is called the 'absorbed dose' D. The dose is measured in joules per kilogram and the resultant unit is the gray (Gy). A unit named the 'rad' is commonly used, where one gray is equivalent to 100 rad. In order to take into account the extent of the biological effects of different types of radiation, the concepts of the relative biological effectiveness (RBE) and the quality factor Q have been introduced. This has led to the definition of the RBE dose and the dose equivalent H, whereby

$$\text{RBE dose (rem)} = D(\text{rad}) \times \text{RBE} \tag{18.1}$$

$$H \text{ (in sievert)} = DQN \tag{18.2}$$

$$1 \text{ sievert} = 1 \text{ J/kg} = 100 \text{ rem} \tag{18.3}$$

N includes all modifying factors, which must be considered in practical radiation protection work (e.g. weighting factors due to radiation sensitivity and physiological importance of different organs). Presently, N is set equal to unity so that the RBE is equal to Q, which ranges from unity (for X-, γ-, and β-radiation at 1 Gy) to 20 (for natural α-radiation at 1 Gy). Radiation doses are usually measured and quoted in millisieverts (mSv), with 1 mSv corresponding to 20 Bq/m^3.

The process of radiation damage involves the breaking of molecular bonds, and thus damage to genetic material or crystal structure. Any radiation, whether α, β or γ, causes radiolysis of the water in a human cell, and the products may react with the DNA to such an extent that the cell dies. For an absorbed dose of 100 Gy (10^4 rad) the number of molecules destroyed is 1.6% for a molecular weight of 10^6 (which is about the weight of a DNA molecule). Since a human chromosome contains some 3×10^9 DNA molecules, such a high dose results in the death of the cell involved (Choppin & Rydberg, 1980).

For external radiation X- and γ-rays are the most dangerous to humans because of their penetrating power. α-Radiation is not dangerous externally as it can only penetrate the outer, rather insensitive, skin layer, but it is the most dangerous radiation as an internal source. α-Particles can cause tissue damage, but as these particles are relatively large and possess a high electrical charge they do not easily travel through clothing or skin. They may, however, enter the body in drinking water and can be inhaled during respiration.

The average amount of radon in household air in Britain is 20 Bq per cubic metre of air. The British Radiation Regulations dictate that annual radon doses should not exceed a radon level of 1000 Bq/m^3. Areas where there is a probability of more than 1% of the households having levels above 200 Bq/m^3 have been designated as 'Affected Areas'. The NRPB published its first assessment of a potentially Affected Area in 1990, which included the counties of Devon and Cornwall. No areas in Cornwall and Devon had concentrations below the 1% probability of homes being above the recommended level (also referred to as the 'Action Level') of 200 Bq/m^3, giving a 3% lifetime risk factor of a fatal cancer. This Action Level represented a compromise by the British government between setting very low risk levels for cancer versus the cost of remedial work which would be necessary to modify existing houses and buildings if an even safer radiation dose level were fixed. Safe concentrations of radon are difficult to assess because the relationships between radon exposure and the incidence of lung cancer are based on studies of workers in uranium mines who have received their doses in a totally different environmental setting.

Exposure to radon gas increases the risk of lung cancer because, once inhaled, the gas sits in the respiratory tract and lungs. Radon-222 has a half-life of 3.8 days, so that it has the opportunity to migrate from rocks and soils into the human environment, in homes and places of work.

While many radioactive wastes are extremely toxic they may be assumed to have become harmless when they have decayed for 10 half-lives. Half-lives and toxicity vary greatly among the radioactive chemical elements. Tables 18.4 and 18.5 show the fission products and chemical elements (of the 'Actinide group') which would be typically found in 1 t of spent liquid reactor fuel. From this table, it can be seen that the half-lives vary by up to millions of years, and typically range over at least tens of thousands of years, far longer than human recorded history. Nevertheless, the pollution potential of nuclear waste needs to be put in context. Mercury is highly toxic and has a half-life of infinity (it does not decay

*Table 18.4. Radioactive products from spent reactor fuel: nuclear fission products**

Element	Half-life (years)	Radioactivity remaining (Ci) After 10 years	After 100 years	After 500 years	After 1000 years
^{144}Ce/^{144}Pr	0.78	300	—	—	—
^{106}Ru/^{106}Rh	1.0	1100	—	—	—
^{155}Eu	1.8	160	—	—	—
^{134}Cs	2.1	8300	—	—	—
^{125}Sb/^{125}Te	2.7	980	—	—	—
^{90}Sr/^{90}Y	28	1.2×10^5	1.3×10^4	0.6	—
^{137}Cs/^{137}Ba	30	1.6×10^5	2.1×10^4	2	—
^{151}Sm	90	1100	520	30	0.4
^{99}Tc	2×10^5	15	15	15	15
^{93}Zr	9×10^5	3.7	3.7	3.7	3.7
^{135}Cs	2×10^6	1.7	1.7	1.7	1.7
^{107}Pd	7×10^6	0.013	0.013	0.013	0.0013
^{128}I	17×10^6	0.025	0.025	0.0025	0.025
Total (approx.)		300 000	35 000	53	22

Ce, cerium; Pa, protactinium; Ru, ruthenium; Rh, rhodium; Eu, europium; Cs, caesium; Sb, antimony; Te, tellerium; Sr, strontium; Y, yttrium; Ba, barium; Sm, samarium; Tc, technetium; Zr, zirconium; Pd, palladium; I, iodine.

* From Pickering & Owen (1994).

*Table 18.5. Radioactive products from spent reactor fuel: typical yields of chemical elements from the actinide group**

Element	Half-life (years)	Decay interval (years) 0.3 Y	0.3 R	10 Y	10 R	500 Y	500 R	10 000 Y	10 000 R	100 000 Y	100 000 R
^{237}Np	2×10^6	760	0.59	760	0.59	786	0.61	810	0.63	790	0.61
^{238}Pu	86	5.8	105	5.5	100	0.1	1.8	—	—	—	—
^{239}Pu	24×10^3	27.5	1.7	27.5	1.7	32	2.0	59	3.6	4.5	0.3
^{240}Pu	6.6×10^3	8.5	2.0	19.2	4.5	38.4	8.8	13.9	3.2	—	—
^{241}Pu	13.2	4	464	2.4	273	—	—	—	—	—	—
^{242}Pu	379×10^3	2	0.009	2	0.009	2	0.009	2	0.009	1.7	0.007
^{241}Am	462	54	189	55.6	198	29.5	103	—	—	—	—
^{243}Am	7×10^3	82	17.0	8.2	17.0	77	16.0	31	6.5	—	0.001
^{244}Cm	17.6	30	2570	19.7	1700	—	—	—	—	—	—
Total (g/t fuel)		974		974		964		916		796	

Am, americium; Cm, curium; Np, neptunium; Pu, plutonium; R, radioactivity output in curies; Y, yield in grammes per tonne of fuel.

* From Pickering & Owen (1994).

with time but is merely redistributed by geomorphological processes), whereas plutonium-242 is not nearly as toxic and has a half-life of only 387 000 years.

The main radioactive elements that are involved in polluting the environment and causing ill-health and death are the isotopes of iodine, strontium, caesium and ruthenium (Paasikallio *et al.*, 1994). Iodine accumulates in the thyroid gland. Strontium, which is chemically similar to calcium, is absorbed through the walls of the intestine and collects in the bones. Caesium behaves chemically in a similar manner to potassium and, therefore, can be distributed throughout the body in much the same way. Ruthenium has no chemical analogue with a biological function.

18.1.3. General aspects of disposal

A common misconception is that there are vast amounts of solid radioactive waste. In fact all the radioactive waste produced over the next 40 years will only be the volume equivalent of about 2–3 weeks domestic refuse production in Britain. Furthermore, not all of the waste poses an equal threat to the environment, and in Britain radioactive waste is designated as belonging to one of four levels, according to its heat-generating capacity and activity content, as indicated in Table 18.6. Most (about 90%) of the waste has a low-level of radioactivity that presents such a small risk that it can be handled by workers wearing normal industrial clothes (gloves and laboratory coats). Wastes with an intermediate level of radioactivity (about 10% of the total amount of radwaste) can be handled safely by using remote tools, cranes or simple shielding. Nevertheless, these wastes may retain their radioactivity for many centuries, and they must therefore be disposed of in a way that isolates them from the environment for very long periods. The remainder of the waste (only about 0.1% of the total) is high-level, or heat generating, and is concentrated in a very small volume. However, it contains over 95% of the radioactivity with which we have to deal.

The overall aim when disposing of radioactive waste is to remove the radionuclides to a position where they will not cause any harm to the environment or biosphere (defined as soils, seas, seabeds, freshwater bodies, the atmosphere, and the organisms they contain) (Figure 18.5). This concern also includes consideration of pathways back to man via the food chain. The fact that radioactivity naturally decays with time means that the waste will eventually render itself radiologically harmless. In theory, the waste need only be contained away from the biosphere until the radioactivity has fallen to levels similar to the background levels found in nature. Forming a judgement about the level of safety afforded by a disposal facility involves an assessment of the means by which radionuclides in the wastes might move, from the wastes through the immediate physical and chemical environment of the facility (and in the case of deep disposal, through the surrounding host rocks), back to the human environment. For deep disposal facilities this involves considering the potential behaviour of radionuclides over extended periods (in excess of thousands of years).

The target for a radioactive waste disposal centre is that it should not subject anybody to a radiation dose greater than the equivalent of 0.1 mSv per year. This

Table 18.6. Categorization of radioactive wastes

Level	Characteristics
High (HLW)	Wastes in which the temperature may rise significantly as a result of their radioactivity, so that this factor has to be taken into account in designing storage or disposal facilities
Intermediate (ILW)	Radioactivity levels exceed the upper boundaries for low-level wastes. Do not require heating to be taken into account in the design of storage or disposal facilities
Low (LLW)	Contain radioactive materials other than those acceptable for disposal with ordinary refuse. Wastes do not exceed 4 GBq/t of α- or 12 GBq/t of β- or γ-activity
Very Low (VLLW)	Can be safely disposed of with ordinary refuse, each tonne of material containing less than about 400 kBq of β- or γ-activity or single items containing less than 40 kBq of β- or γ-activity

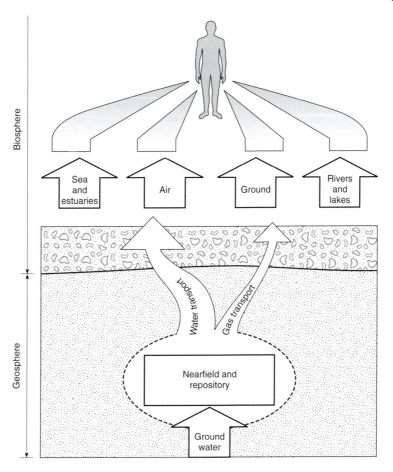

Figure 18.5. Radioactive waste in relation to the environment (from the Royal Society, 1994)

represents an annual risk of death to any individual of about one chance in a million (Beale, 1987). It has been estimated that this radiation dosage is equal to:

- that received by a person flying from Britain to Spain, ten times, on a jet aircraft
- about one-third of the dose each person gets in a year from naturally, occurring radioactive material in their bodies
- one-twentieth of the natural background radiation from the Earth's rocks, the Sun and outer space,
- one two-hundredths of the dose people living in some houses in granitic parts of the country (such as Cornwall) get from their surroundings.

Furthermore, the peak dose from a disposal centre should not be emitted until more than 10 000 years have elapsed.

In general, the disposal of radioactive waste consists of surrounding the material with a man-made container (which may be composed of several different encapsulation 'layers'), which is then sealed within a natural geological regime that also acts as a barrier to radionuclide movement. This concept is termed the *multibarrier approach*. The man-made packaging of the waste is intended to prevent initial leakage and to provide adequate shielding to ensure that there is no likelihood of anyone being irradiated while the radioactivity in the waste decays to a low level. The geological medium is expected, because of its stability, to maintain isolation of the waste until the radiation levels are insignificant. When the man-made packaging decays and fails, as it will do with time, then the geology provides ongoing containment. Clays are a preferred host geology because they have low permeability and they have some capacity for filtering out and retaining many of the radionuclides involved. Many radioactive species move

Table 18.7. Repositories for radioactive wastes*

Country	Wastes	Potential/actual disposal facility
France	LLW/ILW	Two engineered near-surface disposal facilities for LLW and short-lived ILW
	HLW	Studies of deep geological disposal of HLW have been going on since the 1970s. Investigations began at four sites (one in clay, two in granite and one in salt)
Germany	LLW/ILW	One repository which has been in use for some time is an abandoned salt mine with workings at about 400–500 m below ground level. A potential repository is an abandoned iron ore mine with workings at depths of 800 m to 1.3 km
Sweden	LLW/ILW	Wastes are disposed of in a repository at a depth of about 60 m in crystalline rock, under the Baltic Sea. It has been operational since 1988. The existing facility has four rock vaults (one for ILW and three for LLW) and one silo (for the more active ILW), with a total capacity of 60 000 m^3
Switzerland	LLW	Investigation of the suitability of argillaceous rocks (marls and shales) as a host formation for a repository for short-lived radioactive waste. The rocks are suitable by virtue of their low permeability, high clay content (which ensures good sorption properties) and strength (which will allow construction of tunnels and caverns)
	ILW/HLW	Geological studies for the HLW/ILW repository are, as yet, regional and cover both crystalline and sedimentary rocks
UK	LLW	Two main, engineered, shallow trench disposal facilities
	LLW/ILW	Disposal in underground caverns
	HLW	Planned disposal within deeply buried competent rocks
USA	LLW/ILW	LLW and short-lived ILW have been disposed of in near-surface facilities since the start of the nuclear programme. Some of these facilities are now in an unsatisfactory state and leakages of radiation from low-level disposal sites have been monitored
	HLW	At present, even after half a century of generating nuclear waste, no permanent HLW disposal site exists in the USA. The Yucca Mountain repository is mainly for spent fuel from the civil nuclear power programme. The host rock envisaged for the repository is welded tuff at a depth of about 500 m, but some 300 m above the water-table in the area

* Sources: Rodgers (1986), Pickering & Owen (1994), Royal Society (1994), Kappeler (1996).

through clays up to one million times more slowly than the groundwater in which they are dissolved.

The current international situation for radioactive waste disposal is illustrated in Table 18.7. Underground and subsea repositories for radioactive waste storage are being adopted by many governments as the most acceptable option.

18.2. Intermediate and low level wastes

In Britain disposal sites currently exist for low level wastes (LLW), but capacity may be reached in the early decades of the 21st century. It is desirable to dispose of LLW and intermediate level wastes (ILW) as soon as possible after their collection, and to avoid creation of additional accumulations and construction of costly and extensive facilities. ILW is typically a thousand times more radioactive than LLW. For LLW the radioactivity will decay away to trace levels in about 700 years, while the radioactivity for ILW may take thousands of years to decay away to similar trace levels. A repository that can accept both LLW and ILW is thus a priority.

18.2.1. Intermediate level waste

Between 3000 and 5000m^3 of ILW is produced in Britain each year. This waste consists of solid and liquid materials from nuclear power stations, the radioisotopes industry, fuel reprocessing and defence establishments. There are a number of specific waste streams with broadly similar radioactivity:

- fuel element cladding and related debris, which was removed prior to processing or which will remain after dissolution in acid
- various sludges and ion-exchange resins from storage pond water and concentrates of liquid waste streams
- miscellaneous wastes containing β or γ active materials formed by items arising from operations and maintenance
- plutonium contaminated materials
- graphite sleeves and stainless steel components of Advanced Gas-cooled Reactor fuel assemblies.

ILW is highly heterogeneous but may be broadly divided into long-lived wastes, which contain substantial quantities of radionuclides with half-lives of thousands or even millions of years, and short-lived wastes, the principal constituents of which have half-lives of at most a few tens of years. Most ILW will be incorporated in concrete prior to packaging and disposal. At present, a large proportion of existing ILW is stored in 'raw' form at the sites where it arises.

18.2.2. Low level waste

LLW comprises solids, liquids and gases that may be slightly contaminated with traces of radioactive materials. It comes from: the generation of electricity by the nuclear power plants; the preparation and use of radioactive materials in hospitals, where the materials are used for diagnosis and treatment; industry, where such materials are used in measuring and inspection systems; research laboratories; and defence establishments. Between 30 000 and 50 000 m^3 of LLW is produced in Britain each year.

Typical LLW is low in radioactivity (although some wastes will remain significantly radioactive for fairly long periods of time, i.e. 300 years or more) and high in bulk and comprises:

- rubbish (e.g. gloves, packaging, shoes, paper towels)
- worn-out and damaged equipment
- crushed glassware
- protective clothing
- air filters
- chemical sludges from the treatment of liquid LLW (before it is discharged into the environment)
- ash from the burning of low level combustible wastes
- soils and residues from the processing of ores containing naturally occurring radionuclides such as uranium and thorium.

18.2.3. Current disposal methods

Very low level waste (VLLW) (see Table 18.6) can be, and is, disposed of with domestic refuse, via incineration, controlled burial or landfill. In Britain, ILW is mainly stored at the site at which it arises, while LLW is routinely disposed of at a number of designated shallow land facilities. The main two sites are the facility at Drigg in Cumbria, which is owned and operated by British Nuclear Fuels, and the 'burial pits' at Dounreay in Caithness. In the past, the LLW was largely disposed of without any form of treatment and the system relied on massive dilution and slow releases to the atmosphere, ground and sea so that the quantities of radionuclides discharged presented no significant risk to the human population by any pathway (Royal Society, 1994).

The Drigg disposal site occupies about 120 ha. The local geology consists of a thick and variable sequence of sands and gravel, silts, clays and boulder clays overlying sandstone bedrock. The thickness of the glacial deposits is variable, generally ranging between 15 and 45 m. The surface of the bedrock is irregular and the sandstone is about 1000 m thick. The hydrogeology of the boulder clay is complex because of the variable nature of the glacial deposits. Within the glacial sequence the clays represent relatively impermeable horizons that restrict the vertical movement of water. The intervening sand and gravel layers are more permeable and provide routes for both vertical and lateral flow of groundwater (Openshaw *et al.*, 1989). Authorization was granted in 1958 for the shallow burial disposal of solid LLW in trenches (about 6–8 m deep, 25 m wide, up to 700 m long) dug into the underlying boulder clays. It was assumed that the clay would form an effective barrier against contamination of the underlying St Bees sandstone aquifer. Since the land slopes towards the sea, the 'impermeable' boulder clay underlying the burial site ensures that all groundwater movement is to the sea.

The material dumped at Drigg is LLW (see Table 18.6) (e.g. plastic containers, protective clothing, electrical cabling, scrap metal, process wastes, excavation spoil). The radionuclides with the greatest contamination potential are strontium, caesium and ruthenium. Initially the waste was continuously tipped and buried with a minimum of 1 m of soil, which was then covered by a layer of hard core, a geotextile and a surface layer of fine stone and ash. However, some rainwater infiltrated the waste trenches and flowed, via drains and culverts, into a nearby stream and off the site. This provided a pathway for the accidental release of radioactive leachate. Hence a 1 m wide cement–bentonite cut-off wall, which is keyed into the boulder clay of the site, was constructed around part of the site. The wall is designed to limit the movement of water leaching from the trenches and to prevent flow reversal and the inflow of groundwater. In order to achieve these aims the main specified performance criteria for the wall were (Chipp, 1990): permeability not greater than 10^{-9} m/s, strength greater than $100 \, \text{kN/m}^2$, strain at failure greater than 5%.

The current disposal practice is to stack containers within concrete-lined vaults. Highly compacted waste (to minimize future settlement of the cap) is disposed of directly. Non-compacted waste is put into boxes or drums and compressed highly. These containers are placed and grouted into large containers, which are stacked in the concrete-lined vault (BNFL, 1991). The site is capped to reduce the amount of water entering and draining from its trenches. Before the final closure of the site all trench and vault facilities will be capped with a thicker and more durable cap.

18.2.4. Future disposal

A repository for long-lived ILW should be positioned deep underground to ensure that the pathways back to man are very long. It must be emphasized that deep disposal is chosen not because long-lived waste is significantly more radioactive or more dangerous than short-lived waste, it is simply because it remains radioactive for a longer time. Radioactive materials could reach the environment by: dissolving in the groundwater and moving with it to rivers or the sea, diffusion as a result of intrusion (by exploration companies for example), or a major disruptive event such as an earthquake. A repository should be located within a carefully chosen geological structure where the second two eventualities are extremely unlikely. In areas of low permeability and low hydraulic gradient, the groundwater is effectively static, taking hundreds of thousands of years to reach watercourses. However, radionuclides can still move by diffusion. In addition, certain rocks have the property of immobilizing impurities. These natural geological features provide effective containment of radioactivity and are known as the 'farfield' part of the multibarrier approach.

Several scenarios present themselves as being potentially suitable for a deep repository, all of which involve the use of caverns or tunnels that may be natural, man-made or purpose-built. A hard 'host' would permit the use of self-supporting large caverns, whereas for softer rocks engineering support would be required (even for tunnel systems). In terms of their use as a secure repository, natural caverns have the disadvantage that, invariably, they have been created by the flow of water that has removed soluble components. This process may well have produced other, unknown voids, channels and passages within the ground and this, combined with their hydrogeology, means that natural caverns are not generally considered suitable for radioactive waste repositories. Man-made caverns such as mines will have an understood geology, and the history of the site operation can be documented from recent workings. A repository in such a disposal mine would then need to be designed around the space available. Although there are many disused mines in the UK, from mineral workings dating back to Roman times, very few are suitable for development. Coal mines in particular offer little or no potential because they are invariably associated with limestone and other aquifers as part of the carboniferous series of rocks. This means they are prone to water flow.

In 1991 it was announced that the chosen site for the disposal of LLW and ILW in the UK was an on-land underground site in Cumbria. Some of the sites considered for disposal of long-lived ILW are described in Box 18.1. LLW will undergo limited treatment (such as compaction) at source and be packaged in 200-litre carbon steel drums. These will be transported to the repository and stacked in the LLW galleries, which will not be backfilled. Between 0.1 and $1.4\,Mm^3$ of LLW will be contained. The majority of the ILW to be accommodated will be immobilized in a cementitious grout and packed in a variety of iron or steel containers. The containers will be stacked in reinforced concrete bays in the separate ILW waste disposal caverns and all the void space between the drums will be backfilled with a specially formulated, low-strength, enhanced porosity cementitious grout.

The nature of the waste and its packaging will lead to the generation of gases when the repository resaturates after closure. Once ILW has been incorporated in concrete and packaged, a variety of chemical reactions can occur within the packaged waste owing, in particular, to the presence of trace quantities of water. Similarly, supercompacted LLW is not completely dry and, as with ILW, will degrade within its packaging.

18.3. High level waste

High level waste (HLW) is currently vitrified and housed in surface stores to allow much of the radioactivity to decay prior to final disposal. This waste is expected to remain in storage for some 50 years after conditioning. Although it seems certain that HLW will be permanently disposed of in a deep repository, there is at present no such facility.

18.3.1. Origins

HLW is a concentrated, heat-generating waste that results almost entirely from reprocessing of spent nuclear fuel. Reprocessing is carried out to remove a large fraction of the uranium and plutonium spent nuclear fuel, so that these can be re-used, (e.g. in fabricating fresh fuel). It is estimated that when $4\,m^3$ of spent fuel is reprocessed it will produce $2.5\,m^3$ of HLW, $40\,m^3$ of ILW and $600\,m^3$ of LLW. Wastes are currently stored in high-integrity stainless steel tanks fitted with cooling tubes. Some of the constituents of HLW have very long half-lives (Table

Box 18.1. *Sites considered for hosting a repository for long-lived ILW*

Billingham

Anhydrite was formally extracted from the mine using the 'room and pillar' method, which has left a honeycomb of 'rooms' separated by massive anhydride pillars at a depth of 140–280 m and over an area of about 500 ha. The rock is several times stronger than concrete, so the mined cavity is very stable and subsidence is prevented. The total capacity of the mined cavity is about 11 Mm^3. The main anhydrite seam in which the mine lies is about 6 m thick and underlies great thicknesses of permian marl and shale sequences, which themselves are overlain by the Sherwood sandstones. Should radionuclides find their way into the groundwater, both the anhydrite and the permian marls above are very strong absorbers of dissolved radionuclides, and thus makes their effective rate of travel very slow.

Elstow Storage Depot

Located in clay where the Oxford clays outcrop with considerable thickness. The ILW would be disposed of in deep, specially engineered and concrete-lined trenches dug approximately 10–15 m into the clay stratum. The hydraulic conductivity of the clay is low and its absorption capacity is high. However, faults are known to exist in the Oxford clays and have been recorded in worked pits to the south-west of the Elstow site. These faults may form lines of increased permeability.

Bradwell

The site lies on up to 15 m of recent estuarine and marine alluvium, which overlie up to 50 m of London Clay. The London Clay overlies 25 m of lower London tertiary deposits, consisting of sands and hard clays, which themselves overlie chalk. Groundwater movements between the chalk and the London Clays and tertiary deposits are upwards through the clays, driven by the higher hydraulic potential in the underlying chalk.

Fulbeck

The geology of the site is quite complex, being made up a series of interbedded lower Lias clays, mudstones and limestones. The clays outcrop at the site to a maximum depth of 20 m. Beneath the lower Lias rocks is the Penarth group, which overlies the Mercia mudstone group which is believed to be over 200 m thick. Under the existing hydrogeological regime it is probable that groundwater movement is upwards through the lower Lias driven by the potential head in the higher ground of the Lincoln Ridge and the Sherwood Sandstone below.

South Killingholme

The site is covered to a depth of between 13 and 25 m with a mantle of boulder clay (till), the intended disposal medium. This overlies chalk. There are two main layers in the boulder clay. The upper till (stiff silty clay with fine gravel and chalk fragments) has a maximum depth of 5.5 m. This is underlain by stiff silty clay again with gravel and chalk fragments, but with the addition of fine sand horizons about 0.3–0.6 m thick. The overlying boulder clays are relatively impermeable and tend to confine most recharge water to within the chalk. The hydraulic potential within the chalk creates a hydraulic head above the level of the chalk–boulder clay interface, so that any water movement between the two deposits will be upwards into the boulder clay.

18.8). Generally, HLW contains more than 95% of the radioactivity that results from the nuclear industries waste products, and it is necessary to take the heat produced by radioactive decay into account in the design of storage or disposal facilities. Initially, HLW is about a thousand times more radioactive than ILW (per tonne of material), but its radioactivity falls rapidly.

18.3.2. Disposal facility location

Four factors (geology, population density, environmental conservation and site accessibility) are considered in identifying areas that show the greatest potential for the location of a repository. The original approach to the selection of a site for HLW disposal concentrated on the properties of the host rock. This resulted in the definition of a widely accepted group of geological materials (salt, crystalline

Table 18.8. Decay periods for radioactive elements

Nuclide	Half-life (years)	Time to decay to 0.01% of original level (years)
Radon-222	0.01	0.10
Iodine-131	0.02	0.22
Strontium-89	0.14	1.3
Ruthenium-106	1	9.9
Plutonium-241	15	150
Strontium-90	28	279
Caesium-137	30	299
Plutonium-238	87	867
Americium-241	433	4 316
Radium-226	1 600	16 000
Plutonium-240	6 600	66 000
Thorium-229	7 300	73 000
Plutonium-239	24 000	239 000
Thorium-230	80 000	797 000
Uranium-234	240 000	2 392 000
Plutonium-242	387 000	3 858 000
Caesium-135	2 000 000	19 936 000
Iodine-129	17 000 000	169 454 000
Uranium-235	710 000 000	7 077 209 000
Uranium-238	4 500 000 000	44 855 553 000

rocks and clays), largely on the basis of their low permeability and thermal stability (Box 18.2).

The current approach is to consider the features of the complete geological environment. This approach introduces new possibilities for repository sites where the host rock type may be subordinate to the flow regime (e.g. small island environments). The area within which geological information is required once the site has been approximately determined depends on the assessment of threats to the integrity of the geosphere barrier. Groundwater advection is taken to be the major threat so that the area of detailed investigation is determined by:

- the present groundwater catchment within which the site lies, and the zones of recharge and discharge
- the possible future extents of the catchment and its zones of recharge and discharge.

Box 18.2. Deposits suitable for accommodating radioactive waste

Rocksalt
- Low permeability and very low water content.
- Easily mined.
- Self-healing around boreholes and fractures.
- High thermal conductivity.

Anhydrite
- Impermeable and chemically stable.
- Hydrates to generate gypsum, which seals fractures.

Granite and related crystalline rocks
- Great strength and resistance to erosion.
- Low permeability when intact.

Clay
- Plastic nature gives potential for sealing of holes and voids.
- May have low permeability.

Chapman *et al.* (1986) defined five types of hydrogeological environment that are likely to be of interest (Box 18.3). Areas within the UK that satisfy some of the foregoing requirements are indicated in Figure 18.6.

Box 18.3. Hydrogeological environments for disposal of HLW*

Hard rocks in low relief terrain
Low relief means that there is little or no driving potential for groundwater flow and flow that does occur tends to be controlled by major fractures within the rock.

Small islands
These should be sufficiently far from the coast to have their own groundwater flow systems. Should any radioactive material eventually find its way to the seabed then the dilution effect of the sea would be both massive and immediate.

Seaward dipping strata
Groundwater movement is expected to be very slow towards, and under, the coast.

Offshore sediments
In sub-seabed formations there are almost zero flow conditions. Any groundwater flow would be a very slow upwards movement (caused by geothermal heating).

Inland basinal
Deep basins of mixed sedimentary rocks containing a high proportion of impermeable formations such as mudstones and evaporites. Flow is downward, following the dip of the rocks towards the centre of the basin, where near-stagnant conditions have developed.

Low-permeability basement rocks
These rocks (principally hard shales, mudstones, slates, volcanic rocks, etc.) occur under more recent sedimentary cover. Groundwater flows occur predominantly within the cover with little anticipated connection with the basement rocks.

* After Chapman *et al.* (1986).

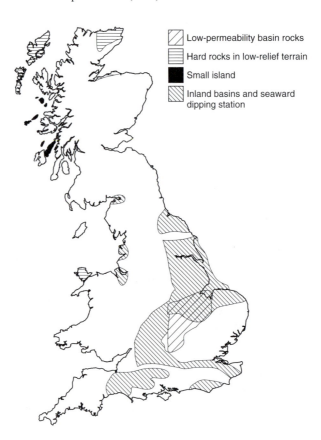

Low-permeability basin rocks

Hard rocks in low-relief terrain

Small island

Inland basins and seaward dipping station

Figure 18.6. Suitable disposal sites (after Openshaw et al., 1989)

Box 18.4. Important properties of a medium for radioactive waste containment

- **Hydrogeological:** to minimize the extent to which the waste is exposed to moving groundwater.
- **Geochemical, mineralogical:** to ensure the radionuclides will be retarded or immobilized before reaching man's environment.
- **Thermomechanical:** to allow for a heat loading to account for waste emplacement without damaging the structural competency of the repository or the host medium.
- **Structural strength, stability:** to ensure that the physical integrity of a repository is not jeopardized during the operational period.

The fundamental requirement of a suitable geological environment for a repository is that it should be relatively stable. For example, it should not be in an area of high seismic–tectonic activity, and its behaviour should be adequately predictable (Miller *et al.*, 1994). The need for predictability arises from the requirement to make scientifically sound evaluations of the long-term radiological safety of a disposal facility. Such evaluations, which are necessary in order to demonstrate that a repository will comply with safety regulations, are generally termed 'performance assessment' or 'safety assessment' exercises. The most important natural events that could affect the performance of the barrier are changes to groundwater flow accompanying earthquakes or climatic change. Earthquakes are known to affect the hydrological regime around faults. The load from ice sheets changes the stress state in the crust, which in turn may control the orientation and distribution of open permeable cracks around the repository. The general nature of the probable climatic changes themselves is not controversial; at some stage in the next 10 000–100 000 years Britain will experience glacial or periglacial conditions with sea levels different from those of today (Royal Society, 1994). The important properties of a medium suitable for waste disposal are outlined in Box 18.4.

Population density is important because it directly affects the number of people who could potentially be affected by any leakage. Areas that have been designated as being of national importance in terms of nature conservation and landscape conservation cannot be utilized. Transport distances from the source of the waste to the site have a significant effect on transport costs and the chances of transport accidents.

18.3.3. Facility design

All repository design studies favour disposal in large caverns or tunnels, since these offer the most economic utilization of the available space. It would appear that hard rocks, such as anhydrite or granite, would be most suitable for large cross-section disposal vaults. All types of engineered repository utilize the 'multibarrier' concept (Figure 18.7), whereby the wastes are placed inside a series of engineered structures and natural barriers that act in concert to control the rate of release of radionuclides over long periods. The components in the multibarrier approach are described in Table 18.9.

The waste is 'solidified' by encasing it in glass (vitrification) or mixing it with some material which coats and hardens (e.g. bitumen). This constitutes the first barrier. The solidified waste is then placed in canisters, which may themselves be placed inside an 'overpack'. Both containers and overpacks (which together comprise the second barrier) will be made of metal or cement. The waste packages are then placed at some depth in an excavated gallery or borehole in a repository. The surrounding space may be filled with grout (the third barrier) to provide long-term structural, hydraulic and chemical stability for the package. A highly alkaline grout mix will help to retard the corrosion rate of the steel can and reduce the solubility of any escaping heavy metals. The main materials that will be used as buffers and seals are bentonite (possibly

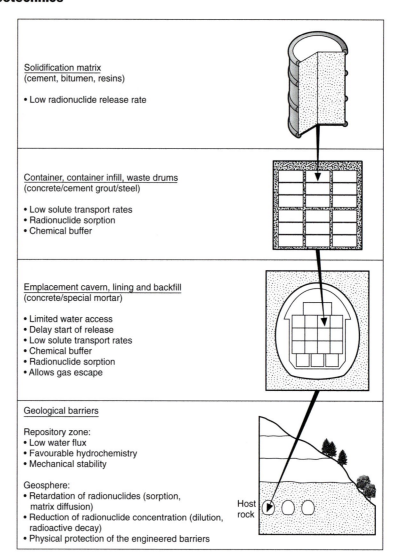

Figure 18.7. The multibarrier concept (after Kappeler, 1996)

Table 18.9. Components in the multibarrier approach

Barrier	Components	Materials	Functions
1	Solidification matrix	Cement, bitumen, resins, glass	Ensures low radionuclide release rate
2	Container, container infill, waste drums	Concrete, cement, grout, steel	Ensures low solute transport rates, provides radionuclide sorption and chemical buffering
3	Emplacement cavern, lining and backfill	Concrete, special mortar	Ensures limited water access, delayed start of nuclide release, low solute transport rates. Provides chemical buffering, radionuclide sorption. Allows gas escape
4	Geological barriers	Intact rock, clay	Ensures low water flux, favourable hydrochemistry, mechanical stability (in repository zone). Provides retardation of radionuclides by sorption and matrix diffusion, reduction of radionuclide concentration by dilution

mixed with various additives) and cement (including concrete). A number of other materials have been proposed for use in other designs, including chemical grouts and synthetic materials. The encapsulated HLW will generate sufficient heat, certainly for the first 500 years of emplacement, to affect the surrounding rock and water regime. Because of the heat generation the canisters will generally be spaced such that the overall heat build up is not excessive and can be accommodated in the design. The fourth barrier is the geological regime surrounding the repository.

Even with the facility sited deep underground (depths of repositories will be in the range of several hundred to one thousand metres below ground level), engineers accept that the rock store cannot be 100% leak proof. A likely design brief is that any radioactivity eventually migrating to the surface in groundwater should contain a radiation no higher than 0.06 mSv, which is equivalent to one-thousandth of the background radioactivity already present in the general environment and given off by rocks, food, sunlight, etc. The major concern for effective containment is the competence of the surrounding rock.

18.3.4. Long-term performance of disposal facilities

It is impossible to predict exactly what will happen in and around a deep repository over the millions of years needed for substantial decay of some of the radionuclides in radioactive wastes. Post-closure performance assessment is thus a matter of identifying possibilities and attempting to quantify, or at least put bounds on, uncertainties. Some uncertainties can be quantified, in the sense that statistical bounds can be placed on them, through model calculations for a range of scenarios for the future evolution of the repository and its surroundings. Other uncertainties can be treated in a semiquantitative fashion through 'scoping calculations', using expert judgement to determine whether particular events and processes will have significant effects on repository performance. There are also uncertainties that, at present, can be treated only qualitatively (Royal Society, 1994).

After all the disposal rooms, tunnels and shafts to the surface have been backfilled and sealed, recharging of the drawdown zone above the repository will begin. During the resaturation period, heat liberated by decay of species and by cement hydration reactions will cause temperatures to rise within the repository. A maximum of 80 °C is envisaged by NIREX, dropping back to 50 °C after 1000 years (Royal Society, 1994). The principal effect of temperature rise will be acceleration of reaction rates (perhaps by an order of magnitude). Because different rock types have different thermal conductivities, the maximum temperature rise will be different for the same heat load whether the host deposit be salt, granite, basalt, shale or clay.

Although the backfill material around a waste canister will restrict entry of the groundwater it will eventually make contact with the canister and the metal will begin to corrode. Current estimates are that this will take between 100 and 1000 years (NIREX). During resaturation the porewater will be very highly alkaline due to traces of hydroxides of sodium and potassium. Oxygen will be progressively consumed in reactions with reducing agents (notably metals and organic matter), and it is believed that reducing conditions will be established within a few tens of years (Royal Society, 1994).

Corrosion will eventually create a path for water to flow through the metal and come into contact with the waste. The time taken for this to occur will depend on the rate of the water flow, its chemical make up and the temperature. The vitrified waste will begin to dissolve, but very slowly (it has been estimated from laboratory tests that it will take about 2000 years to leach about 1 mm from the waste glass). As the radionuclides flow away from the canister some particles will attach themselves to clay particles by the process of sorbtion.

The high initial pH (13) in the ILW nearfield will gradually drop to around 12 over a few thousand years. This hyperalkaline environment will be maintained for some years until the calcium silicate hydrate gels begin to dissolve, at which point the pH will drop to about 10.5 over a period of about one million years. This is much in excess of the expected life-span of the engineered barriers, and indicates that the chemical conditions in the nearfield should act as a buffer even after the physical integrity of the concrete has been lost (Miller *et al.*, 1994).

Some radioactive particles will be transported through the backfill material to the surrounding geological material, and the groundwater will carry radionuclides into fractures and joints around the access passages. The concentrations of the radionuclides along the path to the ground surface will be continually diluted by freshwater continually moving into the areas from the sides and above. Even in the worst situation, involving an early exposure of the waste form to the circulating groundwater, it would appear highly unlikely that any radionuclides could migrate from the repository to the ground surface in less than a few thousands of years. It is much more likely that radionuclide escape to the ground surface will take tens of thousands of years due to the combination of the retardation of radioactive particles and the extremely long paths through the fractures and joints.

According to NIREX, most of the radionuclides in HLW will decay inside the metal drums, with at most 1% escaping. Of this latter volume 99% will be captured by the chemical grout, leaving just 0.01% of the original quantity to be tackled by the undisturbed geology.

Extensive use has been made of mathematical modelling to make long-term predictions of the release of radionuclides and other effects (Royal Society, 1994). Underlying these mathematical models are conceptual models of the disposal system and its evolution with time, and models of radionuclide movement through the nearfield (the repository itself and surrounding disturbed rock), geosphere (the undisturbed rocks and groundwater surrounding the repository) and the biosphere. Construction and validation of models of the current groundwater state require field measurements of heads and fluxes. Development and validation of time-dependent models that seek to predict changes over periods of the order of 10^4 to 10^6 years require knowledge of groundwater states over similar periods in the past and, ideally, simulation of past changes. Unfortunately, we do not have definitive data of this sort. Furthermore, there are gaps in our knowledge of the long-term performance of parts of the containment system:

- Cementation of bentonites may restrict their ability to self-heal following mechanical displacement, thus causing a significant rise in permeability. In extreme cases this mechanism may possibly cause fissuring, allowing direct radionuclide transport through the buffer by advection.
- Leached cations from the canister (copper or iron) may exchange with sodium or calcium in the bentonite, with a subsequent rise in permeability.
- Cement may be adversely affected by interactions with the porewater from clay formations, and hyperalkaline groundwaters may affect the porosity and sorptive capacity of the rock.
- Radionuclide sorption onto cement has been poorly addressed in analogue studies.
- The permeability of cements and concretes to gas produced from steel corrosion may be excessive.

References and further reading

Acar, Y. B. & Alshawabkeh, A. N. (1993). Principles of electrokinetic remediation. *Environmental Science and Technology* **27**, No. 13, pp. 2638–2647.

Acar, Y. B., Hamidon, A., Field, S. D. & Scott, L. (1985). The effect of organic fluids on hydraulic conductivity of compacted kaolinite. In *Hydraulic Barriers in Soil and Rock* (eds, A. I. Johnson, R. K. Frobel, N. J. Cavalli & C. B. Patterson). American Society for Testing and Materials, Philadelphia, *ASTM STP 874*, pp. 171–187.

Acar, Y. B., Alshawabkeh, A. N. & Gale, R. J. (1992). A Review of fundamentals of removing contaminants from soil by electrokinetic soil processing. In *Mediterranean Conference on Environmental Geotechnology* (eds, M. Usmen & Y. B. Acar). Balkema, Rotterdam, pp. 321–330.

Acar, Y. B., Alshawabkeh, A. N. & Gale, R. J. (1994). Cd(u) removal from saturated kaolinite by application of electrical current. *Géotechnique* **44**, No. 3, pp. 239–254.

Adams, M. S. & McManus, F. (1994). *Noise and Noise Law. A Practical Approach.* Wiley Chancery Law, London.

Agg, A. R., Wellstein, N. & Cartwright, N. (1992). *Sewage Sludge: Current Disposal Practice and Future Developments in Selected Companies.* Foundation for Water Research, FWR Report FR0265.

Alker, S. C., Sarsby, R. W. & Howell, R. (1995). The composition of leachate from waste disposal sites, *Proceedings of GREEN'93: Waste Disposal by Landfill* (ed., R. W. Sarsby). Balkema, Rotterdam, pp. 215–223.

All England Law Reports (1873). Rylands and another v. Fletcher. *All England Reports (1861–73)*, pp. 1–15.

All England Law Reports (1922). Hoare and Co Ltd v. Sir Robert McAlpine, Sons & Co. *All England Reports*, Chap. D, pp. 759–764.

Alpan, I. & Meidav, Ts. (1963). Effect of pile driving on adjacent buildings, a case history. In *Proceedings of the RILEM Symposium: Measurement and Evaluation of Dynamic Effects and Vibrations of Constructions, Budapest*, pp. 171–180.

Anandan, J. I. & Piercy, J. R. (1991). A259 Littlehampton-by-pass embankment construction with vertical drains. *Highways & Transportation* **38**, No. 2, pp. 10–15.

Anderson, D. C., Crawley, W. & Zabcik J. D. (1985). Effects of various liquids on clay soil: bentonite slurry mixtures. In *Hydraulic barriers in soil and rock* (eds, A. I. Johnson, R. K. Frobel, N. J. Cavalli & C. B. Petterson). ASTM STP 874, American Society for Testing and Materials, Philadelphia, pp. 93–103.

Anon. (1988a). *The Collection and Disposal of Waste Regulations.* HMSO, London, Statutory Instrument 819, pp. 1870–1881.

Anon. (1988b) Engineering for profit from waste. In *Proceedings of the IME Conference, Coventry.* Institute of Mechanical Engineers, London.

Anon. (1990). *Environmental Protection Act 1990.* HMSO, London, Chap. 43.

Anon. (1993). *Radioactive Substances Act.* HMSO, London.

Anon. (1994). Recycling the rubble. *Interconnect*, 1 December, pp. 6–7.

Anon. (1995a). *Review of Radioactive Waste Management Policy, Final Conclusions.* HMSO, London.

Anon. (1995b). Safely silencing the noise. *Highways and Transportation*, **42**, No. 6, pp. 14–16.

Anon. (1996). Shropshire mines stabilised in £4 m grouting scheme. *The Surveyor*, 8 August, p. 4.

Araya, S., Ishii, T., Sekijima, K., Utsugida, Y., Watanabe, K. & Imazu, M. (1986). Conceptual design of geological repositories for high level wastes. In *Proceedings of an International Symposium on the Siting, Design and Construction of Underground Repositories for Radioactive Wastes, Hannover.* International Atomic Energy Agency,

Vienna, pp. 495–500.

Attewell, P. B. (1993). *Ground Pollution. Environment, Geology, Engineering and Law.* E. & F. N. Spon, London.

Attewell, P. B. & Farmer, I. W. (1973). Attenuation of ground vibrations from pile driving. *Ground Engineering* **6**, No. 4, pp. 26–29.

Bagchi, A. (1994). *Design, construction and monitoring of landfills*, 2nd edn. Wiley, New York.

Baker, K. H. & Herson, D. S. (1994). *Bioremediation.* McGraw Hill, New York.

Barden, L. (1974). *A Comparison of Quay Wall Design Methods.* Part III: *Sheet Pile Wall Design Based on Rowe's Method.* CIRIA, London, Technical Note 54.

Barlaz, A. & Ham, R. K. (1993). Leachate and gas generation. In *Geotechnical Practice for Waste Disposal* (ed. D. Daniel). Chapman & Hall, London, Chap. 6, pp. 113–136.

Barnes, G. E. (2000). *Soil Mechanics Principles and Practice.* 2nd edn. Macmillan, Basingstoke.

Barron, R. A. (1948). Consolidation of fine grained soils by drain wells. *Transactions of the ASCE*, **113**, pp. 718–742.

Barron, J. M. (1989). *An Introduction to Waste Management.* Institute of Water and Environmental Management (IWEM), London, IWEM Booklet 1.

Barry, D. L. (1987). Hazards from methane on contaminated sites. In *Building on Marginal and Derelict Land.* Thomas Telford, London, pp. 323–338.

Barry, D. L. (1991). Hazards in land recycling. In *Recycling Derelict Land* (ed., G. Fleming). Thomas Telford, London, pp. 28–63.

Beale, H. (1987). The assessment of potentially suitable repository sites. In *Proceedings of Seminar on The Management and Disposal of Intermediate and Low Level Radioactive Wastes.* Institute of Mechanical Engineers, London, pp. 11–17.

Beaman, A. L. & Jones, R. D. (1977). *Noise from Construction and Demolition Sites-Measured Levels and their Prediction.* CIRIA, London, Report 64.

Beckett, M. (1993). 'Trigger concentrations: more or less. *Land Contamination and Reclamation* **1**, No. 2, pp. 67–70.

Beeuwsaert, L. (1998). *Construction quality control of compacted clay landfill liners: a laboratory-field investigation.* PhD thesis, Manchester University (Bolton Institute), Manchester.

Beeuwsaert, L. & Sarsby, R. W. (2000). The precision of the compaction-permeability relationship. *Geotechnical Engineering, Proceedings of the Proc ICE (submitted).*

Bell, F. G. (1993). *Engineering Treatment of Soils.* E. & F. N. Spon, London.

Bell, R. M., Gildon, A. & Parry, G. D. R. (1983). Sampling strategy and data interpretation for site investigation of contaminated land. In *Reclamation of former iron and steel works sites* (ed. G. P. Doubleday). Durham and Cumbria County Councils, Durham, pp. 23–31.

Benson, C. (1991). Predicting excursions beyond regulatory threshold of hydraulic conductivity from quality control measurements. In *Proceedings of the 1st Canadian Conference on Environmental Geotechnics*, pp. 447–453.

Berti, G., Villa, F., Dovera, D., Genevois, R. & Brauns, J. (1988). The disaster of Stava/ Northern Italy. In *Proceedings of a Speciality Conference on Hydraulic Fill Structures.* American Society of Civil Engineers, New York, ASCE SP21, pp. 492–510.

Bishop, A. W. (1955). The use of the slip surface in the stability analysis of slopes. *Géotechnique* **5**, No. 1, pp. 7–17.

Bishop, A. W. (1973). The stability of tips and spoil heaps. *Quarterly Journal of Engineering Geology*, **6**, pp. 335–356.

Bishop, A. W. & Bjerrum, L. (1960). The relevance of the triaxial test to the solution of stability problems. In *ASCE Research Conference on Shear Strength of Cohesive Soils, Boulder*, pp. 437–502.

Bishop, A. W. & Morgenstern, N. (1960). Stability coefficients for earth slopes. *Géotechnique*, **10**, No. 4, pp. 129–150.

Bishop, A. W., Green, G. E., Garga, V. K., Andresen, A. & Brown, J. D. (1971). A new ring shear apparatus and its application to the measurement of residual strength. *Géotechnique* **21**, No. 4, pp. 273–328.

Bjerrum, L. (1972). Embankments on soft ground. General report. In *ASCE Speciality Conference on the Performance of Earth and Earth Supported Structures*, Vol. 2.

American Society of Civil Engineers, New York, pp. 1–54.

Blanchet, M. J. (1977). *Treatment and Utilization of Landfill Gas: Mountain View Project Feasibility Study*. Office of Solid Waste, USEPA and Pacific Gas and Electric Co., Cincinnati, OH, EPA SW-583.

Blight, G. E. (1994). The master profile for hydraulic fill tailings beaches. *Proceedings of the ICE, Geotechnical Engineering*, **107**, pp. 27–40.

Blight, G. E. & Bentel, G. M. (1983). The behaviour of mine tailings during hydraulic deposition *Journal of the South African Institute of Mining and Metallurgy*, **83**, No. 4, pp. 73–86.

Blum, H. (1930). *Stress Conditions for Bulkheads*. Diss. Tech., Hochschule Braunschweig.

Blümel, W. & von Bloh, G. (1991) Determination of shear strength of different sludges — methods and effects. In *Proceedings of Sardinia '91, 3rd International Landfill Symposium, Cagliari*. CISA (Environmental Sanitary Engineering Centre), Cagliari, pp. 1433–1442.

BNFL (1991). *The Drigg Low-level Waste Site*. BNFL Waste Management Unit, Risley.

Bogardi, I., Kelley, W. E. & Bardossy, A. (1990). Reliability model for soil liner: post construction. *Journal of Geotechnical Engineering, ASCE*, **116**, No. 10, pp. 1502–1520.

Bowders, J. J. & Daniel D. E. (1987). Hydraulic conductivity of compacted clay to dilute organic chemicals. *Journal of Geotechnical Engineers*, **113**, No. 12, pp. 1432–1448.

Bracci, G., Giardi, M. & Paci, B. (1991). The problem of clay liner testing in landfills. In *Proceedings of Sardinia '91, 3rd International Landfill Symposium*, pp. 679–689.

Bradbury, H. W. (1984). Grouting with PFA. *AshTech '84, 2nd International Conference on Ash Technology and Marketing, London*, pp. 513–518.

Braithwaite, P. (1995). Protective custody. *NCE Magazine*, 26 January, pp. 25–26.

BRE (1983). *Hardcore*. BRE Digest 276. Building Research Establishment, Watford.

BRE (1987). *Measurement of gas emissions from contaminated land*. Building Research Establishment. Garston, Watford.

BRE (1991). *Construction of New Buildings on Gas-contaminated Land*. Building Research Establishment, Garston, Watford, BRE Report 212.

BRE (1996). *Sulfate and Acid Resistance of Concrete in the Ground*. Building Research Establishment, Garston, Watford, BRE Digest 363.

BRE (1997). *Low-rise buildings on fill*. Part 1: Classification. Part 2: Site investigation and foundation design. Part 3: Engineered fill. Building Research Establishment, Garston, Watford, BRE Digest 427.

British Drilling Association (1991). *Guidelines for the Drilling of Landfill, Contaminated Land and Adjacent Areas*. BDA, Brentwood.

Bromhead, E. N. (1987). The treatment of landslides. *Proceedings of the ICE, Journal of Geotechnical Engineering* **125**, pp. 85–96.

Bromhead, E. N. (1992). *The Stability of Slopes*. Blackie Academic and Professional, Glasgow.

Broms, B. (1971). Lateral earth pressures due to compaction of cohesionless soils. In *Proceedings of the 4th Budapest Conference on Soil Mechanics, Budapest*, pp. 373–384.

Brown, K. W. & Anderson D. C. (1983). Effect of organic solvents on the permeability of clay soils. Municipal Environmental Research Laboratory, USEPA, Cincinnati, EPA Report.

Bruce, A. M. (1989). Sewage sludge — making the best of it. *IWEM Yearbook*, Institute of Water and Environmental Management, London, pp. 57–64.

Bruce, A. M., Pike, E. B. & Fisher, W. J. (1990). A review of treatment options to meet the EC sludge directive. *Journal of the IWEM*, **4**, pp. 1–13.

Brunelle, T. M., Dell, L. R. & Meyer, C. J. (1987). Effect of permeameter and leachate on a clay liner. In *Geotechnical Practice for Waste Disposal '87*. (ed. R. D. Woods). American Society of Civil Engineers, New York, ASCE Special Publication 13, pp. 347–361.

BSI (1981, 1999). *BS 5930: British Standard Code of Practice for Site Investigation*. British Standards Institution, London.

BSI (1984). *BS 6472: Evaluation of Human Exposure to Vibration in Buildings (1 Hz to 80 Hz)*. British Standards Institution, London.

BSI (1985). *BS 6543: Use of Industrial By-products and Waste Materials in Building and Civil Engineering*. British Standards Institution, London.

BSI (1988). *DD 175: Code of Practice for the Identification of Potentially Contaminated Land and its Investigation*. British Standards Institution, London.

BSI (1990). *BS 1377: British Standard Methods of Test for Soils for Civil Engineering Purposes*. British Standards Institution, London.

BSI (1994). *BS 8002: Code of Practice for Earth Retaining Structures*. British Standards Institution, London.

BSI (1996). *BS 5228: Noise and Vibration Control on Construction and Open Sites*. Parts 1 to 5. British Standards Institution, London.

Bureau of Mines (1981). Mine waste disposal technology. In *Proceedings of the Bureau of Mines Technology Transfer Workshop, Denver*. US Department of the Interior, Reston.

Burford, D. & Charles, J. A. (1991). *Long-term Performance of Houses Built on Opencast Minestone Mining Backfill at Corby 1975–1990*. Building Research Establishment, Garston, Watford, Paper PD 35/91.

Burland, J. B. & Burbridge, M. C. (1985). Settlements of foundations on sands and gravels. *Proceedings of the ICE*, **78**, pp. 1325–1337.

Burland, J. B., Potts, D. M. & Walsh, N. M. (1981). The overall stability of free and propped embedded cantilever retaining walls. *Ground Engineering*, **14**, No. 5, pp. 28–38.

Burt, A. & Bradshaw, A. (1986). *Transforming our Waste Land: The Way Forward*. Department of the Environment HMSO, London.

Cadwallader, M. W. (1988). Durable liners for hazardous waste containment. In *Hazardous Wastes, Detection, Control and Treatment* (ed. R Abbo). Elsevier Science, London, pp. 1513–1524.

Cairney, T. (ed). (1987). *Reclaiming Contaminated Land*. Blackie, Glasgow.

Cairney, T. (1995). *The Re-use of Contaminated Land. A Handbook of Risk Assessment*. Wiley, London.

Calder, I. (1993). Grand buildings, Trafalgar Square, London WC2. *Proceedings of the ICE, Civil Engineering*, pp. 127–134.

Campbel, D. J. V. (1983). *Understanding Water Balance in Landfill Sites*. Institute of Wastes Management, Cambridge.

Campbell, G. (1997). *A Comparison between Permeability Measurements in the Compaction Permeameter and the Triaxial Cell*. MSc thesis, Bolton Institute.

Campell, J. A., Towner, J. V. & Vallurupalli (1988). Distribution of heavy metals in sewage sludge: the effect of particle size. In *Chemical and Biological Characterisation of Sludges, Sediments, Dredge Spoils and Drilling Muds* (eds, J. Lichten *et al.*). American Society for Testing and Materials, Philadelphia, ASTM STP 976, pp. 93–101.

Caquot A & Kerisel J (1948). *Tables for the Calculation of Passive Pressure, Active Pressure and Bearing Capacity of Foundations*. Gauthier-Villers, Paris.

Card, G. B. (1995). *Protecting Development from Methane*. CIRIA London, Report 149.

Carey, P. J. & Swyka, M. A. (1991). Design and placement considerations for clay and composite clay/geomembrane landfill final covers. *Journal of Geotextiles and Geomembranes* **10**, pp. 133–140.

Carman, P. C. (1956). *Flow of gases through porous media*. Academic Publishers, New York.

Carrier, W. D. & Beckman, J. F. (1984). Correlations between index tests and properties of remoulded clays. *Géotechnique* **34**, No. 2, pp. 211–228.

Carrillo, N. (1942). Simple two and three-dimensional cases in the theory of consolidation of soils. *Journal of Mathematics and Physics* **21**, pp. 1–5.

Carter, M. J. (1985). Hydrogeological assessment and selection of hazardous waste disposal sites. In *Hazardous Waste Management Handbook* (ed., A. Porteous), Butterworths, London, Chap. 3.

Carter, M. & Bentley, S. P. (1991). *Correlations of Soil Properties*. Pentech Press, London.

Cartwright, K. & Hensel, B. R. (1993), Hydrogeology. In *Geotechnical Practice for Waste Disposal* (ed. D. Daniel). Chapman & Hall, London, Chap. 4, pp. 66–93.

Carucci, A., Gabrielli, B. & Grisolia, M. (1991), Stability of sanitary landfill slopes. In *Proceedings of the 3rd International Landfill Symposium, Sardinia,* CISA (Environmental Sanitary Engineering Centre), Cagliari, Vol. II, pp. 1161–1170.

CEC (1980). The protection of groundwater against pollution caused by certain dangerous substances. *Official Journal of the European Communities,* Directive 80/68/EEC.

CEC (1985). The assessment of the effects of certain public and private projects on the environment. *Official Journal of the European Communities,* Directive 85/337/EEC.

Cedergren, H. (1967). *Seepage, Drainage and Flow Nets.* Wiley, New York.

CEGB (1978). *Grouting. PFA Data Book.* Central Electricity Generating Board, London.

CEGB (1985). *Ash Marketing Notes for Information — Specification of Grouts Utilising dry Blended PFA/Cement Mixes and dry PFA for Site Mixing.* Central Electricity Generating Board, London.

CEGB (1986). *Grouting with PFA. PFA in Stabilisation.* Ash Marketing Board, Central Electricity Generating Board, London.

Chakraborty, A. K., Morthy, V. M. S. R. and Jethwa, J. L. (1996). Innovative cautious blasting technique for excavation close to a running hydro-electric powerhouse — a case study. *Proceedings of the ICE, Geotechnical Engineering,* **119**, pp. 57–63.

Chandler, R. J. & Tosatti, G. (1995). The Stava tailings dams failure, Italy, July 1985. *Proceedings of the ICE, Geotechnical Engineering,* **113**, pp. 67–79.

Chapman, N. A., McEwen, T. J. & Beale, H. (1986). Geological environments for deep disposal of intermediate level wastes in the United Kingdom. In *Proceedings of International Symposium on the Siting, Design and Construction of Underground Repositories for Radioactive Wastes, Hannover.* International Atomic Energy Agency, Vienna, pp. 311–328.

Charles, J. A. (1986). Hydraulic fills, colliery discard lagoons — review paper. In *Proceedings of the ICE Conference on Building on Marginal and Derelict Land.* Thomas Telford, London, pp. 95–109.

Charles, J. A. (1991). The causes, magnitudes and control of ground movements in fills. In *Proceedings of the 4th International Conference on Ground Movements and Structures, Cardiff.* Pentech Press, London, pp. 3–9.

Charles, J. A. (1993). *Building on Fill: Geotechnical Aspects.* Building Research Establishment, Garston, Watford.

Charles, J. A. & Burland, J. B. (1982). Geotechnical considerations in the design of foundations for buildings on deep deposits of waste materials. *The Structural Engineer* **60A**, No. 1, pp. 8–14.

Childs, K. A. (1985). In-ground barriers and hydraulic measures. In *Contaminated Land, Reclamation and Treatment* (ed., M. A. Smith). Plenum Press, New York, pp. 145–182.

Chipp, P. N. (1990). Geotechnical containment measures for pollution control. In *Proceedings of the International Conference on Construction in Polluted and Marginal Land, Brunel,* Engineering Technics Press, pp. 103–116.

Choppin, G. R. & Rydberg, J. (1980) *Nuclear Chemistry. Theory and Applications.* Pergamon Press, Oxford.

Christopher, B. R. (1991). Geotextiles in landfill closures. *Journal of Geotextiles and Geomembranes* **10**, pp. 77–88.

CIRIA (1993). *Environmental Issues in Construction. Overview and Summary of Information Needs.* CIRIA, London, Special Publication 93.

CIRIA (1994a). *Environmental Assessment.* CIRIA, London, Special Publication 96.

CIRIA (1994b). *Environmental Handbook for Building and Civil Engineering Projects. Design and Specification.* CIRIA, London, Special Publication 97.

CIRIA (1994c). *Environmental Handbook for Building and Civil Engineering Projects Construction Phase.* CIRIA, London, Special Publication 98.

CIRIA (1995). *Environmental impact of materials. Vol. A: Summary.* CIRIA, London, Special Publication 116.

CIRIA (1995–98). *Remedial Treatment for Contaminated Land.* CIRIA, London.
   Vol. I (1998). *Introduction and guide.* Special Publication 101.
   Vol. II (1995). *Decommissioning, Decontamination and Demolition.* Special Publication 102.
   Vol. III (1995). *Site Investigation and Assessment.* Special Publication 103.

Vol. IV (1995). *Classification and Selection of Remedial Methods.* Special Publication 104.

Vol. V (1995). *Excavation and Disposal.* Special Publication 105.

Vol. VI (1996). *Containment and Hydraulic Measures.* Special Publication 106.

Vol. VII (1995). *Ex-situ Remedial Methods for Soils, Sludges and Sediments.* Special Publication 107.

Vol. VIII (1995). *Ex-situ Remedial Methods for Contaminated Groundwater and Other Liquids.* Special Publication 108.

Vol. IX (1995). *In-situ Methods of Remediation.* Special Publication 109.

Vol. X (1995). *Special situations,* Special Publication 110.

Vol. XI (1995). *Planning and Management.* Special Publication 111.

Vol. XII (1997). *Policy and Legislation.* Special Publication 112.

CIRIA (1996). Environmental assessment: good practice. *Proceedings of Construction Industry Environmental Forum Conference* (ed., J. Petts). CIRIA, London, Special Publication 126.

CIRIA (1998). *Waste Minimisation in Construction — Design Manual.* CIRIA, London, Special Publication 134.

CIRIA (1999). *Waste Minimisation and Recycling in Construction — Boardroom Handbook.* CIRIA, London, Special Publication 135.

Clarke, B. G. (1992). Structural fill. In *National Seminar on the Use of PFA in Construction, Dundee* (eds, R. K. Dhir & M. R. Jones), pp. 21–32.

Clarke, B. G. (1996). Pressuremeter in ground investigation. Part 1: Site operations. *Proceedings of the ICE, Geotechnical Engineering.* **119**, pp. 96–108.

Claydon, J. R., Eadie, H. S. & Harding, C. (1997). Deighton tip — failure, investigation and remedial works. In *Proceedings of the 19th congress of ICOLD, Florence,* pp. 233–245.

Clayton, C. R. I. & Milititsky, J. (1986). *Earth Pressure and Earth-retaining Structures.* Surrey University Press, London.

Clayton, C. R. I., Sissons, N. G. & Matthews, M. C. (1982). *Site Investigation. A Handbook for Engineers.* Granada, St Albans.

Coch, N. K. (1995) Waste disposal. In *Geohazards. Natural and Human.* Prentice Hall, Englewood Cliffs, NJ, pp. 310–343.

Collinge, V. K. & Bruce, A. M. (1981). *Sewage Sludge Disposal: A Strategic Review and Assessment of Resource Needs. Water Research Centre,* Medmenham, Technical Report TR 166.

Collins, R. J. & Atkinson, C. J. (1995). Specifications and the use of wastes in construction in the United Kingdom. Paper presented at *ICE Seminar on Recyclable Building Materials,* London.

Collins R. J. & Ciesielski. S. K. (1992). Highway construction use of wastes and by-products. In *ASCE Conference on Utilisation of Waste Materials in Civil Engineering Construction,* Amiercan Society of Civil Engineers, New York, pp. 140–152.

Construction News:
(1994). Bath undermined by historic past. 4 August, p. 28.
(1995). Delta probes chalk mines. 9 February, p. 21.
(1996). Southern estates home in on pit village bricks. 5 September, p. 3.
(1999). Add a dash of lime. 2 September, pp. 26–27.

Contracts Journal:
(1995). Balfour reuses silt in A13 upgrade. 26 October, p. 8.
(1996). M8 void costs to soar. 23 May, p. 5.

Coppin, N. J. & Richards, I. G. (1990). *Use of vegetation in Civil Engineering,* CIRIA Butterworths, London.

Corbett, B. O. (1983). Contaminated ground. *Building Technology and Management,* May, pp. 22–26.

Cork, J. P. & Hinsley, M. R. (1998). The application of materials recovery and soils washing to a former gas works. In *Land Reclamation: Achieving Sustainable Benefits* (eds, H. R. Fox, H. M. Moore & A. D. McIntosh). Balkema, Rotterdam, pp. 473–482.

Coulomb C. A. (1776). Essai sur une application des règles de maximis et minimis à quelques problèmes de statique, rélatifs à l'architecture. *Memoires de Mathématique et de Physique présentés à l'Academie Royale des Sciences,* **7**, Paris, pp. 343–382.

Coutts, D. A. P., Dunk, M. & Pugh, S. Y. R. (1990). The microbiology of landfills: the Brogborough landfill test cells. In *Landfill Microbiology: R&D Workshop* (eds, P. Lawson and Y. R. Alston), Harwell, Oxford, pp. 133–149.

Cowell, R. (1985). 'Hong Kong and Shanghai Banking headquarters redevelopment — control of noise and vibration from construction/demolition. In *Proceedings of the Conference on Pollution in the Urban Environment* (eds, M. W. Chan, R. W. M. Hoare, P. R. Holmes, R. J. S. Law & S. B. Reed). Elsevier Applied Science, Hong Kong, pp. 351–355.

Coyle, H. M. & Bartoskewitz, R. E. (1976). Earth pressures on a precast, panel retaining wall. *Journal of Geotechnical Engineering*, ASCE, **102**, No. 5, pp. 441–456.

Crawford, J. F. & Smith, P. G. (1985). *Landfill Technology*. Wiley, New York.

Crawshaw, M. D. E. (1993). The construction of Drighlington by-pass through methane generating industrial refuse. *Highways and Transportation* **40**, No. 12, 11–14.

Cripwell, J. B. (1992). Pulverised fuel ash. In *National Seminar on the Use of PFA in Construction* (eds, R. K. Dhir & M. R. Jones). Dundee, pp. 1–20.

Crockett, J. H. A. (1979). Piling vibrations and structural fatigue. In *Proceedings of the Conference on Recent Developments in the Design and Construction of Piles*. ICE Thomas Telford, London.

Crowhurst, D. (1987). *Measurement of Gas Emissions from Contaminated Land.* Building Research Establishment, Garston, Watford.

Crowhurst, D. & Manchester, S. J. (1993). *The Measurement of Methane and other Gases from the Ground.* CIRIA, London, Report 131.

Daniel, D. E. (1990). Summary review of construction quality control for compacted soil liners. In *Waste Containment Systems: Construction, Regulations and Performance* (ed., R. Bonaparte). American Society of Civil Engineers, New York, pp. 175–189.

Daniel, D. E. (1993). Clay liners. In *Geotechnical Practice for Waste Disposal* (ed., D. Daniel), Chapman & Hall, London, Chap. 7, pp. 137–163

Daniel, D. E. (1995). Pollution prevention in landfills using engineered final covers. In *Proceedings of GREEN'93: Waste Disposal by Landfill* (ed., R. W. Sarsby), Balkema, Rotterdam, pp. 73–92.

Daniel, D. E., Anderson, D. C. & Boynton, S. S. (1985). Fixed wall versus flexible-wall permeameters. In *Hydraulic Barriers in Soil and Rock* (eds., A. I. Johnson, R. K. Frobel, N. J. Cavalli & C. B. Pettersson), American Society for Testing and Materials, Philadelphia, ASTM STP 874, pp. 107–126.

Daniel, D. E. & Benson, C. H. (1990). Water content-density criteria for compacted soil liners. *Journal of Geotechnical Engineering, ASCE*, **116**, No. 12, pp. 1811–1830.

Daniel, D. E. & Koerner, R. M. (1993). Final cover systems. In *Geotechnical Practice for Waste Disposal* (ed., D. E. Daniel). Chapman & Hall, London, Chap. 18.

Daniel, D. E. & Liljestrand H. M. (1984). *Effects of landfill leachates on natural liner systems.* Report to Chemical Manufacturers Association, University of Texas, Austin.

Daniel, P., Gyori, Z. & Gaspar, A. (1998). Accumulation of heavy metals within cabbages. In *Proceedings of GREEN2: Contaminated and Derelict Land* (ed., R. W. Sarsby). Thomas Telford, London, pp. 166–174.

Darbyshire, W. (1996). Lining pockets. *Ground Engineering*, July/August, pp. 13–15.

Dass. P., Tamko, G. R. & Stoffel, C. M. (1977). Leachate production at sanitary landfill sites. *Journal of the Environmental Engineering Division, ASCE.* **103**, EE6.

Day, S. & Daniel, D. E. (1985). Hydraulic conductivity of two prototype clay liners. *Journal of Geotechnical Engineering, ASCE* **111**, No. 8, pp. 957–970.

Deegan, J. (1987). Looking back at Love Canal. *Environmental Science Technology Journal* **21**, No. 4, 328–331; **21**, No. 5, pp. 421–426.

Derbyshire County Council (1988). *Report of the Non-statutory Public Inquiry into the Gas Explosion at Loscoe, Derbyshire on 24 March 1986*, Vol. 1 and 2. Matlock, Derbyshire.

DeWalle, F. B., Chian, E. S. & Hammerbeg, E. (1978). Gas production from solid waste in landfills. *Journal of the Environmental Engineering Division, ASCE*, **104**, EE3, pp. 415–432.

DOE (1987). *Guidance on the Assessment and Redevelopment of Contaminated Land.* Department of the Environment, London, ICRCL Guidance Note 59/83, 2nd edn.

DOE (1988). *A Basic Study of Landfill Microbiology and Biochemistry.* Harwell, Oxford,

ETSU B1159.

DOE (1989). *Environmental Assessment — A Guide to the Procedures*. Welsh Office HMSO, London.

DOE (1990). *Landfilling Wastes*. HMSO, London, Waste Management Paper No. 26.

DOE (1991a). *Landfill Gas*. HMSO, London, Waste Management Paper No. 27.

DOE (1991b). *The Building Regulations 1991. C4, Section 3: Floors Next to the Ground*. HMSO, London.

DOE (1991c). *Environmental Effects of Surface Mineral Workings*. HMSO, London, report prepared by Roy Waller Associates Ltd.

DOE (1992). *Environment in Trust — Radioactive Waste Management — a Safe Solution*. Department of the Environment, London.

DOE (1993a). *UK Landfill Practice — Co-disposal*. Department of the Environment, London.

DOE (1993b). *The Householder's Guide to Radon*, 3rd edn., HMSO, London.

DOE (1994a). *Government Response to the 17th Report of the Royal Commission on Environmental Pollution. Incineration of Waste*. Department of the Environment, London.

DOE (1994b). *Environmental Assessment. A Guide to the Procedures*. Welsh Office HMSO, London.

DOE (1994c). *Landfill Completion*. HMSO, London, Waste Management Paper No. 26A.

DOE (1995a). *Landfill Design, Construction and Operational Practice. Part 2: Landfill Engineering*. HMSO, London, Waste Management Paper No. 26B.

DOE (1995b). *Making Waste Work — A Waste Strategy for England and Wales*. HMSO, London.

DOE (1995c). *Slate Waste Tips and Workings in Britain*. HMSO, London, report prepared by Richards, Moorehead and Laing Ltd.

DOE (1996). *Consultation on Draft Statutory Guidance on Contaminated Land, Environmental Protection Act 1990. Part IIA: Contaminated Land*, Vols 1 and 2, Department of the Environment, London.

Don and Low Ltd (1989). *A Geotextiles Design Guide*, 2nd edn., Don & Low, Angus.

DOT (1987). *Reinforced Earth Retaining Walls and Bridge Abutments for Embankments*. Technical Memo (Bridges) BE3/78.

Dowding, C. H. (1992). Frequency based control of urban blasting. In *Proceedings of the ASCE International Convention and Exposion on Excavation and Support for the Urban Infrastructure* (eds, T. D. O'Rourke & A. G. Hobelman). American Society of Civil Engineers, New York, ASCE Geotechnical Special Publication No. 33, pp. 181–211.

Druschel, S. J. & Wardwell, R. E. (1991). Impact of long-term landfill deformation. In *Proceedings of the Engineering Congress, Boulder*, Vol. II. American Society of Civil Engineers, New York, ASCE Geotechnical Special Publication No. 27, pp. 1268–1279.

DTI (1995). *Landfill Gas Microbiology Workshop, Workshop, Solihull*.

Dunnicliff, J & Green, GE (1994). *Geotechnical instrumentation for monitoring field performance*. Wiley, London.

Duplancic, N. (1990). Landfill deformation monitoring and stability analysis. In *Geotechnics of Waste Fills — Theory and Practice* (eds, A. Landva & G. D. Knowles), American Society for Testing and Materials, Philadelphia, ASTM STP 1070, pp. 303–314.

Edgers, L., Noble, J. J. & Williams, E. (1992). A biologic model for long term settlement in landfills. In, *Mediterranean Conference on Environmental Geotechnology* (eds, M. A. Usmen & Y. B. Acar). Balkema, Rotterdam, pp. 177–184.

Edil, T. B. & Erickson, A. E. (1985). Procedure and equipment factors affecting permeability testing of a bentonite-sand liner material. In *Hydraulic Barriers in Soil and Rock* (eds, A. I. Johnson, R. K. Frobel, N. J. Cavalli. & C. B. Petterson) American Society for Testing and Materials, Philadelphia, ASTM STP 874, pp. 155–170.

Edil, T. B., Ranguette, V. J. & Wuellner, W. W. (1990). Settlement of municipal refuse. In *Geotechnics of Waste Fills — Theory and Practice* (eds, A. Landva & G. D. Knowles), American Society for Testing and Materials, Philadelphia, ASTM STP 1070, pp. 225–239.

Edmeades, R. M. & Mangabhai, R. J. (1992). PFA grouts: selection of materials and flow properties. In *National Seminar on the Use of PFA in Construction* (eds, R. K. Dhir & M. R. Jones). Dundee, pp. 75–88.

Edwards, A. T. & Northwood, T. D. (1960). Experimental studies of the effects of blasting on structures. *The Engineer* **210**, pp. 538–546.

Ehrig, H. J. (1988). Quality and quantity of sanitary landfill leachate. *Waste Management and Research* **1**, 53–68.

Eith, A. W., Boschuk, J. & Koerner, R. M. (1991). Prefabricated bentonite clay liners. *Journal of Geotextiles and Geomembranes*, **10**, pp. 192–217.

Eklund, A. G. (1985). A laboratory comparison of the effects of water and waste leachate on the performance of soil liners. In *Hydraulic Barriers in Soil and Rock* (eds, A. I. Johnson, R. K. Frobel, N. J. Cavalli & C. B. Petterson), American Society for Testing and Materials, Philadelphia, ASTM STP 874, pp. 188–202.

Ellis, B. (1992). On site and in-situ treatment of contaminated sites. In *Contaminated Land Treatment Technologies* (ed., J.F. Rees). Elsevier Applied Science, pp. 30–46.

Ellis, D. (1989). *Environments at Risk (Case Histories of Impact Assessment)*. Springer-Verlag, Berlin.

Elsbury, B. R., Daniel, D. E. & Sraders, G. A. (1990). Lessons learned from compacted clay liner. *Journal of Geotechnical Engineering, ASCE* **116**, pp. 1641–1660.

EMCON (1987). *Stability Studies — Waldon Canyon Landfill*. Report to Waste Management Inc. of California, Ventura County, CA.

Engineering Council (1993). *Engineers and the Environment. Code of Professional Practice*. The Engineering Council, London.

Environment Agency (1999). *Interim Guidance on the Geophysical Testing of Geomembranes for Landfill Engineering*. Warrington.

Ervin, M. C. (1993). Specification and control of earthworks. In *Proceedings of Engineered Fills '93* (eds, B. G. Clarke, C. J. F. P. Jones & A. I. B. Moffat). Thomas Telford, London, pp. 18–41.

Esmaili, H. (1975). Control of gas flow from sanitary landfills. *Journal of the Environmental Engineering Division, ASCE*, **101**, EE4, pp. 555–566.

ETSU (1988). *A Basic Study of Landfill Microbiology and Biochemistry*. Report prepared by AFRC Institute of Food Research, Harwell, ETSU B1159.

ETSU (1993). *A Preliminary Assessment of Methane Emissions from UK Landfills*. Department of the Environment, Harwell, DOE Technical Division Report 163/93.

Fadum, R. E. (1948). Influence values for estimating stresses in elastic foundations. In *Proceedings of the 2nd ICSMFE, Rotterdam*, Vol. 3, pp. 77–84.

Fang, H. Y. (1977). *Strength Testing of Bales of Sanitary Landfill*. Department of Civil Engineering, Lehigh University, Pennsylvania.

Fang, H. Y. & Evans, J. C. (1988). *Long-term permeability tests using leachate on a compacted clayey liner material*. ASTM STP 963, ASTM, West Conshohorken, pp. 397–404.

Farquhar, G. J. & Rovers, F. A. (1973). Gas production during refuse decomposition. *Journal of Water, Air and Soil Pollution* **2**, 483–495.

Fellenius, W. (1936). Calculation of the stability of earth dams. In *Transactions of the 2nd Congress on Large Dams, Washington*, Vol. 4, pp. 445–462.

Ferguson, C. C. (1992). The statistical basis for spatial sampling of contaminated land. *Ground Engineering*, June. pp. 34–38.

Ferguson, J. (1995). Practical case studies in construction and demolition waste recycling. In *Seminar on Recyclable Building Materials*. The Institution of Civil Engineers, London.

Ferguson, J., Kermode, N., Nash, C. L., Sketch, W. A. J. & Huxford, R. P. (1995). *Managing and Minimising Construction Waste. A Practical Guide*. Institution of Civil Engineers/Thomas Telford, London.

Finn, L. W. D. (1981). Seismic response of tailings dams. In *Seminar on Design and Construction of Tailings Dams* (ed., D Wilson). Colorado School of Mines Press, Golden, CO, pp. 76–98.

Finno, R. J. & Schubert, W. R. (1986). Clay liner compatability in waste disposal practice. *Journal of Environmental Engineering*, ASCE, **112**, No. 6, pp. 1070–1084.

Fleming, G. (ed.) (1991). *Recycling Derelict Land*. Thomas Telford, London.

Fortlage, C. A. (1990). *Environmental Assessment — A Practical Guide*, Gower Technical, London.

Foss, R. N. (1978). Single screen noise barrier. *Noise Control Engineering*, July–August, pp. 40–44.

Fox, N. H. (1984). Pulverised fuel ash as structural fill. In *AshTech '84, 2nd International Conference on Ash Technology and Marketing, London*, pp. 495–499.

Frantzis, I. (1991). Settlement in the landfill site of Schisto. In *Proceedings of the 3rd International Landfill Symposium*, Sardinia, CISA (Environmental Sanitary Engineering Centre), Cagliari, Vol. II, pp. 1189–1195.

Fraser, C. K. & Lake, J. R. (1967). *A Laboratory Investigation of the Physical and Chemical Properties of Burnt Colliery Shale*. Road Research Laboratory, Crowthorne, Report LR 125.

Frobel, R. K. (1987). *Geosynthetics Terminology, an Interdisciplinary Treatise*. Industrial Fabrics Association International, St Paul, Minnesota.

Frost, R.C., Powlesland, C., Hall, J. E., Nixon, S. C. & Young, C. P. (1980). *Review of Sludge Treatment and Disposal Techniques*. Water Research Centre, Medmenham, Report No. PRD 2306–M/1.

Frost, R. C. (1983). Inter-relation between sludge characteristics. In *Proceedings of the Workshop on Methods of Characterisation of Sewage Sludge*. Reidel, Dordrecht, pp. 106–122.

Gandolla, M., Dugnani, L., Bressi, G. & Acaia, C. (1992). The determination of subsidence effects at municipal solid waste disposal sites. In *Proceedings of the 6th International Solid Wastes Congress, Madrid*, pp. 1–17.

Gardner, N. D. R. & Manley, B. J. W. (1991). Landfill gas extraction from buried refuse — design considerations. In *Proceedings of the Conference on Engineering for Profit from Waste*. Institution of Mechanical Engineers, London, pp. 85–89.

Garvey, D., Davis, R. D., Guarino, C. & Carlton-Smith, C. H. (1992). Treatment and disposal of sewage sludge — current practices. In *Sludge 2000, Conference on Sewage Sludge Use and Disposal*, Cambridge University.

Gay, C. C. W. (1991). Stability of refuse tips containing sludge. In *Proceedings of Third International Landfill Symposium*, CISA (Environmental Sanitary Engineering Centre), Cagliari, pp. 1421–1431.

Genevois, R. & Tecca, P. R. (1993). The tailings dams of Stava (North Italy): an analysis of the disaster. In *Environmental Management, Geo-water and Engineering Aspects*, (eds, P. Choudhury & A. Sivakumar). Balkema, Rotterdam, pp 23–36.

German Geotechnical Society (1991). *Geotechnics of Landfills and Contaminated Land*. Emst and Sohn, Berlin.

Gettinby, J. H. (1999). *The Composition of Landfill Leachate and its Interaction with Compacted Clay Liners*. PhD thesis, Manchester University (Bolton Institute), Manchester.

Geuzens, P. & Dieltjens, W. (1991). Mechanical strength determination of cohesive sludges — a Belgian research project on sludge consistency. In *Recent Developments in Sewage Sludge Processing* (eds, F. Colin, P. J. Newman & Y. J. Puolanne). Elsevier Applied Science, London, pp. 14–23.

Ginniff, M. E. (1987). The characteristics of disposal sites and repositories. In *Proceedings of the Seminar on The Management and Disposal of Intermediate and Low Level Radioactive Waste*. Institution of Mechanical Engineers, pp. 5–10

Giroud, J. P. & Bonaparte, R. (1989). Leakage through liners constructed with geomembranes. Part II: Composite liners. *Geotextiles and Geomembranes*, **8**, No. 2, pp. 71–111.

Giroud, J. P. & Peggs, I. D. (1990). Geomembrane Construction Quality Assurance. In *Waste Containment Systems*. American Society of Civil Engineers, New York, ASCE Special Publication No. 26, pp. 190–225.

Glasson. J., Therivel. R. & Chadwick. A. (1994). *Introduction to Environmental Impact Assessment*. UCL Press, London.

Glegg, G. (1991). A policy for the use of sewage sludge. In *International Conference on Environmental Pollution*. Interscience, Lisbon, pp. 600–607.

Goodman, A. C. (1998). Practical experience of growing trees on derelict and contaminated sites. In *Land Reclamation: Achieving Sustainable Benefits* (eds, H.

R. Fox, H. M. Moore & A. D. McIntosh). Balkema, Rotterdam, pp. 125–130.

Gordon, M. E., Huebner, P. M. & Mitchell, G. R. (1990). Regulation, construction and performance of clay-lined landfills in Wisconsin. In *Waste Containment Systems*. American Society of Civil Engineers, New York, ASCE Geotechnical Special Publication No. 26, pp. 14–24.

Gottheil, K. M. & Brauns, J. (1997). Thermal effects on the barrier efficiency of composite liners — test field measurements. In *Advanced Liner Systems* (eds, H. August, U. Holzlohner & T. Meggyes), Thomas Telford, London, pp. 202–209.

Greenwood, D. A. & Thomson, G. H. (1984). *Ground Stabilisation: Deep Compaction and Grouting*. Thomas Telford, London.

Griffin, R. A., Cartwright, K., Shimp, N. F., Steele, J. D., Ruch, R. R., White, W. A., Hughes, G. M. & Gilkenbon, R. H. (1976). *Attenuation of pollutants in municipal landfill leachate by clay material*. Illinois State Geological Survey, Urbana.

Griffith, A. (1994). Environmental management in the construction process. *Environmental Management in Construction*. MacMillan, Basingstoke, Chap. 5.

Grimes, J. N., Smith, R., Mitchell, I., Walker, C. J. & White, A. J. (1994). Remedial works and hazard management of urbanized land at the old Gunnislake Mine. In 3rd International Conference on the Re-use of Contaminated Land and Landfills, London, pp. 169–178.

Ground Engineering (1992). Clean start. 1st meeting of East Midlands Geotechnical Group, on remediation of contaminated land. December, pp. 24–27.

Guerra, F. (1972). 'Characteristics of tailings from a soil engineer's viewpoint', in *Tailings disposal today* (eds C.L. Aplin & G.O. Argall), University of Arizona, Tucson, pp. 102–136.

Gutenberg, B. & Richter, C. (1956). Earthquake magnitude, intensity, energy and acceleration. *Bulletins of the Seismic Society of America*, **46**, No. 2, pp. 105–145.

Gutowski, T. G. & Dym, C. L. (1976). Propagation of ground vibration: a review. *Journal of Sound and Vibration*, **49**, No. 2, pp. 179–193.

Haines, R. C. (1991). Scale and extent of contaminated land in the UK — an opportunity for the construction industry? In *Proceedings of the Conference on Contaminated Land, A Practical Examination of the Technical and Legal Issues*. IBC Technical Services Ltd, London, Document E-7588.

Haines, R. C. & Harris, M. R. (1987). Main types of contaminant. In *Reclaiming Contaminated Land* (ed., T. Cairney). Blackie, Glasgow, Chap. 3, pp. 39–61.

Hall, J. E. (1995). Sewage sludge production treatment and disposal in the European Union. *Journal of the CIWEM*, **9**, pp. 335–343.

Hansen, T. C. (ed.) (1992). *Recycling of Demolished Concrete and Masonry*. E. & F. N. Spon, London, Report of RILEM Technical Committee 37-DRC.

Harries, C. R., Witherington, P. J. & McEntee, J. (1995). *Interpreting Measurements of Gas in the Ground*. CIRIA, London, Report 151.

Harris, F. (1994). *Modern Construction and Ground Engineering Equipment and Methods*. Longman Scientific and Technical, Harlow.

Harris, M. R. R. (1979). Geotechnical characteristics of landfilled domestic refuse. In *Proceedings of the symposium on Engineering Behaviour of Industrial and Urban Fill, Birmingham*, pp. B1–B10.

Hartless, R. (1992). *Methane and Associated Hazards to Construction: A Bibliography*. CIRIA, London, Special Publication 79.

Haug, R. T. (1979). Engineering principles of sludge composting. *Journal of the Water Control Federation*, **51**, pp. 2189–2206.

Hazen, A. (1892). *Some Physical Properties of Sands and Gravels with Special Reference to their Use in Filtration*. Massachusetts State Board of Health, Boston, MA, 24th Annual Report.

Head, J. M. & Jardine, F. M. (1992). *Ground-borne Vibration Arising from Piling*. CIRIA, London, Technical Note 142.

Heckman, W. S. & Hagerty, D. J. (1978). Vibration associated with pile driving. *Proceedings of the ASCE, Construction Division*, **104**, CD4, pp. 385–394.

Henry, H. F. (1969). *Fundamentals of Radiation Protection*. Wiley Interscience, New York.

Henry, J. G. & Heinke, G. W. (1989). *Engineering Science and Engineering*. Prentice

Hall, Englewood Cliffs, NJ.

Highways Agency (1998). *Specification for Highway Works*. Vol. 1. *Manual of Contract Documents for Highway Works*, HMSO, London.

Hills, C. W. W. & Denby, B. (1996). 'The prediction of opencast backfill settlement. *Proceedings of the ICE, Geotechnical Engineering*, **119**, pp. 167–176.

Hird, C. C., Smith, C. C. & Cripps, J. C. (1998). Issues related to the use and specification of colliery spoil liners. In *Geotechnical Engineering of Landfills* (eds, N. D. Dixon, E. J. Murray & D. R. V. Jones), Thomas Telford, London, pp. 61–79.

Hoek, E. & Bray, J. W. (1974). *Rock Slope Engineering*', Institution of Mining & Metallurgy, London.

Hoeks, J. & Harmsen, J. (1980). Methane gas and leachate from sanitary landfills. *Research Digest*, pp. 132–139.

Holroyd, M. L. & Caunt, P. (1994). Fungal processing: a second generation biological treatment for the degradation of recalcitrant organics in soil. *Land Contamination and Reclamation*, **2**, No. 4, pp. 183–188.

Hooker, P. J. & Bannon, M. P. (1993). *Methane — Its Occurrence and Hazards in Construction*. CIRIA, London, Report 130.

Horner, P. C. (1981). *Earthworks*. Thomas Telford, London.

Howden, C. & Crawley, J. D. (1995). Design and construction of the diaphragm wall. *Proceedings of the ICE* **108**, pp. 48–62.

HSE (1991). *Protection of Workers and the General Public during Development of Contaminated Land*, Health & Safety Executive/HMSO, London.

Hughes, D. (1982). *Notes on Ionizing Radiation*. Science Reviews Ltd, London, Occupational Hygiene Monograph No. 5.

Hunt, R., Dyer, R. H. & Driscoll, R. (1991). *Foundation Movement and Remedial Underpinning in Low-rise Buildings*. Building Research Establishment, Garston, Watford, BRE Report BR 184.

Hurtric, R. (1981). Sanitary landfill settlement rates. *Conference on the prolongation of the capacity of sanitary landfills*. Berlin Technical University, Berlin.

ICE (1990). *Pollution and its Containment*. Infrastructure Policy Group/Thomas Telford, London.

ICE (1993). *Site Investigation in construction*. ICE Site Investigation Steering Group/ Thomas Telford, London, pp. 1–35.

ICE Site Investigation Steering Group, (1993). *Site Investigation in Construction 4: Guidelines for the Safe Investigation by Drilling of Landfills and Contaminated Land*. Thomas Telford, London.

ICOLD (1989). *Tailings Dam Safety. Guidelines*. International Commission on Large Dams, Paris, Bulletin 74.

ICRCL (1983a). *Notes on the Redevelopment of Scrap Yards and Similar Sites*. Interdepartmental Committee on the Redevelopment of Contaminated Land, London, Guidance Note 42/80, 2nd edn.

ICRCL (1983b). *Notes on the Redevelopment of Sewage Works and Farms*. Interdepartmental Committee on the Redevelopment of Contaminated Land, London, Guidance Note 23/79, 2nd edn.

ICRCL (1987). *Guidance on the Assessment and Development of Contaminated Land*. Interdepartmental Committee on the Redevelopment of Contaminated Land, London, Guidance Note 59/83, 2nd edn.

ICRCL (1990a). *Asbestos on Contaminated Sites*. Interdepartmental Committee on the Redevelopment of Contaminated Land, London, Guidance Note 64/85, 2nd edn.

ICRCL (1990b). *Notes on the Development and After-use of Landfill Sites*. Interdepartmental Committee on the Redevelopment of Contaminated Land, London, Guidance Note 17/78, 8th edn.

Indraratna, B. (1992). Problems related to disposal of fly ash and its utilisation as a structural fill. *ASCE Conference on Utilisation of Waste Materials in Civil Engineering Construction, New York*, pp. 274–285.

Ingold, T. S. (1979). Lateral earth pressures on rigid bridge abutments. *The Highway Engineer, Journal of the Institution of Highway Engineers*, December, pp. 2–7.

Ingold, T. S. (1992). Margins of safety in polymeric soil reinforcement. *Journal of Highways and Transportation*, July, pp. 39–44.

ISO (1994). *ISO 4886: 1990/DAM: Vibration of Buildings — Guide Lines for the Measurement of Vibrations and Evaluation of their Effects on Buildings.* International Standards Organisation, Geneva.

Ivanhov, P. L., Kolpachkova, A. B., Trunkov, G. T. & Pavilonsky, V. M. (1989). Methods to estimate and reduce negative impacts of tailings dams on the environment. In *Proceedings of the ICSMFE Rio de Janeiro*, pp. 1877–1880.

James, S. C., Kovalick, W. W. & Bassin, J. (1985). Technologies for treating contaminated land and groundwater. *Chemistry and Industry*, **13**, pp. 492–495.

Janbu, N. (1973). Slope stability computations. In *Embankment Dam Engineering: Casagrande Memorial Volume* (eds, R. C. Hirschfield & S. J. Poulos). Wiley, New York, pp. 47–86.

Jarvis, S. T. (1998). The reclamation of a former engine manufacturing plant in Coventry (UK). In *Proceedings of GREEN2: Contaminated and Derelict Land* (ed. R. W. Sarsby). Thomas Telford, London, pp. 341–347.

Jarvis, S. T. & Braithwaite, P. A. (1993). Infilling Littleton Street mine, Walsall, with colliery spoil paste. In *4th International Symposium on Reclamation, Treatment and Utilisation of Coal Mining Wastes, Krakow*, pp. 661–673.

Jefferis, S. A. (1990a). Contaminated land: significance, sources and treatment. In *Proceedings of the International Conference on Construction on Polluted and Marginal Land*. Engineering Technics Press, London, pp. 59–65.

Jefferis, S. A. (1990b). Cut-off walls: methods, materials and specifications. In *Proceedings of the International Conference on Construction on Polluted and Marginal Land*. Engineering Technics Press, London, pp. 117–125.

Jefferis, S. A. (1992). Remedial barriers and containment. In *Contaminated Land Treatment Technologies* (ed. J. F. Rees). SCI/Elsevier Applied Science, London, pp. 58–82.

Jennings, J. E. (1979). The failure of a slimes dam at Bafokeng. *Die Siviele Ingenieur in Suid-Afrika*, **21**, No. 6, 135–141.

Jessberger, H. L. (1994). Geotechnical aspects of landfill design and construction. Part 2: Material parameters and methods. Part 3: Selected calculation methods for geotechnical landfill design. *Proceedings of the ICE, Journal of Geotechnical Engineering*, **107**, pp. 105–113 and pp. 115–122.

Jessberger, H. L. (1995). Contribution to Session 1 discussion. In *Proceedings of GREEN'93: Waste Disposal by Landfill* (ed., R. W. Sarsby). Balkema, Rotterdam, p. 212.

Jessberger, H. L. & Stone, J. K. L. (1991). Subsidence effects on clay barriers. *Géotechnique* **41**, No. 2, pp. 185–194.

Jewell, R. A. (1996). *Soil Reinforcement with Geotextiles*. CIRIA, London, CIRIA Special Publication 123.

Johnson, S. T. (1993). The role of research, information and demonstration projects. In *Remedial Processes for Contaminated Land* (ed., M. Pratt). Institute of Chemical Engineers, London, pp. 1–17.

Johnston, T. A., Millmore, J. P., Charles, J. A. & Tedd, P. (1990). *An Engineering Guide to the Safety of Embankment Dams in the United Kingdoms*. Building Research Establishment, Garston, Watford BRE Report.

Jones, C. J. F. P., Cripwell, J. B. & Bush, I. (1990). Reinforced Earth trial structure for Dewsbury Road. *Proceedings of the ICE, Part 1*, **88**, pp. 321–345.

Jones, H., Lewis, I. H. & Swindells, C. F. (1993). Legislation and management of mine tailings storages in Western Australia. In *Conference on Environmental Management, Geo-water and Engineering Aspects* (eds, P. Choudhury & A. Sivakumar). Balkema, Rotterdam, pp. 621–628.

Jumikis, A. R. (1962). *Soil Mechanics*. Van Nostrand, New York.

Kappeler, S. (1996). *The Wellenberg Site — Geology and Hydrogeology and Long-term Safety*. Swiss National Cooperative for the Disposal of Radioactive Waste, Nagra Bulletin No. 24, pp. 12–33.

Karol, R. H. (1983). *Chemical Grouting*. Marcel Dekker, New York.

Kelly, R. T. (1979). Site Investigation and materials problems. In *Conference on the Reclamation of Contaminated Land, Eastbourne*. Society of the Chemical Industry, pp. B2/1–B2/14.

Kershaw, K. R. & McCulloch, A. G. (1993). Environmental and planning issues. Channel Tunnel Part 2: Terminals *Proceedings of the ICE, Civil Engineering*, **97**, Special issue 2, pp. 19–31.

Kettle, R. J. (1984). Cement stabilised furnace bottom ash. In *AshTech '84, 2nd International Conference on Ash Technology and Marketing, London*, pp. 483–487.

Kettle, R. J. & Williams, R. I. T. (1978). Colliery shale as a construction material. *Compte Rendus Conf Int Sous-produits et Déchets dans le Génie Civil, Paris*, pp. 475–481.

King, K. S., Quigley, R. M., Fernandez, F., Reades, D. W. & Bacopoulos, A. (1991). Performance monitoring of hydraulic conductivity and diffusion at the Keele Valley landfill site. In *Proceedings of the 1st Canadian Conference on Environmental Geotechnology, Montreal*, pp. 49–56.

Klein, A. & Sarsby, R. W. (1999). Problems in defining the geotechnical behaviour of wastewater sludges. In *Geotechnics of High Water Content Materials*. (eds, T. B. Edil & P. A. Fox). American Society for Testing and Materials, West Conshohoken, ASTM STP 1374.

Kleppe, J. H. & Olson, R. E. (1985). Desiccation cracking of soil barriers. In *Hydraulic Barriers in Soil and Rock* American Society for Testing and Materials, (eds, A. I. Johnson, R. K. Frobel, N. J. Cavalli & C. B. Petterson), Philadelphia, ASTM STP 874 pp. 263–275.

Knocke, W. R. & Wakeland, D. L. (1983). Fundamental characteristics of water treatment plant sludges. *Journal of the American Waterworks Association*, **75**, No. 10, pp. 516–523.

Koenig, A. & Kay, J. N. (1996). The geotechnical characterisation of dewatered sludge from wastewater treatment plants. In *3rd International Conference on Environmental Geotechnology, San Diego*, Vol. 1, pp. 73–82.

Koerner, R. M. (1993). Geomembrane liners. In *Geotechnical Practice for Waste Disposal* (ed. D. Daniel). Chapman & Hall, London, Chap. 8, pp. 164–186.

Kuziemska, I. & Quant, B. (1998). Peat as a sorbent for heavy metal removal from water and waste water. In *Proceedings of GREEN2: Contaminated and Derelict Land* (ed., R. W. Sarsby). Thomas Telford, London, pp. 308–312.

Ladd, C. C., Frott, R., Ishihara, K., Schlosser, F. & Poulos, H. G. (1977). Stress–deformation and strength characteristics. In *Proceedings of the 9th ICSMFE, Tokyo*, Vol. 12, pp. 421–494.

Lageman, R., Wieberen, P. & Seffinga, G. A. (1993). Electro-reclamation: state of the art. In *Proceedings of the 4th conference on Contaminated Land: Policy, Economics and Technology*. IBC Technical Services, London.

Lahti, L., King, K. S., Reades, D. W. & Bacopoulos, A. (1987). Quality Assurance monitoring of a large clay liner. In *Getoechnical Practice for Waste Disposal '87*. (ed., R. D. Woods), American Society of Civil Engineers, New York, ASCE Special Publication 13, pp. 640–654.

Landva, A. O. & Clark, J. I. (1990). Geotechnics of waste fill. In *Geotechnics of Waste Fills Theory and Practice*, (eds, A. Landva & G. D. Knowles). American Society for Testing and Materials, Philadelphia, ASTM STP 1070, pp. 86–103.

Leach, B. A. & Goodger, H. J. (1991). *Building on Derelict Land*. CIRIA, London, CIRIA Special Publication 78/PSA, Civil Engineering Technical Guide 60.

Leckie, J. I., Pacey, J. G. & Halvadakis, C. (1979). Landfill management with moisture control. *Journal of the Environmental Engineering Division, ASCE*, **105**, EE2.

Lee, S. (2000). A Study of the Relationship between Compactive Effort and Permeability for Minestone. MSc thesis, Bolton Institute.

Lentz, R. W., Horst, W. D. & Uppot, J. O. (1985). The permeability of clay to acidic and caustic permeants. In *Hydraulic Barriers in Soil and Rock*, (eds, A. I. Johnson, R. K. Frobel, N. J. Cavalli & C. B. Petterson). American Society for Testing and Materials, Philadelphia, ASTM STP 874, pp. 127–139.

Leonards, G. A., Cutter, W. A. & Holtz, R. D. (1980). Dynamic compaction of granular soils. *Proceedings of the ASCE, Geotechnical Engineering Division*, **106**, GTI, pp. 35–44.

Licence, G. (1993). In situ remediation using vacuum extraction techniques. In *Remedial Processes for Contaminated Land* (ed., M. Pratt), Institution of Chemical Engineers,

London, pp. 113–119.

Littlejohn, G. S. (1979). Surface stability in areas underlain by old coal workings. *Ground Engineering*, **12**, No. 2, pp. 22–30.

Littlejohn, G.S., Cole, K.W. & Mellors, T.W. (1994). Without site investigation ground is a hazard. *Proceedings of the ICE, Civil Engineering*, **102**, pp. 72–78.

Loll, U. (1990). Sludge treatment — state of the art and development. In *Recycling Von Klärschlamme (Sludge Recycling)*. EF-Verlag für Energie- und Umwelttechnik, Berlin, pp. 1–80.

Lord, D. W. (1987). Appropriate site investigations. In *Reclaiming Contaminated Land* (ed., T. Cairney). Blackie, Glasgow, Chap. 4, pp. 62–113.

Lotito, V., Spinosa, L. & Santori, M. (1991). Influence of digestion on sludge characteristics. In *Recent Developments in Sewage Sludge Processing* (eds, F. Colin, P. J. Newman & Y. J. Puolanne). Elsevier Applied Science, London, pp. 32–40.

Loudon, A. G. (1952). The computation of permeability from simple soil tests. *Géotechnique* **2**, No. 3, pp. 165–183.

Lutzic, G. N., Bender, P. A., Wrucke, R. R. & Wainwright, M. E. (1995). New York City sludge management programme. In the *IWEM Yearbook*, Institute of Water and Environmental Management, London, pp. 11–19.

MacFarlane, I. C. (1969). Engineering aspects of peat. In *Muskeg engineering handbook* (ed. I. C. MacFarlane). University of Toronto Press, Toronto, pp. 78–126.

Mair, R. J. & Wood, D. M. (1987). *Pressuremeter Testing, Methods and Interpretation*. Butterworths, London, CIRIA Ground Engineering Report.

Mallet, H. (1996). Case study 1: environmental assessment of a redevelopment scheme on the site of a derelict cement work. In *Environmental Assessment: Good Practice* (ed., J. Petts), CIRIA, London, Special Publication 126, pp. 36–40.

Martin, D. J. (1980). *Ground Vibrations from Impact Pile Driving During Road Construction*. Transport and Road Research Laboratory, Crowthorne, Report SR 544.

Martin, D. J. & Solaini, A. V. (1976). *Noise of Earthmoving at Road Construction Sites*. Transport and Road Research Laboratory, Crowthorne, Report SR 190UC.

Martin, J. D. & Bardis, P. (1994). Recent development in contaminated land treatment technology. *Chemistry and Industry*, 6 June, pp. 411–413.

Matthews, P. J. (1993). *Sewage Sludge Management in Western Europe — A Summer 1993 Perspective*. Anglian Water Services, Cambridge.

McDougall, J. R., Sarsby, R. W. & Hill, N. J. (1996). A numerical investigation of landfill hydraulics using variably saturated flow theory. *Géotechnique* **46**, No. 2, pp. 329–342.

McEntee, J. (1991). Site investigation. In *Recycling Derelict Land* (ed. G. Fleming), Institution of Civil Engineers Thomas Telford, London, Chap. 4, pp. 64–87.

McGrath, S. P., Sidouli, C. M. D., Baker, A. J. M. & Reeves, R. D. (1992). The potential for the use of metal-accumulating plants for in-situ remediation of metal polluted sites. In *Proceedings of the International Conference Eurosol, Maastricht*.

Melent'ev, V. A. (1973) *Hydraulic Fill Structures*. Energy, Moscow.

Miller, W., Alexander, R., Chapman, N., McKinpey, I. & Smellie, J. (1994). *Natural Analogue Studies in the Geological Disposal of Radioactive Wastes*. Elsevier, London, Studies in Environmental Science 57.

Mitchell, J. K. & Jaber, M. (1990). Factors controlling the long-term properties of clay liners. In *Waste Containment Systems* ASCE Geotechnical Special Publication No. 26, American Society of Civil Engineers, New York, pp. 84–105.

Mitchell, J. K. & Madsen, F. T. (1987). Chemical effects on clay hydraulic conductivity. In *Geotechnical practice for waste disposal '87* (ed. R. D. Woods) ASCE Geotechnical Special Publication 13, American Society of Civil Engineers, New York, pp. 87–116.

Mitchell, J. K., Seed, R. B. & Seed, H. B. (1990a). Stability considerations in the design and construction of lined waste repositories. In American Society for Testing and Materials, Philadelphia, ASTM STP 1070, *Geotechnics of Waste Fills — Theory and Practice* (eds, A. Landva & G. D. Knowles). pp. 209–224.

Mitchell, J. K., Seed, R. B. & Seed, H. B. (1990b). Kettleman Hills waste landfill slope failure. I: Liner-system properties. *ASCE Geotechnical Engineering*, **116**, No. 4, pp. 647–668.

Mitchell, R. J. (1983). *Earth Structures Engineering*. Allen & Unwin, Boston.

Mittal, H. & Morgenstern, N. (1975). Parameters for the design of tailings dams. *Canadian Geotechnical Journal*, **12** pp. 285–261.

Moen, J. E. T., Cornet, J. P. & Evers, C. W. A. (1986). Soil protection and remedial actions: Criteria for decision making and standardisation of requirements. In *Contaminated Soil* (eds, J. W. Assink & W. J. van den Brink). Martinus Nijhoff, Rotterdam, pp. 441–448.

Moffatt, A. & McNeill, J. (1994). *Reclaiming Disturbed Land for Forestry*. HMSO, London, Forestry Commission Bulletin 110.

Morgenstern, N. R. & Kupper, A. A. G. (1988). Hydraulic fill structures — a perspective. In *Proceedings of a Speciality Conference on Hydraulics Fill Structures, Fort Collins, CO*. American Society of Civil Engineers, ASCE Geotechnical Special Publication No. 21, pp. 1–31.

Morris, D. V. & Woods, C. E. (1990). Settlement and engineering considerations in landfill and final cover design. In *Geotechnics of Waste Fills — Theory and Practice* (eds, A. Landva & G.D. Knowles). American Society for Testing and Materials, Philadelphia, ASTM STP 1070, pp. 9–21.

Mott, J. G. & Romanow, S. (1992). Sludge characterisation, removal and dewatering. *Journal of Hazardous Materials* **29**, pp. 127–140.

Mugglestone, I. & Hughes, S. (1994). North America — Current technology and development. In *Remedial Processes for Contaminated Land* (ed., M. Pratt). Institution of Chemical Engineers, pp. 19–31.

Mulheron, M. & O'Mahony, M. M. (1990). Properties and performance of recycled aggregates. *Journal of Highways and Transportation*, February, pp. 35–37.

Nagaraj, T. & Murty, B. R. S. (1985). Prediction of the preconsolidation pressure and recompression index for soils. *American Society for Testing and Materials, Geotechnical Testing Journal*, Philadelphia **8**, No. 4, pp. 199–202.

Naylor, J. A., Rowland, C. D., Young, C. P. & Barber, C. (1978). *The Investigation of Landfill Sites*, Water Research Centre, Medenham, Technical Report TR91.

Neal, A. W. (1974). *Formation and Use of Industrial By-products: A Guide*. Business Books, London.

Neden, M. (1996). *The Engineering Properties of Slate Waste and its Potential for Use as a Road Construction Aggregate*. MSc thesis, Bolton Institute.

Nelson, J., Shepherd, T. & Charlie, W. (1977). Parameters affecting the stability of tailing dams. *Proceedings of a Conference on Geotechnical Practice for Disposal of Solid Waste Materials, University of Michigan*, ASCE, New York, pp. 444–460.

Nettleton, A., Robertson, I. & Smith, J. H. (1996). Treatment of soil using lime and PFA to form embankment fill for the new A13. In *Lime stabilisation* (Proceedings of a seminar held at Loughborough University in September 1996), Thomas Telford Ltd, London, pp. 159–175.

New, B. M. (1986). *Ground Vibration Caused by Civil Engineering Works*. Transport Research Laboratory, Crowthorne, TRL RR 53.

New, B. M. (1989). *Trial and Construction Induced Blasting Vibration at the Penmaenbach Road Tunnel*. Transport Research Laboratory Crowthorne, TRL RR 181.

New, B. M. (1992). Construction induced vibration in urban environments. In *Excavation and Support for the Urban Infrastructure* (eds, T. D. O'Rourke & A. G. Hobelman). American Society of Civil Engineers, New York, ASCE Geotechnical Special Publications No. 33, pp. 212–239.

New, B. M. & Hood, R. A. (1989). 'The measurement and control of blasting vibration during civil engineering works. In *Instrumentation in Geotechnical Engineering*. Thomas Telford, London, pp. 253–266.

New Civil Engineer:
(1982). Old mines pose new problems. 9 September, pp. 12–13.
(1984). Bypass gets quick squeeze. 26 January, pp. 18–19.
(1990a). China clay spoil slip alarms experts. 15 February, p. 9.
(1990b). Second china clay spoil heap fails. 22 February, p. 9.
(1990c). Bringing colour to the Black Country. 5 April, pp. 30–34.
(1991a). Poison pen, 21 March, pp. 30–31.

(1991b). Goal break. 1 August, p. 14.

(1991c). Foam foundations. 26 September, pp. 20–21.

(1992a). Race to clear sludge landslide. 20 February, p. 5.

(1992b). Storing up trouble, 2 April, p. 14.

(1992c). Waste line. 25 June, p. 32.

(1992d). The hole truth, 10 September, p. 8 and p. 14.

(1992e). Slip restore. 29 October, p. 6 and p. 8.

(1993). Brick bank. 29 July/5 August, p. 22.

(1994a). South African dam owners were warned about repairs. 3 March, p. 6.

(1994b). Watching the river's flow. Water Supplement, October, pp. 4–6.

(1994c). Clay cleaner. 3 November, p. 19.

(1995a). Barking too loud. 26 January, pp. 28–30.

(1995b). Bury's treasure. 9 February, p. 30.

(1995c). Rock hold. 29 June, pp. 14–15.

(1995d). Greenham Rubble, 24/31 August, p. 8.

(1995e). Industrial blacklist, Black Country new road supplement, November, pp. 14–16.

(1995f). Tough treatment, Black Country new road supplement, November, pp. 30–36.

(1995g). Minesweeping. 7 December, pp. 36–38.

(1996a). Improved Prospect. 25 January, pp. 25–27.

(1996b). Storing up evidence. 25 January, pp. 28–31.

(1996c). Greenwich time switch, 26 September, pp., 24–26.

(1997a). Waste strategy. Hong Kong Handover Supplement, June, p. XLI.

(1997b). Tipping the balance. 12 June, pp. 16–17.

(1997c). On site, not out of site. 25 September, p. 30.

(1997d). Reduced rate, 2 October, p. 18.

(1997e). Restoring pride, Pride Park project study, December, pp. II–VII.

(1998a). Back to work, 29 January, pp. 23–24.

(1998b). Second chance for Felindre, 29 January, pp. 24–25.

(1998c). Core task. 12 March, pp. 22–24.

(1999a). Brown paradise, 28 January, pp. 20–22.

(1999b). Waste not want not, ENTEC project study, January, pp. II–III.

(1999c). Wand wishes settled, Concrete futures, October, pp. 16–18.

(1999d). Poland a la mode, Rail special feature, 25 November, pp. XXXIV–XXXVI.

(2000a). Doing the dirty work, 3 February, pp. 22–23.

(2000b). Pinning hopes, Millenium dome supplement, February, pp. XIV–XV.

New Nordic Technology (1995). Bacterial ground cleansing, **1**, 4–5.

Nicholls, H. R., Johnson, C. F. & Duvall, W. I. (1971). *Blasting Vibrations and their Effects on Structures*. US Bureau of Mines, Pittsburgh, Bulletin 656.

Nixon, P. J., Treadaway, K. & Harrison, W. H. (1979). Durability and protection of building materials in contaminated soil. In *Proceedings of the Conference on the Reclamation of Contaminated Land, Eastbourne*. SCI, London, pp. E4/1–10.

Nobel, J. J., Nunez-McNally, T. & Tansel, B. (1988). The effects of mass transfer on landfill stabilisation rates. In *Proceedings of the 43rd Annual Purdue Industrial Waste Conference, West Lafayette*, pp. 519–532.

North West Regulation Officers Technical Sub-Group (1996). *Earthworks on Landfill Sites. A Technical Note on the Design, Construction and Quality Assurance of Compacted Clay Liners*. Chester, Document NWTECH002.

Obeng, L. E. (1977). Environmental impacts of four African impoundments. In *Proceedings of the ASCE National Convention, Session on Environmental Impact of International Civil Engineering Projects and Practices* (eds, C. E. Gunnerson & J. M. Kalbermatten). American Society of Civil Engineers, New York, pp. 29–43.

O'Connor, M. J. & Mitchell, R. J. (1977). An extension of the Bishop and Morgenstern slope stability charts. *Canadian Geotechnical Journal* **14**, No. 1, pp. 144–155.

O'Flaherty, C. A. (1976). *Highways*. Vol. 2, *Highway Engineering*. Edward Arnold, London.

Ogata, A. (1970). *Theory of Dispersivity in a Granular Medium*. US Geological Survey, Reston.

Okusa, A. & Anma, S. (1980). Slope failures and tailings dam damage in the 1978 Izu Okishima — Kinkai earthquake. *Engineering Geology* **16**, pp. 195–224.

Olsen, H. W. (1962). Hydraulic flow through saturated clays. In *Proceedings of the 9th National Conference on Clays and Clay Minerals*. Pergamon Press, New York, Vol. 9, pp. 131–161.

O'Mahony, M. M. (1990). *Recycling of Materials in Civil Engineering*. PhD Thesis, Oxford University.

O'Mahony, M. M. & Milligan, G. W. E. (1991). Recycling of construction waste. In *Waste Materials in Construction Studies* Environmental Science 48 (eds, J. J. J. R. Goumans, H. A. van der Sloot & Th. G. Aalbers), Elsevier Science, London, pp. 225–231.

Openshaw, S., Carver, S. & Fernie, J. (1989). *Britain's Nuclear Waste — Safety and Siting*. Belhaven Press, London.

O'Riordan, N. J. & Milloy, C. J. (1995). *Risk Assessment for Methane and Other Gases from the Ground*. CIRIA, London, Report 152.

Otte-Witte, R. (1989). Nachbehandlung von entwässerten Schlämmen zum Ziele der Deponierung (After treatment of dewatered sludges destined for deposition). In *Recycling von Klärschlamm*. EF-Verlag für Energie-und Umwelttechnik, Berlin, pp. 165–176.

Overmann, L. K. (1990). Geomembrane seam nondestructive tests: Construction Quality Control (CQC) perspective. In *The Seaming of Geosynthetics* (ed., R. M. Koerner). Elsevier Applied Science, New York, pp. 135–149.

Oweis, I. S. & Khera, R. P. (1990). *Geotechnology of Waste Management*. Butterworths, London.

Oweis, I. S., Smith, D. A., Ellwood, R. B. & Greene, D. S. (1990). Hydraulic characteristics of municipal refuse. Journal of Geotechnical Engineering, ASCE, **116**, No. 4, pp. 539–553.

Paasikallio, A., Rantavaara & Sippola, J. (1994). The transfer of caesium-137 and strontium-90 from soil to food crops after the Chernobyl accident. *Science of the Total Environment*, **155F**, No. 2, pp. 109–124.

Pacey, J. & Augenstein, D. (1990). Modelling landfill methane generation. In *Proceedings of the Internatonal Conference on Landfill Gas: Energy and Environment'90, Bournemouth*, pp. 223–263.

Padfield, C. J. & Mair, R. J. (1984). *Design of Retaining Walls Embedded in Stiff Clay* CIRIA, London. Report 104.

PalRoy, P. (1991). Prediction and control of ground vibration due to blasting. *Colliery Guardian*, November, pp. 215–219.

Parkinson, C. D. (1992). The permeability of landfill liners to leachate. In *The Planning and Engineering of Landfills, Proceedings of a Conference at the University of Birmingham*. The Midland Geotechnical Society, Birmingham, pp. 147–152.

Parry, G. D. R. & Bell, R. M. (1985). Covering systems. In *Contaminated Land, Reclamation and Treatment* (ed., M. A. Smith). Plenum Press, New York, pp. 113–129.

Parry, G. D. R. & Bell, R. M. (1987). Types of contaminated land. In *Reclaiming Contaminated Land* (ed., T. Cairney). Blackie, Glasgow, pp. 30–38.

Parsons, A. W. (1992). *Compaction of Soils and Granular Materials*. Transport Research Laboratory/HMSO, London.

Parsons, A. W. & Boden, J. B. (1979). *The Moisture Condition Test and its Potential Applications*. Transport and Road Research Laboratory, Crowthorne, Supplementary Report SR 522.

Parsons, M. L. (1981). Groundwater aspects of tailings impoundments. *Seminar on Design and Construction of Tailings Dams* (ed., D. Wilson). Colorado School of Mines, Golden, CO, pp. 118–143.

Payne, I. R. (1993). Building a Dartford road embankment on jelly. *Highways and Transportation*, December, pp. 5–10.

Peggs, I. D. (1990). Destructive testing of polyethylene geomembrane seams: various methods to evaluate seam strength. In *The Seaming of Geosynthetics* (ed., R. M. Koerner). Elsevier Applied Science, New York, pp. 125–134.

Peirce, J. J., Sallford, G. & Peterson, E. (1986). Clay liner construction and quality control. *Journal of Environmental Engineering*, ASCE, **112**, No. 1, pp. 13–24.

Penman, A. D. M. (1986). Tailings dams and lagoons. In *Proceedings of the ICE*

Conference on Building on Marginal and Derelict Land. Thomas Telford, London, pp. 37–57.

Perry, J. (1989). *A Survey of Slope Condition on Motorway Earthworks in England and Wales*. Transport and Road Research Laboratory, Crowthorne, Research Report RR199.

Perry, J., MacNeil, D. J. & Wilson, P. E. (1996). The uses of lime in ground engineering: A review of recent work undertaken at the Transport Research Laboratory. In *Lime stabilisation* (Proceedings of a seminar held at Loughborough University in September 1996), Thomas Telford Ltd, London, pp. 27–45.

Peterson, S. R. & Gee, G. W. (1985). Interactions between acidic solutions and clay liners: permeability and neutralisation. In *Hydraulic Barriers in Soil and Rock* (eds, A. I. Johnson, R. K. Frobel, N. J. Cavalli & C. B. Petterson). American Society for Testing and Materials, Philadelphia, ASTM STP 874, pp. 229–245.

Petts, J. (1991). Contaminated land: overview of current issues and concerns. In *Proceedings of a Conference on Contaminated Land: Policy, Regulation and Technology*. IBC Technical Services Ltd, London, Document E7568.

Petts, J. & Eduljee, G. (1994). *Environmental Impact Assessment for Waste Treatment and Disposal Facilities*. Wiley, New York.

Pickering, K. T. & Owen, L. A. (1994). *An Introduction to Global Environmental Issues*. Routledge, London.

Pike, E. B. (1983) Long-term storage of sewage sludge. In *Proceedings of a Workshop on Disinfection of Sewage Sludge: Technical, Academic and Microbiological Aspects* (eds, A. M. Bruce & A. H. Havelaar). Reidel, Dordrecht, pp. 212–225.

Polprasert, C. (1989). Composting. *Organic Waste Recycling*. Wiley, New York, Chap. 3, pp. 63–104.

Portfors, E. A. (1981). Environmental aspects and surface water control. In *Proceedings of a Seminar on Design and Construction of Tailings Dams, Colorado*, pp. 100–117.

Potts, D. M. & Burland, J. B. (1983). *A Parametric Study of the Stability of Embedded Earth Retaining Structures*. Transport and Road Research Laboratory, Crowthorne, Supplementary Report, SR 813.

Potts, D. M. & Fourie, A. B. (1986). Numerical study of effects of wall deformation. *International Journal of Numerical and Analytical Methods in Geomechanics* **10**, No. 4, pp. 383–405.

Powell, A. C. (1992). Use of PFA in highway construction. In *National Seminar on the Use of PFA in Construction* (eds, R.K. Dhir & M.R. Jones), Dundee, pp. 33–43.

PowerGen (1990). *PFA Information. Fill*. PowerGen Ash Products, Coventry.

Powrie, W., Richards, D. J. & Beavan, R. P. (1998). Compression of waste and implications for practice. In *Geotechnical Engineering of Landfills* (eds, N. Dixon, E. J. Murray & D. R. V. Jones). Thomas Telford, London, pp. 3–18.

Prakash, A. J., Bagchi, A., Barman, B. K., Singh, R. B., PalRoy, P., Singh, M. M. & Singh, B. (1991). Structural damage due to blasting in opencast coal mines. *Journal of Mines, Metals and Fuels*, April, pp. 79-86.

Pratt, M. (ed.) (1994). *Remedial Processes for Contaminated Land*. Institution of Chemical Engineers, Rugby.

Privett, K. D., Matthews, S. C. & Hodges, R. A. (1996). *Barriers, Liners and Cover Systems for Containment and Control of Land Contamination*. CIRIA, London, Special Publication 124.

Ramaswany, S.D. & Aziz, M. A. (1992). Some waste materials in road construction. In *ASCE conference on Utilisation of Waste Materials in Civil Engineering Construction, New York*, American Society of Civil Engineers, New York, pp. 153–165.

Rankine, W. J. M. (1857). On the stability of loose earth. *Philosophical Transactions of the Royal Society*, **147**, London, pp. 9–27.

Ray, B. T. (1995). *Environmental Engineering*. PWS, Boston.

Raybould, J. G., Rowan, S. P. & Barry, D. L. (1995). *Methane Investigation Strategies*. CIRIA, London, Report 150.

Rees, J. F. (ed.) (1992). *Contaminated Land Treatment Technology*. SCI/Elsevier Applied Science, London.

Reiher, H. J. & Meister, F. J. (1931). Human sensitivity to vibration. *Forschung auf dem Gebeite der Ingenieurswesens* **2**, No. 11, 381–386.

Richards, I. G., Palmer, J. P. & Barratt, P. A. (1993). *The Reclamation of Former Coal*

Mines and Steelworks. Elsevier, London, Studies in Environmental Science 56.

Richards, K. M. & Aitchison, E. M. (1990). Landfill gas: energy and environmental themes. In *Proceedings of the International Conference on Landfill Gas: Energy and Environment '90, Bournemouth*, pp. 21–40.

Riddell, J. F. (1988). Hydrology of landfill sites. In *Proceedings of Scottish Hydrological Group and British Hydrological Society, Glasgow*, November.

Robb, A. D. (1982). *Site Investigation* Thomas Telford, London, ICE Works Construction Guide.

Robinson, H.D. & Gronow, J. (1992). Groundwater protection in the UK: assessment of the landfill leachate source-term. *Journal of the IWEM* **6**, pp. 229–226.

Robinson, H. D. & Maris, P. J. (1979). *Leachate From Domestic Wastes: Generation, Composition and Treatment: A review*. Water Research Centre, Medenham, Technical Report TR108.

Rodgers, V. C. (1986). Low-level waste: the challenge of disposal. In *Proceedings of the 8th Annual Symposium on Geotechnical and Geohydrological Aspects of Waste Management, Fort Collins*, Balkema, Rotterdam, pp. 51–61.

Rogowski, A. S., Weinrich, B. E. & Simmons, D. E. (1985). Permeability assessment of a compacted clay liner. In *Proceedings of the 8th Annual Madison Conference on Applied Research Practice for Municipal and Industrial Waste*, pp. 315–337.

Routh, C. D. (1984). Civil engineering aspects of China clay and tungsten tailings. In *Conference on Materials for Dams, Monte Carlo*, pp. 1–24.

Rowe, P. W. (1952). Anchored sheet-pile walls. *Proceedings of the ICE* **1**, No. 1, pp. 27–70.

Rowe, P. W. (1962). The stress–dilatancy relation for static equilibrium of an assembly of particles in contact. *Proceedings of the Royal Society, London, Series A*, **269**, pp. 500–527.

Rowe, P. W. (1968). The Influence of Geological Features of Clay Deposits on the Design and Performance of Sand Drains. In *Proceedings of the ICE*, London, Supplementary Paper 7058S.

Rowe, P. W. (1972). The relevance of soil fabric to site investigation practice. 12th Rankine Lecture. *Géotechnique* **22**, No. 2, pp. 195–300.

Rowe, P. W. & Barden, L. (1964). The importance of free ends in triaxial testing. *Journal of Soil Mechanics and Foundations Division, ASCE*, **90**, SM1, 1–27.

Rowe, P. W. & Shields, D. H. (1965). The measured horizontal coefficient of consolidation of laminated, layered or varved clays. *Proceedings of the 6th ICSMFE*, **1**, pp. 342–344.

Rowe, R. K., Quigley, R. M. & Booker, J. R. (1995). *Clayey Barrier Systems for Waste Disposal Facilities*. E. & F. N. Spon, London.

Royal Society (1994). *Disposal of Radioactive Wastes in Deep Repositories*. London.

Rubin, L. S., Burnett, M., Amandson, A., Colaizzi, G. J. & Whaite, R. H. (1981). Disposal of coalmine waste in active underground coal mines. In *Proceedings of the Bureau of Mines Technology, Transfer Workshop, Denver*, July, pp. 8–30.

Rulkens, W. H., Asink, J. W. & van Gemert, W. J. T. (1985). On-site processing of contaminated soil. In *Contaminated Land, Reclamation and Treatment* (ed., M. A. Smith). Plenum Press, New York, pp. 37–90.

Ryan, C. A. (1985). Slurry cut-off walls: applications in the control of hazardous waste. In *Hydraulic Barriers in Soil and Rock* (eds, A. Johnson, R. Fobel, R. K. N. Cavalli & C. Pettersson). STP 874 American Society for Testing and Materials, Philadelphia, pp. 9–23.

Saarela, J. (1987). *Some Factors about Sanitary Landfill Investigations in Helsinki City*. National Board of Waters and the Environment, Helsinki.

Sadlier, M. A. & Christopher, B. R. (1993). Geosynthetic systems in environmental management. In *Environmental Management, Geo-water and Engineering Aspects* (eds, P. Chowdhury and A. Sivakumar). Balkema, Rotterdam, pp. 821–826.

Sanchez-Alciturri J. M., Palma, J., Sagaseta, C. & Canizal, J. (1995). Three years of deformation monitoring at Meruelo landfill. In *Proceedings of GREEN'93: Waste Disposal by Landfill* (ed., R. W. Sarsby). Balkema, Rotterdam, pp. 365–371.

Sanning, D. E. (1995). In-situ treatment. In *Contaminated Land, Reclamation and Treatment* (ed., M. A. Smith). Plenum Press, New York, pp. 91–111.

Sarma, S. K. (1975). Seismic stability of earth dams and embankments. *Géotechnique* **25**,

No. 4, pp. 743–761.

Sarsby, R. W. (1970). *Undrained Stress Path Characteristics of Clay Elements in Plane Strain and Triaxial Compression*. MSc thesis, Manchester University.

Sarsby, R. W. (1982). Noise from sheet-piling operations — M67 Denton Relief Road. *Proceedings of the ICE Part 1* **72**, 15–26.

Sarsby, R. W. (1987a). Conversion of a reservoir into a landfill site. In *Geotechnical Practice for Waste Disposal* (ed., R. D. Wood). American Society of Civil Engineers, New York Special Publication 13, pp. 772–783.

Sarsby, R. W. (1987b). The sliding resistance between grid reinforcement and weathered colliery waste. In *2nd International Symposium on the Reclamation, Treatment and Utilisation of Coal-mining Wastes* (ed., A. K. M. Rainbow). Elsevier, Oxford, pp. 587–596.

Sarsby, R.W. & Cooke, J.E. (1989). 'Treatment of lagoon sludge', *Proc. 2nd International symposium on Environmental Geotechnology*, Vol 2, Bethlehem, pp. 80–90.

Sarsby, R. W. (1991). Environmental pollution in civil engineering training in the United Kingdom. In *Proceedings of the International Conference on Environmental Pollution* (ed. B. Nath) Inderscience Enterprises, Geneva, pp. 432–437.

Sarsby, R. W. (1992a). Reinforced soil retaining walls. In *National Seminar on the use of PFA in Construction* (eds, R.K. Dhir & M.R. Jones), Dundee, pp. 59–73.

Sarsby, R. W. (1992b). *Legislation and Practice on Noise and Vibration Control with Particular Reference to Piling*. British Steel Corus Construction, Scunthorpe.

Sarsby, R. W. & Marshall, C. B. (1987). Field performance of a retaining wall composed of reinforced pulverised fuel ash. *CSIR Conference on Ash — A Valuable Resource*. CSIR Pretoria, Vol. 4.

Sarsby, R. W. & Cooke, J. E. (1989). Treatment of lagoon sludge. In *Proceedings of the 2nd International Symposium on Environmental Geotechnology*, **2**, Bethlehem, pp. 80–90.

Sarsby, R. W. & Williams, M. (1995). Selection of soils for compacted clay lining. In *Proceedings of GREEN'93: Waste Disposal by Landfill* (ed., R. W. Sarsby). Balkema, Rotterdam, pp. 471–475

Sarsby, R. W., Heggie, I. S. & Rainford, S. (1995). Permeability measurements in the triaxial apparatus. In *Proceedings of GREEN'93: Waste Disposal by Landfill* (ed., R. W. Sarsby), Balkema, Rotterdam, pp. 183–187.

Schofield, A. N. & Wroth, C. P. (1968). *Critical State Soil Mechanics*. McGraw-Hill, London.

Schroeder, W. L. & Dickenson, S. E. (1996). Soils in construction. *Embankment Construction and Control*. Prentice Hall, Englewood Cliffs, NJ, Chap. 8, pp. 113–151.

Seed, H. B., Makdisi, F. I. & de Alba P. (1977). The Performance of Dams During Earthquakes. Earthquake Engineering Research Centre, University of California, Berkely, CA, Report No. EERC-77/20.

Seed, R. B., Mitchell, J. K. & Seed, H. B. (1990). Kettleman Hill waste landfill slope failure. II: Stability analyses. *Journal of Geotechnical Engineering, ASCE* **1116**, No. 4, pp. 669–689.

Senior, E. (ed). (1990). *Microbiology of Landfill Sites*. CRC Press, Boca Raton, FL.

Servais, S. E. C. & York, K. I. (1993). *The Precision of Two Laboratory Compaction Tests and a Particle Density Test*. Australian Research Board, Melbourne.

Seymour, K. (1992). Landfill lining for leachate containment. *Journal of IWEM* **6**, pp. 389–396.

Shahabi, A. A., Das, B. M. & Tarquin, A. J. (1984). An empirical relationship for coefficient of permeability of sand. In *Proceedings of the 4th Australian and New Zealand Conference on Geomechanics*, **1**, pp. 54–57.

Sherwood, P. T. (1987). *Wastes for Imported Fill*. ICE Guide, Thomas Telford, London.

Sherwood, P. T. (1995). *Alternative Materials in Road Construction*. Thomas Telford, London.

Siegel, R. A., Robertson, R. J. & Anderson, D. G. (1990). Slope stability investigations at a landfill in southern California. In *Geotechnics of Waste Fills — Theory and Practice* (eds, A. Landva & G. D. Knowles). American Society for Testing and Materials, Philadelphia, ASTM STP 1070, pp. 259–284.

Siegrist, R. L. & Jenssen, P. D. (1990). Evaluation of sampling method effects on volatile organic compound measurements in contaminated soil. *Environmental Science and Technology* **24**, No. 9, pp. 1387–1932.

Sincero, A. P. (1996). *Environmental Engineering*. Prentice-Hall, Englewood Cliffs, NJ: Chap. 8, Sludge treatment and disposal, pp. 358–393; Chap. 11, Solid waste management, pp. 517–574; Chap. 14, Noise pollution and control, pp. 686–723.

Singh, S. & Murphy, B. (1990). Evaluation of the stability of sanitary landfills. In *Geotechnics of Waste Fills — Theory and Practice* (eds, A. Landva & G. D. Knowles). American Society for Testing and Materials, Philadelphia, STP 1070, pp. 240-258.

Siriwardane, H. J. (1988). Mine induced subsidence. In *Proceedings of a Symposium Sponsored by the ASCE, Nashville*.

Skarzynska, K. M. (1995a). Reuse of coal mining wastes in civil engineering. Part 1: Properties of minestone. *Waste Management* **15**, No. 1, pp. 3–42.

Skarzynska, K. M. (1995b). Reuse of coal mining wastes in civil engineering. Part 2: Utilisation of minestone. *Waste Management* **15**, No. 2, pp. 83–126.

Skempton, A. W. (1954). The pore pressure coefficients *A* and *B*. *Géotechnique* **4**, No. 4, pp. 143–147.

Skempton, A. W. (1957). Discussion: Planning and design of the new Hong Kong airport. *Proceedings of the ICE* **7**, pp. 306.

Skempton, A. W. (1964). Long-term stability of clay slopes. 4th Rankine Lecture. *Géotechnique* **14**, No. 2, pp. 75-102.

Skempton, A. W. (1990). Historical development of British embankment dams to 1960. In *Clay Barriers for Embankment Dams*. Thomas Telford, London, pp. 15–52.

Skempton, A. W. & Bjerrum, L. (1957). A contribution to settlement analysis of foundations on clay. *Géotechnique* **7**, No. 4, pp. 168–178.

Skempton, A. W. & Coats, D. J. (1985). Carsington dam failure. In *Symposium on Failures in Earthworks*. Thomas Telford, London, pp. 203–220.

Sleeman, W. (1990). Environmental effects of the utilisation of coal mining wastes. In *Proceedings of the 3rd International Symposium on the Reclamation, Treatment and Utilisation of Coal Mining Wastes*. Balkema, Rotterdam, pp. 65–76.

Slocombe, B. C. (1993). Dynamic compaction. In *Ground Improvement* (ed., M. P. Mosley). Blackie Academic and Professional, Glasgow, Chap. 2, pp. 20–39.

Smith, D. M. (1996). *The use of Geomembranes for Containment of Landfilled Waste Materials*. MSc Thesis, Bolton Institute.

Smith, M. A. (1988). International study of technologies for clearing-up contaminated land and groundwater. In *Proceedings of Contaminated Land Reclamation 88*. Durham County Council, pp. 259–266.

Smith, M. A. (1991a). Data analysis and interpretation. In *Recycling Derelict Land* (ed., G. Fleming). Institution of Civil Engineers/Thomas Telford, London, Chap. 5, pp. 88–144.

Smith, M. A. (1991b). Investigation of contaminated sites and buildings prior to development. In *IBC Conference on Contaminated Land, London*.

Smith, M. A. & Ellis, A. C. (1986). An investigation into methods used to assess gas works sites for reclamation. In *Reclamation and Revegetation Research*, Vol. 4., Elsevier, Amsterdam, pp. 183–209.

Smith, M. R. & Collis, L. (1993). *Aggregates. Sand, Gravel and Crushed Rock Aggregates for Construction Purposes*. The Geological Society, London, Special Publication No. 9.

Somerville, S. M. (1986). *Control of Groundwater for Temporary Works*. CIRIA, London, Report 113.

Sowers, G. F. (1973). Settlement of waste disposal fills. In *Proceedings of the 8th ICSMFE, Moscow*, pp. 207–210.

Spencer, E. (1967). A method of analysis of the stability of embankments assuming parallel inter-slice forces. *Géotechnique* **17**, pp. 11–26.

Spinosa, L. (1985). Technological characterization of sewage sludge. *Waste Management and Research* **3**, pp. 389–398.

Spinosa. L, Santori, M. & Lotito, V. (1989). Rheological characterisation of sewage sludges. In *Recycling von Klärschlamm (Sludge Recycling)*. EF-Verlag für Energie und Umweltlebark, GmbH, Berlin, pp. 177–184.

Stamp, R. J. & Smith, I. E. (1984). Pulverised fuel ash — waste material or mineral resource. In *AshTech '84 — 2nd International Conference on Ash Technology and Marketing*, London, pp. 67–71.

Stedman, L. (1992). Driller killers. *Ground Engineering*, June, pp. 17–18.

Steeds, J. E., Shepherd, E. & Barry, D. L. (1996). *A Guide for Safe Working on Contaminated Sites*. CIRIA, London, Report 132.

Steffens, R. J. (1966). Some aspects of structural vibration. In *Proceedings of the Symposium on Vibration in Civil Engineering*. (ed., B. O. Skip) Butterworths, London, pp. 1–30.

Stegman, R. & Spendlin, H. H. (1989). Enhancement of degradation: German experiences. In *Sanitary Landfilling: Process Technology and Environmental Impact* (ed., T. H. Christensen). Academic Press, London.

Steilen, N. (1994). Remediation of former gasworks sites. In *3rd International Conference on Re-use of Contaminated Land and Landfills*. Brunel University, London, pp. 419–426.

Stevens, C. (1984). *Landfill Lining and Capping*, UKAEA/HMSO, Harwell/London, AERE-R11506.

Stief, K. (1985). Long-term effectiveness of remedial measures. In *Contaminated Land, Reclamation and Treatment* (ed., M. A. Smith). Plenum Press, New York, pp. 13–36.

Stoll, U. W. (1971). Mechanical properties of milled domestic trash. In *National Water Resources Meeting*, ASCE, Phoenix.

Street, A. (1987). The Portworthy china clay tailings disposal scheme. *Proceedings of the ICE, Part 1* **82**, pp. 551–566.

Street, A. (1994). Landfilling-the difference between Continental European and British practice. *Proceedings of the ICE, Geotechnical Engineering*, **107**, pp. 41–46.

Task Committee of the Committee on Air Resources and Environmental Effects Management (Environmental Engineering Division) (1979). Nuclear facilities siting. *Journal of the Engineering Division, ASCE* **LO5**, EE3.

Taylor, D. W. (1937). Stability of earth slopes. *Journal of the Boston Society of Civil Engineers* **24**, pp. 197–246.

Taylor, D. W. (1966). *Fundamentals of Soil Mechanics*, 2nd edn., Wiley, London.

Taylor, G. & Simpson, A. (1995). Aspects of contaminated land reclamation. *Institute of Water and Environmental Management (IWEM) Yearbook 1995*. IWEM, London, pp. 29–39.

Taylor, M. R. G. & McLean, R. A. N. (1992). Overview of clean-up methods for contaminated sites. *Journal of the IWEM* **6**, No. 4, p. 40.

Taylor, R. K. & Cobb, A. E. (1977). Mineralogical and geotechnical controls on the storage and use of British coal-mining wastes. In *Proceedings of the 9th ICSMFE, Tokyo*, pp. 373–388.

Technical Committee on the Experimental Disposal of House Refuse in Wet and Dry Pits (1960). *Pollution of Waste by Tipped Refuse*. HMSO, London.

Terzaghi, K. (1966). *Theoretical Soil Mechanics*. Wiley, New York.

Terzaghi, K. & Peck, R. B. (1967). *Soil Mechanics in Engineering Practice*, 2nd edn., Wiley, London.

Tesarik, D. R. & McWilliams, P. C. (1981). Factor of Safety charts for estimating the stability of saturated and unsaturated tailings pond embankments. *Proceedings of the Bureau of Mines Technology Transfer Working, Denver*, pp. 21–34.

Tesarik, D. R. & McWilliams, P. C. (1981). *Factor of Safety Charts for Estimating the Stability of Saturated and Unsaturated Tailings Pond Embankments*. Bureau of Mines, US Department of the Interior, Pittsburgh, Report of Investigations 8564.

Tillotson, S. (1993). The costs of clean-up. In *Remedial Processes for Contaminated Land* (ed., M. Pratt), Institution of Chemical Engineers, London, pp. 87–112.

Tomlinson, M. J. (1995). *Foundation Design and Construction*, 6th edn, Pitman, London.

Torrance, Y. K. (1975). On the role of chemistry in the development and behaviour of the sensitive marine clays of Canada and Scandinavia. *Canadian Geotechnical Journal*, **12**, pp. 326–335.

Towell, D. L. (1994). The Practicalities and Expense of Contaminated Land Remediation — A Comparison of UK and European Experiences. MSc thesis, Bolton Institute.

Trautwein, S. & Williams, C. (1990). *Performance evaluation of earthern liners, in Waste Containment Systems: Construction, regulations and performance* (ed., R.

Bonaparte). ASCE, New York, pp. 30–51.

Tsonis, P., Christoulas, S. & Kolias, S. L. (1983). Soil improvement with coal ash in road construction. In *Proceedings of the 8th European Conference in Soil Mechanics and Foundation Engineering*, **2**, pp. 961–964.

Tubey, L. W. (1978). *The Use of Waste and Low-grade Materials in Road Construction. 5. China Clay Sand*. Transport and Road Research Laboratory, Crowthorne, Report LR 817.

Turnbull, D. (1984). The role of the Minestone Executive in British mining and civil engineering. In *2nd International Symposium on the Reclamation, Treatment and Utilisation of Coalmining Wastes, Durham*. National Coal Board Minestone Executive, pp. 1.1–1.12.

UCO Technical Fabrics (1992). *Geosynthetics Manual*. Lokeren, Belgium.

United Nations (1987). *Our common future*. World Commission on Environment and Development. Oxford University Press.

USEPA (1983). *Process Design Manual for Land Application of Municipal Sludge*. US Environmental Protection Agency, Cincinnati, Ohio. Report EPA-625/1-83-016.

USEPA (1993a). *Solidification/Stabilization of Organics and Inorganics*. US Environmental Protection Agency, Cincinnati, OH, Report EPA/540/5-92/015.

USEPA (1993b). *Life-cycle Assessment: Inventory guidelines and Principles*. US Environmental Protection Agency, Cincinnati, OH, Report EPA/600/R-92/245.

USEPA (1994). *SITE Technology Capsule. Geosafe Corporation In-situ Vitrification Technology*. US Environmental Protection Agency, Cincinnati, OH, Report EPA/540/R-94/520a.

Van den Berg, J. J., Geuzens, P. & Otte-Witte, R. (1991). Physical aspects of landfilling of sewage sludge. In *Proceedings of the Conference on Alternative Uses for Sewage Sludge*, Pergammon Press, Oxford, pp. 263–276.

Van den Berg, J. W. (1991). Quality and environmental aspects in relation to the application of pulverised fuel ash. In *Waste Materials in Construction* (eds, J. J. J. R. Goumans, H. A. van der Sloot & Th. G. Aalbers). Elsevier, Oxford, pp. 441–450.

Van Zyl, D. (1993). Mine waste disposal. In *Geotechnical Practice for Waste Disposal* (ed., D. E. Daniel). Chapman and Hall, London, Chap. 12, pp. 267–276.

Verheul, J., de Bruijn, P. & Herbert, S. (1993). Biological remediation: a European perspective. In *Remedial Processes for Contaminated Land* (ed., M. Pratt). Institution of Chemical Engineers, London, pp. 53–85.

Versperman, K. D., Edil, T. B. & Berthoux (1985). Permeability of fly ash and fly ash–sand mixtures, In *'Hydraulic Barriers in Soil and Rock* (eds, A. I. Johnson, R. K. Frobel, N. J. Cavalli & C. B. Pettersson), American Society for Testing and Materials, Pennsylvania, ASTM STP 874, pp. 289–298.

Vick, S. G. (1983). *Planning, Design and Analysis of Tailings Dams*. Wiley, New York.

Visser, W. J. F. (1995). Contaminated land policies in Europe. In *Chemistry and Industry* **13**, pp. 496–499.

Volpe, R. (1979). Physical and engineering properties of copper tailings. In ASCE Symposium on *Geotechnical Practice in Mine Waste Disposal*. American Society of Civil Engineers, New York, pp. 242–260.

Voss, T. (1993). *Shear Strength Development of Wastewater Sludges in Mono-fill Disposal Sites* (in German). PhD thesis, Kassel University.

Waldron, S., Hall, A. J., Fallick, A. E., Gilmour, R. A. & MacDonald, M. (1995). Stable isotope analysis as a means of identifying the source of methane. In *Proceedings of GREEN'93: Waste Disposal by Landfill* (ed., R. W. Sarsby). Balkema, Rotterdam, pp. 665–671.

Walker, W. H. (1969). Ilinois groundwater pollution. *Journal of the American Waterworks Association* **61**, pp. 31–40.

Waltham, A. C. (1989). *Ground Subsidence*. Blackie, London.

Wang, M. C., Hung, M. H. & Dempsey, B. (1996). pH effect on plasticity and thixotropy of a water treatment sludge. In *3rd International Conference on Environmental Geotechnology, San Diego*, **1**, pp. 122–131.

Warwickshire Environmental Protection Council (1995). *Landfill Gas from Closed Sites in Coventry and Warwickshire*. Nuneaton.

Wathern, P. (1988). *Environmental Impact Assessment*, Unwin Hyman, London.

Watson, K. L. (1980). *Slate Waste Engineering and Environmental Aspects*. Applied Science, London.

Watts, K. S. & Charles, J. A. (1990). Settlement of recently placed domestic landfill, *Proceedings of the ICE, Part 1*, **88**, pp. 971–993.

Wheeler, P. (1992). Washing cycle. *Ground Engineering*, December, pp. 10–11.

Wheelton, I. S. & Wood, E. (1986). Design of a shallow land repository for final disposal of low level and short-lived intermediate level radioactive wastes on a clay site. In *Proceedings of the International Symposium on the Siting, Design and Construction of Underground Repositories for Radioactive Waste*, International Atomic Energy Agency, London, pp. 201–210.

Whitbread, M., Marsay, A. & Tunnell, C. (1991). *Occurrence and Utilization of Mineral and Construction Wastes*. Department of the Environment, Geological and Minerals Planning Research Programme. HMSO, London.

Whittaker, B. N. & Reddish, D. J. (1989). *Subsidence. Occurrence, Prediction and Control*. Elsevier, London, Developments in Geotechnical Engineering, p. 56.

Whyley, P. J. & Sarsby, R. W. (1992). Ground borne vibration from piling. *Ground engineering*, May, pp. 32–37.

Wiedemann, H. U. (1990). Deep mine disposal of hazardous waste. Presented at *ISWA conference*, Jonkoping.

Williams, G. M. & Aitkenhead, N. (1991). Lessons from Loscoe: the uncontrolled migration of landfill gas. *Quarterly Journal of Engineering Geology*, pp. 191–207.

Williams, G. M. & Harrison, I. B. (1983). *Case Study of a Containment Site — The Hooton Landfill, Cheshire*. Institute of Geological Sciences/NERC, Harwell, Report FLPU 83–8.

Williams, N. D. & Houlihan, M. (1986). Evaluation of friction coefficients between geomembranes, geotextiles and related products. In *Proceedings of the International Conference on Geotextiles, Vienna, IGS*, pp. 891–896.

Wise, D. L. & Trantolo, D. J. (eds) (1994). *Remediation of Hazardous Waste Contaminated Soils. Part II: Specific Case Studies in Hydrocarbon Remediation*. Marcel Dekker, New York, pp. 149–401.

Wiss, J. F. (1967). *Damage Effects of Pile Driving Vibration*. National Research Council/ National Academy of Science, Washingon, Highway Research Report No. 155, pp. 14–20.

Wiss, J. F. (1981). Construction vibrations: state-of-the-art. *Proceedings of the ASCE, Geotechnical Engineering Division* **107**, GT2, pp. 167–181.

Wiss, J. F. & Parmalee, R. A. (1974). Human perception of transient vibrations. *ASCE, Journal of the Structural Division* **100**, ST4, pp. 773–787.

Woods, R. D. (1968). Screening of surface waves in soils. *Proceedings of the ASCE, Journal of Soil Mechanics and Foundation Engineering* **SM4**, pp. 951–977.

Wroth C. P. & Hughs, J. M. O. (1973). An instrument for the in-situ measurement of the properties of soft clays. *Proceedings of the 8th ICSMFE*, Moscow, **1**, pp. 487–494.

Wu, L. (1999). *Geotechnical Properties of Papermill Sludge and its Utilisation in Landfill Cover Systems*. PhD thesis, Manchester University (Bolton Institute), Manchester.

Yen, B. C. and Scanlon, B. (1975). Sanitary landfill settlement rates. *ASCE, Geotechnical Engineering Division*, **101**, GT5, pp. 475–487.

Yong, R. N. (1986). Selective leaching effects on some mechanical properties of a sensitive clay. Engineering Geology. **21**, pp. 279–299.

Yong, R. N. & Warkentin, B. P. (1973). Soil properties and behaviour. In *Developments in Geotechnical Engineering*. Elsevier Scientific, Amsterdam.

Young, P. J., Pollard, S. & Crowcroft, P. (1997). Overview: context, calculating risk and using consultants. In *Contaminated Land and its Reclamation* (eds, R. E. Hester & R. M. Harrison), Royal Society of Chemistry, London, pp. 1–24.

Zimmie, T. F., LaPlante, C. M. and Bronson, D. (1992). The effect of freezing and thawing on the permeability of compacted clay landfill covers and liners. In *Mediterranean Conference on Environmental Geotechnology* (eds, M. A. Usmen & Y. B. Acar). Balkema, Rotterdam, pp. 213–217.

Index

Page references in *italics* refer to figures and tables where located away from relevant text.